The Mathematical Biology of Diatoms

Scrivener Publishing
100 Cummings Center, Suite 541J
Beverly, MA 01915-6106

Diatoms: Biology and Applications

Series Editors: Richard Gordon dickgordoncan@protonmail.com and Joseph Seckbach Joseph.Seckbach@mail.huji.ac.il

Scope: The diatoms are a single-cell algal group, with each cell surrounded by a silica shell. The shells have beautiful attractive shapes with multiscalar structure at 8 orders of magnitude, and have several uses. 20% of the oxygen we breathe is produced by diatom photosynthesis, and they feed most of the aquatic food chain in freshwaters and the oceans. Diatoms serve as sources of biofuel and electrical solar energy production and are impacting on nanotechnology and photonics. They are important ecological and paleoclimate indicators. Some of them are extremophiles, living at high temperatures or in ice, at extremes of pH, at high or low light levels, and surviving desiccation. There are about 100,000 species and as many papers written about them since their discovery over three hundred years ago. The literature on diatoms is currently doubling every ten years, with 50,000 papers during the last decade (2006-2016). In this context, it is timely to review the progress to date, highlight cutting-edge discoveries, and discuss exciting future perspectives. To fulfill this objective, this new Diatom Series is being launched under the leadership of two experts in diatoms and related disciplines. The aim is to provide a comprehensive and reliable source of information on diatom biology and applications and enhance interdisciplinary collaborations required to advance knowledge and applications of diatoms.

Publishers at Scrivener
Martin Scrivener (martin@scrivenerpublishing.com)
Phillip Carmical (pcarmical@scrivenerpublishing.com)

The Mathematical Biology of Diatoms

Edited by

Janice L. Pappas

Department of Mathematics, University of Michigan, Ann Arbor, MI, USA

Scrivener Publishing

WILEY

This edition first published 2023 by John Wiley & Sons, Inc., 111 River Street, Hoboken, NJ 07030, USA and Scrivener Publishing LLC, 100 Cummings Center, Suite 541J, Beverly, MA 01915, USA

For more information about Scrivener publications please visit www.scrivenerpublishing.com.

Wiley Global Headquarters
111 River Street, Hoboken, NJ 07030, USA

For details of our global editorial offices, customer services, and more information about Wiley products visit us at www.wiley.com.

Library of Congress Cataloging-in-Publication Data

ISBN 978-1-119-75043-7

Cover images: Sisyphus pictured rolling a colossal sized marine fossil diatom, *Brightwellia hyperborea*, a colored scanning electron micrograph (SEM) by Dr. Mary Ann Tiffany, Biology Department, San Diego State University, USA. Sample taken from Oamaru, New Zealand.
Cover design by Russell Richardson

Set in size of 11pt and Minion Pro by Manila Typesetting Company, Makati, Philippines

Printed in the USA

10 9 8 7 6 5 4 3 2 1

Contents

List of Figures xiii

List of Tables xxxi

Preface xxxv

Part I: Diatom Form and Size Dynamics 1

1 **Modeling the Stiffness of *Diploneis* Species Based on Geometry of the Frustule Cut with Focused Ion Beam Technology** 3
Andrzej Witkowski, Romuald Dobosz, Tomasz Płociński, Przemysław Dąbek, Izabela Zgłobicka, Horst Lange-Bertalot, Thomas G. Bornman, Renata Dobrucka, Michał Gloc and Krzysztof J. Kurzydłowski
1.1 Introduction 4
1.2 Material and Methods 6
1.2.1 Focused Ion Beam (FIB) Milling 6
1.2.2 Modeling 6
1.3 Results 8
1.3.1 FIB Processing 8
1.3.2 Modeling 11
1.4 Discussion 14
1.4.1 Practical Meaning of the Frustule Geometric Characters 14
1.4.2 Documenting the Mechanical Strength of the Diatom Frustule 14
Acknowledgments 16
References 16

2 **Size-Resolved Modeling of Diatom Populations: Old Findings and New Insights** 19
Jonas Ziebarth, Werner M. Seiler and Thomas Fuhrmann-Lieker
2.1 Introduction 19
2.2 The MacDonald–Pfitzer Rule and the Need for Matrix Descriptions 20
2.3 Cardinal Points and Cycle Lengths 21
2.3.1 Considered Cardinal Parameters 21
2.3.2 Factors Determining Cardinal Points 22
2.3.3 Experimental Data 24
2.4 Asymmetry, Delay and Fibonacci Growth 26
2.4.1 The Müller Model 26
2.4.2 The Laney Model 28
2.5 Continuous vs. Discrete Modeling 28

2.5.1 Discrete Dynamical Systems 29
2.5.2 The Perron-Frobenius Theorem 33
2.5.3 Continuous Dynamical Systems 35
2.5.4 Extensions and Generalizations 37
2.5.5 Individual-Based Models 39
2.6 Simulation Models 41
2.6.1 The Schwarz *et al.* Model 41
2.6.2 The D'Alelio *et al.* Model 43
2.6.3 The Hense–Beckmann Model 45
2.6.4 The Terzieva–Terziev Model 48
2.6.5 The Fuhrmann-Lieker *et al.* Model 49
2.7 Oscillatory Behavior 52
2.7.1 Reproduction of Experimental Data 52
2.7.2 Coupling to External Rhythms 53
2.8 Conclusion 55
Acknowledgment 56
References 56

3 On the Mathematical Description of Diatom Algae: From Siliceous Exoskeleton Structure and Properties to Colony Growth Kinetics, and Prospective Nanoengineering Applications 63
Alexey I. Salimon, Julijana Cvjetinovic, Yuliya Kan, Eugene S. Statnik, Patrick Aggrey, Pavel A. Somov, Igor A. Salimon, Joris Everaerts, Yekaterina D. Bedoshvili, Dmitry A. Gorin, Yelena V. Likhoshway, Philipp V. Sapozhnikov, Nikolai A. Davidovich, Olga Y. Kalinina, Kalin Dragnevski and Alexander M. Korsunsky
3.1 Introduction 64
3.2 Hierarchical Structuring of Matter: Diatom Algae and the Bio-Assisted Nanostructured Additive Manufacturing Paradigm 64
3.3 Structural Design of Diatom Frustules 65
3.4 Mechanical Performance of Diatom Frustules – Experimental Characterization 73
3.4.1 Nanoindentation Testing of Diatom Frustules 75
3.4.2 AFM Studies of Diatom Frustules 77
3.5 Engineering Applications of Diatomaceous Earth 80
3.6 NEMS/MEMS Perspective 85
3.7 On the Mathematical Description of Self-Organized Diatom Frustule Growth 87
3.8 On the Kinetics of Diatom Colony Growth 90
3.9 Advanced Pattern Analysis of the Hierarchical Structure of Diatom Frustules 92
3.10 Concluding Remarks 95
Acknowledgement 96
References 96

Part II: Diatom Development, Growth and Metabolism 103

4 Ring to the Linear: Valve Ontogeny Indicates Two Potential Evolutionary Pathways of Core Araphid Diatoms **105**
Shigeki Mayama and Momoko Kushida
4.1 Introduction 106
4.2 Material and Methods 107
4.2.1 *Fragilaria mesolepta* 107
4.2.2 *Staurosira binodis* 108
4.2.3 Induction of Synchronous Division 109
4.2.4 Electron Microscopy 110
4.3 Results 110
4.3.1 *Fragilaria mesolepta* 110
4.3.2 *Staurosira binodis* 112
4.4 Discussion 114
4.5 Conclusion 116
References 117

5 Mathematical Basis for Diatom Growth Modeling **121**
Dariush Sardari
5.1 Introduction 121
5.2 General Physiology of Diatoms 122
5.3 Mathematical View of Diatom Growth 123
5.4 Physical Basis for Diatom Modeling 127
5.4.1 Diatom Dimensions 127
5.4.2 Ambient Temperature 129
5.4.3 Light Intensity and Duration 129
5.5 Review of Existing Mathematical Models 130
5.5.1 Gompertz Model 130
5.5.2 Monod Model 131
5.5.3 Michaelis-Menten Model 132
5.5.4 Droop Model 133
5.5.5 Aquaphy Model 134
5.5.6 Mechanistic Model 134
5.6 Results 135
5.7 Conclusion 135
5.8 Prospects 136
References 136

6 Diatom Growth: How to Improve the Analysis of Noisy Data **141**
Olga Kourtchenko, Kai T. Lohbeck, Björn Andersson and Tuomas Rajala
6.1 Introduction 142
6.1.1 What is a Growth Curve? 142
6.1.2 Why Measure Growth? 142
6.1.3 Diatoms and Their Growth 143
6.1.4 Growth Data Analysis and Growth Parameter Estimation 147
6.2 Simulation Trials 150

6.2.1 Methodology for the Simulation Trials 150
6.2.2 Candidate Methods for Estimating the Specific Growth Rate 152
6.2.3 Simulation Trials Results 153
6.2.3.1 Results with Only the Noise Challenge 153
6.2.3.2 Results when Crashing Occurs 155
6.2.3.3 Results when Censoring Occurs 156
6.2.3.4 Overall Results and Ranking of the Methods 157
6.3 Empirical Example 158
6.4 Conclusions and Recommendations 159
References 161

7 Integrating Metabolic Modeling and High-Throughput Data to Characterize Diatoms Metabolism **165**
Juan D. Tibocha-Bonilla, Manish Kumar, Karsten Zengler and Cristal Zuniga
7.1 Introduction 166
7.2 Characterization of Diatom Genomes 166
7.2.1 Available Genomics Data 166
7.2.2 Computational Tools to Allocate Gene Functions by Subcellular Localization 169
7.3 Metabolic Modeling of Diatoms: Data and Outcomes 172
7.3.1 Using Genomic Information to Build Genome-Scale Metabolic Models 172
7.3.2 Comprehensive Diatom Omic Datasets Are Useful to Constrain Metabolic Models 173
7.3.3 Unraveling New Knowledge About Central Carbon Metabolism of Diatoms 178
7.3.4 Light-Driven Metabolism that Enables Acclimation to High Light Intensities 178
7.4 Modeling Applications to Study Bioproduction and Genome Changes in Diatoms 180
7.4.1 Predicting Diatom-Heterotroph Interactions and Horizontal Gene Transfer Using Community Metabolic Models 180
7.4.2 Optimization and Scale-Up of the Production of Valuable Metabolites 181
7.4.3 Potential for the Study of Proteome Allocation in Diatoms 182
7.5 Conclusions 183
References 183

Part III: Diatom Motility 193

8 Modeling the Synchronization of the Movement of *Bacillaria paxillifer* by a Kuramoto Model with Time Delay **195**
Thomas Harbich
8.1 Introduction 195
8.2 Materials and Methods 198
8.3 Time Dependence of the Relative Motion of Adjacent Diatoms 198

8.4 Modeling Interacting Oscillators of a *Bacillaria* Colony 203
8.4.1 Observation of the Movement Activity at Uncovered Rhaphes 203
8.4.2 Interaction of Neighboring Diatoms 204
8.4.3 Coupled Oscillators 205
8.5 Kuramoto Model 207
8.5.1 Adaptation of the Kuramoto Model for a *Bacillaria* Colony 207
8.5.2 Analyses and Approximations 208
8.5.3 Critical Coupling 212
8.5.3.1 Uncoupled Oscillators 212
8.5.3.2 Two Oscillators 213
8.5.3.3 Chains without Retardation 214
8.5.3.4 Identical Oscillator Frequencies and Sufficiently Small Delay 214
8.5.3.5 Remarks on the General Case 214
8.5.4 Statistical Considerations and Monte Carlo Simulations 215
8.5.4.1 Expected Value and Standard Deviation 215
8.5.4.2 Distribution of Critical Coupling 216
8.5.5 Simulation of Non-Synchronous States 218
8.5.5.1 Numerical Integration 218
8.5.5.2 Visualization of the Transient 218
8.5.5.3 Discrete Fourier Transform 219
8.5.6 Coupling to a Periodic Light Source 221
8.6 Discussion 223
Acknowledgment 225
References 226

9 The Psychophysical World of the Motile Diatom *Bacillaria paradoxa* **229**
Bradly Alicea, Richard Gordon and Jesse Parent
Abbreviations 230
9.1 Introduction 230
9.1.1 Aneural Architecture of *Bacillaria* 232
9.1.2 Aneural Cognition in a Broader Context 233
9.1.3 Psychophysics as Diatom Information Processing 235
9.1.4 Information Processing and Aneural Cognition 236
9.1.5 Hebbian Intelligence and Predictive Processing 237
9.2 Measurement Techniques 238
9.2.1 Weber-Fechner Law 238
9.2.2 Connectionist Network 240
9.2.3 Algorithmic Information 240
9.2.4 Collective Pattern Generator 241
9.2.5 Dynamical States of the CoPG 242
9.3 CPGs vs. CoPGs 242
9.3.1 Potential of Predictive Processing 247
9.3.2 Phase Transitions in *Bacillaria* Movement 247
9.4 Aneural Regulation 248
9.5 Broader Picture of Intelligence and Emergence 249

9.5.1 Pseudo-Intelligence 249
9.6 Discussion 250
Acknowledgments 252
References 252

10 Pattern Formation in *Diatoma vulgaris* Colonies: Observations and Description by a Lindenmayer-System 265
Thomas Harbich
10.1 Introduction 265
10.2 Materials and Methods 268
10.2.1 Cultivation and Recording of Images 268
10.2.2 Formal Notation of Types of Concatenation and Splitting Processes 269
10.2.3 Methods to Observe the Processes 272
10.2.3.1 Basic Options 272
10.2.3.2 Long-Term Observations 272
10.2.3.3 Analysis of Single Images 273
10.3 Results 273
10.3.1 Observation of Elementary Splitting Processes 273
10.3.2 Observation of Synchronism 274
10.3.3 Observation of the Processes and Appearance of Colonies 275
10.3.3.1 Splitting of Elements of Types c and d 275
10.3.3.2 Splitting of Elements of Types a and b – Dynamic Analysis 276
10.3.3.3 Separation of Elements of Types a and b – Static Analysis 277
10.3.3.4 Dependence on Environmental Parameters 278
10.3.4 Theory Formation 278
10.3.4.1 Description of the Asymmetry 278
10.3.4.2 Lindenmayer System 281
10.3.5 Outer Shape of the Colonies 284
10.4 Discussion 285
Acknowledgment 287
Appendix 10A: Calculation Scheme 287
Appendix 10B: Accordance with the D0L-System 288
References 289

11 RAPHE: Simulation of the Dynamics of Diatom Motility at the Molecular Level – The Domino Effect Hydration Model with Concerted Diffusion 291
Shruti Raj Vansh Singh, Krishna Katyal and Richard Gordon
11.1 Introduction 292
11.2 Parameters 293
11.3 Ising Lattice Modeling 295
11.4 Allowing Bias 298

11.5 Computer Representation 299
11.6 The Roles of the Cell Membrane, Canal Raphes, and the Diatotepum 300
11.7 Raphan and the Raphe 301
11.8 The Jerky Motion of Diatoms 301
11.9 Diffusion and Concerted Diffusion of Raphan 302
11.10 Shear and Janus-Faced Causation: Motility and Raphan Tilting 303
11.11 The Domino Effect Causes Size Independence of Diatom Speed 304
11.12 Quantitating the Swelling of Raphan in the Diatom Trail 306
11.13 A Schematic of Raphan Discharge 307
11.14 Transitions of Raphan 308
11.15 The Roles of the Diatom Trail 310
11.16 Outline of the Simulation 311
11.17 Results 312
11.18 Discussion 315
11.19 Conclusion 316
Dedication 318
Appendix 11.1 318
Appendix 11.2 318
References 328

Part IV: Diatom Ecological and Environmental Analysis 343

12 Following the Photons Route: Mathematical Models Describing the Interaction of Diatoms with Light 345
Edoardo De Tommasi, Alessandra Rogato, Diego Caratelli, Luciano Mescia and Johan Gielis
12.1 Introduction 346
12.2 The Underwater Light Field 347
12.2.1 The Travel of Light from the Sun into Water Bodies 347
12.2.2 Numerical Computation of the Underwater Optical Field 349
12.3 Novel Geometrical Models for Diatoms 352
12.3.1 Gielis Transformations 352
12.3.2 Laplace and Fourier Revisited 355
12.4 Going Through the Wall: Simulating Light Propagation in the Frustule 356
12.4.1 Plane Wave Expansion (PWE) Method 359
12.4.2 Finite Difference Time Domain (FDTD) Method 362
12.4.3 Wide-Angle Beam Propagation Method (WA-BPM) 364
12.4.4 Fast Fourier Transform (FFT) Approach 368
12.5 Fractional Calculus for Diatoms 368
12.5.1 Fractional Calculus Based Dielectric Dispersion Model 370
12.5.2 Basic Time–Marching Scheme 370
12.5.3 Uniaxial Perfectly Matched Layer Boundary Conditions 374
12.6 Beyond the Glass Cage: The Fate of Light Inside the Cell 376
12.6.1 The Diatom Chloroplast and its Evolution 377
12.6.2 The Photosynthetic and Electron Transport Chain 378

12.6.3 The Photoprotection Mechanism 379
12.6.4 The Diatom Photoreceptors 380
12.6.5 Chlorophyll Optical Signals for Satellite Population Monitoring 380
12.7 Conclusions 383
References 384

13 A Generalized Model for the Light Response of the Nonphotochemical Quenching of Chlorophyll Fluorescence of Diatoms 393
João Serôdio and Johann Lavaud
13.1 Introduction 394
13.2 Model Formulation 395
13.2.1 Nonphotochemical Quenching Indices NPQ and Y(NPQ) 395
13.2.2 Standard Model for NPQ LCs 397
13.2.3 Generalized Model for NPQ LCs 397
13.2.4 Model Fitting and Parameter Estimation 398
13.3 Results 403
13.4 Discussion 406
13.4.1 Model Assumptions 406
13.4.2 Fitting to Experimental Data 407
13.4.3 Application 407
Acknowledgments 409
References 409

14 *Coscinodiscus wailesii* as Biogenic Charge-Based Sensors for Heavy Metal Ion Contamination Detection 413
Rajeshwari Taruvai Kalyana Kumar, Diem-Thuy Le, Antra Ganguly and Shalini Prasad
14.1 Introduction 413
14.2 Materials and Methods 416
14.2.1 Chemicals and Reagents 416
14.2.2 Cell Culture 416
14.2.3 Heavy Metal Doping and Characterization 416
14.2.4 Electrophoretic Measurements 417
14.3 Results and Discussion 417
14.3.1 Effect of Heavy Metal Doping on Cell Culture 417
14.3.2 Effect of Heavy Metal Doping on Zeta Potential 418
14.3.3 Dependency of pH on Surface Charge Potential 419
14.3.4 FTIR Characterization 421
14.4 Conclusion 424
Acknowledgments 425
References 425

Index 427

List of Figures

Figure 1.1 Reference models: (a) Model v1_k – the valve with a constant thickness and a raphe sternum. (b) Model v1_p – the valve with a constant thickness. (c) Model v_o – the contour of v1 and v2 models. 7

Figure 1.2 Panduriform *Diploneis* aff. *chersonensis* cut with FIB along transapical axis and the geometric model v1: (a) External view of the whole specimen (frustule) prior to FIB processing. Note the plain and robust girdle (white arrow) and the raphe system with central nodule (black arrow) and two series of poroids marking the external run most typical for *Diploneis* as a genus character – the longitudinal canals (arrowheads). (b) The same frustule with FIB milled apical part. (c) Close up of the frustule imaged in Figure b. Note the longitudinal canals (white arrow) separated by a raphe slit (black arrow) and poroids positioned in canal exterior (white arrowhead) and in the internal surface of the canal (black arrowhead). (d) The same specimen cut through the central nodule. Note the longitudinal canals (arrows) separated by solid silica and internal structure of the transapical striae connected through the small opening (arrowhead) to the canal. (e) The v1 model with valve surface. (f) The v1 model with longitudinal canal uncovered. Scale bars in Figures a–c = 10 μm and in d = 5 μm. 9

Figure 1.3 Panduriform *Diploneis* spp. cut with FIB along the transapical axis and along the valve surface and the geometric model v2: (a) *Diploneis didyma* cut with FIB along transapical axis. Note the presence of two longitudinal canals (arrows) parallel to the raphe system and the solid transapical striae composed of a single silica layer (arrowheads). Figure b–d shows *Diploneis* aff. *chersonensis* milled with FIB along apical axis: (b) half of the frustule prior to FIB cuts. Note the transapical striae composed of two silica layers (arrows). (c) Same frustule with FIB milled superficial valve layer. Note the open system of longitudinal canals. Note the remnants of the transapical ribs upon which the longitudinal canal cover is based (arrowheads). (d) Close up of the specimen illustrated in Figure c. Note the poroids connecting the longitudinal canal with the cell lumen (arrows). (e) The v2 model with valve surface. (f) The v2 model with short ribs uncovered. Scale bars in Figures a–c = 10 μm and in d = 5 μm. 10

Figure 1.4 Panduriform *Diploneis* aff. *chersonensis* valve interior cut with FIB along the apical axis: (a) Valve interior illustrated prior to FIB milling. Note the elevated longitudinal canals (black arrows) expanding at the central nodule (white arrows) parallel to the raphe system (black arrowheads), robust transapical ribs (white arrowheads). (b–e) Consecutive stages of FIB processing: (b) Valve nanocut through the transapical ribs (white arrowhead) with a row of areola distinguishable (black arrowhead). (c) The FIB cuts of the longitudinal margin, still completely closed by the transapical ribs (arrowhead). (d) The first cut of the longitudinal canal through its expansion within central nodule. Note the open space (arrowhead). (e) FIB processing of the longitudinal canal. Note the free space in the valve middle (arrowhead). Scale bars in Figures a–e = 10 μm. 11

Figure 1.5 Mapping of the *Diploneis* aff. *chersonensis* valve stiffness in response to downward oriented stress: (a) Model v1 and (b) Model v2. 12

Figure 1.6 Normalized displacement compared between models: (a) Comparison among v1, v1_p and v1_k models. (b) Comparison between v2, v2_p and v2_k models. (c) Comparison between the v_o, v1 and v2 models, including normalized displacement (blue bars) and normalized displacement with mass factor (orange bars). 13

Figure 2.1 (a) The diatom division scheme according to the MacDonald–Pfitzer rule. From left to right: vegetative process, from right to left: closing the life cycle by sexual reproduction. (b) Change of size distribution from generation to generation according to Pascal's triangle. 21

Figure 2.2 Temperature dependence on the doubling rate for *Haslea ostrearia*. Strains were adapted to 10°C and 26°C and then exposed to different temperatures. (Constructed with data from [2.25]) 23

Figure 2.3 The modified MacDonald–Pfitzer scheme of *Ellerbeckia arenaria* according to Müller. New thecae are enclosed by a girdle band in the first generation which is faced by a thickening or step in the valve of the next cell (a). The detailed formation of this peculiar arrangement was described in more recent publications [2.43] [2.75]. The reference point for the size of the theca when considering the MacDonald–Pfitzer rule is the thickening of the valve if there is one (b). In Müller's division pattern nascent cells always skip one generation before dividing again. This leads to Fibonacci growth and formation of twin and triplet groups. 27

Figure 2.4 Stability and oscillatory regions (yellow and purple area, resp.) in the complex plane for (a) discrete and (b) continuous dynamical systems. The square and diamond-shaped marks show possible locations of the eigenvalues for an asymptotically stable system and for a system with sustained oscillations, respectively. 32

Figure 2.5 Three cumulative distributions functions for generation times ("lifetimes" of a Bellman-Harris particle): (a) fixed lifetime,

(b) exponentially decreasing lifetime distribution, (c) lifetime with Gaussian distribution. 41

Figure 2.6 Size-dependent growth parameter described by a parabolic function. 44

Figure 2.7 Direct comparison of the Fuhrmann-Lieker *et al.* model with and without aging. (a) Development of the population in a small model with 20 size classes in which cells die only due to overpopulation (saturation by Ricker term) (b) Trajectory of the size distribution in (a), represented by mean size and variance. (c) Same model as in (a), but thecae live for only 7 generations as suggested by Jewson. Long-term oscillations are induced. (d) Trajectory of the size distribution in (c). 51

Figure 2.8 Cell size dynamics of *Pseudo-nitzschia multistriata* according to D' Alelio *et al.* [2.19]. Upper graph: cell abundance at the sampling period. Center: Division of the population according to cell size into subpopulations (cohorts, dotted ellipses) and size classes (solid circles). The mean cell size of the cohorts is indicated in pink. The dashed line indicates the postulated upper size threshold for sex. Lower graph: Comparison with the model. Different model cohorts are emphasized by different colors. Changes of cell size reduction speed are attributed to different growth phases: normal growth (gray in the bar below), bloom phases (black), and stationary phases (white). (Reproduced with permission). 53

Figure 2.9 Temporal development of population parameters and dynamics in the Fuhrmann-Lieker *et al.* model with *zeitgeber* rhythm. (a) Population size and *zeitgeber* modulation of the parameter σ with a period of 40 generations, multiplied by 100 for better scaling (red), (b) Number of auxospores, showing years of low and high auxosporulation (c) Fourier transform of the total cell number for the first 8000 generations. Temporal frequency is given as 1/generation (d) Fourier transform of the total cell number for the first 256000 generations. 54

Figure 3.1 Soil species of diatoms capable of living on the surface of wet rocks, high in the mountains: *Hantzschia amphioxys* (a–c), *Pinnularia borealis* (d, e) and *Luticola mutica* (f). SEM images by Helga Kemnitz, GeoForschungsZentrum, Germany. The images of diatom frustules presented in this chapter were taken using three different SEMs: Hitachi TM 1000 (MISIS), Tescan LYRA (IO RAS), and Tescan MAIA3 (Lomonosov Moscow State University). 66

Figure 3.2 Treelike colonies of diatoms from different evolutionary groups: (a) *Licmophora gracilis* (West Antarctica, Powell Basin), (b) *Gomphonema olivaceum* (Moscow-River, Central Russia), (c) *Didymosphenia* sp. (Lake Baikal, East Siberia). (Light microscopy by Ph.V. Sapozhnikov, O.Y. Kalinina, SIO RAS). 67

Figure 3.3 Habitual similarity of the shells of diatoms from different evolutionary groups: (a) *Donkinia carinata*, (b) *Plagiotropis longa*, (c) *Auricula intermedia*, (d) *Plagiotropis gibberula*, (e) *Pl. lepidoptera*, (f) *Trachyneis aspera*. (Light microscopy images by Ph. V. Sapozhnikov, O.Y. Kalinina, SIO RAS). 67

Figure 3.4 Diatoms with a radial symmetry of the frustule are mostly adapted to life conditions within planktonic populations (a–c, e–g); diatoms with a bilateral symmetry are more prevalent in benthos and periphyton (d, h–l). (SEM images by M.V. Kravchishina, Ph. V. Sapozhnikov, O.Y. Kalinina, SIO RAS). 68

Figure 3.5 Fragilarioids are feathery diatoms without sutures. They are incapable of independent movement, but build colonies in the form of bundles, stars and ribbons. (SEM images by Ph. V. Sapozhnikov, O.Y. Kalinina, M.V. Kravchishina, SIO RAS). 69

Figure 3.6 Active wanderers among diatoms: (a, c) naviculoids, (e, f) nitzschioids, (b, g) surirelloids, (d) diploneoids and (h) fallacioids. (SEM images by Ph. V. Sapozhnikov, O.Y. Kalinina, M.V. Kravchishina, SIO RAS). 69

Figure 3.7 Semi-mobile forms of the class Bacillariophyceae, which have a pronounced tendency to build widespread colonial settlements ("shell coverings") on the surface of solid substrates: (a–d) cocconeoids, amphoids (lying "belly up", e–h) and various achnantoids (i, j, l - from the dorsal side of the shell, k - from the ventral). (SEM and light microscopy images by Ph. V. Sapozhnikov, O.Y. Kalinina, M.V. Kravchishina, SIO RAS). 70

Figure 3.8 Different representatives of thalassiosiroids (a–f) and coscinodiscoids *Coscinodiscus radiatus* (g, h). (SEM images by M.V. Kravchishina, SIO RAS, taken through a light microscope by Ph. V. Sapozhnikov, SIO RAS). 71

Figure 3.9 Widespread life forms of planktonic and periphyton-planktonic diatoms: (a–d) biddulphioids, (e–g) skeletonemoids and (h–j) chaetoceroids. (Images by Ph. V. Sapozhnikov, O.Y. Kalinina, SIO RAS). 71

Figure 3.10 (a, b) Colonial form of naviculoids (*Navicula smithii*), a colony living in biopolymer tubes. (c–f) Colonial forms of nitzschioids: ribbon-like colonies (d, f) and separate valves (c, e) of different species from the genus *Fragilariopsis*. (Images by Ph. V. Sapozhnikov, O.Y. Kalinina, SIO RAS). 72

Figure 3.11 SEM images of *Achnanthidium sibiricum* diatoms showing the basic structure of the cell wall (frustules). Insets: close-up of individual sections of the frustules (scale bar: 500 nm). (Reproduced with permission from [3.37].) 74

Figure 3.12 Ashby plot of strength vs. density for naturally occurring biological materials. (Reproduced with permission from [3.60].) 77

Figure 3.13 SEM and AFM structural characterization of porous layers of the silica frustule of *Coscinodiscus* sp. (a) SEM and (b) AFM images of the cribellum surface (z scale 500 nm). (c) SEM and (d) AFM images of the cribrum surface (z scale, 500 nm). (e) SEM and (f) AFM images of the internal plate (z scale, 700 nm). (g) AFM image of the central part of the internal surface (z scale 700 nm) with inset showing channels and ribs. (h) SEM image of a cross section of the diatom frustule showing the internal shape of pores connecting cribellum and cribrum. (i) SEM and (j) AFM images of girdle band area of surfactant-cleaned diatom in side-on position (z scale, 200 nm) and (k) SEM image of a cross section of the girdle band. Numbers within white circles denote the three porous layers of the diatom frustule: (1) cribellum, (2) cribrum, and (3) internal plate. Cross-section image (h) is shown before images of the internal layer (e, f, g). (Reproduced with permission from [3.69].) 79

Figure 3.14 (a) Low-resolution AFM image of the internal plate after nanoindentation. (b) Series of indentation points (solid arrows) obtained using a force of 1.5-3 μN. (c) Typical force penetration curve showing loading and unloading parts (dashed arrows). (d) AFM image showing nanoindentation performed on an areola rib at higher applied load (12 μN). White scale bar is 10 μm. (Reproduced with permission from [3.69].) 80

Figure 3.15 Mode shapes and natural frequencies of *Coscinodiscus* sp. (centric) and *Ulnaria acus* (pennate) observed in the simulation using COMSOL Multiphysics. (Reproduced with permission from [3.112].) 85

Figure 3.16 The example of FIB-SEM milling of diatom species (destructive method). 86

Figure 3.17 Ptychography of *Aulacoseira* diatom cells. (Images courtesy of T.E.J. Moxham, University of Oxford/Diamond Light Source, U.K.) 87

Figure 3.18 A classification scheme for pore occlusions, provided with Latin descriptive terminology. Listed at the bottom are lengths of the major axes of the pores. The pores in the first five panels from the left are characterized with wavelengths of 60–100 nm. For panel six, nanopores have a wavelength of 10 nm, and for panel seven one of 220 nm. (Reproduced from [3.114].) 87

Figure 3.19 (a)-(g) Patterns produced by tuning negative long-range feedback of a lateral dependence stochastic algorithm (LDSA) compared to similar-looking SEM micrographs of pore occlusions of various diatom algae. (Reproduced from [3.114].) 89

Figure 3.20 Probabilistic patterns of self-organized matter calculated for different long and short order parameters and factors: (a) Long-range parameter 4, long-range factor -1, short-range parameter 1, short-range factor 30; (b) Long-range parameter 4, long-range factor -2, short-range parameter 1, short-range factor 10; (c) Long-range parameter 4, long-range factor -1, short-range parameter 2, short-range factor 30; (d) Long-range parameter 4, long-range factor -2, short-range parameter 1, short-range factor 10; (e) Long-range parameter 5, long-range factor -1, short-range parameter 2, short-range factor 5; (f) Long-range parameter 5, long-range factor -1, short-range parameter 3, short-range factor 3. 89

Figure 3.21 (a) IVIS fluorescence images of *A. sibiricum* in cell culture flasks; (b) the total radiant efficiency of diatoms measured during 100 days of cultivation; (c) fluorescence spectra of diatoms; (d) comparison of normalized fluorescence intensity and average radiant efficiency during the cultivation period. (Adapted from [3.122].) 91

Figure 3.22 SEM image of the *Coscinodiscus oculus-iridis* frustule (Ehrenberg). 92

Figure 3.23 (a) Fourier transform of the selected circular area in Figure 3.22. (b) A close-up view of the central region of the Fourier transform in Figure 3.23a. 92

Figure 3.24 Prominent peaks in the radial variation of FFT corresponding to the features at different length scales (reverse logarithmic scale used to correspond to the peak order outward from pattern center). 93

Figure 3.25 Reconstructed patterns corresponding to different peaks of Fourier Transform signal: (a) Initial picture provided for reference; (b) Features that correspond to 120 nm FFT peak; (c) Features that correspond to 460 nm FFT peak; (d) Features that correspond to 860 nm FFT peak. 94

Figure 3.26 Radial distribution function (RDF) $g(r)$ plotted for convenience using the same inverted logarithmic distance scale used in Figure 3.24. Prominent peaks are observed corresponding to structural order across several orders of length scale. 94

Figure 3.27 Spatial distribution function (SDF) illustration. 95

Figure 4.1 *Fragilaria mesolepta.* (a) A whole view of the two valves connected with linking spines. Upper valve showing the inside portion with the linear sternum (dashed line). (b) Enlarged valve end with the apical pore field (large arrow) and the linking spine (arrow head). Note the rimoportula (small arrow) placed in the border of the valve face and mantle. Scale bar = 5 μm (a), 0.5 μm (b). 107

Figure 4.2 *Fragilaria mesolepta.* A part of a chain colony. (a) Each cell has one H-shaped chloroplast. (b) Divided chloroplast shortly before cell division. (c) After cell division. Note the newly forming valves (arrows). Scale bar = 5 μm. 108

Figure 4.3 *Staurosira binodis.* (a) Valve exterior showing the narrow linear sternum (dashed line) and the linking spines (arrow) on the valve face and mantle juncture. (b) Valve interior with the spindle-shaped axial area (*). Scale bar = 5 μm. 109

Figure 4.4 *Staurosira binodis.* A part of a chain colony. (a) Cells in three phases, with one chloroplast (*), two chloroplasts (**), and after cell division (no asterisk). (b) The same colony as (a). Green color indicates newly forming valves with PDMPO fluorescence. Scale bar = 5 μm. 109

Figure 4.5 *Fragilaria mesolepta.* The series of forming valves arranged sequentially by the degree of development. As the valves get longer, the valve components become thicker and more developed. Scale bar = 5 μm. 111

Figure 4.6 *Fragilaria mesolepta.* A forming valve in the earliest stage. (a) A whole view showing a chain of small rings with short virgae (arrow heads). (b) Enlargement of the upper half of (a). The ring chain is composed of silica spherules. The uppermost ring extends the branching virgae. (c) Enlargement of the lower half of (a). At the bottom end, two longer ribs extend multiple virgae. Scale bar = 2 μm (a), 0.5 μm (b, c). 111

Figure 4.7 *Fragilaria mesolepta.* Sequential valve development after the earliest stage. The ring chain gradually becoming obscure and united into a single linear sternum. Note the initiation of the linking spine (c, circled) and the vimens (d, arrows). Scale bar = 2 μm. 112

Figure 4.8 *Staurosira binodis.* Forming valves in the earliest stage. (a–c) Valve composed of a pair of threads arranged as arcs facing each other. Drawing (c) was prepared from multiple images observed with scanning electron microscope equipped with a transmission device (STEM) (a) and SEM (b). A couple of threads, one (TLV, red) with longer virgae (LV) and the other (TSV, blue) with shorter virgae (SV) make a loop at one end and fuse into a linear structure at the opposite end. (d, e) Sibling forming valves side by side. Two pairs of threads are partly broken and overlapped. Note the unorganized silica between two threads (*) and the down feather-like structure (circled). STEM image (d) and drawing (e). Scale bar = 2 μm (a, b, d). 113

Figure 4.9 *Staurosira binodis.* Forming valves in the earliest stage. (a, b) Enlarged sibling valves side by side. Each valve comprises a thread with longer virgae (TLV) and a thread with shorter virgae (TSV). The densities of the longer virgae (arrow heads) and the shorter virgae (arrows) are the same. Note the unorganized silica (*). STEM (a) and SEM (b) images. (c–e) A pair of threads demonstrates both closed ends. Enlargement of top of (c) shows a rounded looping end (d). Enlargement of bottom of (c) shows a linearly fused end of two threads (e). Scale bar = 2 μm (a, b), 5 μm (c), 1 μm (d, e). 113

Figure 4.10 *Staurosira binodis.* Sequential valve development after the earliest stage. (a) An early forming valve with short virgae extending from a sternum. (c) Enlarged top of (b) showing the looping end. (d) A valve with the remaining of hairpin-like loop in the upper half (arrow). (e) A valve with early extending vimens. Note the initiation of linking spine (arrow). (f) Enlargement of (e). Note the virgae near the sternum bearing vimens (large arrows) and denser projections (small arrows). (g) A valve with forming axial area. (h) A valve with completed axial area. Scale bar = 5 μm (a, b, d, e, g, h), 0.5 μm (c, f). 114

Figure 5.1 Exponential population growth. 124

Figure 5.2 Simple growth curve depicting the saturable behavior of Eq. 5.5. 125

Figure 5.3 Gompertzian growth [5.40] [5.41] [5.29]. 126

Figure 5.4 Gompertzian lag time. See Eq. (5.12) for parameters a, b, c. [5.54]. 126

Figure 5.5 Cell population exceeds the limit that may be supported by nutrients leading to negative growth rate and reduction in population [5.37]. 126

Figure 5.6 Small growth for large cells [5.18]. 128

Figure 5.7 Growth vs. cell size variation is not monotonic [5.51]. 128

Figure 5.8 Silicon limited uptake rate in Michaelis-Menten model. 133

Figure 6.1 Growth curve examples: (a) A high data density yeast (*Saccharomyces cerevisiae*) growth curve; (b) a comparatively low data density diatom (*Skeletonema marinoi*) growth curve. Biomass is either recorded by optical density (light absorbance at 600 nm) or chlorophyll fluorescence (excitation and fluorescence emission wavelengths at 425 and 680 nm, respectively). h = hours; OD600 = optical density at 600 nm; RF = relative fluorescence. 142

Figure 6.2 Examples of diatom growth curves: Data was retrieved from an arbitrary selection of published studies without considering data quality or research aims. The selection includes common model species and rarely cultured species, from both benthic and pelagic marine environments. Cell densities are based on cell counts, optical density, and chlorophyll fluorescence, and each species dataset has been scaled to the maximum observed density to enable a unit-free comparison (y-axis values, shown here in the natural logarithm scale). All curves represent an average of multiple replicates, and we have omitted error bars due to the inconsistency of variance reporting between studies. The data is presented on four panels based on the duration of a growth experiment (a) four days, (b) 12 days, (c) 16 days, (d) 27 days. Additional information about each species' growth curve and the respective study can be found in Table 6.1. 144

Figure 6.3 An example of a *Skeletonema marinoi* growth curve. Cells were counted twice a day, at the start and the end of the light "day" phase under laboratory conditions (authors own data, unpublished). 146

Figure 6.4 Examples of fitting a nonparametric curve to empirical *Skeletonema marinoi* data (black dots) using cubic splines (a) and local polynomial regression (b), with different smoothness parameter *s* values.

The empirical data values represent relative *in-vivo* chlorophyll *a* fluorescence, log-transformed. 149

Figure 6.5 Three randomly chosen example growth series for 6 out of 27 challenge groups from the simulation trial. Each series was generated using the Baranyi-Roberts function with a random lag and initial value (shown), fixed maximum 100, and rate 1. Only the dots are taken as observations, the lines connecting the dots are merely part of the illustration and do not necessarily follow the true function. t.a.d. = times a day (sampling frequency). 151

Figure 6.6 Quality of estimating the specific growth rate with only the noise challenge added, consisting of three levels of noise variance and three frequencies of sampling (t.a.d.). (a) The loess GCV error distributions. (b) Quality metrics relative to loess GCV (=0x; lower values indicate better performance). 153

Figure 6.7 Quality of estimating the specific growth rate when the crash is introduced into the series after day 7. (a) The loess GCV error distributions. (b) Quality metrics relative to loess GCV (=0x). 155

Figure 6.8 Quality of estimating the specific growth rate when the experiment is halted at day seven. (a) The loess GCV error distributions. (b) Quality metrics relative to loess GCV (=0x). 156

Figure 6.9 Ranking of the quality metrics over all 27 challenge sets. The small transparent symbols are the rankings per individual challenge set, while the big dark symbols represent the average score. 157

Figure 6.10 Analysis of a temperature treatment growth experiment with 8 strains (3 replicates per strain) of the chain forming diatom *S. subsalsum*, comparison of mean relative growth rates extracted by Gompertz, ad hoc, and spline GCV methods. (a) Empirical growth curves; (b) Specific growth rate estimates by each of the chosen methods. 158

Figure 7.1 State-of-the-art genomics and metabolic modeling of diatoms: (a) Neighbor-joining tree based on full-length of the ribulose-1,5-bisphosphate carboxylase/oxygenase large subunit (chloroplast) gene sequences, showing phylogenetic relationships between diatoms with complete genome sequence available. NCBI or UniProt accession numbers are given in parentheses. GEMs properties (i.e., number of reactions, metabolites, and genes) are given under the diatom name for those that have an available genome-scale metabolic model). (b) Timeline of first genome annotations and published metabolic reconstructions of *T. pseudonana*, *P. tricornutum*, and *F. cylindrus*. The straight line represents the timeline from the first diatom sequenced until the present. Dotted lines connect models that were used as templates (line start) for other models (line end). 167

Figure 7.2 Shared reactions across the latest *T. pseudonana* and *F. cylindrus* GEMs and their template *i*LB1027_lipid. (a) Shared reactions with the same ID across models *i*LB1027_lipid, *i*Tps1432, and Fc_GEM. (b) Shared homologous reactions including: 1) same reaction with a different ID, 2) the same reaction at a different compartment, 3) the same reaction with a different associated enzyme. 173

Figure 8.1 Colored plate from the German translation of the description of *Bacillaria* colonies by Müller [8.28]). Shown are various typical configurations such as an extended colony seen from two perspectives, S-shaped configurations that usually change dynamically, and some compact forms that are found statically or with little movement. 196

Figure 8.2 Image of a moving small colony on a microscope slide with coverslip. Diatom #1 was permanently attached to the slide, so only the paths of the other diatoms had to be tracked. Since the colony did not change its orientation in the plane of observation, it was sufficient to determine the coordinates of the diatoms with respect to a coordinate axis in apical orientation. 199

Figure 8.3 Time curve of the differences of the positions of neighboring diatoms, where the numbering corresponds to the identifiers from Figure 8.2. The duration of one cycle is 62.5 seconds. 200

Figure 8.4 Path-time diagrams, velocity-time diagrams, and a phase diagram of the relative motion of diatoms covering about 2 periods. (a) Position and velocity from a measurement, where the velocity was obtained by numerical differentiation. The inaccuracies in the tracking of positions contribute to the noise of the velocity curve. (b) Figure obtained by removing the rest periods. The adjustment of the velocity curve by straight lines was done manually. (c) Idealization of the curves assuming linearly increasing or decreasing velocity and exactly maintained rest periods. For clarity of presentation, the parameters were not taken from the measured data shown. (d) In addition to the two curves from Figure 8.1a, fitted curves have been plotted based on the above assumptions. Such a good fit based on one set of parameters usually succeeds only over a few periods, presumably because environmental parameters and internal parameters of the diatoms do not remain constant over a long time. (e) Velocity as a function of displacement according to observation and model. (f) Exemplary plot of five displacements (six diatoms) versus phase, with identical parameters chosen except for amplitudes and phase shifts ($t_b/T = 0.1$). 201

Figure 8.5 Trajectory of a single diatom relative to the substrate. As the motion had a constant orientation during the observation period, only one coordinate had to be considered. The period of the fitted theoretical curve (dashed) is 59.5 seconds. 203

Figure 8.6 Control-oriented modeling of the interaction of two adjacent diatoms. The synchronization of two incomplete oscillators by separate controllers is assumed. The function of the controller is included in the controlled oscillators. 204

Figure 8.7 Modeling of a small colony with oscillators and their communication links, as far as it is necessary for the emergence of synchronous motion. The communication within the diatoms, is shown with dark arrows, the hypothetical coupling to the outer incomplete oscillators with light gray arrows. 206

Figure 8.8 Illustration of phases and chronological sequence for identical and nearly identical oscillator frequencies in the case of 12 oscillators. (a) Numerator of Eq. (8.20), which is proportional to the phases ϑ_j. For an even number of oscillators there are two identical minimal phases of equal size, for an odd number there is exactly one minimal phase. (b) elongated colony at rest and appearance of the colony shortly after the onset of motion. The relative displacements are indicated by arrows. It does not matter whether the colony is fixed at one end. (c) Circulating phases from a simulation with 12 oscillators and identical oscillator frequencies. Only 6 marks for the phases are seen, as each mark covers another. In units related to $\bar{\omega}$, a delay of $0.04T'(T' = 2\pi / \bar{\omega})$ and a coupling constant $0.5\bar{\omega}$ of were used. (d) Circulating phases from a simulation with different oscillator frequencies. The parameters correspond to those of c, but with an empirical standard deviation of the angular frequencies of $0.1\bar{\omega}$. 211

Figure 8.9 Histograms as approximation of the probability density function and corresponding cumulative distribution from a Monte Carlo simulation for different numbers of coupled oscillators. Their counts are noted next to the curves. The units are related to the standard deviation μ. Here, $\sigma = 0.064\,\mu$ was always chosen. (a) Histograms without delay. (b) Cumulative probability without delay. (c) Histograms with delay $\tau = 0.16\,T'$ with ($T' = 2\pi/\mu$). If μ corresponds to an oscillation period of 60 seconds, then the delay is 9.6 s. (d) Cumulative probability with delay. 216

Figure 8.10 Illustration analogous to Figure 8.9, but for different values of the standard deviation. Their values in units of μ are noted next to the curves. Here, a chain consisting of 12 oscillators and $\sigma = 0.064\,\mu$ was always chosen. (a) Histograms without delay. (b) Cumulative probability without delay. (c) Histograms with delay $\tau = 0.16\,T'$ ($T' = 2\pi/\mu$). (d) Cumulative probability with delay $\tau = 0.16\,T'$. 217

Figure 8.11 Illustration of the transient of 12 oscillators with normally distributed frequencies. Here the delay is $0.01T'(T' = 2\pi / \bar{\omega})$, the coupling constant is $0.48\bar{\omega}$ and the standard deviation of the frequencies ω_j is $0.034\,\bar{\omega}$. (a) Sine of the phase differences versus time measured in units of T'. (b) Coupling term for each phase. 218

Figure 8.12 Illustration of the transient oscillation of a colony of 24 diatoms consisting of two groups of 12 oscillators with the same frequency. The distance of the frequencies from their mean is $\pm 0.032\,\bar{\omega}$. (a) Coupling constant $0.5\,\bar{\omega}$. (b) Coupling constant $0.032\,\bar{\omega}$. 219

Figure 8.13 Results of a discrete Fourier transform, where $N = 2048$, $\tau = 0.04T'(T' = 2\pi/\bar{\omega})$ and empirical standard deviation of the sample $0.044\bar{\omega}$. (a) $k = 10^{-7}\bar{\omega}$. (b) $k = 0.03\bar{\omega}$. (c) $k = 0.05\bar{\omega}$. (d) $k = 0.2\bar{\omega}(k > k_c)$. 220

Figure 8.14 Short period of time from a simulation with $\tau = 0.04T'$ $(T' = 2\pi/\bar{\omega})$, k = $0.48\bar{\omega}$, $L = 0.64\bar{\omega}$, standard deviation of the angular frequencies is $0.07\bar{\omega}$ and the period of illumination 0.9 $T'(T' = 2\pi/\bar{\omega})$. The phase differences are constant as a periodic light source is switched on. After about two periods, the chain has settled and follows the periodic illumination. (a) Settling of the phase differences (b) Settling of the frequency correction term. 223

Figure 9.1 An image of the *Bacillaria* colony, captured in its elongated (a) and contractile (b) state. (Thanks to Dr. Thomas Harbich.) 231

Figure 9.2 Sensory thresholds (just noticeable differences). Differences in light intensity measured by output movement weights ($W_{mm,nn}$). Network shown as edges and nodes that connect all cells in the colony. 241

Figure 9.3 Deviations from the collective pattern generator (CoPG). Difference between a central pattern generator (CPG), characterizing a neural circuit, and collective pattern generator (CoPG), characterizing collective behavior. 242

Figure 9.4 Movement trace of a single cell that halts at different points in a single oscillation. (top) Sine wave with tangent at quarter-phase; (bottom) Sine wave with tangent at three-quarter phase. 243

Figure 9.5 Movement trace of a single cell that halts at different points in a single oscillation. (top) Two adjacent and overlapping pairs of oscillatory cell movements (sine waves, three-quarter phase); (bottom) Two adjacent and overlapping pairs of oscillatory cell movements (sine waves, quarter-phase). 245

Figure 9.6 Single oscillations with noise and compared with another cell. (top) Sine wave with random (white) noise; (bottom) Attractor map (based on sine waves, quarter-phase). 246

Figure 10.1 *Diatoma vulgaris* after a drawing by Bory de Saint-Vincent [10.2]. Colonies in epiphytic growth are shown at three different scales. In addition to the girdle view, the valve view is also sketched in each partial image. 266

Figure 10.2 Tree structure of binary divisions starting from a pair of diatoms in a chain. Each diatom forms another pair after an individual doubling time. Diatom pairs are represented as adjacent dots. The diatoms of each generation are on a horizontal straight line. Shown is the logical structure, not the doubling times. Thus, the cells of one of a selected generation can be formed at different times. 268

Figure 10.3 (a) The four types shown with their identifiers form the basis for a formal notation of the transitions after cell division. In the "diagonal" types *a* and *b*, the contact points are located at the most distant apices. Types *c* and *d* are referred to as "non-diagonal." (b) Example of a small colony in which each type appears once. 269

Figure 10.4 The diagonal types split into the same type and a non-diagonal type. 270

Figure 10.5 The non-diagonal types can split into two different diagonal types or two identical non-diagonal types. 270

Figure 10.6 Effect of rotation and mirroring on processes I to IV. The execution of S = T*M interchanges processes I and IV, as well as II and III. 271

Figure 10.7 Effect of rotation and mirroring on the processes V to VIII. There is no geometric transformation that interchanges V and VI or VII and VIII. 272

Figure 10.8 A short section from a colony in the DIC. EPS pads are only visible at the connecting points of adjacent diatoms. 274

Figure 10.9 Images (a), (b) and (c) show phases of colony development in synchronous sections in chronological order. In image (d), a long chain is displayed with different stages of division, a section with overlap, an area in valve view, and an area outside the focal plane. In many cases, a focus stack helps. 275

Figure 10.10 A few images from the development of a colony. Image 5 was taken 130 hours after the first image. The colony was exposed to a light-dark cycle with 13 hours of light phase. For analysis, the chains are viewed from the right. 276

Figure 10.11 Section of a colony where an indication of pairing is recognizable. As the types of diatoms are determinable at the image edge, these are included, although they are cut. 278

Figure 10.12 Real types that contain information about asymmetry. 279

Figure 10.13 Elements created by mirroring on the horizontal line. 279

Figure 10.14 Illustration of the processes of splitting $\bar{a}$. The process identifiers have been kept. 280

Figure 10.15 Processes derived by the operation of mirroring M. 280

Figure 10.16 Processes starting from non-diagonal elements. The processes merge into each other by mirroring M. 280

Figure 10.17 Exemplary colony projected into the plane. 285

Figure 10.18 Random structure not formed according to the production rules of the D0L system. In a planar structure and with the angles chosen, a sequence of three consecutive non-diagonal elements would lead to a collision of parts of the chain with itself. 286

Figure 11.1 Scanning electron micrograph showing a single file row of secreted fibrils in *Pinnularia viridis*. Scale bar 500 nm. (Reproduced from [11.93] with permission of John Wiley & Sons/Blackwell Publishing). 295

Figure 11.2 Temperature dependence of average speed for four genera. (Reproduced from [11.30] with permission of Taylor & Francis). 300

Figure 11.3 A pencil balanced on its tip could fall in any direction. A perfect balance of raphan can be designated by the symbol |, perpendicular to the axis of the raphe, whereas tilting left or right can be indicated by \ or /. On hydration, such raphans would push the diatom right or left, respectively (with the pivot point corresponding to the "eraser" end). We assume tilting occurs instantaneously, upon injection from the membrane into the raphe. Groups are allowed to flip back and forth (variable: bias). (Reproduced from [11.171] with open access permission from SlideServe). 303

Figure 11.4 Falling dominoes. (From [11.208] under a Creative Commons Attribution License). 304

Figure 11.5 Speed of diatoms (μm/s) in Table 11.3 versus length (μm) for various diatoms of many species. The linear fits have R2 values of 60.7% for the maximum speed (blue) and 4.1% for average speed (green). While these low values, despite being below or near the standard cutoff for significance of 50%, might contradict our and other's assumptions of independence of speed from length (see text), we hope this will inspire more definitive experiments. Given the decreasing size vs. generation in the vegetative phase of diatoms, a more definitive test would be to compare speeds of diatoms of different sizes from the same clone. 305

Figure 11.6 A sampling of patterns generated by the Wolfram rules. Rule 30 is chaotic. (Reproduced from [11.238] with permission of Wolfram Research). 309

Figure 11.7 Top: Details of the first 50 events for bias = 0.8. Bottom: Wolfram-like representation of the consecutive states, ignoring tilt of the raphans. This emphasizes the diffusion events, which do not directly contribute to motility, as the hydration events do this. 311

Figure 11.8 Sample runs with the bias parameter incremented from 0.5 (no bias) to 1 (maximum bias). Note: 1) even with no bias, the source of the raphan being highest at one end is sufficient, so that the net motion is forwards; 2) in all cases, except bias = 1, some backwards motion occurs. More or less steady forward progress of the diatom may require bias > 0.7. All other variables were set to the defaults (Table 11.1). For intuitive clarity, in this figure the mean motion of the simulated diatom is to the right. Note that in all cases with bias > 0.5 there are long periods of nearly constant speed. 312

Figure 12.1 Visual representation of the quantities used in the definition of radiance $L = L(\theta, \phi)$: polar angle defining photon travel (θ), elemental surface (dS), its projection on the direction of propagation of light ($dS \cos \theta$), and elemental solid angle surrounding the same direction ($d\Omega$). 351

Figure 12.2 Schematic representation of the connections between frustule morphogenesis, geometry, and physical functions. (Reproduced with permission from De Tommasi *et al.* [12.27].) 353

Figure 12.3 The solution of the Robin problem for the Laplace equation in a shell S with fourfold symmetry in XY direction. Left: Boundary behaviour along the inner shell surface of the partial sum U_N of order $N = 11$. Right: relative boundary error e_N as function of the order N of the truncated spherical harmonic expansion for the super-shaped shell on the left. (Reproduced with permission from De Tommasi *et al.* [12.27].) 355

Figure 12.4 Artificial PhC (a) and a detail of a *Coscinodiscus wailesii* valve (b). (Reproduced with permission from De Tommasi *et al.* [12.27].) 359

Figure 12.5 SEM image of a *C. granii* whole frusule (a) and two-dimensional band structure of a valve immersed in water as retrieved by effective-index method calculations (b). Detail of the girdle (c) and relative, numerically reconstructed three-dimensional band structure obtained by PWE method, considering air as surrounding medium (d). Reproduced with permission from Fuhrmann *et al.* [12.45].) 362

Figure 12.6 (a) Cross-sectional view of a *C. granii* girdle (SEM image). (b) 3D CAD reconstruction over four unit cells of the girdle. Numerical values derived by SEM characterization: $a_1 = 285 \pm 5$ nm; $a_2 = 279 \pm 11$ nm; $d = 124 \pm 21$ nm; $h_1 = 225 \pm 21$ nm; $h_2 = 250 \pm 32$ nm; $C = 290 \pm 18$ nm; $D = 745 \pm 43$ nm. (c) Reflectance of the girdle immersed in water as a function of incidence angle (for in-plane propagation) and for different wavelengths as retrieved by FDTD simulations (left) and measured by micro-scatterometry (right). (Reproduced with permission from Goessling *et al.* [12.56].) 363

Figure 12.7 (a) Single valve of an *Arachnoidiscus* sp. diatom illuminated by radiation at $\lambda = 640$ nm; (b) refractive index map (XY plane) of the valve immersed in air retrieved from a SEM characterization of the frustule; (c) experimental transmitted intensity in XZ plane; (d) transmitted intensity in XZ plane as evaluated by WA-BPM; (e) hot-spot acquired at $z = 320$ μm; (f) transmitted intensity evaluated by WA-BPM in XY plane at $z = 307$ μm. 367

Figure 12.8 (a) SEM image of a detail of a *C. granii* valve with indication of the hexagonal symmetry of the lattice (top) and corresponding mask used in numerical simulations (bottom). (b) transmitted intensity as a function of the distance from the frustule for λ = 532 nm (top) and λ = 633 nm (bottom). Dots represent experimental data, while continuous (top) and dashed (bottom) lines refer to numerical simulations. (Reproduced with permission from Maibohm *et al.* [12.82].) 368

Figure 12.9 Origin and structure of the diatom chloroplast. The upper panel shows the schematic representation of the secondary (a) and tertiary (b) endosymbiont hypothesis of diatom evolution. The lower panel (c) shows a schematic diagram of the four membranes surrounding the diatom chloroplast. Abbreviations are as follows: cERM; chloroplast endoplasmic reticular membrane; ER, endoplasmic reticulum; iEM, inner envelope membrane; IMS, intermembrane space; oEM, outer envelope membrane; PPC, periplastid compartment; PPM, periplastid membrane. (Reproduced with permission from [12.92].) 377

Figure 12.10 Schematic diagrams of the photosynthetic electron transport chain. (Created with BioRender.com.) 379

Figure 12.11 Schematic of photoprotection mechanism. The figure shows the chloroplastic membranes, the light-harvesting complexes (LCH) of the photosystem II (PSII) containing pigments that absorb light for photosynthesis and the effectors implicated in the non-photochemical quenching of chlorophyll fluorescence (NPQ). (Created with BioRender.com.) 380

Figure 12.12 Schematic representation of a simplified view of the phototransduction cascades mediated by photoreceptors. Light in the visible spectrum colors is absorbed by specific photoreceptors which trigger signal transduction cascades that generate physiological responses. (Created with BioRender.com.) 381

Figure 12.13 This map shows the average spatial concentration of chlorophyll *a* in the oceans. (Adapted from Rogato *et al.*, [12.102].) 381

Figure 12.14 Jewels of the sea. https://maxseeger.de/jewels-of-the-sea. (Reproduced with permission of the author.) 384

Figure 13.1 Schematic description of the model proposed to describe light curves of chlorophyll fluorescence NPQ index (Eq. 13.6) and graphical depiction of its parameters. Lines and letters in blue and red refer to the functions describing relaxation of dark-induced NPQ (NPQ_{Dark}; Eq. 13.5) and build-up of high light-induced NPQ (NPQ_{HL}, Eq. 13.4), respectively. Black line represents the sum of the two components NPQ_{Dark} and NPQ_{HL} following Eq. 13.6. For clarity, the value of C was exaggerated relative to the values typically obtained from fitting the model to experimental data. The figure and the parameters refer to NPQ but are readily applicable to light curves of Y(NPQ). 398

Figure 13.2 Fit of the model (Eq. 13.6) to light curves of NPQ (a, b) and Y(NPQ) (c, d) of contrasting magnitude of NPQ_{Dark}, as induced by different photoacclimation states (light step: 120 s). Data are unpublished results from cultures of the benthic diatom *Navicula* sp. (see Table 13.2 for details), grown in f/2 medium photoacclimated to 15 μmol m^{-2} s^{-1} (LL) and 160 μmol m^{-2} s^{-1} (HL). 403

Figure 13.3 Further examples illustrating the fit of the model (Eq. 13.6) to experimental data. Three Y(NPQ) light-response curves show contrasting relationships between NPQ_{Dark} and NPQ_{HL} components. While having similar magnitude of NPQ_{HL} (as quantified by NPQ_m estimates), the three light curves show very different contributions of NPQ_{Dark}. (a) High light-acclimated natural diatom-dominated biofilm [13.35] showing a high dark-induced NPQ_{Dark} (NPQ_0), dissipating gradually with light increase, as well as high E_{50}, resulting in very high E_{NPQmin} (light step: 90 s). (b) High light-acclimated natural diatom-dominated biofilm [13.37] showing a lower NPQ_0 level, while maintaining light saturation at high light levels (high E_{50}) and resulting high E_{NPQmin} (light step: 90 s). (c) Culture of the low light-adapted artic diatom *Fragilaria* sp. (J. Lavaud, unpublished data) showing NPQ_0 and NPQ_m similar to panel b, but dramatically lower values of E_{50} and E_{NPQmin} (light step: 30 s; note x-axis scale). 404

Figure 13.4 Dependence of NPQ_{Dark} on light history, as measured on the diatom *Thalassiosira weissflogii*. Variation of parameter $\boldsymbol{F'_m}$ (a) and NPQ (b) during sequential increase (up) and decrease (down) of light intensity (light step: 120 s). Lines in panel b represent the fit of the proposed model (Eq. 13.6). Arrows represent the direction of increasing light intensity. 405

Figure 14.1 Results of diatoms doped with arsenic and lead heavy metal ions: (a) cell density after 24-hour incubation, (b) cell density after 1-week incubation. 418

Figure 14.2 Zeta potential of diatoms at varying concentrations of heavy metal salts after (a) 24-hour incubation and (b) 1-week incubation with control sample as diatoms without any dopant along with dead diatoms (silica frustule only). (c) SEM micrographs of diatom frustules. 419

Figure 14.3 Schematic representation of Stern layer model of electric double layer on (a-b) silica and anionic charges and (c-d) silica and cationic charges. (e-f) Protonated and deprotonated silanol groups are located at the interface layer at higher and lower pH values respectively. (g) Zeta potential of diatoms treated with arsenic- and lead-treated diatoms at 100 ppm salt concentration with varying solution pH post treatment. 420

Figure 14.4 FTIR spectra of (a) arsenic- and (b) lead-treated diatoms with 100 ppm salt concentration at 24 hours and 1 week exposure time. (c) Full range FTIR spectra of heavy metal-doped diatom samples. 422

List of Tables

Table 2.1 Reported life cycles and cardinal points of selected species. Cycle lengths marked with a star are rough estimates based on the reported size diminution speeds and the size range. The size in brackets marks the size at which cells were observed to become sexually inducible. 25

Table 2.2 Size clusters for *P. delicatissima* according to Schwarz *et al.* [2.76]. Given is mean μ_i, standard deviation σ_i (from graph), lower boundary, cluster range and holding time τ for the Gaussian functions and their derived clusters, respectively. 42

Table 2.3 Parameters used in the model of D'Alelio *et al.* Given is the original denotation in the paper [2.19] as well as an abbreviated version used here in the mathematical expressions. 44

Table 2.4 Selected parameters used in the Hense–Beckmann PPND model. Parameters for the second, bulk phytoplankton species (index *P* instead of *B*) are omitted here. 46

Table 2.5 Selected variables and parameters used in the Hense–Beckmann DiaLCM model as far as mentioned in the text. For a complete list please refer to the appendix of the original article [2.36]. 47

Table 2.6 Parameters used in the Fuhrmann-Lieker *et al.* model. 50

Table 2.7 Comparison of the models. C: class model treated with differential equations, D: class model treated with difference equations, I: individual-based branching model. 55

Table 3.1 Summary of various experiments of diatom algae by nanoindentation and bending tests. 78

Table 6.1 Study details for each of the diatom species shown in Figure 6.2. The species are given in alphabetic order; the study details include species-specific information (symmetry type, natural habitat type), the growth experiment-specific metadata (type of treatment, biomass proxy, the length of the experiment, total number of observations, approximate number of observations in the exponential phase, light cycle conditions), and a reference. *Indicates unpublished growth data from the authors presented in Subsection 6.1.3. 145

Table 6.2 Average ranking of estimators in terms of quality metrics, averaged over challenge sets with varying noise and sampling frequency. Mean of ranks over challenge sets. Lower values signify better performance. 154

Table 6.3 Average ranking of estimators in terms of quality metrics, averaged over challenge sets with a crash event on day 7 and varying noise and sampling frequency. Means of ranks over challenge sets. Lower values signify better performance. 155

Table 6.4 Average ranking of estimators in terms of quality metrics, averaged over challenge sets with experiment stopped at day 7 and varying noise and sampling frequency. Means of ranks over challenge sets. Lower values signify better performance. 156

Table 6.5 Estimated difference of mean growth rates in two temperatures. Growth rates are derived from different methods. 159

Table 7.1 Comparison of annotation methods used on diatom genomes. 168

Table 7.2 Tools for predicting subcellular localization of the proteome of diatoms. This table contains information about the tools that have previously been applied to diatoms' genomes. 170

Table 7.3 Omics datasets that will facilitate the metabolic modeling of diatoms. 175

Table 9.1 A demonstration of how autonomy can be partitioned into generated and observed components. 250

Table 10.1 Observed processes and their frequencies. 276

Table 10.2 Measured angles and standard deviations. 284

Table 11.1 The set of parameters that require values for a run of the RAPHE computer program. Except for the number of events simulated, we have 9 independent input variables that could vary between individual diatoms and species. A variable that changes with each event is labeled "none" under Default value. 294

Table 11.2 States and rates. All rates in columns 4 through 7 (variable: total_columns-1) are set to zero at the beginning of each computational loop for an event. For those events that can happen, the rates are then calculated and become positive real numbers, indicated here by a + in those columns for this example of a raphe configuration. These are recalculated for each event after concatenation. The actual numbers depend on the values for the parameters in Table 11.1. If groups contain oppositely directed raphans, they are first concatenated, as described in the text. Only already concatenated groups are shown here. States of lattice sites are shown both numerically and via symbols. -1 / means hydrates to the left, 0 means empty, +1 \ means hydrates to the right. The diatom is presumed to move in the opposite direction (variable: directionForDiatom) of the hydration when hydration occurs, by a fixed amount (variable: distanceIncrementContributedBySwelling). For readability of the code, the column names correspond to variable names, which are assigned the column number. 296

Table 11.3 Published data on pennate diatom speeds and their lengths. Speeds in three-dimensional substrates are not compiled here. As in our simulation, data for the typical smooth, clean microscope slide is assembled here and all other properties related to the substrate are ignored, such as adhesion, roughness, gaps, or obstacles. 305

Table 12.1 Symbols, units and definitions of the quantities in Eq. 12.8. 350

Table 12.2 Low-order Padé approximants expressed in terms of the operator *P* defined in Eq. (12.40). 366

Table 13.1 Notations. 396

Table 13.2 Results of fitting of Eq. (13.5) to NPQ vs. *E* and Y(NPQ) vs. *E* curves measured in diatoms or diatom-dominated communities. Growth conditions/Treatment: sampling dates, conditions applied during growth or before the measurement of the light curves, when more than one LC available for the same taxon (HL: high growth light; LL: low growth light). LC: light curve protocol (LCx: steady state LC; RLCx: rapid light curve, x seconds of light step; UP, Down: sequential increase or decrease light levels, respectively). 399

Table 14.1 Toxicity limits, common sources, sensing modality, and limit of detection. 415

Table 14.2 FTIR spectra and band assignment for their corresponding functional groups. 423

Preface

Mathematics is important in all fields of study, whether implicitly as it is in the arts, or explicitly as it is in the sciences. There comes a time when the interface of the arts and sciences emerge, and that emergence is in the form of applied mathematics. Diatoms are both works of art and as unicellular organisms, biological entities. Diatoms exhibit microscopic external glass scaffolded structures that are at once exquisite in their beauty and geometrical in their symmetry and complexity, exhibiting art and science. Such a structural form is the starting point of the application of mathematics to diatoms.

More precisely, diatoms are unicellular algae that are mostly autotrophic with a few members that are heterotrophic, attached to abiotic substrate or free-living, and found in freshwater or marine or brackish waters. Diatoms can exist in a variety of habitats from Arctic and Antarctic ice to oceanic hydrothermal vents, from acidic peat bogs to soda lakes, in aerial habitats, or hitching a ride on an animal, plant or other unicell. Diatoms were first viewed by C.G. Ehrenberg from aeolian samples collected by Charles Darwin aboard the H.M.S. *Beagle* in 1833. The samples originated from the Bodélé Depression in the southern Sahara Desert, known as the dustiest place on Earth. Currently, diatoms are thought to have been in existence since the Jurassic period.

On an evolutionary basis, diatoms originated via secondary, or potentially, tertiary endosymbiosis and play an important role in the global processes that drive our dynamic planet. They are major drivers of the global carbon and silica cycles, account for a large percentage of global primary productivity, and are the primary basis for marine food webs. During nutrient depleted times, diatoms induce carbon sequestration, and during nutrient enriched times, diatoms produce blooms. As the highly ornamented unicells that they are, diatoms are unique in their structure and influence in the biological well-being of Earth. Because of this, diatoms are important indicators of the environmental and climate health of our planet.

Even with all of this, diatoms are important to human beings. They are used in technology, forming the basis of nanostructure replication in nanoengineering applications, nano-delivery devices in medicine and the health sciences, and in the production of biofuels. Diatoms have been found in archaeological clay-based artifacts and are used in the forensic sciences to solve cases involving the less-than-admirable side of humans. Fossil diatoms in diatomite production, also known as diatomaceous earth, form the basis of a multimillion-dollar industry in products used for filtration, polishes, cleaning up spills via absorption or adsorption, as an environmentally-friendly insecticide, and in food supplements. Humans are rather dependent on diatoms.

Why have a volume dedicated to the mathematical biology of diatoms now? The application of mathematics to diatoms is a natural progression in the study of organisms. Using shape analysis or other ways to quantify the geometry of diatom shape, multivariate statistics is used to understand the relation among a group of diatom phenotypes. Using phylogenetic analyses, combinatorics and graphs (trees) along with probability-based methods produce structures that are used to interpret the evolutionary relation among a group of diatom species. For many disciplines in the biological sciences, mathematics has provided a way to make very complicated and almost intractable problems tractable. Mathematics provides a unifying way to interpret results, both descriptively and prescriptively. Unlike studies of many other organisms, we have neither a precise number of species tallied nor a definitive phylogeny of all species, which is why the work of taxonomists, systematists and phylogeneticists is so important. Mathematical applications in diatom research are an aid to such basic studies of diatoms, as the utility of modeling and simulation via analytical and algorithmic tools has become increasingly apparent throughout the biological sciences.

Mathematical biology entails the development of theoretical principles as well as methodological procedures in which patterns and processes are analyzed and expressed. The interdisciplinary nature of mathematical biology borrows from many fields to provide the impetus to elucidate otherwise heretofore unanalyzable phenomena. The net result is to find practical methodological applications or theoretical understanding in such fields as diatom research. When applied to diatoms, this results in the studies you will find in this volume. Application of mathematics to topics covered in the volume are morphological and structural analyses, studies on cellular growth and metabolism, motility, population biology, nanoengineering, and applications in ecological and environmental studies. In some chapters, the mathematics is more extensive than in others, and in some cases, the mathematics is implicit rather than explicit. The goal of the volume was to present mathematically-based studies without the formalism of proofs, theorems and lemmas so that all diatom researchers, regardless of background or ability in mathematics, could glean something useful from the text.

This volume, *The Mathematical Biology of Diatoms*, is the fifth in the series Diatoms: Biology & Applications, series editors: Richard Gordon & Joseph Seckbach. Other volumes in the series include *Diatoms: Fundamentals and Applications*, *Diatom Gliding Motility*, *Diatom Morphogenesis and Diatom Microscopy*. Much gratitude goes to the authors who diligently produced their works for the volume, the reviewers who all too often do not get recognition or kudos, the series editors, Richard Gordon and Joseph Seckbach, for the idea and opportunity to produce the volume, the typesetters who tirelessly perfect the art of satisfying the authors and publisher alike, and Martin Scrivener for his expertise and support in facilitating the publishing of this volume.

Janice L. Pappas
University of Michigan
February, 2022

Part I

DIATOM FORM AND SIZE DYNAMICS

1

Modeling the Stiffness of *Diploneis* Species Based on Geometry of the Frustule Cut with Focused Ion Beam Technology

Andrzej Witkowski[1*], Romuald Dobosz[2], Tomasz Płociński[2], Przemysław Dąbek[1], Izabela Zgłobicka[3], Horst Lange-Bertalot[4], Thomas G. Bornman[5,6], Renata Dobrucka[2], Michał Gloc[2] and Krzysztof J. Kurzydłowski[3]

[1]*Institute of Marine and Environmental Sciences, University of Szczecin, Szczecin, Poland*
[2]*Warsaw University of Technology, Faculty of Materials Science and Engineering, Warszawa, Poland*
[3]*Bialystok University of Technology, Faculty of Mechanical Engineering, Białystok, Poland*
[4]*Goethe University, Institute of Ecology, Evolution and Diversity, Frankfurt am Main, Germany*
[5]*South African Environmental Observation Network, Elwandle Coastal Node, Gqeberha (Port Elizabeth), South Africa*
[6]*Institute for Coastal and Marine Research, Nelson Mandela University, Gqeberha (Port Elizabeth), South Africa*

**Corresponding author*: andrzej.witkowski@usz.edu.pl
Andrzej Witkowski: https://orcid.org/0000-0003-1714-218X, https://scholar.google.pl/citations?user=v_1sQtYAAAAJ&hl=pl, https://www.researchgate.net/profile/Witkowski-Andrzej
Romuald Dobosz: https://orcid.org/0000-0002-8975-4153, https://www.scopus.com/authid/detail.uri?authorId=22233905800
Tomasz Płociński: https://orcid.org/0000-0003-3325-3904, https://scholar.google.pl/citations?user=96YltxQAAAAJ, https://www.scopus.com/authid/detail.uri?authorId=55396936000
Przemysław Dąbek: https://www.researchgate.net/profile/Przemyslaw-Dabek, https://orcid.org/0000-0002-3736-3011, https://www.facebook.com/marinebenthicdiatomproject/
Izabela Zgłobicka: https://www.researchgate.net/profile/Izabela-Zglobicka, https://orcid.org/0000-0002-4432-9196, https://scholar.google.pl/citations?user=UlpniH0AAAAJ&hl, https://www.scopus.com/authid/detail.uri?authorId=55816999000
Horst Lange-Bertalot: https://diatoms.org/contributor/HorstLange-Bertalot, https://worldcat.org/identities/lccn-n85165948/
Thomas G. Bornman: http://orcid.org/0000-0003-1868-479X, https://www.researchgate.net/profile/Thomas-Bornman, https://nmmu.academia.edu/ThomasBornman, https://scholar.google.com/citations?user=bmE5ehAAAAAJ&hl=en
Renata Dobrucka: https://orcid.org/0000-0003-1486-9412, https://scholar.google.pl/citations?user=Zo22iv4AAAAJ&hl=pl, https://www.scopus.com/authid/detail.uri?authorId=55886435400
Michał Gloc: https://repo.pw.edu.pl/info/author/WUT1178774685814270 89f84201e756ace8/Micha%25C5%2582%publication&tab=publications&conversationPropagation=begin&sort=&lang=en
Krzysztof J. Kurzydłowski: https://www.researchgate.net/profile/Kj-Kurzydlowski, https://orcid.org/0000-0003-3875-4820, https://scholar.google.pl/citations?user=sHuyq58AAAAJ&hl=en, https://www.scopus.com/authid/detail.uri?authorId=7006335771

Janice L. Pappas (ed.) The Mathematical Biology of Diatoms, (3–18) © 2023 Scrivener Publishing LLC

Abstract
Siliceous exoskeletons of diatoms (frustules) provide protection against environmental and biological factors, including waves and currents in the surf and coastal zone, osmotic pressure in the tidal flats and the grazing activity of zooplankton and zoobenthos. With its intricate 3D patterns, diatomaceous biosilica is an interesting research subject of applied sciences, including materials science and chemistry. Biological, environmental, and applied aspects of diatomaceous exoskeletons pose questions concerning their mechanical strength and protective role against the destructive natural forces and the stiffness of the frustules once the applied issues are considered. To answer such questions, frustules of marine benthic diatoms have been subjected to nano cuts with focused ion beam (FIB) sequential milling along the transapical (cross sectioning) and apical (longitudinal sectioning) axis. However, in a few cases, the whole external valve surface of the *Diploneis* aff. *chersonensis* and *Diploneis* sp. have been FIB milled. FIB milling image series were applied in the modelling of the frustule response to the mechanical deformations. The geometric and numerical models were constructed with Ansys software based on the finite element method. Assumed mechanical properties involved Young's Modulus E = 22.4 GPa and Poisson constant ν = 0.3, while the identical load for all cases was the force of 500 μN acting on the upper surface of the raphe sternum—the frustule's backbone. Simulations show maps of displacement within the diatom frustule model under downward oriented force. The weakest point is located in the middle of the valve resulting from directional deformation along the Y axis (pervalvar axis). In addition to deformation along the acting force, a smaller displacement is observed as directional deformation of the girdle (Z axis). The consequence of the frustule's biological and mechanical response to such experiments is discussed.

Keywords: 3D shape, diatoms, siliceous exoskeleton, mechanical deformations, geometric modeling, numerical modeling

1.1 Introduction

Diatoms are unicellular eukaryotic photoautotrophic microalgae abundant in various aquatic and terrestrial habitats, which play a major role as primary producers on the global scale in the phytoplankton component [1.1] but also as microphytobenthos in particular habitats, e.g., in the littoral and sublittoral zones of the world's oceans (e.g., [1.2] [1.3]). The most characteristic feature of diatoms is the presence of a siliceous (opaline) exoskeleton bearing a species-specific ornamentation (e.g., [1.4]). This exoskeleton, called a frustule, encases the organelles, and after the death of the cells is deposited in the sediments to form a natural archive of the climatic and environmental conditions over the geological past. In the most productive zones of ancient lakes and nutrient-rich upwelling zones in the oceans, large amounts of diatomaceous biosilica formed extensive deposits known as diatomaceous earth or diatomites [1.5]. Diatoms and their siliceous exoskeletons are primarily of interest to biologists and Earth and environmental scientists involved in taxonomy, biodiversity, ecology and paleoecology [1.4] [1.6]. Over the last few decades there has been an increasing interest in diatomaceous biosilica in applied research due to its unique physical and chemical properties, including a highly ordered 3D structure and enormous porosity (e.g., [1.7] [1.8]). Despite the interest in the physical and chemical properties of pure or doped biosilica (e.g., [1.9]), little attention has been paid to the mechanical strength of diatom frustules, which is crucial for diatom well-being when their role in environmental services is taken into account. Recently, Cox *et al.* [1.3] estimated the role of diatoms in the coastal ecosystem (seagrass meadows) to be 80%, where diatoms apparently rely on the mechanical strength of their siliceous exoskeleton against grazer activity. In fact, diatoms, as one of the most important groups of primary producers seems to counter-adapt in some way the protective role of the siliceous frustule also against filter feeders, cf. [1.10].

To test the physical strength of the diatomaceous frustules we selected the benthic marine diatom *Diploneis* Ehrenberg *ex* Cleve which can range in size from ca. 20 to 200 μm and usually forms very robust frustules [1.11] [1.12]. The taxa belonging to the genus *Diploneis* Ehrenberg *ex* Cleve are common components of benthic diatom assemblages in both freshwater and marine littoral habitats [1.4] [1.12] [1.13]. There are 419 species names in the data base at present, as well as 322 infraspecific names (https://www.algaebase.org). Of the species names, 395 have been flagged as accepted taxonomically on the basis of the listed literature under the species name [1.14]. Numerous *Diploneis* taxa are easy to observe with a light microscope due to their large size and very robust frustules combined with their characteristic shape, being either linear, broad elliptical or panduriform. In fact, it would be hard to find another diatom genus with so many panduriform valves as in *Diploneis,* including the generitype, i.e., *D. didyma* (Ehrenberg) Ehrenberg [1.12]. Although their specific habitats are widespread among marine and freshwater ecosystems, usually *Diploneis* taxa are common sediment dwellers and included in motile epipelon and epilithon. Particular *Diploneis* species rarely attain increased abundance and dominant species status, but due to their size and biomass they are important in the local trophic chain of marine benthic habitats, e.g., in coral reef habitats [1.15]–[1.17]. Although *Diploneis* from marine (cf. [1.13] [1.18]) or freshwater habitats in North America [1.19] and Europe [1.12] have a prominent size and relative importance in the food chain for marine benthic grazers, the genus is still awaiting a thorough study in terms of its role in diatomaceous assemblages and biogeography.

In the most recent review, Lange-Bertalot *et al.* [1.12] characterized the construction of the valves and named the lateral canals as the most important characteristic of the taxa belonging in *Diploneis.* Medianally located lateral canals are unique and known only in *Diploneis* thus far. At both sides of the raphe are two internal 3D canals running through the whole length of the frustule's apical axis [1.4] [1.12]. Longitudinal canals are composed of two cylindrical structures (usually flattened), which are irregular in shape. They have acute apical ends and expand in width towards the valve middle where they decrease in size. Located in the middle of the valve, the raphe sternum involves two raphe slits interrupted by the central nodule (CN). In terms of mechanical strength, the raphe sternum and the two longitudinal canals are considered the valve's backbone.

Research on the mechanical properties of siliceous frustules built from glass-like biosilica aims at answering the questions related to the frustule's mechanical properties, e.g., stiffness and elasticity (e.g., [1.20]–[1.22]). To the best of our knowledge, the presented results are the first of its kind using the frustule of *Diploneis,* and certainly the first on the diatom frustule originating from a natural sample and not from a culture. Hamm *et al.* [1.23] published results of experiments with planktonic diatoms from the Southern Ocean, i.e. two centric (*Thalassiosira punctigera* and *Coscinodiscus granii*) and one pennate species (*Fragilariopsis kerguelensis*). *Navicula pelliculosa,* another pennate diatom, has been studied by Almqvist *et al.* [1.20] with an experimental set involving atomic force microscopy (AFM). Here we present the FIB technique, which has already been used in sectioning and imaging diatoms (e.g., [1.24] [1.25]), but this is the first time that results of FIB sequential cuts are used for the modeling of *Diploneis* aff. *chersonensis* (Cleve) Grunow and *Diploneis* sp.

Unlike the studies on the mechanical properties of diatom frustules referred to, the diatom we studied originated from a natural sample and not from a clonal strain isolated and maintained in a culture. The diatom frustules of *Diploneis* were subjected to FIB nanosectioning along transapical (Z) and apical (X) axes. The models built on sequential FIB cuts were used in the calculations of the deformational displacements. For the first time, the results of our research illustrate the mechanical properties of the diatom frustule with the raphe sternum system positioned within the valve middle.

1.2 Material and Methods

Diatom valves processed with FIB all originated from natural samples collected from coral reef habitats in the Western Indian Ocean. Species identified as *Diploneis* aff. *chersonensis* and *Diploneis* sp. were selected for this study. Prior to valve processing with FIB, the sediment sample was cleaned with 10% HCl to remove carbonates and with 37% hydrogen peroxide to remove organic matter. Thereafter, samples were washed several times with deionized water until neutral (7) pH. Suspension of the cleaned sample was spread on the SEM stub and coated with gold for FIB analysis.

1.2.1 Focused Ion Beam (FIB) Milling

The FIB-based sample preparation was performed by means of FIB/SEM Hitachi NB5000 integrated system. This system consists of an ultra-high performance focused Ga^+ ion beam gun (40 kV) and high-resolution field emission scanning electron microscope (30 kV). This dual-beam system enabled high-throughput specimen preparation, high-resolution imaging and analysis and precision nanosectioning. Manipulation of FIB samples was carried out under parameters of ion beam conditions selected to minimize their damage. Prior to cutting, a tungsten (W) layer was applied to the sample for protection. The diatom valves selected for the FIB processing were subject to sequential milling (sectioning) and consecutive SEM imaging immediately after cuts. The gallium ion beam gun is strongly invasive and destroys the processed valves. Particular transapical sectioning ranged from a few nanometers to a few micrometers depending on the valve structure. Except for transapical sectioning, we also performed milling of the diatom valves along their external surface, which along with the SEM imaging made insights into the depth of the valve structure along a determined surface possible.

1.2.2 Modeling

The purpose of the study was to analyze the stiffness of the diatom frustule. Two models of the frustule geometry was built on the basis of the FIB produced SEM images: the first one treated the raphe sternum—the frustule backbone—as a hollow body (v1); and in the second the inner part of the frustule's backbone is made up of short ribs (v2). This distinction in the structure of the diatom was considered due to observations obtained with FIB processing and SEM imaging.

The preparation of geometric models and calculations was carried out using Ansys software based on the finite element method (FEM) modeling. The result of the calculations is the deformation of the diatom frustule geometry. In order to simplify the construction of geometric and numerical models, the unit system applied was expressed in μm, μN and MPa. The finite element mesh contained 32.5 and 34 thousand quadratic finite elements respectively.

The deformations of the v1 and v2 models were compared with the deformations of the reference models: v1_p and v2_p where the valve is considered to have a constant thickness (Figure 1.1a) and v1_k and v2_k where the valve had a constant thickness and a raphe sternum, a sort of backbone (Figure 1.1b). The valves of models v1_p and v2_p were characterized with the same mass as the v1 and v2 models, respectively. The same assumption applies to the v1_k and v2_k models. The analysis was complemented with model v_o (Figure 1.1c), where the geometry was based on the contour of the v1 and v2 models.

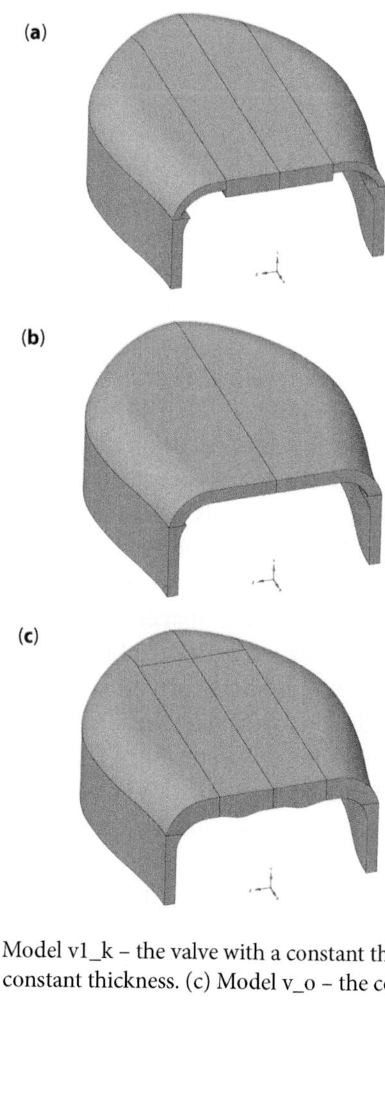

Figure 1.1 Reference models: (a) Model v1_k – the valve with a constant thickness and a raphe sternum. (b) Model v1_p – the valve with a constant thickness. (c) Model v_o – the contour of the v1 and v2 models.

The assumed mechanical properties are Young's Modulus $E = 22.4$ GPa and Poisson constant $\nu = 0.3$ [1.23], while the identical load for all cases was a force of 500 μN acting on the upper surface of the raphe sternum.

1.3 Results

Processing of the *Diploneis* valves with FIB is illustrated in Figures 1.2 to 1.4. The valves have been subjected to nanocuts along the two axes: Z (transapical) and X (apical). Prior to the FIB sectioning, care was taken that all the selected valves possessed well-developed structures and are representative in terms of characterized structures. Externally the valves exhibit the striae composed of a single row of areolae. Positioned along the middle part of the valve is a raphe — a slit cutting through the valve surface composed of two branches separated by a solid valve part called the central nodule. The two slits of the raphe are positioned within the so-called raphe sternum, which is parallel to the longitudinal canals connected through the system of pores with the cell lumen and with the frustule external environment.

1.3.1 FIB Processing

The frustules of *D.* aff. *chersonensis* and *Diploneis* sp. in this study were panduriform and connected with very high girdle composed of a few open and plain copulae (bands, Figures 1.2a–c). The valve exterior is characterised by two perpendicular ornamentation systems. In the middle of the valve external view, a large and mostly hyaline raphe-sternum is present with the raphe composed of two branches. Sectioning along the Z axis revealed that the raphe branches are longitudinal slits within the valve face, in the shape of a "<", and not a simple cut through (Figures 1.2c-d). In the middle part of the valve, a small bridge in the raphe branches, called the central nodule, is present. As illustrated in cuts performed along the Z axis, the bridge in the raphe slit is filled with solid silica (Figure 1.2d). In all three images with nanocuts along this axis, two longitudinal canals parallel to the raphe system are visible. Longitudinal canals extend from the top of the valves (apices) through the mid position valve and tend to expand towards the central nodule. In images documenting nanocuts through the longitudinal canals it is clearly visible that they are connected through tiny poroids to the cell ambient environment and the cell lumen (Figures 1.2b-c), which are absent from the central nodule area. The spaces between the transapical ribs are filled with biosilica. The valve face is constructed from robust arch-like ribs named transapical striae composed of two silica layers. The space formed between the two silica layers appears to be connected to the longitudinal canal (arrowhead in Figure 1.2d). The first model of the *Diploneis* aff. *chersonensis* with the raphe sternum (frustule backbone) as a hollow body originated from FIB sequential sectioning (v1) with complete valve surface as illustrated in Figure 1.2e, whereas the same v1 model with the longitudinal canal uncovered is illustrated in Figure 1.2f.

Figure 1.3, illustrates the use of FIB milling of the whole valve external surface (axis X) to show the interior of the longitudinal canals and the internal surface of the striae forming transapical ribs. Longitudinal canals are illustrated in Figures 1.3c-d as distinct depressions and their margin bears remnants of the rib system supporting the hyaline external cover perforated with solitary pores. The pores are positioned between two neighboring short ribs (Figures 1.3c-d). The removal of the external valve layer without nanocuts would be

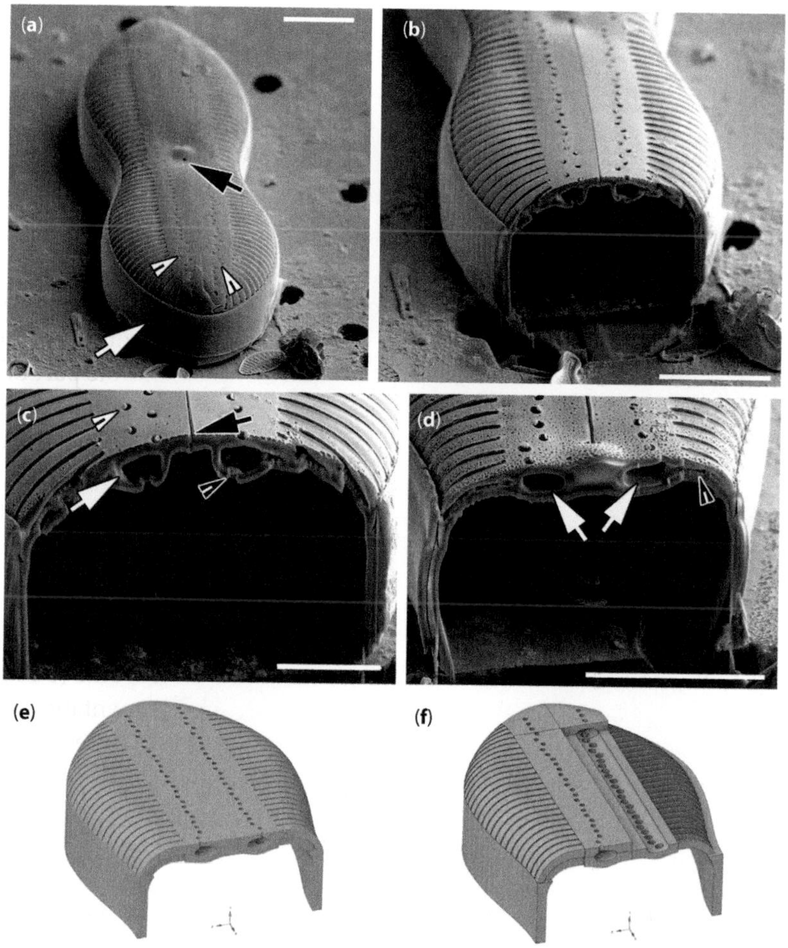

Figure 1.2 Panduriform *Diploneis* aff. *chersonensis* cut with FIB along the transapical axis and the geometric model v1: (a) External view of the whole specimen (frustule) prior to FIB processing. Note the plain and robust girdle (white arrow) and the raphe system with central nodule (black arrow) and two series of poroids marking the external run most typical for *Diploneis* as a genus character – the longitudinal canals (arrowheads). (b) The same frustule with FIB milled apical part. (c) Close up of the frustule imaged in Figure b. Note the longitudinal canals (white arrow) separated by a raphe slit (black arrow) and poroids positioned in canal exterior (white arrowhead) and in the internal surface of the canal (black arrowhead). (d) The same specimen cut through the central nodule. Note the longitudinal canals (arrows) separated by solid silica and internal structure of the transapical striae connected through the small opening (arrowhead) to the canal. (e) The v1 model with valve surface. (f) The v1 model with longitudinal canal uncovered. Scale bars in Figures a–c = 10 μm and in d = 5 μm.

impossible to perform. Likewise, expanded pores connecting the longitudinal canals to the cell lumen (Figure 1.3d) are also illustrated. In addition, short apically oriented ribs appear between the transapical ribs, which would not have been detected without FIB milling of the external valve surface (Figures 1.3c-d). When compared with the appearance of the longitudinal canals observed from nanocuts along the Z axis it is evident that these short ribs are an artefact and result from the FIB milling process (Figures 1.3a–d). In the second model of the species we studied, the inner part of the frustule's backbone is made up of short ribs – (v2),

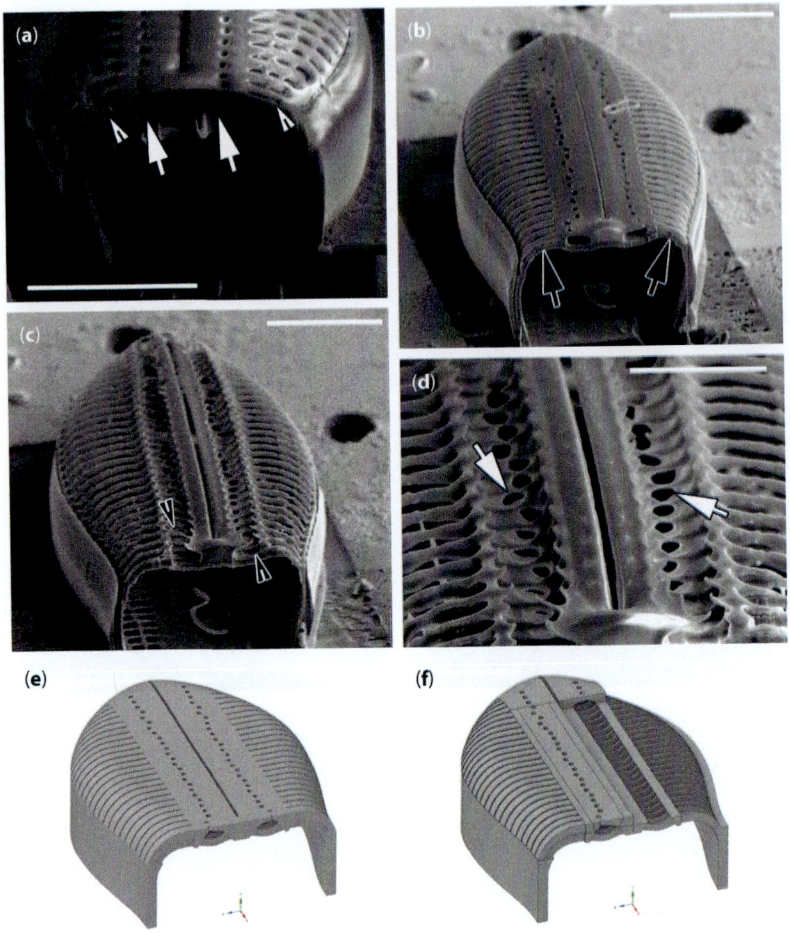

Figure 1.3 Panduriform *Diploneis* spp. cut with FIB along the transapical axis and along the valve surface and the geometric model v2: (a) *Diploneis didyma* cut with FIB along the transapical axis. Note the presence of two longitudinal canals (arrows) parallel to the raphe system and the solid transapical striae composed of a single silica layer (arrowheads). Figure b–d shows *Diploneis* aff. *chersonensis* milled with FIB along the apical axis: (b) half of the frustule prior to FIB cuts. Note the transapical striae composed of two silica layers (arrows). (c) Same frustule with FIB milled superficial valve layer. Note the open system of longitudinal canals. Note the remnants of the transapical ribs upon which the longitudinal canal cover is based (arrowheads). (d) Close up of the specimen illustrated in Figure c. Note the poroids connecting the longitudinal canal with the cell lumen (arrows). (e) The v2 model with valve surface. (f) The v2 model with short ribs uncovered. Scale bars in Figures a–c = 10 μm and in d = 5 μm.

as illustrated in Figures 1.3e–f. In Figure 1.3e, the v2 model with valve surface is presented; while in Figure 1.3f illustrated the v2 model with short ribs and longitudinal canal uncovered.

The third row of FIB processed valve of a panduriform *Diploneis* sp. illustrates nanocuts of the valve internal side along the apical (X) axis. In a series of images, the extent of the longitudinal canal to the apices is illustrated. The valve interior illustrated prior to FIB milling clearly shows the elevated longitudinal canals expanding at the central nodule and parallel to the raphe system (Figure 1.4a). Figure 1.4b confirms that the longitudinal canal is framed by transapical ribs and with the next sequential cut one can easily observe the open canal

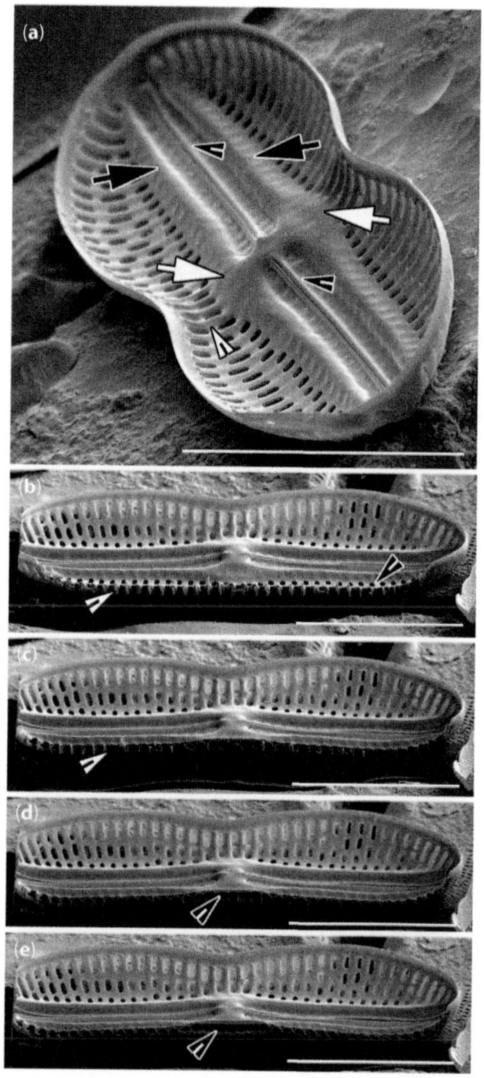

Figure 1.4 Panduriform *Diploneis* aff. *chersonensis* valve interior cut with FIB along the apical axis: (a) Valve interior illustrated prior to FIB milling. Note the elevated longitudinal canals (black arrows) expanding at the central nodule (white arrows) parallel to the raphe system (black arrowheads), robust transapical ribs (white arrowheads). (b–e) Consecutive stages of FIB processing: (b) Valve nanocut through the transapical ribs (white arrowhead) with a row of areola distinguishable (black arrowhead). (c) The FIB cuts of the longitudinal margin, still completely closed by the transapical ribs (arrowhead). (d) The first cut of the longitudinal canal through its expansion within the central nodule. Note the open space (arrowhead). (e) FIB processing of the longitudinal canal. Note the free space in the valve middle (arrowhead). Scale bars in Figures a–e = 10 μm.

space in the central nodule area, which is still framed by transapical ribs, which gradually open the longitudinal canal to the observer (Figures 1.4c–e).

1.3.2 Modeling

Based on the FIB, sequential sectioning diatom frustule models were built on which simulation experiments were carried out. Two types of valves are presented in both models. The first type of model shows the complete valve with well-developed solid longitudinal canals and the raphe slit, but the bridging central nodule is absent (Figure 1.2e); the valve with the upper layer removed with FIB milling and the raphe slit is missing (Figure 1.2f). In the second model, the first valve has longitudinal canals with the central nodule bridging both halves of the valve (Figure 1.3e); and in the next valve, the upper layer is removed with FIB milling, the longitudinal canals are formed with the transapical ribs and the bridging central nodule remained (Figure 1.3f).

The basic model v1 is characterized by the highest stiffness and the maximum displacement along axis Y of 0,642 μm (Figure 1.5a). The zone of the highest displacement is located within the valve middle and is distributed equally throughout the frustule backbone (raphe sternum with longitudinal canals). The v2 model is characterized by the lowest stiffness and amounts to 2,231 μm (Figure 1.5b). The distribution of the maximum displacement is connected with the valve backbone and corresponds to the middle part of the raphe sternum disconnected by the central nodule, which is solid.

As shown in FIB milling processed images, the *Diploneis* species we used for diatom frustule modeling is characterized by a significant girdle composed of a few open bands. Interestingly, a single copula has the width of ca. 8 μm, which is at least two times higher than the valve itself (Figure 1.3). From the two illustrated models we can see that in the v1 model (with the highest

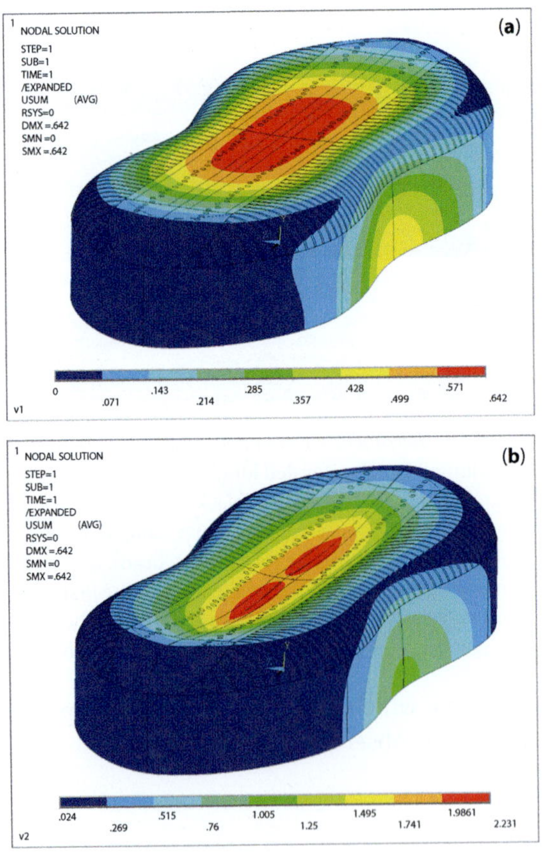

Figure 1.5 Mapping of the *Diploneis* aff. *chersonensis* valve stiffness in response to downward oriented stress: (a) Model v1 and (b) Model v2.

stiffness) the gridle shows low displacement along the Z axis and amounts to ca. 0,5 μm (Figure 1.5a). This is best illustrated in the maximum constriction of the frustule. In the v2 model, the maximum displacement along the Y axis is received by the raphe sternum (without central nodule,) and hence the displacement of the girdle bands is less than ca. 1,0 – 1,5 μm (Figure 1.5b).

The results of the simulation for the v1, v1_p and v1_k models in the form of normalized displacements by the displacement value of the v1 model are presented diagrammatically in Figure 1.6. The maximum deflections for the v1_p and v1_k models are 10 and 20% larger, and amount to 0,75 and 0,77 μm (Figure 1.6a). In the case of the v2 model, the situation is different, and the displacements of the v2_p and v2_k models are approximately 50% smaller (Figure 1.6b). This results from a significant weakening of the valve for longitudinal bending due to the presence of ribs (i.e., material voids between them). The loss of material resistant to stretching or compression in the longitudinal direction results in a reduction in bending stiffness. The comparison of the v1 and v2 models with the v_o model being their outline is shown in Figure 1.6c. The normalized displacements in relation to the displacements of the v_o model show the lowest stiffness in the v2 model—about 11 times with about 3 times lower stiffness compared to the v1 model. If we take into account the masses of the models by multiplying the normalized displacement by the ratio of the analyzed model mass to the mass v_o, these differences are smaller, and the stiffness is 2.4 and 7.7 times lower, respectively (Figure 1.6c). The appearance of the rib structure in the case of the v1 model gives a significant increase in stiffness (under the condition of constant mass). The addition of ribs in the lower portion of the longitudinal canal dramatically reduces the stiffness (Figure 1.6c). The loss of material resisting stretching or compression in the longitudinal direction results in a reduction in bending stiffness. The model with more flexibility, on the other hand, may be characterized by higher fracture toughness.

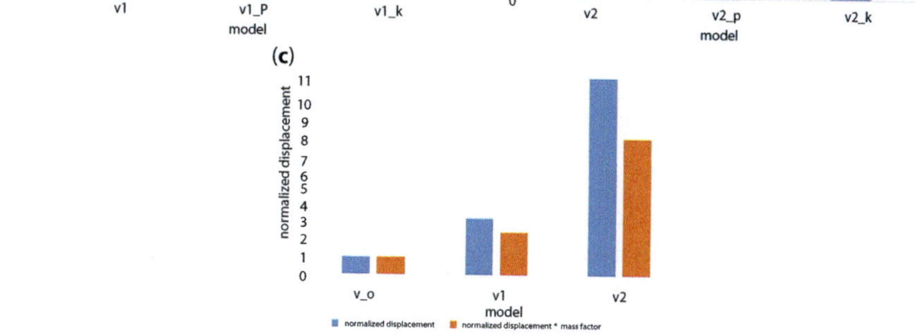

Figure 1.6 Normalized displacement compared between models: (a) Comparison among v1, v1_p and v1_k models. (b) Comparison between v2, v2_p and v2_k models. (c) Comparison between the v_o, v1 and v2 models, including normalized displacement (blue bars) and normalized displacement with mass factor (orange bars).

1.4 Discussion

1.4.1 Practical Meaning of the Frustule Geometric Characters

The most important feature of *Diploneis* taxa is its characteristic lateral canal system. This is a unique bilateral internal construction parallel to the apical axis. The median raphe complex as an integrated part consists of raphe branches lying in a particular narrow and shallow internal sternum, which is mirrored by distinct depressions (grooves) on both sides [1.12]. From a diatomologists point of view, this complex is interrupted by a characteristic central nodule forming a bridge in the continuum to the unperforated walls of both canals and projects into the cell as an "axial plate." However, according to Lange-Bertalot *et al.* [1.12] this term is less appropriate for the micromorphology and function of this construction element present in all *Diploneis* species. This H-shaped complex consists of two strong and hollow apical supporting beams connected by a short crossbeam (in German: "Doppel-T-Träger"). Such a construction is well known in architecture and guarantees very effective stability of the cell skeleton, prevailing against structural weakness due to the raphe slits.

However, the results of our simulations on *Diploneis* aff. *chersonensis* in both types of models, with and without raphe sternum and ribs in longitudinal canals, show that models without the raphe sternum, central nodule and the ribs are more resistant to deformations than those with a central nodule and ribs. What we interpret as much higher elasticity is due to the presence of longitudinal slits in the form of the raphe that cut through the valve surface. This could be an advantage of *Diploneis* as a representative of diatoms grazed by zooplankton (e.g., [1.26]), by suspension feeders in the littoral and sublittoral zones [1.10] or by benthic invertebrates grazing on epiphytic diatoms in freshwater streams (e.g., [1.27]). As shown by combined molecular and morphologic approaches, shallow water marine habitats abound with diatom populations (cf. [1.28]). Research on feeding behaviors of zooplankton grazing on diatoms showed that Copepods are potentially able to crush the frustules using their mandibles armored with opaline blades (e.g., [1.29]; cf. review [1.7]). However, suspension feeders may use more sophisticated ways to obtain their lipid drops from tightly sealed diatomaceous cells. As shown by Houki *et al.* [1.10], an efficient mechanism of the Manila clam that crushes the diatom frustules in the gut involves attrition of the densely packed cells. However, the experimentally measured stiffness of the diatom taxa depended on the degree of silicification of the exoskeleton, with robust *Navicula ramosissima* being the most resistant to the ruptures and delicate *Cylindrotheca closterium* being unmeasurable as it was already breaking down prior to measurements [1.10].

1.4.2 Documenting the Mechanical Strength of the Diatom Frustule

Research on the mechanical properties of diatom siliceous frustules was initiated a few decades ago and involved experiments on both centric and pennate forms originating from cultures. Experiments involved either AFM (e.g., [1.20]) or the use of nanoindentation facilities (e.g., [1.21]–[1.23]). Except for Almqvist *et al.* [1.20], in all of these experiments the diatom frustules involved cultured strains of marine planktonic diatoms [1.21], [1.22], sometimes from the most productive waters of the Southern Oceans, where diatoms are grazed by zooplankton represented by the massive occurrence of Copepods and Euphausiids (e.g., [1.26] [1.29]–[1.33]. Since both zooplankton groups can actively crush the diatom frustules with special mechanisms, their prey must adapt to these conditions by improving the strength of the siliceous protection of their cells. It has been documented that those marine planktonic diatoms, e.g., *Fragilariopsis* spp. and *Thalassiosira* spp., constitute the major food resource of marine zooplankton [1.31], but they are not exclusive feed. Frequently, zooplankton utilizes the so-called marine snow, which also includes marine planktonic diatoms [1.30]. The first experiments on the mechanical strength of diatoms from natural samples were conducted by Houki *et al.* [1.10] and involved marine planktonic and benthic diatom species which are preyed on by the bivalve Manila clam, a tidal flat suspension feeder. The above authors studied the mechanical strength of the diatom frustules using a microscope and nanoindentation set.

The measured strength of the most resistant species, *Navicula ramosissima,* was fractured with the force of 353.5 ± 32.3 μN, which is lower than the 730 μN measured by Hamm *et al.* [1.23] for *Fragilariopsis kerguelensis,* but significantly higher than the 158-250 μN measured by Wilken *et al.* [1.32] for the latter species. Taking into account the fact that frustules of *Fragilariopsis kerguelensis* are much more strongly silicified than *N. ramosissima* and have a different raphe system positioned diagonally as in most Bacillariceae, not medianally as in *Navicula* and Naviculaceae, the only explanation for the relatively lower force necessary to break the more robust species may be the effect of culture conditions. Experiments carried out by Wilken *et al.* [1.33] in iron-manipulated cultures have shown that planktonic diatom cultures induce thickening of their siliceous frustules in response to iron-depleted culture conditions. The latter authors suggest iron depletion induced valve thickening as an adaptation against increased grazing from marine zooplankton (cf. [1.33]). This can be true in the case of large marine planktonic diatoms of genera *Coscinodiscus* and *Thalassiosira*; however, not necessarily true of the planktonic pennate diatoms representing the biraphid genera of *Fragilariopsis, Nitzschia* and *Pseudo-nitzschia* dominant in the Southern Ocean and in the Arctic [1.34]. From the experiments performed by Hamm *et al.* [1.23] and Wilken *et al.* [1.33] it looks as though *Fragilariopsis kerguelensis* frustules only partly require very high pressures to be broken in nanoindentation experiments. In fact, *F. kerguelensis* has very robust frustules but, more importantly, the geometry of the valves is composed of strongly silicified ribs oriented perpendicularly to each other, which adsorb the tension from weaker areolated transapical striae positioned in between. As simulated by Hamm *et al.* [1.23], removal of the transapical ribs decreased the force necessary to break the frustule of *F. kerguelensis* by 60%. In *Fragilariopsis* belonging to Bacillariaceae, the raphe (longitudinal slit) is positioned eccentrically, which means the potential breakage of the frustule is mitigated by the fact that the raphe is located over the transapical ribs present on the valve mantle, which dissipate the stress [1.23]. However, *Fragilariopsis kerguelensis* is just an example of a very robust and stress-resistant species in Bacillariaceae, while numerous species in other genera are weakly silicified. From other experiments on the hardness of taxa belonging to Bacillariaceae, it can be seen that the thickness and general morphology of siliceous frustules is an important factor for their hardness. In hardness measured *in situ* by Houki *et al.* [1.10], one of the most delicate species of Bacillariaceae, *Cylindrotheca closterium,* was unmeasurable due to its fragile frustules. This is completely different in the diatom we studied belonging to the order Naviculales, where the two raphe slits are positioned medianally. In our experiment with *Diploneis* aff. *chersonensis,* the frustules weakest area is located along the raphe slit and its weakest point is the siliceous bridge—the central nodule. In our v1 model with high stiffness, the entire CN area plus an associated part of the longitudinal canal showed equal displacement of 0.642 μm with quite high displacement in the middle part of the girdle.

Apparently, in natural frustules of *Diploneis* aff. *chersonensis*, the valve center reacts against the force as a whole surface. It is different in the valve milled with FIB in our v2 model, where the short ribs in our interpretation reduce the stiffness against bending along the valve X (apical) axis. Therefore, the distribution of displacement over the *Diploneis* valve seems to not support the "Doppel-T-Träger" hypothesis for the central nodule (CN) and the longitudinal canals. The "Doppel-T-Träger" hypothesis is quite well illustrated in the v2 model, which is the weakest in terms of displacement. However, this model was impacted by the earlier removal of valve surface with FIB and separation of the short ribs, which significantly reduced the valve stiffness. In light of the above comparison of models v1 and v2, an answer to the question of the mechanical role of the raphe in diatoms is difficult to answer. In a robust diatom exemplified by *Diploneis*, the overall valve stiffness is high with the geometric center of the frustule outlined by its position in it: part of the raphe sternum and raphe slits, central nodule and part of the longitudinal canals as the weakest area. However, it does not allow us to conclude whether the raphe in the *Diploneis* we studied is only limited in its function to gliding motility [1.35]. Experiments illustrated by Hamm *et al.* [1.23] with *Fragilariopsis kerguelensis* show a completely different situation. In the latter species the highest stiffness is characteristic of the transapical ribs which support the entire frustule. Unlike in *Diploneis*, the raphe in the latter species has no CN, is strongly eccentric, and is positioned at the junction of the valve face and the mantle [1.4].

Acknowledgments

This research was carried out within the frame of the project "Advanced Biocomposites for Tomorrow's Economy BIOG-NET" funded by the Foundation for Polish Science from the European Regional Development Fund (POIR.04.04.00-00-1792/18-00).

The authors acknowledge the sampling from the South Africa-Poland Bilateral Collaboration Project facilitated by the National Research and Development Centre, Warsaw, Poland (NCBiR- project number PL-RPA/TemBioDiaSA/02/2016 granted to AW) and by the National Research Foundation of South Africa (NRF - Grant UID 102283 to TGB). The authors also wish to express their appreciation to Dr. Janice Pappas, the editor of the monograph, for her valuable comments and encouragement.

References

[1.1] Field, C.B., Behrenfeld, M.J., Randerson, J.T. and Falkowski, P. (1998) Primary production of the biosphere: integrating terrestrial and oceanic components. *Science* 281(5374), 237–240.

[1.2] Costanza, R., Groot, R., Sutton, P., Van der Ploeg, S., Anderson, S., Farber, S. and Turner, R. (2014) Changes in the global value of ecosystem services. *Global Environmental Change* 26, 152–158.

[1.3] Cox, T., Cebrian, J., Tabor, M., West, L. and Krause, J. (2020) Do diatoms dominate benthic production in shallow systems? A case study from a mixed seagrass bed. *Limnology and Oceanography Letters*.

[1.4] Round, F.E., Crawford, R.M. and Mann, D.G. (1990) *The Diatoms. Biology and Morphology of the Genera.* Cambridge University Press, Cambridge, 747 pp.
[1.5] Armbrust E. V. (2009) The life of diatoms in the world's oceans. *Nature* 459(7244), 185–192.
[1.6] Smol, J. and Stoermer, E. (eds.) (2010) *The Diatoms: Applications for the Environmental and Earth Sciences* (2nd ed.). Cambridge University Press.
[1.7] Hamm C.E. (2005) The evolution of advanced mechanical defenses and potential technological applications of diatom shells. *Journal of Nanoscience and Nanotechnology* 5(1), 108–119.
[1.8] Sterrenburg, F.A., Tiffany, M.A. and del Castillo, M.E. (2005) Valve morphogenesis in the diatom genus Pleurosigma w. Smith (Bacillariophyceae): nature's alternative sandwich. *Journal of Nanoscience and Nanotechnology* 5(1), 140–145.
[1.9] Sprynskyy, M., Szczyglewska, P., Wojtczak, I., Nowak, I., Witkowski, A., Buszewski, B. and Feliczak-Guzik, A. (2021) Diatom Biosilica Doped with Palladium(II) Chloride Nanoparticles as New Efficient Photocatalysts for Methyl Orange Degradation. *International Journal of Molecular Sciences* 22, 6734.
[1.10] Houki, S., Kawamura, T., Ogawa, N. and Watanabe, Y. (2018) Efficient crushing of hard benthic diatoms in the gut of the Manila clam *Ruditapes philippinarum* – Experimental and observational evidence. *Journal of Experimental Marine Biology and Ecology* 505, 35–44.
[1.11] Witkowski, A., Lange-Bertalot, H. and Metzeltin, D. (2000) *Diatom Flora of Marine Coasts I.* Iconographia Diatomologica 7. Koeltz Botanical Books, Glashütten, Germany, 925 pp.
[1.12] Lange-Bertalot, H., Fuhrmann, A. and Werum, M. (2020) Freshwater *Diploneis*: Species Diversity in the Holarctic and Spot Checks from Elsewhere. *Diatoms of Europe: Diatoms of the European Inland Waters and Comparable Habitats. Volume 9.* Lange-Bertalot, H. (ed.). Ruggell: A.R.G. Gantner Verlag K.G., 526 pp.
[1.13] Pennesi, C., Caputo, A., Lobban, C., Poulin, M. and Totti, C. (2017) Morphological discoveries in the genus *Diploneis* (Bacillariophyceae) from the tropical west Pacific, including the description of new taxa. *Diatom Research* 32.
[1.14] Guiry, M.D. and Guiry, G.M. (2022) AlgaeBase. World-wide electronic publication, National University of Ireland, Galway.
[1.15] Lobban, C., Schefter, M., Jordan, R., Arai, Y., Sasaki, A., Theriot, E., Ashworth, M., Ruck, E. and Pennesi, C. (2012) Coral-reef diatoms (Bacillariophyta) from Guam: New records and preliminary checklist, with emphasis on epiphytic species from farmer-fish territories. *Micronesica* 43, 237–479.
[1.16] Gottschalk, S., Uthicke, S. and Heimann, K. (2007) Benthic diatom community composition in three regions of the Great Barrier Reef, Australia. *Coral Reefs* 26, 345–357.
[1.17] Risjani, Y., Witkowski, A., Kryk, A., Yunianta, Y., Górecka, E., Krzywda, M., Safitri, I., Sapar, A., Dąbek, P., Arsad, S., Gusev, E., Rudiyansyah, Peszek, Ł. and Wróbel, R.J. (2021) Indonesian coral reef habitats reveal exceptionally high species richness and biodiversity of diatom assemblages. *Estuarine Coastal and Shelf Science* 261, 107551.
[1.18] Mann, D.G. and Droop, S.J.M. (1996) Biodiversity, Biogeography and Conservation of Diatoms. *Hydrobiologia* 337, 19–32.
[1.19] Spaulding, S. and Metzeltin, D. (2011) *Diploneis.* In: Diatoms of North America. https://diatoms.org/genera/diploneis
[1.20] Almqvist, N., Delamo, Y., Smith, B. L., Thomson, N. H., Bartholdson , A., Lal, R., Brzezinski, M. and Hansma, P. K. (2001) Micromechanical and structural properties of a pennate diatom investigated by atomic force microscopy. *Journal of Microscopy* 202(Pt 3), 518–532.
[1.21] Subhash, G., Yao, S., Bellinger, B. and Gretz, M.R. (2005) Investigation of mechanical properties of diatom frustules using nanoindentation. *Journal of Nanoscience and Nanotechnology* 5(1), 50–56.

[1.22] Lu, J., Sun, C. and Wang, Q. (2015) Mechanical Simulation of a Diatom Frustule Structure. *Journal of Bionic Engineering* 12.

[1.23] Hamm, C., Merkel, R., Springer, O., Jurkojc, P., Maier, C., Prechtel, K. and Smetacek, V. (2003) Architecture and material properties of diatom shells provide effective mechanical protection. *Nature* 421, 841–843.

[1.24] Witkowski, A., Płociński, T., Grzonka, J., Zgłobicka, I., Bąk, M., Dąbek, P., Gomes, A. and Kurzydłowski, K.J. (2019) Application of Focused Ion Beam Technique. In: *Diatoms: Fundamentals and Applications. Taxonomy-Oriented Research on Ultrastructure of Diatoms.* J. Seckback and R. Gordon (eds.) Wiley-Scrivener Publishing LLC: 115–127.

[1.25] Zgłobicka, I., Gluch, J., Liao, Z., Werner, S., Guttmann, P., Li, Q., Bazarnik, P., Plocinski, T., Witkowski, A. and Kurzydłowski, K.J. (2021) Insight into diatom frustule structures using various imaging techniques. *Scientific Reports* 11(1), 14555.

[1.26] Jansen, S. and Bathmann, U. (2007) Algae viability within copepod faecal pellets: Evidence from microscopic examinations. *Marine Ecology-Progress Series* 337, 145–153.

[1.27] Tall, L., Cloutier, L. and Cattaneo, A. (2006) Grazer-diatom size relationships in an epiphytic community. *Limnology and Oceanography* 51(2), 1211–1216.

[1.28] Perez-Burillo, J., Valoti, G., Witkowski, A., Prado, P., Mann, D.G. and Trobajo, R. (2022) Assessment of marine benthic diatom communities: insights from a combined morphological–metabarcoding approach in Mediterranean shallow coastal waters. *Marine Pollution Bulletin,* 174, 113, 183, https://doi.org/10.1016/j.marpolbul.2021.113183

[1.29] Michels, J., Vogt, J. and Gorb, S. (2012) Tools for crushing diatoms – opal teeth in copepods feature a rubber-like bearing composed of resilin. *Scientific Reports* 2, 465.

[1.30] Dilling, L., Wilson, J., Steinberg, D. and Alldredge, A. (1998) Feeding by the euphausiid *Euphausia pacifica* and the copepod *Calanus pacificus* on marine snow. *Marine Ecology-Progress Series* 170, 189–201.

[1.31] Gurney, J.L., Froneman, P.W., Pakhomov, E. and McQuaid, C.D. (2001) Trophic positions of three euphausiid species from the Prince Edward Islands (Southern Ocean): Implications for the pelagic food web structure. *Marine Ecology Progress Series* 217, 167–174.

[1.32] Grønning, J. and Kiørboe, T. (2020) Diatom defence: Grazer induction and cost of shell-thickening. *Functional Ecology* 34.

[1.33] Wilken, S., Hoffmann, B., Hersch, N., Kirchgessner, N. Dieluweit, S., Rubner, W., Hoffmann, L.J., Merkel, R. and Peeken, I. (2011) Diatom frustules show increased mechanical strength and altered valve morphology under iron limitation. *Limnology and Oceanography* 56, 1399–1410.

[1.34] Hasle, G.R. and Syvertsen, E.E. (1997) Marine Diatoms. In: *Identifying Marine Phytoplankton.* C.R. Tomas (ed.) Academic Press, San Diego: 5-385.

[1.35] Poulsen, N.C., Spector, I., Spurck, T.P., Schultz, T.F. and Wetherbee, R. (1999) Diatom gliding is the result of an actin-myosin motility system. *Cell Motility Cytoskeleton* 44(1), 23–33.

2

Size-Resolved Modeling of Diatom Populations: Old Findings and New Insights

Jonas Ziebarth[1], Werner M. Seiler[2] and Thomas Fuhrmann-Lieker[1]*

[1]*Institute of Chemistry and Center for Interdisciplinary Nanostructure Science and Technology (CINSaT), University of Kassel, Kassel, Germany*
[2]*Institute of Mathematics, University of Kassel, Kassel, Germany*

Abstract
The unique size reduction and restitution cycle of diatoms requires special attention in modeling their population dynamics. Matrix models based on distinct size classes are generally necessary to fully address periodic changes in cell size distribution and understand infra-annual algae blooms. Here, different approaches for such size-resolved models are compared, with their conclusions about oscillatory behavior in diatom populations. Particularly, the coupling of model parameters with environmental signals that may act as *zeitgebers* are discussed.

Keywords: Diatoms, biological clocks, population dynamics, dynamical systems, matrix models

2.1 Introduction

The simulation of ecosystems by population dynamics of contained species has a long tradition [2.12]. Normally, coupled differential equations are used that describe the interdependence of species or the reaction to changing environmental conditions. Mostly, a species is considered as a single compartment characterized by the number of individuals, but in some models age classes such as juvenile, adult, and senescent are distinguished [2.50]. This idea can be transferred for treating different size classes such as in diatoms. Among unicellular organisms, diatoms are unique due to their size reduction-restitution life cycle that defines an intrinsic long-period rhythm, described by William Lewis as a "sex clock" [2.52]. For modeling this means that the size variance has to be taken into account if real scenarios have to be described properly. Only a few contributions in literature have been devoted to treating this issue in some mathematical depth, despite the ecological significance of diatoms for carbon fixation and therefore the world climate [2.8] [2.83]. It should be noted that some diatoms do not undergo or have lost size-diminution in their life cycle, which could have an impact on their ecological role and might offer insights into the evolutionary relevance of the diatoms' life cycle [2.85]. This review will examine the very different approaches for size-resolved population

**Corresponding author*: th.fuhrmann@uni-kassel.de
Jonas Ziebarth: 0000-0002-6765-0003, https://www.researchgate.net/profile/Jonas_Ziebarth
Werner M. Seiler: 0000-0002-0565-1334, https://www.researchgate.net/profile/Werner-Seiler
Thomas Fuhrmann-Lieker: 0000-0003-3473-534X, https://www.researchgate.net/profile/Thomas-Fuhrmann-Lieker

Janice L. Pappas (ed.) The Mathematical Biology of Diatoms, (19–62) © 2023 Scrivener Publishing LLC

dynamics described in the literature. After an introduction of the size reduction-restitution cycle and the important parameters that need to be considered in models, we discuss experimentally detected deviations of the simple scheme. For instance, we will meet asynchronization of the two halves that changes the growth pattern and the timing of the proliferation process. Having set the biological background behind modeling, we introduce the basic mathematical concepts that are needed for understanding the various models. The main section is devoted to a detailed description of the different modeling approaches. Finally, some conclusions are made concerning the dependence of model parameters and thus the internal timing on external variables reflecting fluctuations in the environment.

2.2 The MacDonald–Pfitzer Rule and the Need for Matrix Descriptions

The size-diminution and restitution life cycle of diatoms was first described independently by John MacDonald and Ernst Pfitzer in 1869 [2.54] [2.68]. Cyclic changes in size distribution and mean size are rarely found in nature. In fact, protists over multiple orders of magnitude in size have been observed to have stable size distributions which define their ecological niche [2.33]. The mechanism is a consequence of the rigid cell wall structure that prevents a further growth in diameter after its solidification. Some species that have been studied frequently, e.g., *Thalassiosira pseudonana* or *Phaeodactylum tricornutum*, seem to be able to bypass this restriction by a yet unknown mechanism [2.16]. However, these cases will not be considered here. The two cell halves of the silicified frustule overlap like the halves of a Petri dish. The larger half is denoted as epitheca, the smaller one as hypotheca. When a cell divides vegetatively, the missing half is replenished inside, thus forming a new hypotheca in each of the parental thecae [2.72]. This leads to a decreasing mean cell size of the diatom population. On the other hand, the cell size distribution increases since the individuals inheriting the larger epitheca also continue to divide. If the generation time for all individuals is equal and no cell is removed, a distribution of size classes according to Pascal's triangle is formed. If cells become too small for further vegetative division they change from mitosis to meiosis and switch to sexual reproduction. There are variants of sexual reproduction involving gametes and auxospores (see below) but at the end of the sexual process new initial cells of large size are generated, starting the vegetative process from the beginning (Figure 2.1). In field studies, different cohorts derived from initial cells that were formed at different times (in different years) can be clearly distinguished, which demonstrates the cyclic character of the process in natural populations. In order to keep the sizes within each cohort together, however, there seem to be mechanisms counteracting the broadening of the size distributions. We will discuss this important point further below.

It should have become evident that the state and further evolution of a diatom population strongly depends on the size distribution. For each individual cell, the probability between further vegetative division or sexual reproduction is decoded in its size. Conveniently, distinct size classes can be defined for mathematical modeling, so the state of the population is expressed as a vector. The transformation over time can then be implemented as a matrix acting on this vector and changing the cell size distribution. Different transition or retention parameters in this matrix represent asymmetric division, replenishment of initial cells, and cell

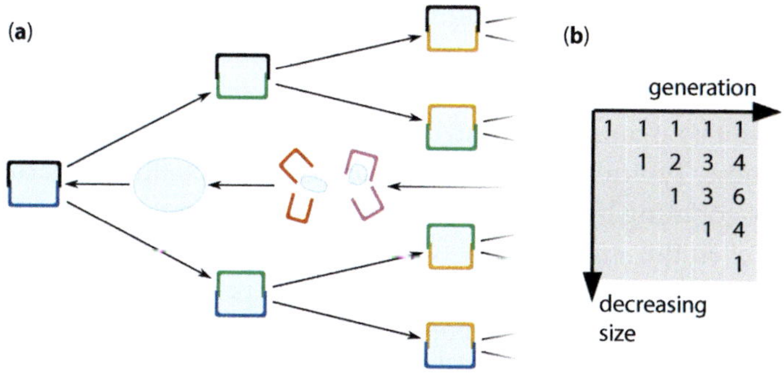

Figure 2.1 (a) The diatom division scheme according to the MacDonald–Pfitzer rule. From left to right: vegetative process, from right to left: closing the life cycle by sexual reproduction. (b) Change of size distribution from generation to generation according to Pascal's triangle.

death. This bears some resemblance to age classes in other population models but the number of size classes needed here is typically much larger. So also in mathematical modeling diatom populations turn out to be unique and justify a separate treatment as in this review.

2.3 Cardinal Points and Cycle Lengths

2.3.1 Considered Cardinal Parameters

For the different sizes diatom cells may attain in the course of their life cycle, so-called cardinal points were first defined by L. Geitler of the University of Vienna. The cardinal points constitute the most important parameters needed for any size-resolved modeling of the diatom life cycle. These four cardinal points are:

(i) the size of the initial cells,
(ii) the upper sexual size threshold, i.e., the upper size limit below which cells are able to produce gametes for sexual reproduction,
(iii) a lower sexual size threshold, i.e., a lower size limit below which cells are not able any more to produce gametes,
(iv) a final minimum size, at which cells are not able to further divide and eventually die.

That means that the cells of a culture can be divided into three subpopulations, limited by the respective cardinal points: Large cells that proliferate only by vegetative division between (i) and (ii), medium-sized cells that are capable of both vegetative and sexual reproduction between (ii) and (iii), and small cells between (iii) and (iv) that divide vegetatively until they are not viable any more. This last category is not relevant for closing the life cycle by the sexual phase, and therefore may be neglected in studies aiming at cycle lengths. On the other hand, they contribute to the size distribution of a population as well and have to be taken into account when analyzing experimental distributions.

Since there are many cell divisions within each of these categories, the incremental size diminution in one division step, or alternatively the total number of steps between two cardinal points, is a further parameter that has to be specified. It should be mentioned that size could be defined by diameter of centric diatoms, respectively the long axis in pennate diatoms, or by volume. Typically, lengths are more easily detectable by microscope, whereas other methods like counting by electrical conductivity (Coulter counter) address rather the volume [2.65], or even the surface (fluorescence staining of cell walls) [2.69]. The scaling between length and volume is not always clear, since it depends on the species [2.6]. In long pennate diatoms like *Synedra tabulata*, length and volume are directly proportional [2.20]. Changes in size can be accompanied by allometric changes which could give an opportunity for deeper analysis of the population [2.47]; however, they are not considered in this introduction of population size-distribution modeling.

These parameters mentioned so far affect only the sizes. For describing temporal events, the time between two mitotic cell divisions τ is an important parameter. It surely depends on environmental conditions and may be subjected to a certain distribution function (see also Section 2.4 for a discussion on asymmetry) [2.1] [2.59]. Typically, this time has to be derived from the growth rate from counting of cells or turbidimetric measurements. The cell division rate, based on binary logarithm, can be calculated from two population numbers X_1 and X_2 with a time leap Δt by

$$\mu_D = \tau^{-1} = \frac{\log(X_2/X_1)}{\Delta t \log 2} \tag{2.1}$$

and is obtained with higher reliability by fitting log X versus time. Attention should be paid to the fact that sometimes the rate is given as growth rate $\mu = \mu_D \ln(2)$ on the basis of natural logarithms.

The sexual process involving gametogenesis and auxosporulation takes a certain amount of time that may be modeled as a multiple of τ.

2.3.2 Factors Determining Cardinal Points

The formation of initial cells shall be considered as the first step in the cycle. For simplicity in modeling, it may be reasonable to assume a certain fixed size for the initial cells. However, it can be expected that the size of initial cells is subjected to some distribution. For example, Bethge reported a ratio of 2-4 for initial cell size relative to the parental cell size for species in the *Melosira* genus [2.10]. For string-forming species of the genus *Aulacoseira,* a plot of initial cell diameter versus maternal filament diameter revealed an almost linear dependence, with a ratio of sizes between initial cells and their parental cells in the range of 2-3 [2.27]. More recently, Davidovich added further data on linear dependence for *Nitzschia lanceolata, Synedra tabulata* and *Licmophora ehrenbergii,* finding a ratio between maximum and minimum size of initial cells of 1.5 to 2 [2.20]. Overall, the size range of diatom species also appears to be related to marine and freshwater conditions, indicating an evolutionary link between size and the respective nitrogen and phosphorous availability.

In the vegetative phase, the timing for the sequence of generations follows the cell cycle. The cell cycle is controlled by some checkpoints that ensure that the right conditions for proliferation are met [2.39]. For instance, nutrients and silica required for the formation of a new theca have to be available. Notably, there is also a checkpoint for blue light that is sensed by an aureochrome photoreceptor [2.41]. This enables the synchronization of the cell cycle with the diurnal rhythm and can also be applied to synchronize cell cultures, as demonstrated for *Seminavis robusta* [2.32] and *Phaeodactylum tricornutum* [2.40]. Davidovich *et al.* [2.25] investigated the influence of temperature on the rate of cell divisions for *Haslea ostrearia* and *Haslea karadagensis*. Data are given in Figure 2.2 and show division times between two days at low and one day at higher temperature [2.25]. Williams characterized the influence of salinity for 14 species of pennate diatoms and measured maximum rates between 0.6 and 3.2 divisions/day that were found to correlate with cell volume and area/volume ratio [2.87]. Much slower rates are reported for *Stephanodiscus neoastreaea* in an northern Irish lake with only 3-7 divisions in the spring growing period [2.42]. The variability in growth rates can be high as these data show, but often it cannot be taken into account in modeling because continuous data are missing. The problem is circumvented in models in which not the real time, but the generation time, i.e., the time between two cell divisions as unit, defines the time axis. If a mapping of generation time to real time can be made, with a corresponding stretching or compression of the time axis, the models fully retain their validity.

The checkpoint for the switch to sexual reproduction is one of the mysteries that is not fully understood. By artificially altering the cell size using nutritional as well as operative methods, Von Stosch at the University of Marburg was able to show in the 1960s that this threshold is indeed determined by the cell size and not by the age of a population [2.80] [2.56]. For *Stephanodiscus neoastreaea*, nitrate was found to be a prerequisite to start sexual reproduction in fall [2.42]. For *Seminavis robusta*, it was recently found that the light quality (color) also influences sexual reproduction [2.11].

Typically, new initial cells are formed within auxospores. The kinetics of auxospore formation depends on the detailed mechanism of sexual reproduction of which a large variety is known [2.44] [2.15]. Some bypasses exist that allow single cells to form new enlarged initial cells without the need of combining genetic material stemming from another cell. One example is vegetative enlargement [2.80], another is automixis in which gametes produced in one parental cell immediately fuse to form the auxospore [2.15]. In these uniparental cases there is no need

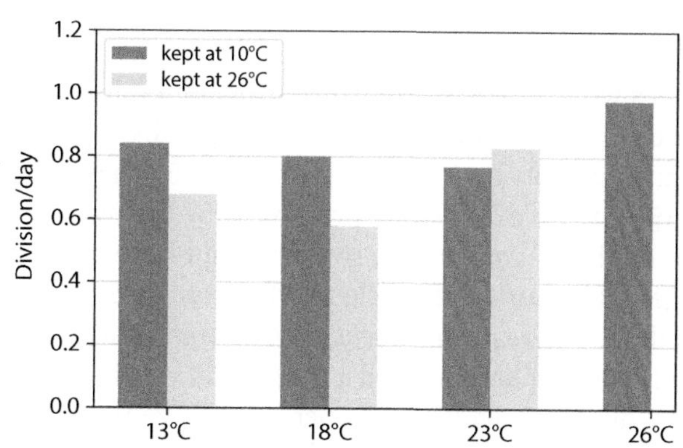

Figure 2.2 Temperature dependence on the doubling rate for *Haslea ostrearia*. Strains were adapted to 10°C and 26°C and then exposed to different temperatures. (Constructed with data from [2.25]).

to consider encounter dynamics of gametes by diffusion. In biparental auxosporulation we distinguish between homothallic strains in which one clonal strain produces two types of gametes, and heterothallic strains, in which different mating types exist in separate strains. Since gametes from different parents have to combine it is no wonder that the success rate of auxosporulation depends on the cell density [2.74].

The lower limit for sexual reproduction is not described for all diatom species. If such a lower limit exists, the life cycle is called a "closed cycle," if not, it is referred to as an "open cycle" [2.44].

Finally, aging and natural cell death shall be discussed. Obviously, there is some difficulty in maintain the silica pattern with possibly self-organized pores in very small frustules [2.72]. Also, there would be a crowding of organelles, so it is understandable that there is some size limit for viability, and very small cells will eventually die. Jewson addressed the important question whether some aging exists in the vegetative phase of larger cells, i.e., whether there is a maximum number of divisions an epitheca can undergo or whether the cells will leave the population after a certain number of smaller offspring [2.42] [2.43]. For *Stephanodiscus neoastraea* he deduced a lifespan of six to eight generations and extrapolated a similar value for *Aulacoseira subarctica*. By the removal of older cells from the scheme, an asymmetry is induced that changes the size-resolved population dynamics. The occurrence of such asymmetries and consequences for modeling will be further treated in Section 2.4.

2.3.3 Experimental Data

Experimental data normally stem from investigations of the size distribution of natural diatom populations in some time intervals. Such data may have some gaps due to the organization of the field study and are influenced by strong weather fluctuations and singular natural events. Nevertheless, they give important insight into the cycles of real ecosystems. Notable studies include the investigation of populations in lakes, such as *Tabellaria fenestrata* and *Aulacoseira* (formerly *Melosira*) *islandica* in the Zürichsee [2.63], *Nitzschia sigmoidea* in Blackford Pond [2.55], and *Aulacoseira subarctica* and *Stephanodiscus* sp. in Lough Neagh [2.42] [2.43]. Investigations of sediments that made the assignment of distinct years possible might open a window into the past [2.63]. Only in some cases in which cultures can be held successfully in the laboratory, parameter data under controlled conditions are available. The most conspicuous finding of these investigations are multimodal distributions of cell sizes which are used for calculating life cycle length, or more precisely the periodic occurrence of initializations events by large cells. For plotting and detection of deviations from Gaussian distributions, Rankit plots [2.67] have been favorably used as visualization tool [2.55]. In Table 2.1, some proposals for life cycle lengths are compiled. Interpretation of data might be ambiguous, so some data have been reanalyzed later. For example, Jewson suggested a shorter life cycle for *Aulacoseira islandica* than the one originally reported by Nipkow, whereas for *Tabellaria fenestrata* Mann suggested a longer life cycle. For *Aulacoseira subarctica* he proposed a theoretical maximum cycle of 15 years corresponding to over 100 divisions, while the upper sexual size threshold is already reached after 4-6 years.

Table 2.1 Reported life cycles and cardinal points of selected species. Cycle lengths marked with a star are rough estimates based on the reported size diminution speeds and the size range. The size in brackets marks the size at which cells were observed to become sexually inducible.

Species	Cycle length	Size range	Ref.
Seminavis robusta	<1 year	25(50)–73 μm	[2.14] [2.16]
Eucampia zodiacus	~1 year	10(15)–80 μm	[2.64]
Cocconeis scutellum	1 year	15–49 μm	[2.57]
Fragilariopsis kerguelensis	~1 year	10(31)–76 μm	[2.48] [2.5]
Pseudo-nitzschia delicatissima	<1 year	8(80)–94 μm	[2.3]
Pseudo-nitzschia multistriata	2 years	30(55)–82 μm	[2.19]
Tabularia fasciculata	1–1.5 years	10(113)–248 μm	[2.23]
Haslea karadagensis	<1–2 years	22(52)–97 μm	[2.24]
Stephanodiscus sp.	1–2 years	25–66 μm	[2.70]
Haslea ostrearia	1–2.5 years*	17(68)–140 μm	[2.22]
Melosira helvetica	3–4 years	4–31 μm	[2.10]
Pseudo-nitzschia australis	2 years	75(121)–170 μm	[2.38]
Pseudo-nitzschia pungens	3 years	75(115)–185 μm	[2.38]
Pseudo-nitzschia multiseries	3 years	39–146 μm	[2.21]
Chaetoceros dichaeta	>3 years*	6(15)–44 μm	[2.6]
Stephanodiscus sp.	3–4 years	20(34)–74 μm	[2.42]
Coscinodiscus wailesii	<4 years*	90(200)–350 μm	[2.62]
Nitzschia sigmoidea	>5 years	200(300)–424 μm	[2.55]
Tabellaria fenestrata	2–8 years	40(40)–90 μm	[2.63] [2.55]
Aulacoseira subarctica	4–6 years	3(8)–14 μm	[2.43]
Aulacoseira islandica	40 years	4–22 μm	[2.63] [2.55]

2.4 Asymmetry, Delay and Fibonacci Growth

2.4.1 The Müller Model

Since the introduction of the MacDonald–Pfitzer rule in 1869, it has been evident that the theory cannot represent full reality. The rule is commonly used as a simple description of the diatoms life cycles but over time diatomists have repeatedly disputed its simplicity. Some have specifically rejected the idea that diatoms divide uniformly and would thus generate a binomial distribution of cell sizes.

In 1883, Otto Müller proposed an approach to the cell division scheme which involves a delay [2.60]. He defined a process of simultaneous division as the vegetative cell division in which cells divide into two daughter cells which then in turn continue to divide simultaneously again. Müller, however, found that this was not in agreement with his observations of some diatom species. Instead, if the larger or older daughter cells would divide first and the younger daughter would delay division by one generation, his observations could be explained. Furthermore, a simultaneous division would lead to a large number of auxospores at one point. As this cannot be found in nature, a delay of some kind should slow down the formation of auxospores and the progression of the size reduction.

The diatom *Ellerbeckia* (formerly *Melosira*) *arenaria* [2.71] possesses particular modifications of the girdle band that allow the deduction of the age of the cells and thecae. Müller first found evidence of a delay when investigating diatoms of the genus *Terpsinoë*. These chain-forming diatoms often form groups of three (triplets), hinting at a faster division of one cell. Lacking the proper tools to assign the age of the thecae, he could not prove his assumption. However, in *Ellerbeckia arenaria* the girdle bands can indicate the age of the theca. Because of its division mechanism, this species has girdle bands only on one theca opposite to a thickening [2.61] or a step [2.18] on the next cell (inset B Figure 2.3). During division, a new girdle band is formed and smaller thecae are enclosed. Again, the girdle band ends at a step which is formed on the adjacent new valve of the next cell (inset A Figure 2.3). Hence, two parameters can be assigned to each theca in the chain, thereby creating four types: the existence or absence of a step and whether the thecae are enclosed or free. Every cell is then found to be part of a twin or triplet group which has free thecae as its boundary. If the knowledge about the parameter sequence in the chain and the grouping of the cells is combined, the age of each theca can be deduced and the formation of the chain could be traced back until the initial cell. Applying this method, Müller concluded that the younger cell which contains the smaller half of the mother cell and a fresh even smaller theca, must delay its division for one generation. He hypothesized that the older daughter would inherit richer material, by quality or quantity, from the mother.

In 1884, following up on his first publication, Müller presented a deeper analysis of the composition of the *Ellerbeckia arenaria* chains [2.61]. As the motivation of the analysis was to find the influence of the delay on auxosporulation, it is important to understand the number of cells and their corresponding sizes first. The delayed division scheme generates a chain whose total number of individuals follows the Fibonacci sequence (Figure 2.3). The Fibonacci sequence dominates the division process in all aspects of single theca generation as well as frustule grouping. When transitioning into the next generation, a twin of an older and a younger cell produces one additional cell which turns it into a triplet with a fresh triplet-twin and an undivided triplet cell. A triplet reproduces another triplet from the triplet-twin and a twin

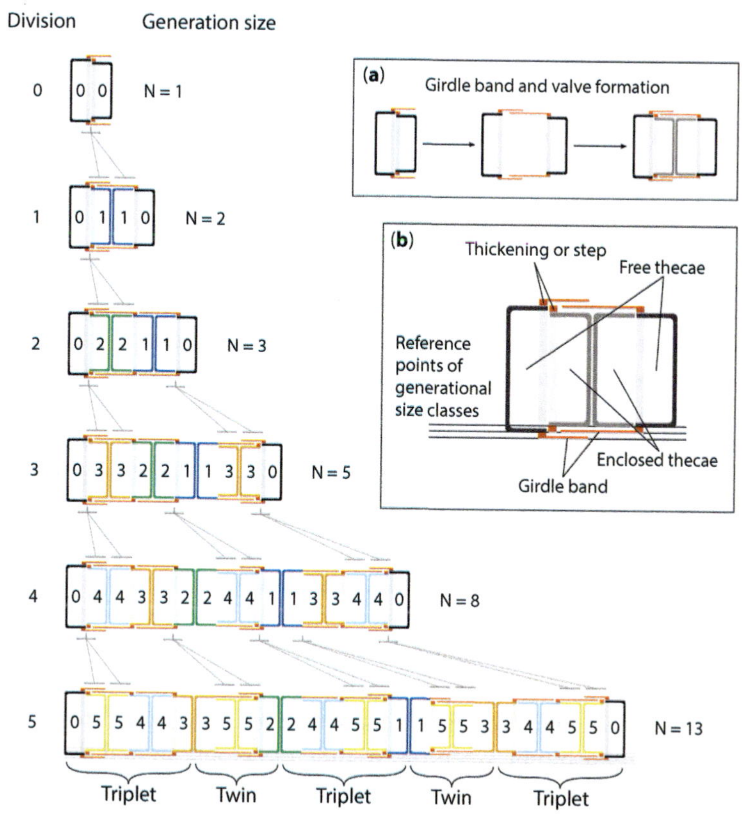

Figure 2.3 The modified MacDonald–Pfitzer scheme of *Ellerbeckia arenaria* according to Müller. New thecae are enclosed by a girdle band in the first generation which is faced by a thickening or step in the valve of the next cell (a). The detailed formation of this peculiar arrangement was described in more recent publications [2.43] [2.75]. The reference point for the size of the theca when considering the MacDonald–Pfitzer rule is the thickening of the valve if there is one (b). In Müller's division pattern nascent cells always skip one generation before dividing again. This leads to Fibonacci growth and formation of twin and triplet groups.

from the triplet cell. After the first few divisions, this leads to Fibonacci growth of twin and triplet numbers. Furthermore, Fibonacci growth also occurs with the numbers of the four different theca types. However, there is an exception from this rule for the fiber-initiating cell, which can cause a difference of the numbers by two depending on the theca types. After establishing the total number of cells, the number of cells in a certain size after a random number of generations are essential since the cells of the smallest size can auxosporulate. These numbers are obtained with arithmetic sequences of the corresponding order. Thus, the onset of auxosporulation in the nth generation in a simultaneous division scheme is delayed to the $(2n + 1)$th generation. However, the numbers of auxospores that are produced from this generation on are the same as with simultaneous division.

Further investigation of the chain provides insights into the symmetry of the chains. Here, symmetry is defined as the symmetry of the arrangement of the four types of theca from each end of the chain. The individuals can be grouped in different levels. The first level would be the twins and triplets, which in turn can be regrouped on multiple levels. The analysis and breakdown of this grouping leads to three key findings. Firstly, the previous elements of the chain are always reproduced until the point where the chain ended in the previous generation. Secondly, the entire chain is symmetrical with the exception of up to four cells in the center and the outmost thecae. Lastly, the chain can be broken down into smaller groups which are in turn symmetrical except for their outmost thecae. The length of these groups is also derived from Fibonacci growth.

2.4.2 The Laney Model

The previous section discussed asymmetric division and favoring of the older daughters. Laney *et al.* noted that much of recent research has been focusing on the phenotypic differences of daughter cells in diatoms, but not on the timing of theca inheritance. In their experiments, contrary to Müller's results, diatoms seem to favor their younger daughters [2.49].

Laney *et al.* investigated the division rates and cell volume of *Ditylum brightwellii* using a modern time-lapse imaging approach [2.49]. The pictures were taken at low density using infrared microscopy. Because of higher density after only a few divisions and close proximity of the cells, the analysis was mostly performed manually and therefore had to be limited. Nevertheless, daughters inheriting the hypotheca were found to need about 4% less total light dosage to divide. The smaller cells also received more of the mother's cytoplasm. Interestingly, the division speed did not seem to accumulate over multiple generations. Nonetheless, results were comparable to observations of growth speed in *E. coli* [2.79], which suggested a common behavior across kingdoms. A linear regression of the received additional cytoplasm and the light dosage supported a relationship between the volume of cytoplasm and division rate as well. Moreover, the analysis indicated that the younger daughters have an advantage independent of cytoplasm volume. This would be in agreement with the previously introduced theory that the quality of inherited material is better. Specifically, it is assumed that the smaller daughter receives material that is fresh and received less oxidative stress. Asymmetric distribution of oxidative stress products is also known from other kingdom, e.g., for oxidatively damaged proteins in *Saccharomyces cerevisiae* [2.2]. Overall, the higher quality of the smaller cell's material in diatoms could improve the chance of success in sexual reproduction at the end of a size-diminution cycle.

The authors propose that the asymmetrical division of diatoms could also play a role in tuning the diatom sex clock. This would be an important modeling parameter for the clock besides the cardinal points that depend on the diatom life cycle. Ecologically, a faster division of smaller cells would lead to more copies of a mother cell's genome and permit more of its progeny to reach sexual reproduction. Müller had already concluded that delays in cell division fundamentally alter the growth speed and size distribution of a population. Laney *et al.* remark that it would be essential to further investigate the influence of different division speeds and subsequently improve modeling approaches. As the delays in both cases are assigned to different halves, a wide spectrum of systematic asymmetry in diatom division is suggested.

2.5 Continuous vs. Discrete Modeling

Population dynamics represents a typical *evolution problem*: we are interested in the evolution of certain quantities over time. Mathematically, such problems are described by *dynamical systems* which may abstractly be considered as rules describing how we can derive from the knowledge of the values of the involved quantities at some time t_0 their values at any later time $t_1 > t_0$. A fundamental question in the design of mathematical models is then whether time is considered as a continuous or as a discrete quantity, as these options lead to rather different type of equations. A general overview over various mathematical approaches to population dynamics can be found in the textbook [2.12].

In everyday life, we perceive time as a continuous quantity, i.e., mathematically it is identified with the real numbers $\mathbb{R}$. In typical models, we then also treat quantities like the number X of diatoms in a given population as real numbers, although they are actually integers. Since X is usually a rather large number, this does not lead to relevant errors and allows us to consider X as a function $\mathbb{R} \to \mathbb{R}$ for which we can derive *differential equations* representing a *continuous dynamical system*. Some classical introductions into their basic theory are the books [2.37] [2.82] [2.86].

In population dynamics, it is often also natural to think in "generations": one combines all births within a certain time span into one generation and enumerates the different generations, i.e., mathematically time is now identified with the integers $\mathbb{Z}$. In principle, this makes it also natural to describe X by integers, i. e., as a function $\mathbb{Z} \to \mathbb{Z}$. However, most models contain real parameters like birth rates or survival probabilities and thus one must consider X as a real-valued function $\mathbb{Z} \to \mathbb{R}$ satisfying *difference equations* representing a *discrete dynamical system*. Here, some introductory textbooks are [2.13] [2.28] [2.45] [2.73].

In practice, one usually deals with systems and not with scalar equations. Thus, one deals with vector-valued functions $\mathbf{X}$ from either $\mathbb{R}$ or $\mathbb{Z}$ to $\mathbb{R}^n$. The meaning of the components of $\mathbf{X}$ can be rather different. They may represent different species or compartments, but vectors also naturally appear in age-structured models [2.50] or in the case of diatoms in size-structured models. If one uses a model with both age and size structure, then it may be useful to consider $\mathbf{X}$ actually as a matrix-valued function from $\mathbb{R}$ or $\mathbb{Z}$ to $\mathbb{R}^{m \times n}$. For notational simplicity, in this section we will consider such models still as vector-valued with vectors in $\mathbb{R}^{mn}$.

In this section, we will briefly survey some basic mathematical facts for both types of equations with an emphasis on *linear* systems where much information can be gleaned with the help of elementary linear algebra. This concerns in particular the stability of equilibria and the existence of periodic solutions (i.e., oscillations). In many situations, it is relevant for the understanding of a linear system to have information about the dominant eigenvalue of the system matrix. In the context of population dynamics, such information can often be obtained rather easily with the help of the Perron-Frobenius theorem. Therefore, we will also briefly discuss this theorem and its application.

2.5.1 Discrete Dynamical Systems

Let $\mathbf{X}$ be a function $\mathbb{Z} \to \mathbb{R}^n$ whose values are n-dimensional vectors decoding different species or size classes. Alternatively, we may consider $\mathbf{X}$ as an infinite sequence of vectors

$$(\ldots, \mathbf{X}(-2), \mathbf{X}(-1), \mathbf{X}(0), \mathbf{X}(1), \mathbf{X}(2), \ldots).$$

As in evolutionary problems one is typically only interested in the future, one considers **X** only from some starting index n_0 on, which is usually chosen to be 0. Hence we will consider from now **X** as a function $\mathbb{N}_0 \to \mathbb{R}^n$ or as a sequence $(\mathbf{X}(0), \mathbf{X}(1), \mathbf{X}(2), \ldots)$.

An *(autonomous) first-order system of difference equations* for **X** is of the form

$$\mathbf{X}(t+1) = \mathbf{f}(\mathbf{X}(t)) \tag{2.2}$$

with some function $\mathbf{f}: \mathbb{R}^n \to \mathbb{R}^n$. Equation (2.2) may be considered as a rule describing the population in generation $t + 1$, if the population in generation t is known. Consequently, it is common to consider *initial value problems*: prescribing the values $\mathbf{X}(0) = \mathbf{X}_0 \in \mathbb{R}^n$, the rule (2.2) uniquely determines the population $\mathbf{X}(t)$ in all generations $t > 0$.

An *equilibrium* (or *steady state*) *solution* of an autonomous system is defined by a vector $\mathbf{V} \in \mathbb{R}^n$ such that $\mathbf{f}(\mathbf{V}) = \mathbf{V}$ (a *fixed point* of the function **f**) [2.28, Sect. 1.3]. Indeed, in this case the constant function $\mathbf{X}(t) = \mathbf{V}$ for all $t \in \mathbb{N}_0$ obviously solves (2.2). An equilibrium **V** is called *stable*, if for any $\epsilon > 0$ we can choose a $\delta > 0$ such that for any initial value $\mathbf{X}_0$ with $||\mathbf{X}_0 - \mathbf{V}|| < \delta$ the estimate $||\mathbf{f}^k(\mathbf{X}_0) - \mathbf{V}|| < \epsilon$ holds for all $k > 0$; otherwise **V** is called *unstable* [2.28, Sect. 1.3]. Here $|| \cdot ||$ denotes an arbitrary norm on $\mathbb{R}^n$.

A *periodic solution with period* ℓ (for some $\ell > 1$) is defined by a vector $\mathbf{V} \in \mathbb{R}^n$ such that $\mathbf{f}^\ell(\mathbf{V}) = \mathbf{f}(\mathbf{f}(\cdots(\mathbf{f}(\mathbf{V})))) = \mathbf{V}$ and $\mathbf{f}^j(\mathbf{V}) \neq \mathbf{V}$ for any exponent $1 \leq j < \ell$ [2.28, Sect. 1.6]. Indeed, setting $\mathbf{X}(0) = \mathbf{V}$, $\mathbf{X}(1) = \mathbf{f}(\mathbf{V})$, $\mathbf{X}(2) = \mathbf{f}^2(\mathbf{V})$, ..., $\mathbf{X}(\ell - 1) = \mathbf{f}^{\ell-1}(\mathbf{V})$, we can construct a solution by continuing with $\mathbf{X}(t) = \mathbf{X}(t - \ell)$ for all indices $t \geq \ell$.

A *second-order system of difference equations* is of the form

$$\mathbf{X}(t+2) = \mathbf{f}(\mathbf{X}(t), \mathbf{X}(t+1)), \tag{2.3}$$

i.e., the population in generation $t + 2$ depends not only on the population in the last generation $t + 1$ but also on the last but first generation t. This may be interpreted as a model for delay effects. In an *initial value problem*, it is now necessary to prescribe both the values $\mathbf{X}(0) = \mathbf{X}_0$ and $\mathbf{X}(1) = \mathbf{X}_1$, as now the rule (2.3) is able to determine the population $\mathbf{X}(t)$ only in all generations $t > 1$. One can extend this scheme in an obvious manner to higher-order systems.

The simplest difference equations are the linear ones and we will concentrate here on these. A *linear, first-order system of difference equations with constant coefficients* has the form

$$\mathbf{X}(t+1) = A\mathbf{X}(t) \tag{2.4}$$

where $A \in \mathbb{R}^{n \times n}$ is a real matrix with constant entries. In the context of population dynamics, A is often called the *propagator matrix* or population projection matrix [2.13]. Obviously, the zero vector (or more generally any vector **V** with $A\mathbf{V} = \mathbf{V}$) defines an equilibrium for such a system. Furthermore, such systems can be solved in closed form via elementary linear algebra, namely the theory of eigenvalues and -vectors [2.66].

Here, in the simplest case A is a diagonalizable matrix which is also the most relevant case for applications [2.28, Sect. 3.3.1]. For such a matrix, we can find n linearly independent eigenvectors $\mathbf{V}_1, \ldots, \mathbf{V}_n$ with corresponding eigenvalues $\lambda_1, \ldots, \lambda_n$ (i. e., solving the matrix equation $A\mathbf{V}_k = \lambda_k \mathbf{V}_k$; note that we do not require that $\lambda_i \neq \lambda_j$ for $i \neq j$). If we furthermore assume that all these eigenvalues and -vectors are real, then a fundamental system of real solutions is given by the functions

$$\mathbf{X}_k(t) = \lambda_k^t \mathbf{V}_k, \quad 1 \leqslant k \leqslant n \tag{2.5}$$

(this follows simply from the fact that $\mathbf{X}(t) = A^t\mathbf{X}_0$ and that for $\mathbf{X}_0 = \mathbf{V}_k$ we have $A^t\mathbf{V}_k = \lambda_k^t \mathbf{V}_k$). The general solution is the linear combination $\mathbf{X}(t) = \sum_{k=1}^{n} c_k \mathbf{X}_k(t)$ with arbitrary coefficients $c_k \in \mathbb{R}$. The unique solution of an initial value problem with $\mathbf{X}(0) = \mathbf{X}_0$ is then obtained by solving the linear system of equations $\sum_{k=1}^{n} c_k \mathbf{V}_k = \mathbf{X}_0$ for the coefficients c_k (i.e. one expresses the vector $\mathbf{X}_0$ in the eigenbasis $(\mathbf{V}_1, \ldots, \mathbf{V}_n)$).

If the matrix A also possesses complex eigenvalues, then (2.5) still represents a fundamental system of solutions, but at least some of them are complex. It is, however, easily possible to derive an alternative fundamental system consisting entirely of real solutions. If $\lambda = \rho e^{i\omega} \in \mathbb{C}$ is a complex eigenvalue of A, then its complex conjugate $\bar{\lambda} = \rho e^{-i\omega}$ is also an eigenvalue. Furthermore, if $\mathbf{V} \in \mathbb{C}^n$ is an eigenvector for λ, then $\bar{\mathbf{V}}$ is an eigenvector for $\bar{\lambda}$. Decomposing the eigenvector in its real and imaginary part, $\mathbf{V} = \mathbf{W}_{\text{re}} + i\mathbf{W}_{\text{im}}$, we can replace in our fundamental system the two complex solutions corresponding to λ and $\bar{\lambda}$ by the following two real ones:

$$\mathbf{X}_{\text{re}}(t) = \rho^t[\cos(\omega t)\mathbf{W}_{\text{re}} - \sin(\omega t)\mathbf{W}_{\text{im}}], \tag{2.6}$$

$$\mathbf{X}_{\text{im}}(t) = \rho^t[\sin(\omega t)\mathbf{W}_{\text{re}} + \cos(\omega t)\mathbf{W}_{\text{im}}].$$

The expressions (2.5) and (2.6) describing a real fundamental system of (2.4) easily allow us to make certain statements about the asymptotic behavior of solutions. If there exists an eigenvalue λ_k with $|\lambda_k| > 1$, then for any initial condition with $c_k \neq 0$ the solution will be unbounded for $t \to \infty$, i.e., the population of at least one species will become arbitrarily large; i.e., the equilibrium $\mathbf{V} = \mathbf{0}$ is unstable. Conversely, if all eigenvalues satisfy $|\lambda_k| < 1$, then any solution will converge for $t \to \infty$ towards $\mathbf{0}$, i.e., all species become extinct; i.e., the equilibrium $\mathbf{V} = \mathbf{0}$ is even *asymptotically stable.* If the absolute value is greater than 1 for some eigenvalues and less than 1 for some others, then $\mathbf{V} = \mathbf{0}$ is a *saddle point.* (2.6) indicates that complex eigenvalues are related to *oscillatory behavior*: if $\rho < 1$, the oscillations are damped, whereas for $\rho > 1$, the amplitude grows towards infinity.

The situation is more complicated if eigenvalues λ_k exist with $|\lambda_k| = 1$. If λ_k is real, this means that $\lambda_k = \pm 1$. In the case $\lambda_k = +1$, any eigenvector $\mathbf{V}_k$ defines an equilibrium; in the case $\lambda_k = -1$, a periodic solution with period 2 is given by $\mathbf{X}(t) = (-1)^t\mathbf{V}_k$ for any eigenvector $\mathbf{V}_k$. If λ_k is complex, then one can observe for appropriate initial conditions sustained oscillations. One should note that such oscillations do not necessarily lead to periodic solutions. For $\lambda_k = e^{i\omega_k}$, periodic solutions will only arise if ω_k is a rational multiple of 2π. Otherwise, one finds *almost periodic* solutions which never exactly repeat themselves, but come arbitrarily close to doing so.

Figure 2.4a shows in yellow the region in the complex plane where all eigenvalues must be located for an asymptotically stable equilibrium, namely inside the unit circle. For sustained oscillations, one must also have at least some eigenvalues on the unit circle itself (marked with a diamond symbol). The figure also shows that complex eigenvalues always appear in complex conjugated pairs, i.e., the eigenvalues lie symmetric with respect to the real axis. The *Jury test* [2.13] [2.51] allows for effectively determining the number of eigenvalues outside of the unit circle (indicating instability) without explicitly computing the eigenvalues which is quite useful for higher-dimensional systems.

We remark that the case that the propagator matrix A is not diagonalizable can also be handled in closed form [2.28, Sect. 3.3.2]. The situation is more complicated then and one needs the so-called *Jordan normal form* of A. In this case, further possibilities for unstable behavior exist, namely multiple eigenvalues of absolute value 1 without the corresponding number of linearly independent eigenvectors. In such situations, one no longer has an exponential instability but only a polynomial one.

We thus conclude that using linear algebra allows us to solve any system of the form (2.4) with a fundamental system composed of the solutions (2.5). This explicit form furthermore permits statements about the asymptotic behavior of the solution. Assume for simplicity that one eigenvalue, say λ_1, has the greatest absolute value, i.e., $|\lambda_1| > |\lambda_k|$ for all $k > 1$. We know from the discussion above that for the initial data $\mathbf{X}_0 = \sum_{k=1}^{n} c_k \mathbf{V}_k$ the solution is given by $\mathbf{X}(t) = \sum_{k=1}^{n} c_k \lambda_k^t \mathbf{V}_k$. It is now easy to see that we obtain for large times t

$$\lim_{t\to\infty} \frac{\mathbf{X}(t)}{|\lambda_1|^t} = c_1 \mathbf{V}_1 . \tag{2.7}$$

Thus, in the long run the solution mode corresponding to the eigenvalue λ_1 dominates all other modes, which consequently effectively disappear, as they become negligibly small in

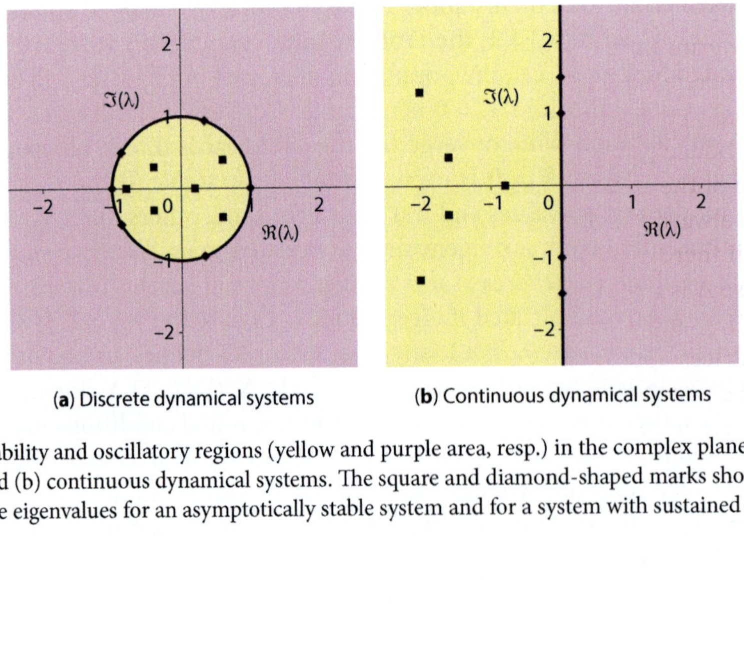

Figure 2.4 Stability and oscillatory regions (yellow and purple area, resp.) in the complex plane for (a) discrete and (b) continuous dynamical systems. The square and diamond-shaped marks show possible locations of the eigenvalues for an asymptotically stable system and for a system with sustained oscillations, respectively.

comparison. If several eigenvalues have the same maximal absolute value, then the long-time behavior is controlled exclusively by the corresponding modes.

For nonlinear first-order systems of the form (2.2), some of the above sketched out ideas about linear systems can be reused [2.28, Sect. 4.6]. Assume that **V** is an equilibrium of (2.2) and that the right-hand side **f** is partially differentiable with respect to all of its arguments. Then we can form the *Jacobian* of **f** at **V**

$$J = \begin{pmatrix} \frac{\partial f_1}{x_1}(\mathbf{V}) & \cdots & \frac{\partial f_1}{x_n}(\mathbf{V}) \\ \vdots & & \vdots \\ \frac{\partial f_n}{x_1}(\mathbf{V}) & \cdots & \frac{\partial f_n}{x_n}(\mathbf{V}) \end{pmatrix} \in \mathbb{R}^{n \times n} \tag{2.8}$$

and in many cases the local behavior of (2.2) near **V** can by analyzed by studying its *linearization* $\mathbf{Y}(t+1) = J\mathbf{Y}(t)$. If, for example, all eigenvalues of the Jacobian J have an absolute value less than 1, then the equilibrium **V** of (2.2) is asymptotically stable. Such a simple relationship between linear and nonlinear stability exists, however, only for *hyperbolic* equilibria, i.e., if none of the eigenvalues of J has absolute value 1. Otherwise, more sophisticated methods must be used.

2.5.2 The Perron-Frobenius Theorem

It has become apparent from the discussions in the previous section that information about the size of the eigenvalues of a matrix are crucial for assessing the stability of solutions or equilibria on the one hand and for the search for periodic solutions on the other hand. For small systems, it is often possible to explicitly determine all eigenvalues by hand or with the help of a computer algebra system. For larger systems, one usually fails here. The *Perron-Frobenius Theorem* provides information for certain matrices about the dominant eigenvalue which determines the long-time behavior of a dynamical system as discussed above [2.73] [2.13]. We refer to the monograph [2.9] for an in-depth treatment of all concepts and results discussed in this section.

We first introduce some necessary terminology. Let $A = (a_{ij}) \in \mathbb{R}^{n \times n}$ be a square matrix. The maximum of the absolute values $|\lambda|$ of the eigenvalues is the *spectral radius* $\rho(A)$ of A (note that it is always a real number, even if all the eigenvalues are complex). We say that A is *reducible* if there exists a permutation matrix P such that PAP^{-1} is in block triangular form; otherwise A is *irreducible.* A *positive* matrix A has only positive entries, i.e., if $a_{ij} > 0$ for all indices i, j, and in a *non-negative* matrix also zero entries are allowed, i.e., $a_{ij} \geq 0$ now. Such matrices appear naturally in stochastic problems, but also in many biological applications like population dynamics.

A property like irreducibility is not easy to check directly from its definition. A practically useful approach is based on a directed graph $G(A)$ associated with any non-negative matrix $A \in \mathbb{R}^{n \times n}$. It has n vertices $1, \ldots, n$ and a directed edge goes from i to j, if $a_{ij} > 0$. Thus, this graph encodes where the positive entries are located in A. A *path* of length ℓ connecting two vertices is a sequence of edges $i_0 \to i_1 \to \cdots i_\ell$; it is called a *loop*, if $i_\ell = i_0$. The graph $G(A)$ is *strongly connected*, if for any pair (i, j) there exists a path from i to j. One can now prove that the matrix A is irreducible, if and only if $G(A)$ is strongly connected.

We consider as a simple example the following two non-negative matrices

$$A = \begin{pmatrix} 1 & 0 & 1 & 0 \\ 0 & 0 & 0 & 1 \\ 0 & 1 & 0 & 0 \\ 1 & 0 & 0 & 1 \end{pmatrix}, \qquad B = \begin{pmatrix} 1 & 0 & 1 & 0 \\ 0 & 1 & 1 & 1 \\ 1 & 0 & 1 & 0 \\ 1 & 1 & 0 & 1 \end{pmatrix} \tag{2.9}$$

which possess the following associated directed graphs

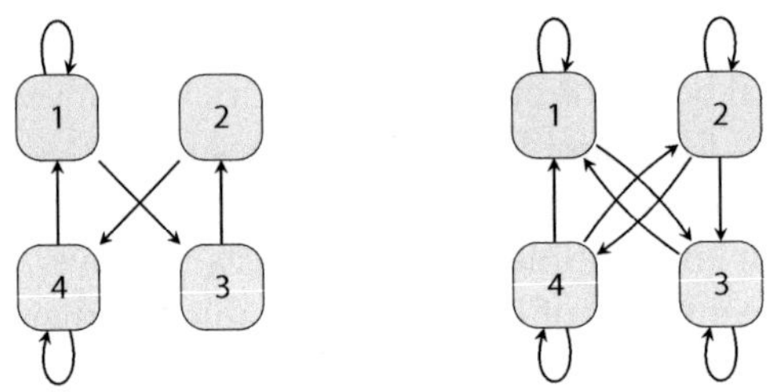

It is easy to see that in the graph $G(A)$ on the left we have a loop $1 \to 3 \to 2 \to 4 \to 1$ which connects all vertices. Hence, $G(A)$ is strongly connected and the matrix A is irreducible. In the graph $G(B)$ on the right, there exists no possibility to reach 2 or 4 from 1. Hence, the graph is not strongly connected and the matrix B is reducible.

The Perron–Frobenius theorem makes a number of statements about matrices with specific properties. The first assertion is that the spectral radius $\rho(A)$ of a non-negative matrix A is always an eigenvalue and that it always possesses an eigenvector with non-negative entries. Thus, we always find a non-negative *real* eigenvalue of maximal absolute value and in applications like population dynamics where negative values make no sense in solutions it is very important to have in addition an eigenvector without negative entries.

If the matrix A is in addition irreducible, then $\rho(A)$ is a simple eigenvalue and its one-dimensional eigenspace can be generated by a positive eigenvector (and $\rho(A)$ is the only eigenvalue with such an eigenvector). In the context of population dynamics, this means that no species is extinct. Furthermore, all other eigenvalues λ with $|\lambda| = \rho(A)$ are simple, too. The total number of such eigenvalues is called the *index of cyclicity* of A. If $\rho(A)$ is the only eigenvalue with this property, then A is a *primitive* matrix. The index of cyclicity can again be read off the graph $G(A)$; it is the greatest common divisor of the lengths of the different loops existing in $G(A)$. For the concrete irreducible matrix A defined in (2.9), the situation is very easy: we already noted above that there exists a loop of length 4 passing through all vertices; in addition, we obviously have loops of length 1 like $4 \to 4$ and hence the greatest common divisor of all lengths can only be one (there is no need to check whether further loops exist), so that A is a primitive matrix.

2.5.3 Continuous Dynamical Systems

The theories of differential and difference equations, respectively, share many similarities, but also differ in crucial aspects. We now consider functions $\mathbf{X}: \mathbb{R} \to \mathbb{R}^n$ whose values are n-dimensional vectors and always tacitly assume that $\mathbf{X}$ is sufficiently often differentiable. An *(autonomous) first-order system of differential equations* for $\mathbf{X}$ is of the form

$$\dot{\mathbf{X}}(t) = \mathbf{f}(\mathbf{X}(t)) \tag{2.10}$$

with $\dot{\mathbf{X}}$ denoting the first derivative of $\mathbf{X}$ (taken componentwise) and with some function $\mathbf{f}: \mathbb{R}^n \to \mathbb{R}^n$ (the *vector field* corresponding to the equation (2.10) [2.82, Sect. 6.2]). Thus at each time t, the system allows us to compute the rate of change $\dot{\mathbf{X}}(t)$ from the current state $\mathbf{X}(t)$. Again it is common to consider *initial value problems* where one prescribes the initial state $\mathbf{X}(0) = \mathbf{X}_0 \in \mathbb{R}^n$, as normally any differential equation possesses infinitely many solutions and such an initial condition then selects a particular one.

An *equilibrium* (or *steady state*) *solution* of an autonomous system is defined by a vector $\mathbf{V} \in \mathbb{R}^n$ such that $\mathbf{f}(\mathbf{V}) = 0$, i.e., by the zeros of the vector field $\mathbf{f}$ [2.82, Sect. 6.3]. Indeed, in this case the constant function $\mathbf{X}(t) = \mathbf{V}$ for all $t \in \mathbb{R}$ obviously solves (2.10). An equilibrium $\mathbf{V}$ is called *stable*, if for any $\epsilon > 0$ there exists a $\delta > 0$ such that any solution $\mathbf{X}(t)$ for an initial vector $\mathbf{X}_0$ with $||\mathbf{V} - \mathbf{X}_0|| < \delta$ satisfies $||\mathbf{V} - \mathbf{X}(t)|| < \epsilon$ for all $t > 0$; otherwise $\mathbf{V}$ is *unstable* [2.82, Sect. 6.5]. A *periodic solution with period* $T > 0$ is a solution $\mathbf{X}(t)$ satisfying $\mathbf{X}(t + T) = \mathbf{X}(t)$ for all $t \in \mathbb{R}$ (with T the smallest value with this property) [2.82, Sect. 6.3]. It is generally a hard task to decide whether or not a system of the form (2.10) possesses any periodic solution. One classical tool for studying such questions is the *Poincaré map* [2.82, Sect. 6.4], which is, however, beyond the scope of this article.

While it was obvious for systems of difference equations in the form (2.2) that any initial value problem has a unique solution, this represents for differential equations a theorem: the *Picard–Lindelöf Theorem* [2.82, Sect. 2.2] asserts that, given $\mathbf{X}(0)$, the system (2.10) has a unique solution $\mathbf{X}(t)$ for t in some open interval I containing 0 – under mild conditions on $\mathbf{f}$ which are in applications almost always satisfied. In general, it is a non-trivial task to decide whether I is infinite or whether the solution $\mathbf{X}(t)$ becomes infinite already for some finite time $t_s > 0$. Only for linear systems, it is guaranteed that one may choose $I = \mathbb{R}$.

As in the discrete case, we concentrate on *linear first-order systems of differential equations with constant coefficients* which have the form

$$\dot{\mathbf{X}}(t) = A\mathbf{X}(t) \tag{2.11}$$

where $A \in \mathbb{R}^{n \times n}$ is a real matrix with constant entries [2.82, Sect. 3.2]. Any vector $\mathbf{V}$ with $A\mathbf{V} = 0$ (and thus in particular the zero vector) defines an equilibrium. As in the discrete case, any such system can be explicitly solved with elementary linear algebra and we restrict our discussion to diagonalizable matrices. Thus, we assume again that $\mathbf{V}_1, \ldots, \mathbf{V}_n$ are n linearly independent eigenvectors with corresponding eigenvalues $\lambda_1, \ldots, \lambda_n$. A fundamental system of solutions is then given by

$$\mathbf{X}_k(t) = e^{\lambda_k t}\,\mathbf{V}_k, \quad 1 \leqslant k \leqslant n. \tag{2.12}$$

For solving the initial value problem with $\mathbf{X}(0) = \mathbf{X}_0$, we must again express the initial vector $\mathbf{X}_0 = \sum_{k=1}^{n} c_k \mathbf{V}_k$ as a linear combination of the eigenvectors with coefficients c_k and obtain the unique solution $\mathbf{X}(t) = \sum_{k=1}^{n} c_k \mathbf{X}_k(t)$.

If some eigenvalues are complex, then the same is true for the corresponding eigenvectors. But as in the discrete case, we can rewrite complex members of the fundamental system (2.12) as real ones. Again this is based on the fact that with $\lambda = \rho + i\omega$ also its complex conjugate $\bar{\lambda} = \rho - i\omega$ is an eigenvalue of A and if $\mathbf{V} \in \mathbb{C}^n$ is an eigenvector for λ, then $\bar{\mathbf{V}}$ is an eigenvector for $\bar{\lambda}$. Decomposing the eigenvector in its real and imaginary part, $\mathbf{V} = \mathbf{W}_{re} + i\mathbf{W}_{im}$, we can replace in our fundamental system the two complex solutions corresponding to λ and $\bar{\lambda}$ by two real ones:

$$\mathbf{X}_{re}(t) = e^{\rho t}[\cos(\omega t)\mathbf{W}_{re} - \sin(\omega t)\mathbf{W}_{im}], \tag{2.13}$$

$$\mathbf{X}_{im}(t) = e^{\rho t}[\sin(\omega t)\mathbf{W}_{re} + \cos(\omega t)\mathbf{W}_{im}].$$

One key difference to the discrete case is that here ρ and ω are the real and the imaginary part of λ, respectively, whereas in the discrete case ρ was the modulus and ω the argument of λ.

Based on the explicit expressions (2.12) and (2.13), we can again make statements about the asymptotic behavior of solutions. If the real part $\Re(\lambda_k)$ of some eigenvalue λ_k (for $\lambda_k \in \mathbb{R}$ obviously $\Re(\lambda_k) = \lambda_k$) is positive, then the solution will be unbounded for $t \to \infty$ for any initial condition with $c_k \neq 0$. Thus, the population of at least one species/size class will become arbitrarily large and the system is *unstable.* On the other hand, if all eigenvalues have a negative real part, then any solution will converge for $t \to \infty$ to the origin, so that all species/size classes become extinct. This represents an *asymptotically stable* system. If positive and negative real parts exist among the eigenvalues, then the origin is a *saddle point.* As in the discrete case, complex eigenvalues indicate an *oscillatory behavior* which is damped for $\rho < 0$ and unbounded for $\rho > 0$.

The situation is more complicated if eigenvalues λ_k exist with a vanishing real part. If λ_k is real, this means $\lambda_k = 0$ and the matrix A is singular. In this case, we have not only one equilibrium at the origin, but at least a whole line of equilibria. If $\lambda_k \neq 0$, it is purely imaginary, $\lambda_k = i\omega_k$ and periodic solutions with frequency ω_k exist for appropriate initial conditions. They correspond to sustained oscillations.

Using the theory of the Jordan normal form, these results can be extended also to non-diagonalizable matrices [2.82, Sect. 3.2]. While the fundamental system above was composed of pure exponential functions, now products of polynomials with exponential functions also appear as components. Assume now for simplicity that one eigenvalue, say λ_1, has the real part with the largest absolute value, i.e., $|\Re(\lambda_1)| > |\Re(\lambda_k)|$ for all $k > 1$. Now we find that for long times the solution $\mathbf{X}(t) = \sum_{k=1}^{n} c_k e^{\lambda_k t} \mathbf{V}_k$ for the initial data $\mathbf{X}_0 = \sum_{k=1}^{n} c_k \mathbf{V}_k$ satisfies

$$\lim_{t\to\infty} e^{-\lambda_1 t}\mathbf{X}(t) = c_1\mathbf{V}_1. \tag{2.14}$$

Hence, the fundamental mode corresponding to the eigenvalue λ_1 dominates the long-term behavior of solutions.

These results can to some extent be applied to study equilibria of nonlinear systems of the form (2.10), too [2.82, Sect. 2.6.5, Chapt. 2.9]. Assume that $\mathbf{V}$ is an equilibrium of (2.10). As in the discrete case, we use the Jacobian J of the right-hand side $\mathbf{f}$ at $\mathbf{V}$ defined by (2.8). We call $\mathbf{V}$ a *hyperbolic* equilibrium, if no eigenvalue of J has a vanishing real part. For such equilibria, the local solution behavior of (2.10) near $\mathbf{V}$ is the same as for the linear system $\dot{\mathbf{Y}} = J\mathbf{Y}$. Hence, the equilibrium is asymptotically stable if all eigenvalues have a negative real part.

Figure 2.4b shows on the right hand side in yellow the region in the complex plane where all eigenvalues must be located for an asymptotically stable equilibrium, namely in the left half plane. The *Routh-Hurwitz criterion* [2.82, Sect. 2.3.2] allows for detecting this without explicitly computing of the eigenvalues and requires only calculation of certain determinants. For sustained oscillations, one must also have at least some eigenvalues on the imaginary axis (marked with a diamond).

2.5.4 Extensions and Generalizations

We mentioned above that higher-order difference equations may be considered as a model for delay effects. This is not the case for continuous models: a higher-order differential equation has nothing to do with a delay; one has to generalize to *delay differential equations.* In the simplest case an autonomous first-order system of delay differential equations takes the form

$$\dot{\mathbf{X}}(t) = \mathbf{f}(\mathbf{X}(t), \mathbf{X}(t-\tau)) \tag{2.15}$$

where $\tau > 0$ denotes the delay. One may also consider several delays $\tau_1, \ldots, \tau_r$ by allowing that $\mathbf{f}$ depends on $\mathbf{X}(t - \tau_1), \ldots, \mathbf{X}(t - \tau_r)$. In addition, delay models frequently incorporate integrals (one then speaks of a *distributed delay*).

The theory of delay differential equations is considerably more involved than that of ordinary differential equations. This starts already with the formulation of initial value problems. Whereas for an ordinary differential equation like (2.10) it suffices to prescribe a single value $\mathbf{X}(0)$ (implying that the solution space is finite-dimensional), an initial condition for (2.15) has typically the form $\mathbf{X}(t) = \mathbf{X}_0(t)$ for $-\tau \leqslant t \leqslant 0$, i. e. $\mathbf{X}_0$ is no longer a constant vector but a function defined on the interval $[-\tau, 0]$ (implying that the solution space is now infinite-dimensional like for most partial differential equations). Using the method of steps, i. e. looking at intervals of the form $[k\tau, (k + 1)\tau]$ for $k = 0, 1, 2, \ldots$, one can use for each interval the Picard–Lindelöf Theorem to prove the (local) existence of a unique solution for the described initial value problem.

The simplest form of a linear delay system with constant coefficients is

$$\dot{\mathbf{X}}(t) = A\mathbf{X}(t) + B\mathbf{X}(t - \tau) \tag{2.16}$$

with two matrices A, $B \in \mathbb{R}^{n \times n}$. If one searches for exponential solutions as in the case of linear systems of ordinary differential equations, then the equation for the "eigenvalues" λ takes the form $h(\lambda) = \det(\lambda E_n - A - e^{-\tau\lambda}B) = 0$ and thus is no longer simply a polynomial in λ. One can show that h is an entire function entailing many special properties of the equation $h(\lambda) = 0$, i. e. that there are at most countably many solutions and that these do not possess a finite accumulation point. Most remarkably, for any real number $r \in \mathbb{R}$ there exist only finitely many solutions with $\Re(\lambda) > r$ and again the dominating solution modes are those corresponding to the solutions λ with maximal real part.

Delays can lead to oscillations even in trivial linear models, if the delay τ is greater than some threshold. For example, in the simplest scalar delay equation $\dot{X}(t) = -aX(t-\tau)$, the solutions will start to oscillate whenever $\alpha\tau \geqslant e^{-1}$. The oscillations will be damped, as long $a\tau \geqslant \pi/2$; for even greater values of $\alpha\tau$, the amplitude of the oscillations will grow with t.

We refrain from delving deeper into the theory of delay differential equations and refer instead to textbooks like [2.77] for an introduction or to [2.26] [2.34] for in-depth treatments of different aspects.

So far, we have always assumed that $\mathbf{X}$ depends only on the time t. Of course, in reality populations will also show a spatial distribution or in the case of marine species like diatom a dependence on the water depth. This leads to models with *partial differential equations* for functions $\mathbf{X}(t, \mathbf{r})$ where the spatial variable $\mathbf{r}$ may again be vector-valued. The dimension of $\mathbf{r}$ depends on the modeling. If one considers simply a water column to include the effect of the water depth, then $\mathbf{r}$ would be a scalar quantity, namely the depth. If one studies the distribution of a population over some region on the surface of the earth, $\mathbf{r}$ would be two-dimensional (cartesian coordinates or longitude and latitude). If the region is part of an ocean and water depth is taken into account, $\mathbf{r}$ would be even three-dimensional. As a general rule of thumb, both the theoretical analysis and the numerical simulation of a system of partial differential equations become rapidly more complicated and expensive with each additional dimension. As it suffices for our purposes here, we will restrict to the case that $\mathbf{r}$ is a scalar quantity x.

In the context of population dynamics (and many other biological applications), a frequent approach to formulating distributed models consists of simply adding a *diffusion term* to model the spatial spreading. This leads then to *reaction-diffusion equations* of the form

$$\frac{\partial \mathbf{X}}{\partial t}(t,x) = \frac{\partial}{\partial x}\left(D(t,x)\frac{\partial \mathbf{X}}{\partial x}(t,x) \right) + \mathbf{f}(\mathbf{X}(t,x)) \tag{2.17}$$

where D is a diagonal matrix containing the diffusion coefficients for the different species (which may depend on t and x, although in many models they are assumed to be constant) and $\mathbf{f}$ represents the reaction part of the equation. Thus (2.17) may be considered as a distributed refinement of the simpler model $\dot{\mathbf{X}}(t) = \mathbf{f}(\mathbf{X}(t))$. Such equations appear in many biological, ecological and chemical applications, but also in geology, physics and many other fields of science. Some classical textbooks describing their mathematical theory and some applications are [2.30] [2.35] [2.78].

From a mathematical point of view, (2.17) represents a parabolic system. As the only nonlinear part is the reaction term $\mathbf{f}$, many results from the theory of linear partial differential equations can be exploited in its analysis. Nevertheless, this nonlinearity leads to many interesting phenomena like traveling waves solutions or pattern formation as discovered by Turing [2.84]. One usually studies *initial boundary value problems* for a system like (2.17). Here, the spatial variable x is constrained to some closed interval $[a, b]$ and one prescribes an initial condition $\mathbf{X}(t_0, x) = \mathbf{X}_0(x)$ for all $x \in [a, b]$ with a given function $\mathbf{X}_0(x)$ and two boundary conditions $\mathbf{X}(t, a) = \mathbf{X}_a(t)$, $\mathbf{X}(t, b) = \mathbf{X}_b(t)$ for all $t > t_0$ with two given functions $\mathbf{X}_a(t)$, $\mathbf{X}_b(t)$. Using methods from functional analysis, one can show that under modest conditions on the reaction term $\mathbf{f}$ and the prescribed functions $\mathbf{X}_0(x)$, $\mathbf{X}_a(t)$ and $\mathbf{X}_b(t)$ a unique solution exists.

2.5.5 Individual-Based Models

In the approaches discussed so far, individuals are grouped into classes in order to define a state vector on which propagating matrices are applied. Stochastic branching processes constitute a different approach by following the fate of individual cells during proliferation [2.46]. On the population level, stochastic methods are also available in the form of time-continuous, Markovian birth-death-models [2.17], but in the context of diatom models we meet branching processes as the corresponding individual-based theory. The general idea is as follows: At a certain time, a cell or generally a particle, is generated. This particle has a lifetime that underlies a certain distribution. After the end of the lifetime, a number of new cells are generated, each starting a new process. There are also variants in which new particles are generated during the lifetime, but these are mainly applicable for higher organisms, so we will not discuss them further in this context. Important, however, is that the newly generated progeny may occur in different type classes. This type space can be attributed to different size classes, so the treatment leads to matrix equations just as in the case of differential or difference equation systems. An important point is that also non-Markovian processes can be easily written down due to the introduction of finite lifetimes. In a Markov process, the future development of a system only depends on the present state, in non-Markovian processes also states in the past contribute to the development. Actually, a system in which the previous generation contributes may be classified as a second-order Markov chain, and so on, so a true non-Markovian process has to take the complete history into account.

In a multitype branching process, the state is characterized by a vector $\mathbf{X}(t)$ with the number of particles (cells) of the different types (for diatoms: size classes). To make it simple, we first discuss a one-dimensional system (i. e. one type) with one initial particle at $t = 0$ with lifetime realization $\tau(\omega)$ and progeny number $Y(\omega)$. Then, after time t, the number of particles is given by

$$X(t) = \begin{cases} 1, & t < \tau \\ \sum_{i=1}^{Y} X^{(i)}(t-\tau), & t \geqslant \tau, \end{cases} \tag{2.18}$$

which is a recurrence relation because each individual starts a new process.

Now, different models can be considered for the lifetime distribution. The simplest case is that each particle has the same fixed lifetime of one generation (Galton-Watson process). In this case the process is Markovian and a stochastic version of a discrete system is reproduced, with all the properties given above. In the second special case of a lifetime distribution that falls as a single exponential, also a Markov system results, which can be proved analytically [2.46]. The general case is the Bellman-Harris process in which τ is an arbitrary non-negative random variable from another distribution. Here, a true non-Markovian, age-dependent process results.

The key for treating such problems is to rewrite it with probability generating functions. Such probability generating functions apply to non-negative random variables such as the number of particles or the number of progeny and are polynomials on a symbolic new variable s with the possible values as exponents and the probabilities p_i as coefficients. For instance, the probability generation function for the number of offspring is defined as

$$f(s) = \sum_i p_i s^i \tag{2.19}$$

in which i are the possible values for Y. In the case of exact two daughters, $f(s)$ is simply s^2 since the probability for the value $Y = 2$ is exactly 1. For the total number of particles we write $F(s, t)$ with the same definition and take i as possible values for X. The mean (expectation value $\mathbb{E}$) of the particle number after time t is given then conveniently by the first derivative by s at $s = 1$

$$\mathbb{E}[Z(t)] = F'(1,t) = \left.\frac{\partial F(s,t)}{\partial s}\right|_{s=1}. \tag{2.20}$$

Higher moments of $X(t)$ are represented by higher derivatives, so the variance is given by $\mathbb{V}(X) = F''(1) + F'(1) - (F'(1))^2$.

With these two probability functions we can rewrite the above expression and bring it into a single line by introducing a cumulative distribution function for the lifetimes $G(\tau)$, resulting in

$$F(s,t) = \begin{cases} s, & t < \tau \\ f[F(s,t-\tau)], & t \geqslant \tau \end{cases} \tag{2.21}$$

$$= s[1-G(t)] + \int_0^t f[F(s,t-u)]\mathrm{d}G(u). \tag{2.22}$$

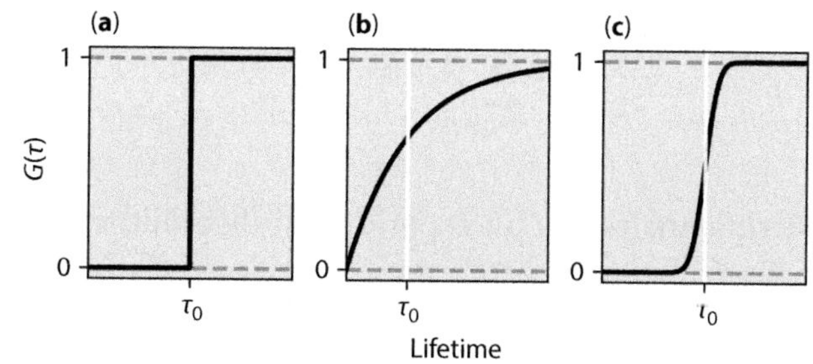

Figure 2.5 Three cumulative distributions functions for generation times ("lifetimes" of a Bellman-Harris particle): (a) fixed lifetime, (b) exponentially decreasing lifetime distribution (c) lifetime with Gaussian distribution.

Examples for cumulative distribution functions are a step function at the fixed lifetime τ_0 for the Galton-Watson process, the function $G(\tau) = 1 - \exp(-\tau/\tau_0)$ for an exponentially decreasing lifetime distribution, or the integral of the Gaussian function for a normal distribution with τ_0 as mean value (Figure 2.5). The computational task in solving Bellman-Harris processes is to treat the resulting integral equation to obtain self-consistent solutions, representing steady-states, or follow the temporal evolution.

In multitype branching processes, the offspring may occur in different types (classes), so $X(t)$ and $F(s, t)$ become vectors and a progeny matrix M is needed that sorts the offspring according to some rules to these classes. This matrix is also non-negative and similar to the propagation matrices of difference equations; thus the evolution is also subjected to the Perron–Frobenius theorem [2.7]. In the case of the Galton-Watson process, the description is just an alternative to the time-discrete class-based approach by difference equations since the eigenvalues of the same matrices have to be solved in order to obtain analytical solutions [2.13]. The strength of individual-based approaches, however, is the treatment of time-delay since it is easier to implement and can be subjected to distributions via the Bellman–Harris formulae. On the other hand, the recurrence equations cannot be solved analytically, and picking random numbers for many individuals causes substantial computational cost. The implementation of stochastic branching can be time-discrete, especially for simulations, or time-continuous in analytical expressions.

2.6 Simulation Models

2.6.1 The Schwarz *et al.* Model

Schwarz *et al.* developed a stochastic model based on experimental data to find optimal values for Markov-type propagation matrices [2.76]. They reduced the number of size classes from around 40 as expected for the modeled species *Pseudo-nitzschia delicatissima* to seven. The size distribution was described as a superposition of seven normal distributions with respective means and widths (Gaussian mixture model). The number of seven size classes has been found as optimal number according to the Bayesian information criterion preventing overfitting with insignificant parameters added by further Gaussian functions.

$$\frac{dX(d)}{dd} = \sum_{i=1}^{7} A_i \exp \frac{(d-\mu_i)^2}{2\sigma_i^2} \tag{2.23}$$

A univariate clustering analysis of an experimental distribution (with cell size as the single variable) led to the definition of the seven classes with respective boundaries, as given in Table 2.2. It should be noted that in this procedure for the determination of the boundaries, the contained size ranges may differ substantially from cluster to cluster.

Having defined the distinct size classes, the propagation matrix for continuous time modeling was obtained by fitting experimental data for a laboratory strain of *Pseudo-nitzschia delicatissima* [2.3] with size distributions followed for 265 days. Choosing a monoclonal strain of a single mating type, sexual reproduction was prevented, so that the vegetative phase could be modeled in isolation. Allowing only a transition into the next size class with smaller size and abandoning the reentry by the sexual phase, the continuous, time-independent propagation matrix was constructed as

$$Q = \begin{pmatrix} -q_{12} & q_{12} & & 0 \\ & \ddots & \ddots & \\ & & -q_{67} & q_{67} \\ 0 & & 0 & 0 \end{pmatrix}. \tag{2.24}$$

The diagonal elements correspond to the negative inverse holding times that were as an initial guess extracted from the scaling law of the size distribution versus time plot (values in Table 2.2). These holding times vary significantly for the different size classes and do not scale with the size range of an individual class. For instance, they are conspicuously long for size class 6 without an identifiable reason.

The fitted entries were tested with parallel data of two other strains by predicting the time evolution of the size distribution. In order to estimate the confidence intervals for being in

Table 2.2 Size clusters for *P. delicatissima* according to Schwarz *et al.* [2.76]. Given is mean μ_i, standard deviation σ_i (from graph), lower boundary, cluster range and holding time τ for the Gaussian functions and their derived clusters, respectively.

	1	2	3	4	5	6	7
μ_i in μm	80.8	75.4	64.5	43.8	51.6	28.7	17.9
σ_i in μm	0.5	0.5	7.1	4.0	4.2	3.6	2.0
limits (μm)	>80.0	>74.5	>57.0	>48.0	>37.5	>21.0	>14.0
range in μm	5.5	5.5	17.5	9.0	10.5	16.5	7.0
τ (days)	5.1	3.0	15.6	15.0	31.9	168.7	∞

the experimentally determined states after certain amounts of time, the Fisher information approach was used. The Fisher information $I(\theta)$ expresses the significance of parameter values θ that are contained in a probability distribution function $f(X; \theta)$ and is defined as the expectation value of the second derivative – log f versus the parameter.

$$I(\theta) - -\int\left(\frac{\partial^2 \log f(X;\theta)}{\partial\theta^2}\right) f(X;\theta)\mathrm{d}X. \tag{2.25}$$

In this case, the parameter is time with n different distributions at certain time intervals. Schwarz *et al.* calculated 95% confidence intervals for the time estimates as

$$\bar{t} \pm 1.96\sqrt{1/(nI(t))} \tag{2.26}$$

and showed that the predicted times for the observed size distributions match the experimental time within these limits, thus verifying the model.

In extension of the model, an 8th class was introduced representing the sexual phase, being accessible only from size class 6 and 7. The average holding time was set to 2 days after which class 1 is repopulated (not explicitly stated in the paper). In this case, a stationary distribution with population maximum in the 5th class was obtained.

2.6.2 The D'Alelio *et al.* Model

D'Alelio *et al.* developed a parametrized model that aimed at explaining experimental data for *Pseudo-nitzschia multistriata* taken over a time span of 10 years in the Gulf of Naples [2.19]. It is formulated as a continuous model with a sampling time of one day. From the original paper it is not completely conclusive how the size classes are defined, but the largest, initial cells were set to 80 μm, the smallest size to 30 μm, and the upper sexual threshold to 55 μm. The resolution of the output was 2.5 μm which would correspond to 20 or 21 size classes. Even if cells were collected to size classes and hence the model is denoted as matrix model, there seems to be a certain tracing of individual cells. This can be deduced from a size-depending rate of cell size decrease, given as

$$R(d) = 10^{-7}\ \mu\text{m/division}\ (d/\mu\text{m})^{2.5} \tag{2.27}$$

(units added here, d represents cell size), and the statement that the total number of cells was kept below 10^4 by randomly removing cells in each time step. The size-dependent expression for $R(d)$ exhibits a larger reduction for larger cells and was justified by comparing the output of the model with the observed data.

Similarly, the growth and death rates were taken as cell size dependent (Figure 2.6):

$$g(d) = g_0[0.25 + 0.04(d/\mu\text{m}) - 5 \cdot 10^{-4}\ (d/\mu\text{m})^2] \tag{2.28}$$

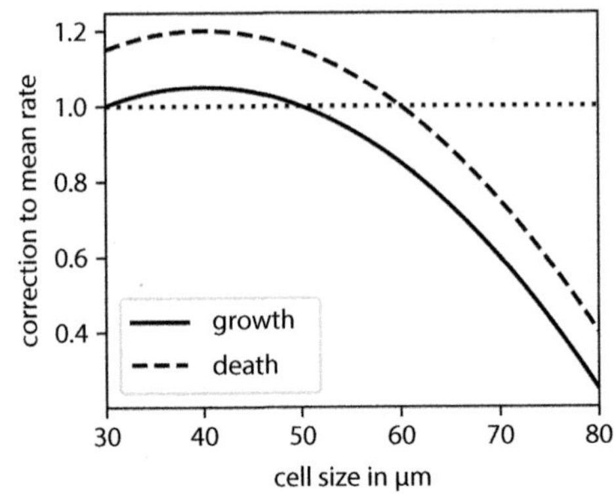

Figure 2.6 Size-dependent growth parameter described by a parabolic function.

$$k(d) = k_0[0.4 + 0.04(d/\mu\text{m}) - 5 \cdot 10^{-4}\,(d/\mu\text{m})^2)] \tag{2.29}$$

Also for these equations, the choices of prefactors in the second order polynomial were derived from comparison with experimental data. The basic rates were modulated according to the seasons: g_0 was set tentatively from 0.69 to 1.40 per day during the bloom season and from 0.07 to 0.20 per day during the rest of the year. Coupled to g_0, k_0 varied accordingly from from 0.05 to 1.00 per day.

The probability for auxosporulation was 2%, with instantaneous reentering into the process (an experimentally determined delay of 2-3 days was neglected deliberately). The authors compared different timing scenarios for sexual reproduction with a gating function $\theta(t)$ between 0 and 1. While the assumption of a continuous sexual reproduction irrespective of the seasons and an annual sex time window during the bloom failed to reproduce

Table 2.3 Parameters used in the model of D'Alelio *et al.* Given is the original denotation in the paper [2.19] as well as an abbreviated version used here in the mathematical expressions.

Parameter	**Code**	**Here**
Size	*s*	*d*
Size of initial cells	Init size	80 μm
Size of smallest cells	Small size	30 μm
Gametangia threshold size	Sex size	55 μm
Parameterized growth rate	Pgr	$g(d)$
Parameterized death rate	Pdr	$k(d)$
Rate of cell size decrease	Red	$R(d)$
Probability of forming gametes	Sex prob	2%
Timing of sexual reproduction	Sex time	$\theta(t)$

the experimental observations, the assumption of biennial sexual periods displayed the best matching. If sex would occur less than every 4th year, the model predicted local extinction of a population.

The parameters used in the model are compiled in Table 2.3 for reference.

2.6.3 The Hense–Beckmann Model

Hense and Beckmann [2.36] developed a rather complex model of the population dynamics of diatoms and embedded it into the context of a complete ecosystem, which leads to a system of partial differential equations combined with a discrete system. The basis is a compartmental model PPND of a water column consisting of four compartments: a diatom species B, a bulk phytoplankton P, a limiting nutrient N and detritus D. This model takes the form of a reaction-diffusion system as in (2.17) where the vector $\mathbf{X}$ consists of the four component (P, B, N, D), given as biomass in nitrogen equivalents. The equations for B and D are augmented each by a term describing a sinking along axis z, whereas P and N are assumed to be buoyant. More precisely, one obtains the following model

$$\begin{aligned}
\frac{\partial B}{\partial t} &= \mu_B B-(\gamma_B+\delta_B(B+P))B+A_v\frac{\partial^2 B}{\partial z^2}-w_B\frac{\partial B}{\partial z},\\
\frac{\partial P}{\partial t} &= \mu_P P-(\gamma_P+\delta_P(B+P))P+A_v\frac{\partial^2 P}{\partial z^2},\\
\frac{\partial N}{\partial t} &= -\mu_P P-\mu_B B+\tau D+A_v\frac{\partial^2 N}{\partial z^2},\\
\frac{\partial D}{\partial t} &= (\gamma_B+\delta_B(B+P))B+(\gamma_P+\delta_P(B+P))P-\tau D+A_v\frac{\partial^2 D}{\partial z^2}-w_D\frac{\partial D}{\partial z}.
\end{aligned} \tag{2.30}$$

The parameters and typical values are listed in Table 2.4. Here, the growth rates $\mu_{B/P}$ of the species B and P, respectively, depend on the temperature T, the nutrient N and the light I in the water column:

$$\mu_{B/P} = \mu_{0,B/P}\mathcal{F}_{B/P}(T)\frac{N}{k_{N,B/P}+N}\frac{I}{k_{I,B/P}+I}. \tag{2.31}$$

Here, the temperature dependency for the species P follows an empirical law of a constant times 1.066^T [2.29], whereas for diatoms B an additional Gaussian term is used defining an optimal temperature [2.58]. The temperature itself plays the role of an external forcing with seasonal variation. For the light, self-shading is taken into account by modeling its depth dependency as

$$I(z) = I_0 \exp\ k_w z + k_c\int_0^z (P(z') + B(z') + D(z'))\mathrm{d}z'. \tag{2.32}$$

Table 2.4 Selected parameters used in the Hense–Beckmann PPND model. Parameters for the second, bulk phytoplankton species (index P instead of B) are omitted here.

Parameter	Symbol	Value
Diatom maximum growth rate	$\mu_{0,B}$	0.98 day^{-1}
Diatom optimal temperature	T_{opt_B}	8 °C
Diatom nutrient half saturation	k_{N_B}	0.16 mmol N m^{-3}
Diatom light half saturation	k_{IB}	33.6 Wm^{-2}
Attenuation coefficient for water	k_w	0.08 m^{-1}
Attenuation coefficient for biomass	k_c	0.07 m^2 mmol N^{-1}
Diatom cell lysis rate	γ_B	0.01 day^{-1}
Diatom grazing rate	δ_B	0.10 m^3 day^{-1} mmol N^{-1}
Subsurface vertical mixing coefficient	A_v	10^{-5} m^{2}s^{-1}
Diatom sinking rate	w_B	0.12 m day^{-1}
Nutrient remineralization rate	τ	0.1 day^{-1}
Detritus sinking rate	w_D	1.5 m day^{-1}

In this relatively classical form, the size reduction and restitution cycle of the diatoms is not yet taken into account. For this purpose, Hense and Beckmann developed a rather detailed individual based model of the diatom life cycle called DiaLCM. It aggregates a fixed number c of diatoms with identical status into one entity called agent, so that a smaller number of individual cells has to be followed in total. The biomass of the individually modeled diatoms b_i constitutes the total biomass, i.e.

$$B = c\sum_i b_i. \tag{2.33}$$

For each cell, five processes are taken into account:

(i) Cell growth: Depending on temperature and limited by nutrient and light availability as described above but with size dependent parameters, individual cylindrical cells uptake biomass and change their shape by extending into the pervalvar direction due to a fixed diameter.

(ii) Cell division: When a size dependent maximum biomass is reached, the biomass is split evenly between epivalvar and hypovalvar daughter, but the hypovalvar daughter gets a smaller diameter, calculated with a constant factor.

(iii) Cell mortality: Cells die due to size independent cell lysis and size dependent grazing.

(iv) Auxosporulation: If cells are below the upper sexual threshold (after 80% of the maximum number of generations) and meet additional conditions on available energy and cell concentration, they form new cells of maximum size after an developmental period.

(v) Vertical movements in the water column: This is modeled by size dependent sinking and turbulent mixing in the surface region. Since cells change their shape, an equivalent hydrodynamic radius is calculated for this and other processes.

The precise modeling of these processes is too complex to be presented here in detail. For example, in size dependent aspects the authors have to recalculate between biomass, volume, surface area and equivalent radius. Empirical functional relationships between the various relevant quantities are assumed and taken from the literature. On the whole, the DiaLCM model depends on almost 20 parameters for which values must be chosen. Some variable ranges and parameters are compiled in Table 2.5. The authors report a comparison of the results of simulations of the DiaLCM model with the results of laboratory experiments conducted by Armbrust and Chisholm [2.4] where they obtained a good agreement. In particular, they could observe periodic changes.

For the coupling of the two models, the diatom compartment B is removed from the PPND model, i.e. the equation for B is removed from (2.30) and replaced by the DiaLCM model. This approach allows determination of all B dependent terms in the remaining equations

Table 2.5 Selected variables and parameters used in the Hense–Beckmann DiaLCM model as far as mentioned in the text. For a complete list please refer to the appendix of the original article [2.36].

Quantity	**Symbol**	**Value**
Number of cells per agent	c	10^4
Number of agents	M	$2 \cdot 10^5$–$3.1 \cdot 10^6$
Cell biomass	b_{i}	0.18–1.82 pmol N cell^{-1}
Equivalent radius	r_{e_i}	2.61–6.89 μm
Pervalvar length	h_i	3.85–17.69 μm
Number of generations	G	100
Biomass ratio epivalve:hypovalve	$\delta_e : \delta_h$	50:50
Maximum age of epivalve	a_{max}	7 divisions
Generation threshold for sex	g_a^{crit}	0.8 G
Daily mean energy threshold for sex	E_{crit}	12 Wm^{-2}
Minimum cell concentration for sex	c_{crit}	10^7 cells m^{-3}
Developmental period	τ_{dev}	1 day

in (2.30). Hense and Beckmann report that in the coupled model the strict periodicity of the DiaLCM model under "laboratory conditions" disappears. They observed that under the influence of environmental factors achieved through the coupling with the PPND model, the size reduction and restitution cycle can be significantly accelerated or decelerated. Due to the complexity of their model with its many parameters and assumptions, it is, however, difficult to see which factors are mainly responsible for these observations.

2.6.4 The Terzieva–Terziev Model

Terzieva and Terziev followed a non-Markovian approach and applied the Bellman-Harris stochastic branching concept to the cell division scheme [2.81]. This allows them to model a flexible time between cell divisions. Explicitly, the time between two divisions, constituting the lifetime of a particle in the multitype Bellman–Harris scheme, is drawn from a Gaussian normal distribution with mean τ_0 and variance σ^2. Originally, for the size range between 30 μm and 150 μm of a generic diatom, five size classes (types) have been defined which are reduced to four by uniting the two lowest classes (i. e. 1 from 30 to 80 μm, 2 from 80 to 105 μm, 3 from 105 to 130 μm, 4 from 130 to 150 μm). For the individual cells of these four types, the following individual generation functions are constructed:

$$f_1(\mathbf{s}) = 1 - \rho + \rho s_1, \tag{2.34}$$

$$f_i(\mathbf{s}) = 1 - \rho + \rho s_{i-1} s_i \quad \text{for } i = 2, 3, 4 \tag{2.35}$$

forming together the four-dimensional vector $\mathbf{f}$ with the four-dimensional symbolic variable $\mathbf{s}$ as argument. ρ denotes the number of offspring after dividing in each of the daughter types. ρ is not subjected to a distribution and represents the probability of survival until the next division. The lack of asymmetry between the daughter types reveals that one type essentially represents one distinct size, otherwise the number of progeny would be less for the smaller type. Also, as a probability, ρ is meant to be not larger than 1. In the multitype Bellman–Harris variant, the number of progeny is expressed as matrix (particle production mean matrix M) with elements given as

$$m_{i,j} = \left. \frac{\partial f_i}{\partial s_j} \right|_{s=\mathbf{1}} \tag{2.36}$$

performing the differentiation at the vector of ones $\mathbf{1} = (1, 1, 1, 1)$, i.e.

$$M = \begin{pmatrix} \rho & 0 & 0 & 0 \\ \rho & \rho & 0 & 0 \\ 0 & \rho & \rho & 0 \\ 0 & 0 & \rho & \rho \end{pmatrix}. \tag{2.37}$$

Note that a sexual size restoring mechanism is not implemented in this matrix.

For the ensemble of particles with individual lifetimes, a vectorial generation function was then constructed, with integrals over the cumulative lifetime distribution functions $G(\tau)$ as in the single type case with components

$$F_i(\mathbf{s},t) = s_i(1 - G_i(t)) + \int_0^t f_i(\mathbf{F}(\mathbf{s},t - \tau))\mathrm{d}G_i(\tau). \tag{2.38}$$

Note that in the integral the generating function vector $\mathbf{F}$ is used as argument for the individual generating functions f to define its own components, hence the recursion.

A matrix of means for the ensemble $U(t)$ was then similarly constructed, in which the last row gives the expectation of the population $\mathbf{X}(t)$ for the starting condition of one initial cell

$$\mathbf{X}(t) = \begin{pmatrix} u_{11}(t) & 0 & 0 & 0 \\ u_{21}(t) & u_{22}(t) & 0 & 0 \\ u_{31}(t) & u_{32}(t) & u_{33}(t) & 0 \\ u_{41}(t) & u_{42}(t) & u_{43}(t) & u_{44}(t) \end{pmatrix} \begin{pmatrix} 0 \\ 0 \\ 0 \\ 1 \end{pmatrix}.$$

The respective elements of $U(t)$ are given by

$$u_{ij}(t) = 1 - G_i(t) + \rho\int_0^t (1 - G_i(t - u))\mathrm{d}\sum_{n=1}^{\infty} \rho^{n-1}(G_i(t))^{*n} \tag{2.39}$$

in which $*n$ denotes the nth convolution [2.7].

The paper of Terzieva and Terziev, which is not easily accessible at first glance, demonstrate how multitype Bellman–Harris procedures may be applicable to size-resolved population dynamics of diatoms in principle, but numerical simulations are still lacking.

2.6.5 The Fuhrmann-Lieker *et al.* Model

Fuhrmann-Lieker *et al.* understood the diatom population as a discrete dynamical system [2.31]. The authors varied parameters and expanded their model to investigate whether a linear matrix model could produce the diatom clock's oscillations in cell and population size. For that, a basic model was introduced which assumed the population to consist of discrete size classes according to the MacDonald–Pfitzer scheme. Consequently, cells would divide simultaneously at a discrete periodical rate. The number of cells in a certain size class is denoted in a population vector $\mathbf{X}(t)$ with elements x_i. The distribution of the next generation $t + 1$ is obtained by multiplication of the vector with the propagator matrix (as introduced in (2.4)):

$$P = \begin{pmatrix} \alpha p & p & 0 & \cdots & 0 & 0 \\ 0 & \beta p & p & \cdots & 0 & 0 \\ \vdots & \vdots & \vdots & & \vdots & \vdots \\ 0 & 0 & 0 & \cdots & \beta p & p \\ s & 0 & 0 & \cdots & 0 & \beta p \end{pmatrix}. \quad (2.40)$$

The propagator matrix contains the parameters of survival and auxosporulation rate as well as an optional asymmetry factor according to Table 2.6. For simplification of P and later analytical treatment, the parameter σ was introduced.

In addition to the basic model, four different modifications were examined. Asymmetry between two daughter cells was considered according to the observations of Müller and Laney [2.60] [2.49]. In a second modification the authors applied two different methods to introduce a nonlinearity. First, to account for a biparental sexual reproduction, the auxosporulation probability was modified:

$$x_{n-1,t+1} = \frac{s x_{0,t}^2}{(1 + 2 s x_{0,t})}. \quad (2.41)$$

Second, the natural saturation was included by addition of a Ricker function which limits the maximal size of the population by modifying the survival probability depending on the population size:

Table 2.6 Parameters used in the Fuhrmann-Lieker *et al.* model.

Parameter	**Meaning**	**Model**
N	number of size classes	all models
i	index for size class ($0 \leqslant i < N$)	all models
p	survival probability of theca	all models
s	probability of auxosporulation	all models
α	survival factor of smallest cells	all models
β	asymmetry factor between daughters	all models
σ	$(s/p)^{1/n}$	all models
t	generation as time scale	all models
γ	inverse carrying capacity	saturation, aging models
m	number of age classes	aging models
j	index for age class ($0 \leqslant j < m$)	aging models

$$p = p_0 e^{-\gamma\left(\sum_i x_i\right)_t}. \tag{2.42}$$

The third modification involves the aging of daughter cells as it has been observed in various microorganisms to control population development. The authors follow Jewson's findings that the daughter cells containing the epitheca would age and subsequently die after a certain number of divisions [2.42]. To include this mechanism, another dimension is added to the population vector, thus turning it into a matrix with a size and an age dimension. Upon division, the cell containing the epitheca moves along the age axis of the matrix into the older class while the hypothecal cell moves into the smaller size class along the size axis. The last variation of the model introduced a sinusoidal oscillation of the survival and auxosporulation parameters to mimic seasonal changes.

The authors evaluated their models by generational development of total cell number, number of auxospores, mean size and the variance of the sizes. Furthermore, eigenvalues and eigenvectors of the systems were analyzed (as described in section 2.5). The authors conclude that a continuous oscillation as in natural environments could only be obtained by including temporal modulations into the model. With certain combinations of parameters and variations of the model a very long decay time of millions of generations for initially occurring inherent oscillations was found. Under natural conditions it could be assumed that these

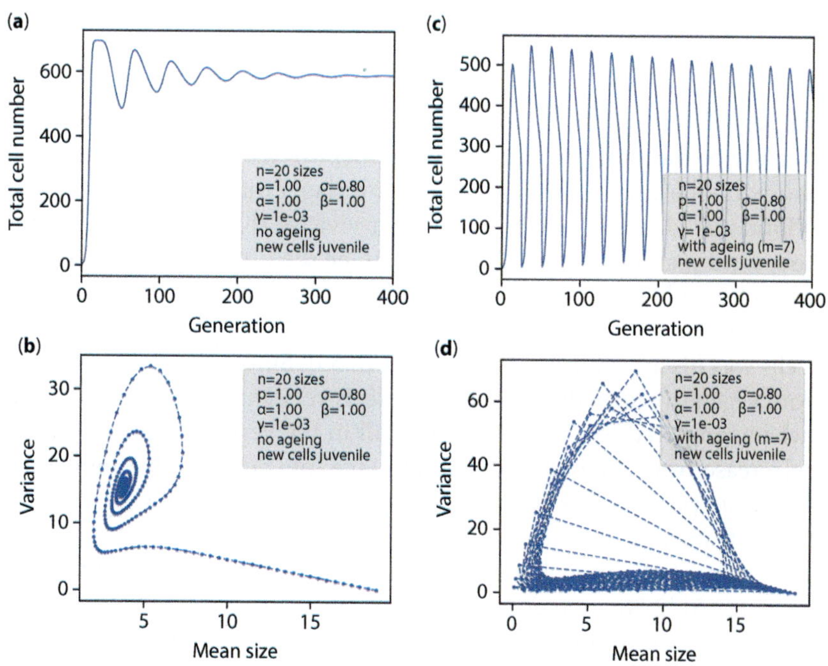

Figure 2.7 Direct comparison of the Fuhrmann-Lieker *et al.* model with and without aging. (a) Development of the population in a small model with 20 size classes in which cells die only due to overpopulation (saturation by Ricker term) (b) Trajectory of the size distribution in (a), represented by mean size and variance. (c) Same model as in (a), but thecae live for only 7 generations as suggested by Jewson. Long-term oscillations are induced. (d) Trajectory of the size distribution in (c).

oscillations would appear stable since no system could stay undisturbed for such a long period. Nevertheless, an equilibrium is always reached in theory as long as no fluctuation is included.

As an example of numerical results from the model, in Figure 2.7 the situation for two different scenarios is compared. On the left side the population dynamics and the trajectory of size distribution is followed for the original MacDonald–Pfitzer scheme, showing an approach towards an equilibrium population and size distribution after a few life cycles. On the right side, the assumption of Jewson about a limited lifespan of individual cells is added. Here, the oscillations persist for a long time and decay only slowly.

2.7 Oscillatory Behavior

2.7.1 Reproduction of Experimental Data

Models have to be capable of describing present data and giving predictions for the future, also in alternative scenarios with modified parameters. In this section we show for the few existing long-term studies agreements and difficulties in the application of the models. Comparison mainly concentrates on the observation of distinct cohorts, each started by a synchronous auxosporulation event which starts the reduction-restitution cycle.

Figure 2.8 can be taken as a representative example for long-term studies. For 10 years, the size distribution in the population of *Pseudo-nitzschia multistriata* in the Gulf of Naples was monitored by taking samples in regular intervals [2.19]. During bloom phases *P. multistriata* was abundant, so sampling was easy, but the bloom did not occur every year with the same cell concentration. In samples in the rest of the year, the diatoms occurred less frequently, so the statistic significance of the data is also less. A bimodal distribution of cell sizes was observed in many of the samples, but the assignment of a certain cell size to one of the assumed cohorts can be ambiguous if the modes overlap. From the bimodal distribution, a total of 7 cohorts was identified, each of which survive for three years and generate a new cohort in the second year, so a biannual rhythm can be seen. The first phase from the initial cell to midsize cells could only be postulated with no experimental data because of the low frequency of occurrence. The problem of observing auxosporulation and the generation of new initial cells was discussed in depth by Mann [2.55]. Since the total percentage of cells switching from vegetative growth to gametogenesis is very small, auxosporulation is a very rare event, when compared to all cells that are still in the vegetative phase. In reading the further size decrease from the graph, one has to take into account that model parameters change in the bloom phase (black bar at the bottom) and in the following stationary phase (white bar). This feature is further discussed in the next section. As this example shows, finding total agreement of a model and experimental data is not easy. Similar studies confirm this picture [2.55] [2.56]. Sometimes multimodal behavior can be seen, but sometimes the size classes merge, and cohorts are not easily distinguished. Therefore, the only hope is to be able to deduce some of the parameters that would represent an ideal life cycle, neglecting stochastic fluctuations of real populations. In the laboratory, chances are better, but also here unforeseen events and discontinuity in the culture may obstruct the analysis.

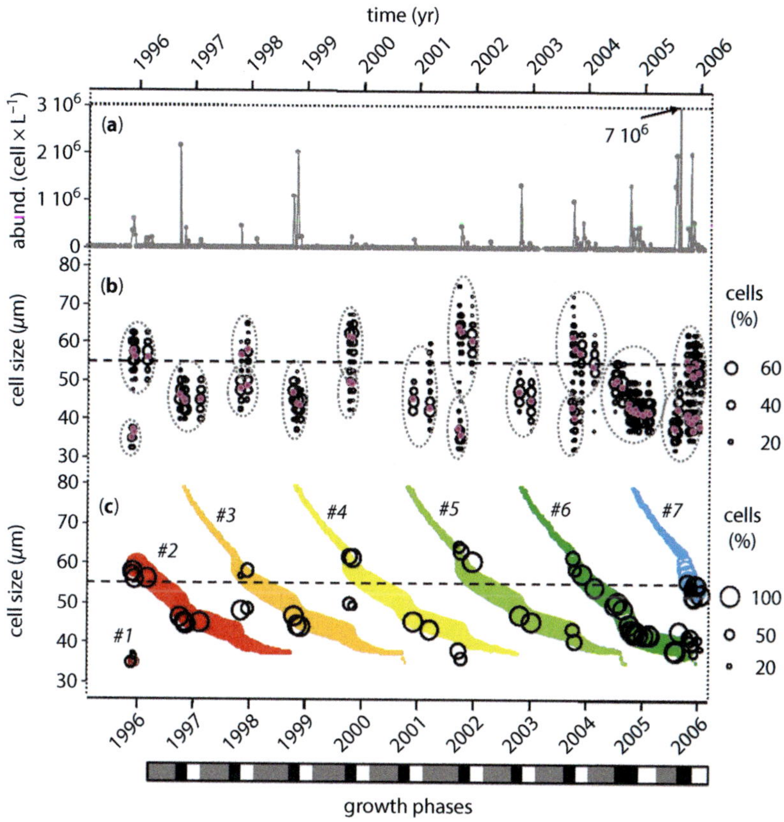

Figure 2.8 Cell size dynamics of *Pseudo-nitzschia multistriata* according to D'Alelio *et al.* [2.19]. Upper graph: cell abundance at the sampling period. Center: Division of the population according to cell size into subpopulations (cohorts, dotted ellipses) and size classes (solid circles). The mean cell size of the cohorts is indicated in pink. The dashed line indicates the postulated upper size threshold for sex. Lower graph: Comparison with the model. Different model cohorts are emphasized by different colors. Changes of cell size reduction speed are attributed to different growth phases: normal growth (gray in the bar below), bloom phases (black), and stationary phases (white). (Reproduced with permission.)

2.7.2 Coupling to External Rhythms

In all models the question arises if the parameters used in the differential equations and matrices can be taken as constant or if they may be subjected to external influences. Wherever doubling rates are given, they are subjected to fluctuations in nutrient and light conditions. Mann and others suggested that auxosporulation occurs only within a few weeks each year [2.55]. A major reason might be the influence of cell density and accompanying pheromone activity, but these are ultimately coupled to annual changes in the environmental conditions. So a coupling of growth rate and/or auxosporulation probability to photo- or thermoperiod should be taken into account in appropriate modeling. For the life cycle it means that cell size and season simultaneously decide about the switch to sexual reproduction.

Typically external influences vary roughly with some typical period that can be defined as *zeitgeber* rhythm, in the terms of chronobiology. *Zeitgeber* can be implemented in the models as gating functions (see discussion of the D'Alelio *et al.* model) or, as all periodic signals can be described as Fourier series of the fundamental frequency and its higher harmonics, as sine functions. Then, the intrinsic frequency given by the size reduction and restitution cycle (the "sex clock") and the annual cycle of the *zeitgeber* interact.

In Figure 2.9 the behavior is shown when the auxosporulation parameter σ is varied sinusoidally in the Fuhrmann-Lieker *et al.* model. Since the auxosporulation probability is given by $s = p\sigma^n$, it mimics a confined time window with high probability for sexual reproduction and a longer phase with a strongly reduced probability. In the resulting plot of the population dynamics, both periodicities are involved (a), with the result that auxosporulation does not take place significantly in every year but with a burst in the years when sexually productive cells are abundant during the mating season (b). Fourier transformation of the cell number reveal the frequency coupling on short and long terms. At the beginning of the oscillations, the Fourier spectrum shows the signals for the size cycle at 0.018/generation and the annual cycle at 0.025/generation, but also mixed frequencies by nonlinear coupling. For example, the dominant Fourier peak at 0.07/generation is just the difference frequency of both underlining rhythms (c). In the long term however, the *zeitgeber* rhythm controls the dynamic behavior (d) and becomes the only surviving mode. This again is a

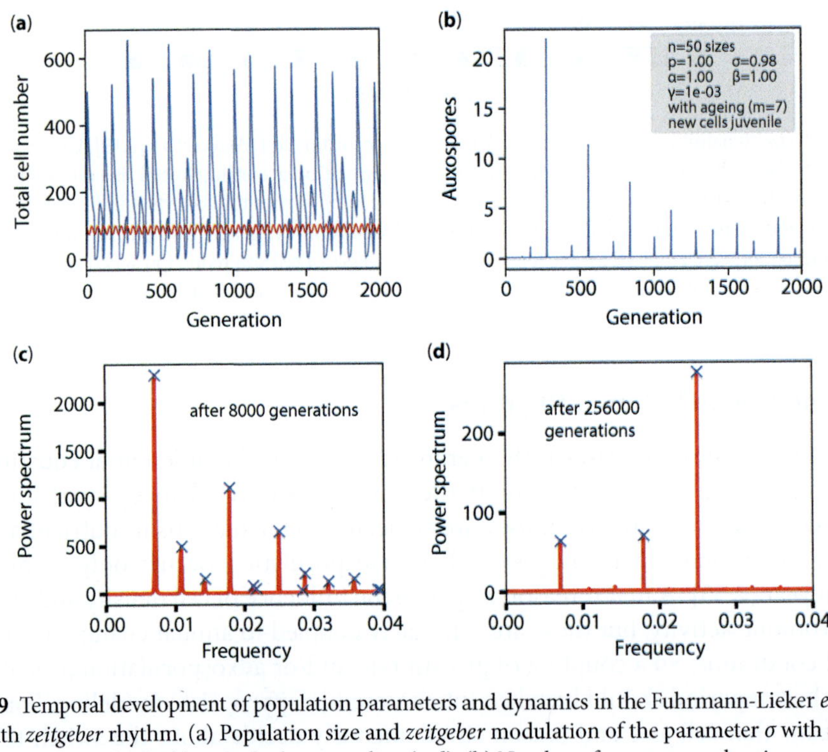

Figure 2.9 Temporal development of population parameters and dynamics in the Fuhrmann-Lieker *et al.* model with *zeitgeber* rhythm. (a) Population size and *zeitgeber* modulation of the parameter σ with a period of 40 generations, multiplied by 100 for better scaling (red), (b) Number of auxospores, showing years of low and high auxosporulation (c) Fourier transform of the total cell number for the first 8000 generations. Temporal frequency is given as 1/generation (d) Fourier transform of the total cell number for the first 256000 generations.

result of the Perron–Frobenius theorem that forbids a dominant oscillating eigenvector in the matrices describing the size class structure to which the external signal is not subjected.

2.8 Conclusion

We have seen that there are many different approaches possible in modeling the size resolved population dynamics of diatoms. They all have in common that the population is sorted into classes of distinct cell sizes in order to describe the vegetative size reduction process.

Then the similarity ends. Under the proposed models, there are systems of differential equations, difference equations, as well as individual-based stochastic processes. The models are either constructed to describe an experimentally obtained data set for a distinct species, or are developed as rather generic models to understand the principles of size-dependent population dynamics (Table 2.7). Often the choice whether to use class-based or individual-based models, or time-discrete or time-continuous approaches is a matter of taste, since more or less the same numerical results should result. However, we can see that in the different papers the strengths of the methods are favorably used for the respective aims of the studies. To investigate non-Markovian behaviour in which the individual past is important, stochastic branching is a very suitable approach, as applied in the Terzieva–Terziev paper. They focused on the mathematical theory but it remains to be seen whether this approach supplies useful results in numerical simulations. Time-discrete models are the obvious method of choice for synchronized populations and the transition of the time axis from real time to generation time. This was the choice in the description of Fuhrmann-Lieker *et al.* Also, advantages can be taken of the properties of non-negative matrices via the Perron–Frobenius theorem which is not straightforward for the matrices in systems of differential equations with their negative entries. In desynchronized populations, however, a time-continuous description seems more natural, hence the choice of Schwarz *et al.*

The parametrization is of much higher importance as the choice of the method. In models with few parameters such as in the Fuhrmann-Lieker *et al.* model, it is easier to see the general properties of the model, but they cannot be applied easily to experimental data sets.

Table 2.7 Comparison of the models. C: class model treated with differential equations, D: class model treated with difference equations, I: individual-based branching model.

Model	Method	Aim
Schwarz *et al.*	C	Finding optimum size classes and parameters from experimental data sets
D'Alelio *et al.*	D	Reproducing experimental observations
Hense–Beckmann	C + I	Mapping mechanisms as realistic as possible
Terzieva–Terziev	I	Developing formulae for individual-based non-Markov approach
Fuhrmann-Lieker *et al.*	D	Analyzing generic models with regard to oscillations

If many parameters are used, such as in the models of D'Alelio *et al.* and Hense–Beckmann, then a fine-tuning to reproduce experimental results is easy, but there is the danger of over-parametrization. Sensitivity or elasticity analyses ([2.13]) may help in parameter reduction, but they are rarely done. Of course, for replicability a good model should give full information about the governing equations and initial conditions, which is unfortunately not always the case.

In any case, a good model for population dynamics of diatoms takes cell size and environmental parameters into account. Such models will be important for simulating whole ecosystems that contain diatoms as one of the most influential primary production sources and carbon dioxide sequestration agents.

Acknowledgment

Funding by the Research Training Group "Biological Clocks on Multiple Time Scales" (GRK 2749) of the German Research Foundation (DFG), the Graduate School "Biological Clocks" (P/979) of the University of Kassel and the bilateral project ANR-17-CE400036/DFG-391322026 SYMBIONT is gratefully acknowledged.

We thank Nico Kubetschek and Giagkos-Ion Chlomoudis for proofreading and help in formatting formulae and figures.

References

[2.1] W. Admiraal. Influence of light and temperature on the growth rate of estuarine benthic diatoms in culture. *Marine Biology*, 39(1):1–9, 1977.

[2.2] H. Aguilaniu, L. Gustafsson, M. Rigoulet, and T. Nyström. Asymmetric inheritance of oxidatively damaged proteins during cytokinesis. *Science (New York, N.Y.)*, 299(5613): 1751–1753, 2003.

[2.3] A. Amato, L. Orsini, D. D'Alelio, and M. Montresor. Life cycle, size reduction patterns, and ultrastructure of the pennate planktonic diatom *Pseudo-nitzschia delicatissima* (Bacillariophyceae). *Journal of Phycology*, 41:542–556, 2005.

[2.4] E. Armbrust and S. Chisholm. Patterns of cell size change in a marine centric diatom: Variability evolving from clonal isolates. *Journal of Phycology*, 28:146–156, 1992.

[2.5] P. Assmy, J. Henjes, V. Smetacek, and M. Montresor. Auxospore formation by the silica-sinking, oceanic diatom *Fragilariopsis kerguelensis* (Bacillariophyceae). *Journal of Phycology*, 42(5):1002–1006, 2006.

[2.6] P. Assmy, D. U. Hernández-Becerril, and M. Montresor. Morphological variability and life cycle traits of the type species of the diatom genus *Chaetoceros*, *C. dichaeta*. *Journal of Phycology*, 44(1):152–163, 2008.

[2.7] K. Athreya and P. Ney. *Branching processes*. Die Grundlagen der mathematischen Wissenschaften in Einzeldarstellungen 196. Springer, New York, 1972.

[2.8] A.-S. Benoiston, F. M. Ibarbalz, L. Bittner, L. Guidi, O. Jahn, S. Dutkiewicz, and C. Bowler. The evolution of diatoms and their biogeochemical functions. *Philosophical Transactions of the Royal Society B: Biological Sciences*, 372(1728):20160397, 2017.

[2.9] A. Berman and R. Plemmons. *Nonnegative Matrices in the Mathematical Sciences.* Classics in Applied Mathematics 9. SIAM, Philadelphia, 1994.

[2.10] H. Bethge. *Melosira und ihre Planktonbegleiter.* Salzwasser Verlag, Paderborn, reprinted from 1925, 2012.

[2.11] G. Bilcke, L. van Craenenbroeck, A. Castagna, C. M. Osuna-Cruz, K. Vandepoele, K. Sabbe, L. de Veylder, and W. Vyverman. Light intensity and spectral composition drive reproductive success in the marine benthic diatom *Seminavis robusta. Scientific Reports,* 11(1):17560, 2021.

[2.12] F. Brauer and C. Castillo-Chavez. *Mathematical Models in Population Biology and Epidemiology.* Springer, New York, 2nd edition, 2010.

[2.13] H. Caswell. *Matrix Population Models.* Sinauer Associates, Sunderland, 2nd edition, 2001.

[2.14] V. A. Chepurnov, D. G. Mann, W. Vyverman, K. Sabbe, and D. B. Danielidis. Sexual reproduction, mating system, and protoplast dynamics of *Seminavis* (Bacillariophyceae). *Journal of Phycology,* 38(5):1004–1019, 2002.

[2.15] V. A. Chepurnov, D. G. Mann, K. Sabbe, and W. Vyverman. Experimental Studies on Sexual Reproduction in Diatoms. volume 237 of *International Review of Cytology,* pages 91–154. Academic Press, 2004.

[2.16] V. A. Chepurnov, D. G. Mann, P. von Dassow, P. Vanormelingen, J. Gillard, D. Inzé, K. Sabbe, and W. Vyverman. In search of new *tractable* diatoms for experimental biology. *BioEssays: news and reviews in molecular, cellular and developmental biology,* 30(7):692–702, 2008.

[2.17] F. Crawford, L. Ho, and Suchard, M.A. Computational methods for birth-death processes. *WIREs Comput. Stat.,* 10:e1423, 2018.

[2.18] R. M. Crawford and Y. Likhoshway. The frustule structure of original material of *Aulacoseira distans* (Ehrenberg) Simonsen. *Diatom Research,* 14(2):239–250, 1999.

[2.19] D. D'Alelio, M. R. d'Alcalà, L. Dubroca, D. Sarno, A. Zingone, and M. Montresor. The time for sex: A biennial life cycle in a marine planktonic diatom. *Limnology and Oceanography,* 55(1):106–114, 2010.

[2.20] N. A. Davidovich. Factors controlling the size of initial cells in diatoms. *Russian Journal of Plant Physiology,* 41:220–224, 1994.

[2.21] N. A. Davidovich and S. S. Bates. Sexual reproduction in the pennate diatoms *Pseudo-nitzschia multiseries* and *P. pseudodelicatissima* (Bacillariophyceae). *Journal of Phycology,* 34(1):126–137, 1998.

[2.22] N. A. Davidovich, J.-L. Mouget, and P. Gaudin. Heterothallism in the pennate diatom *Haslea ostrearia* (Bacillariophyta). *European Journal of Phycology,* 44(2):251–261, 2009.

[2.23] N. A. Davidovich, I. Kaczmarska, and J. M. Ehrman. Heterothallic and homothallic sexual reproduction in *Tabularia fasciculata* (Bacillariophyta). *Fottea,* 10(2):251–266, 2010.

[2.24] N. A. Davidovich, R. Gastineau, P. Gaudin, and O. I. Davidovich. Sexual reproduction in the newly-described blue diatom, *Haslea karadagensis. Fottea,* 12(2):219–229, 2012.

[2.25] O. Davidovich, N. Davidovich, and J.-L. Mouget. The effect of temperature on vegetative growth and sexual reproduction of two diatoms from the genus *Haslea* Simonsen. *Russian Journal of Marine Biology,* 44(1):8–13, 2018.

[2.26] O. Diekmann, S. van Gils, S. Verduyn Lunel, and H.-O. Walther. *Delay Equations.* Applied Mathematical Sciences 110. Springer-Verlag, New York, 1995.

[2.27] M. Edlund and R. Bixby. Intra- and inter-specific differences in gametangial and inital sizes in diatoms. Proceedings of the 16th international diatom symposium, pages 169–190. Amvrosiou Press, 2001.

[2.28] S. Elaydi. *An Introduction to Difference Equations.* Undergraduate Texts in Mathematics. Springer-Verlag, New York, 3rd edition, 2005.

[2.29] R. Eppley. Temperature and phytoplankton growth in the sea. *Fish. Bull.*, 70:1063–1085, 1972.

[2.30] P. Fife. *Mathematical Aspects of Reacting and Diffusing Systems.* Lecture Notes in Biomathematics 28. Springer-Verlag, Berlin, 1979.

[2.31] T. Fuhrmann-Lieker, N. Kubetschek, J. Ziebarth, R. Klassen, and W. Seiler. Is the diatom sex clock a clock? *Journal of the Royal Society, Interface*, 18(179):20210146, 2021.

[2.32] J. Gillard, V. Devos, M. J. Huysman, L. De Veylder, S. D'Hondt, C. Martens, P. Vanormelingen, K. Vannerum, K. Sabbe, V. A. Chepurnov, D. Inzé, M. Vuylsteke, and W. Vyverman. Physiological and transcriptomic evidence for a close coupling between chloroplast ontogeny and cell cycle progression in the pennate diatom *Seminavis robusta. Plant Physiology*, 148:1394–1411, 2008.

[2.33] A. Giometto, F. Altermatt, F. Carrara, A. Maritan, and A. Rinaldo. Scaling body size fluctuations. *Proceedings of the National Academy of Sciences*, 110(12):4646–4650, 2013.

[2.34] K. Gopalsamy. *Stability and Oscillations in Delay Differential Equations of Population Dynamics.* Kluwer, New York, 1992.

[2.35] P. Grindrod. *The Theory and Applications of Reaction-Diffusion Equations: Patterns and Waves.* Oxford Applied Mathematics and Computing Science Series. Clarendon Press, Oxford, 2nd edition, 1996.

[2.36] I. Hense and A. Beckmann. A theoretical investigation of the diatom cell size reduction–restitution cycle. *Ecological Modelling*, 317:66–82, 2015.

[2.37] M. Hirsch, S. Smale, and R. Devaney. *Differential Equations, Dynamical Systems, and an Introduction to Chaos.* Academic Press, Amsterdam, 3rd edition, 2013.

[2.38] K. E. Holtermann, S. S. Bates, V. L. Trainer, A. Odell, and E. Virginia Armbrust. Mass sexual reproduction in the toxigenic diatoms *Pseudo-nitzschia australis* and *P. pungens* (Bacillariophyceae) on the Washington coast, USA. *Journal of Phycology*, 46 (1):41–52, 2010.

[2.39] M. J. Huysman, W. Vyverman, and L. De Veylder. Molecular regulation of the diatom cell cycle. *Journal of Experimental Botany*, 65(10):2573–2584, 2013.

[2.40] M. J. Huysman, C. Martens, K. Vandepoele, J. Gillard, E. Rayko, M. Heijde, C. Bowler, D. Inzé, Y. Van de Peer, L. De Veylder, and W. Vyverman. Genome-wide analysis of the diatom cell cycle unveils a novel type of cyclins involved in environmental signalling. *Genome Biology*, 11:R17, 2010.

[2.41] M. J. J. Huysman, A. E. Fortunato, M. Matthijs, B. S. Costa, R. Vanderhaeghen, H. van den Daele, M. Sachse, D. Inzé, C. Bowler, P. G. Kroth, C. Wilhelm, A. Falciatore, W. Vyverman, and L. de Veylder. AUREOCHROME1a-mediated induction of the diatom-specific cyclin dsCYC2 controls the onset of cell division in diatoms (*Phaeodactylum tricornutum*). *The Plant cell*, 25(1):215–228, 2013.

[2.42] D. H. Jewson. Life cycle of a *Stephanodiscus* sp. (Bacillariophyta). *Journal of Phycology*, 28(6):856–866, 1992.

[2.43] D. H. Jewson. Size reduction, reproductive strategy and the life cycle of a centric diatom. *Philosophical Transactions of the Royal Society of London. Series B: Biological Sciences*, 336(1277):191–213, 1992.

[2.44] I. Kaczmarska, A. Poulíčková, S. Sato, M. B. Edlund, M. Idei, T. Watanabe, and D. Mann. Proposals for a terminology for diatom sexual reproduction, auxospores and resting stages. *Diatom Research*, 28(3):263–294, 2013.

[2.45] W. Kelley and A. Peterson. *Difference Equations: An Introduction with Applications.* Academic Press, Boston, 1991.

[2.46] M. Kimmel and D. Axelrod. *Branching Processes in Biology.* Interdisciplinary Applied Mathematics. Springer-Verlag, New York, 2002.

[2.47] C. P. Klingenberg. Size, shape, and form: concepts of allometry in geometric morphometrics. *Development Genes and Evolution*, 226(3):113–137, 2016.

[2.48] M. Kloster, A. S. Rigual-Hernández, L. K. Armand, G. Kauer, T. W. Trull, and B. Beszteri. Temporal changes in size distributions of the southern ocean diatom *Fragilariopsis kerguelensis* through high-throughput microscopy of sediment trap samples. *Diatom Research*, 34(3):133–147, 2019.

[2.49] S. R. Laney, R. J. Olson, and H. M. Sosik. Diatoms favor their younger daughters. *Limnology and Oceanography*, 57(5):1572–1578, 2012.

[2.50] P. H. Leslie. On the use of matrices in certain population mathematics. *Biometrika*, 33(3):183–212, 1945.

[2.51] E. Lewis. *Network Models in Biology*. Biomathematics 7. Springer-Verlag, Berlin, 1977.

[2.52] W. M. Lewis. The diatom sex clock and its evolutionary significance. *The American Naturalist*, 123(1):73–80, 1984.

[2.53] E. Litchman, C. A. Klausmeier, and K. Yoshiyama. Contrasting size evolution in marine and freshwater diatoms. *Proceedings of the National Academy of Sciences*, 106 (8):2665–2670, 2009.

[2.54] J. D. Macdonald. I.—On the structure of the Diatomaceous frustule, and its genetic cycle. *Annals and Magazine of Natural History*, 3(13):1–8, 1869.

[2.55] D. G. Mann. Why didn't Lund see sex in *Asterionella*? A discussion of the diatom life cycle in nature. *Algae and the Aquatic Environment*, pages 383–412. 1988.

[2.56] D. G. Mann. Size and Sex. The Diatom World, pages 145–166. Springer Netherlands, Dordrecht, 2011.

[2.57] M. Mizuno and K. Okada. Seasonal change in the distribution of cell size of *Cocconeis scutellum* var. *ornata* (Bacillariophyceae) in relation to growth and sexual reproduction. *J. Phycol.*, 21:547–553, 1985.

[2.58] J. Moisan, T. Moisan, and M. Abbott. Modelling the effect of temperature on the maximum growth rates of phytoplankton populations. *Ecol. Model.*, 153:197–215, 2002.

[2.59] C. M. Moore, M. M. Mills, K. R. Arrigo, I. Berman-Frank, L. Bopp, P. W. Boyd, E. D. Galbraith, R. J. Geider, C. Guieu, S. L. Jaccard, T. D. Jickells, J. La Roche, T. M. Lenton, N. M. Mahowald, E. Marañón, I. Marinov, J. K. Moore, T. Nakatsuka, A. Oschlies, M. A. Saito, T. F. Thingstad, A. Tsuda, and O. Ulloa. Processes and patterns of oceanic nutrient limitation. *Nature Geoscience*, 6(9):701–710, 2013.

[2.60] G. F. O. Müller. Das Gesetz der Zelltheilungsfolge von *Melosira* (*Orthosira*) *arenaria* Moore. *Berichte der Deutschen Botanischen Gesellschaft*, v. 1:35–44, 1883.

[2.61] G. F. O. Müller. Die Zellhaut und das Gesetz der Zelltheilungsfolge von *Melosira (Orthosira* Thwaites) *arenaria* Moore. *Jahrbücher für wissenschaftliche Botanik*, 14: 232–290, 1884.

[2.62] S. Nagai and T. Manabe. Auxospore formation of a giant diatom, *Coscinodiscus wailesii* (Bacillariophyceae), in culture. *Bulletin of Plankton Society of Japan (Japan)*, 40(2):151–167 1994.

[2.63] F. Nipkow. Über das Verhalten der Skelette planktischer Kieselalgen im geschichteten Tiefenschlamm des Zürich- und Baldeggersees. *Zeitschrift für Hydrologie*, 4(1):71–120, 1928.

[2.64] T. Nishikawa. Cell size reduction and restoration of the diatom *Eucampia zodiacus* in culture. *Plankton and Benthos Research*, 12(2):139–142, 2017.

[2.65] E. T. Olivieri. Feasibility of estimating phytoplankton size and biomass in fresh and preserved samples from the Benguela current with a Coulter counter. *South African Journal of Marine Science*, 3(1):99–110, 1985.

[2.66] P. Olver and C. Shakiban. *Applied Linear Algebra*. Pearson Prentice Hall, 2006.

[2.67] P. H. Petersen, O. Blaaberg, M. Andersen, L.G.M. Jørgensen, K. Schousboe, and E. Jensen. Graphical interpretation of confidence curves in rankit plots. *Clin. Chem. Lab. Med.*, 42(7):715–724, 2004.

[2.68] E. Pfitzer. Über Bau und Zellteilung der Diatomeen. *Botanische Zeitung*, 27:774–776, 1869.

[2.69] L. Roselli, E. Stanca, F. Paparella, A. Mastrolia, and A. Basset. Determination of *Coscinodiscus* cf. *granii* biovolume by confocal microscopy: comparison of calculation models. *Journal of Plankton Research*, 35(1):135–145, 2013.

[2.70] F. Round. Auxospore structure, initial valves and the development of populations of *Stephanodiscus* in Farmoor Reservoir. *Ann. Bot.*, 49:447–459, 1982.

[2.71] F. E. Round, editor. *Algae and the aquatic environment: Contributions in honour of J. W. G. Lund*. Biopress, Bristol, 1988.

[2.72] F. E. Round, R. M. Crawford, and D. G. Mann. *The diatoms: Biology & morphology of the genera*. Cambridge Univ. Press, Cambridge, reprinted edition, 1990.

[2.73] J. Sandefur. *Discrete Dynamical Systems*. Clarendon Press, Oxford, 1990.

[2.74] E. Scalco, K. Stec, D. Iudicone, M. I. Ferrante, and M. Montresor. The dynamics of sexual phase in the marine diatom *Pseudo-nitzschia multistriata* (Bacillariophyceae). *Journal of Phycology*, 50(5):817–828, 2014.

[2.75] A.-M. M. Schmid and R. M. Crawford. *Ellerbeckia arenaria* (Bacillariophyceae): formation of auxospores and initial cells. *European Journal of Phycology*, 36(4):307–320, 2001.

[2.76] R. Schwarz, M. Wolf, and T. Müller. A probabilistic model of cell size reduction in *Pseudo-nitzschia delicatissima* (Bacillariophyta). *Journal of Theoretical Biology*, 258 (2):316–322, 2009.

[2.77] H. Smith. *An Introduction to Delay Differential Equations with Applications to the Life Sciences*. Texts in Applied Mathematics 57. Springer-Verlag, New York, 2011.

[2.78] J. Smoller. *Shock Waves and Reaction Diffusion Equations*. Grundlehren der Mathematischen Wissenschaften 258. Springer-Verlag, New York, 2nd edition, 1994.

[2.79] E. J. Stewart, R. Madden, G. Paul, and F. Taddei. Aging and death in an organism that reproduces by morphologically symmetric division. *PLOS Biology*, 3(2):e45, 2005.

[2.80] H. A. V. Stosch. Manipulierung der Zellgrösse von Diatomeen im Experiment. *Phycologia*, 5(1):21–44, 1965.

[2.81] A. Terzieva and G. Terziev. Model diatom population by branching stochastic processes. *Applied and Computational Mathematics*, 8(1):3–8, 2019.

[2.82] G. Teschl. *Ordinary Differential Equations and Dynamical Systems*. Graduate Studies in Mathematics 140. American Mathematical Society, Providence, 2012.

[2.83] P. Tréguer, C. Bowler, B. Moriceau, S. Dutkiewicz, M. Gehlen, O. Aumont, L. Bittner, R. Dugdale, Z. Finkel, D. Iudicone, O. Jahn, L. Guidi, M. Lasbleiz, K. Leblanc, M. Levy, and P. Pondaven. Influence of diatom diversity on the ocean biological carbon pump. *Nature Geoscience*, 11(1):27–37, 2018.

[2.84] A. Turing. The chemical basis of morphogenesis. *Phil. Trans. Roy. Soc. B*, 237(641): 37–72, 1952.

[2.85] P. von Dassow and M. Montresor. Unveiling the mysteries of phytoplankton life cycles: patterns and opportunities behind complexity. *Journal of Plankton Research*, 33(1): 3–12, 2011.

[2.86] S. Wiggins. *Introduction to Applied Nonlinear Dynamical Systems and Chaos.* Texts in Applied Mathematics 2. Springer-Verlag, New York, 2nd edition, 2003.
[2.87] R. Williams. Division rates of salt marsh diatoms in relation to salinity and size. *Ecology*, 45(4):877–880, 1964.

3

On the Mathematical Description of Diatom Algae: From Siliceous Exoskeleton Structure and Properties to Colony Growth Kinetics, and Prospective Nanoengineering Applications

Alexey I. Salimon[1*], Julijana Cvjetinovic[2], Yuliya Kan[1], Eugene S. Statnik[1], Patrick Aggrey[1], Pavel A. Somov[1], Igor A. Salimon[2], Joris Everaerts[3], Yekaterina D. Bedoshvili[4], Dmitry A. Gorin[2], Yelena V. Likhoshway[4], Philipp V. Sapozhnikov[5], Nickolai A. Davidovich[6], Olga Y. Kalinina[7], Kalin Dragnevski[8] and Alexander M. Korsunsky[1,8]

[1]*Center for Energy Science and Technology, Skolkovo Institute of Science and Technology, Moscow, Russia*
[2]*Center for Photonic Science and Engineering, Skolkovo Institute of Science and Technology, Moscow, Russia*
[3]*Institute of Materials, École Polytechnique Fédérale de Lausanne, Lausanne, Vaud, Switzerland*
[4]*Limnological Institute Siberian Branch of the Russian Academy of Sciences, Irkutsk, Russia*
[5]*Shirshov Institute of Oceanology of the Russian Academy of Sciences, Moscow, Russia*
[6]*Karadag Scientific Station-Natural Reserve of the Russian Academy of Sciences, Crimea, Russia*
[7]*Lomonosov Moscow State University, Moscow, Russia*
[8]*Department of Engineering Science, University of Oxford, Oxford, United Kingdom*

Abstract

Diatoms, the unicellular algae encased within fascinating nanostructured siliceous exoskeletons, continue to capture the interest and attention of mechanical engineers and materials scientists due to the outstanding strength and fracture resistance of their frustules (shells) achieved at minimal weight. The high mechanical performance appears to be a beneficial consequence of the sophisticated hierarchical structural design of the valves and girdles that sustain the complex biological and photochemical processes playing out within diatom microorganisms under vastly diverse, frequently hostile environmental conditions.

The superlative mechanical performance of diatom frustules paves the way for a wide range of engineering applications in NEMS, MEMS, nanofluidics, and many other applications.

Groundbreaking new technologies require exploring the classical research chain "structure-properties-function-applications," so that the underlying relationships can be systematically identified, elucidated, adapted and exploited via the most relevant theoretical, instrumental, computational (mathematical) and technological developments.

In this chapter, we overview and discuss the background knowledge and fundamental issues of the mechanics of thin perforated shells for application in hierarchically structured diatom exoskeletons,

**Corresponding author*: a.salimon@skoltech.ru

Janice L. Pappas (ed.) The Mathematical Biology of Diatoms, (63–102) © 2023 Scrivener Publishing LLC

overview the relevant theoretical and experimental approaches, and present some recent results of SEM and X-ray characterization accompanied by critical evaluation and discussion.

Keywords: Photoacoustics, nanostructuring, Fourier transform, AFM, SEM, vibration, MEMS/NEMS

3.1 Introduction

The expression "Mathematical Biology" relates two fundamental clusters of modern science: mathematics, which nowadays refers to the study of the full breadth of the quantitative, logical, often abstracted relationships between arbitrary objects and their properties; and biology, which explores diverse objects ranging from cell to tissue to organisms and populations characterized by their ability for reproduction. The overlap and interaction between these two disciplines give birth to numerous synthetic methods and approaches that yield new fundamental knowledge and practical benefits.

Diatom algae have been the focus of scientific interest of biologists for centuries, and represent an area for the application of diverse mathematical methods, from the description of their structure and growth to statistics and beyond. In particular, nanostructured siliceous exoskeletons of diatom algae are capturing the interest of mechanical engineers and materials scientists due to their suitability as mimetic prototypes for minimal weight and vibrational design, and as a platform for fine-scale manipulations in nanotechnologies. Moreover, the neat organization and space ordering of pore-like discontinuities has a direct impact on light-matter interactions applicable in photonics, photochemistry, and photosensors.

We present a compendium of landscape notes devoted to several particular aspects of the science of diatom algae where mathematics and biology converge and interact, namely:

- differential and integral equations to describe the diatom colonies growth kinetics;
- finite element modeling and evaluation of the static, dynamic and vibrational mechanical response of diatom frustule;
- combinatorial modeling to simulate the appearance of ordering in quasi-random systems.

The advances in mathematical treatment of the problems of diatom algae biology have clear implications for applications in aquaculture production of diatoms as a renewable material and energy resource with synergetic potential to counteract the tendency of increasing atmospheric carbonization.

3.2 Hierarchical Structuring of Matter: Diatom Algae and the Bio-Assisted Nanostructured Additive Manufacturing Paradigm

The paradigm of hierarchical structuring of matter [3.1] is the deliberate and purposeful arrangement of material building blocks relevant to each dimensional level (from molecules through nanometer clusters to micrometer and millimeter entities) ultimately to construct *multifunctional* organisms, structures, or mechanisms. Nature provides numerous examples of successful hierarchical structuring in soft and hard tissues to satisfy the diverse and frequently contradictory requirements posed by the varying environments of living organisms [3.2].

Generally speaking, the fundamental reason for hierarchical structuring, i.e., scale-specific arrangement of parts and components, lies in the alteration of the predominant physical interaction mechanisms with the significant change of the dimensional scale of consideration (by one to two orders of magnitude) that entails the dominance of certain physical forces acting at the characteristic lengths comparable with the size of material entities (atoms and molecules of organic monomers, DNA spirals, nanoparticles, crystallites and spherulites, grain boundaries and phase interfaces, etc.). It is crucial that material entities interact to create hierarchical structures in the presence of external force fields and energy flows. The non-equilibrium thermodynamics of Onsager and Prigogine provides a means of describing the self-organization of matter to highly ordered complex structures lying far from the global minimum of Gibbs free energy.

In the following sections we discuss the structural design of diatom frustules in pursuit of the idea that the diversity of diatom appearance and key features of frustule structure are firmly linked to their specific living conditions that require different mechanical and optical properties to ensure optimal performance of the frustules. Certain hierarchical multifunctional architecture of a diatom frustule is DNA-coded and biologically driven. However, the fine local mechanisms of phase separation in silica deposition vesicles at the nanometer scale (the precipitation of silica globules from (over)saturated silicate water solutions) still need to be systematically studied to be understood. This would open the pathways for these processes to be directed and customized, and potentially used to implement the paradigm of Bio-Assisted Nanometer Size Additive Manufacturing (BANSAM). BANSAM assumes the transition to nature-like fabrication technologies of the 21st century that are simultaneously green, sustainable, non-carbon, and possess nanometer tolerance.

3.3 Structural Design of Diatom Frustules

At present, diatoms are considered to be an extensive taxonomic group of unicellular microorganisms at the type or division level (Biota: Chromista, Kingdom: Harosa, Subkingdom: Heterokonta, Infrakingdom: Bacillariophyta, Phylum). Despite the fact that many species of diatoms are capable of forming colonies of considerable size, at the individual level they remain unicellular.

Figures on the origination period of diatoms as an evolutionary group of marine organisms differ depending on the method of study. There are several specific assumptions about the lifetime of diatoms depending on the various studies on their age. Thus, according to the estimation using the molecular clock method, they may be as old as 240 million years [3.3]. However, paleontological approaches—revolving around an age analysis of ancient microfossils—indicate that diatoms were separated as a group in the world ocean about 145 million years ago [3.4]. The paleontological approach with respect to the diatoms should now be considered as more reliable. These organisms formed in vast lagoons with variable salinity. Lagoons formed at the bottom of the receding ocean, in shallow waters. The earliest diatoms lived in conditions where the rains filled the lagoons with water, and then the sun dried them up again. They formed dense silicon shells around them, through which they interacted with the habitat, where temperature and mineralization often changed. Since then, gradually evolving and assimilating new ecological niches, diatoms have spread across the world.

They populated a variety of aquatic and semi-aquatic biotopes: seas, rivers, lakes, streams, wet rocks and wet salt crusts, swamps, moss clumps, tree bark and soil. There are several types of diatoms that live on the surface of rocks, at an altitude of over 3 km, as shown in Figure 3.1. Some species have found stable ecological niches on the skin of whales [3.5]–[3.7]. In the areas of water bodies, diatoms live both in the water column (in plankton) and at the bottom (as part of benthos, to depths of several hundred meters), as well as on macroalgae, on the covers of invertebrates and fish, and on floating objects (as part of communities of periphyton). In recent decades, when a large amount of different types of plastic ended up in the water bodies of the planet as garbage, some species of diatoms have successfully adapted to life and on artificial polymers. They even formed very special communities on them, capable of surviving in a rather aggressive environment.

In order to live, carry out metabolic functions and reproduce successfully in any biotope, diatoms have developed a complex adaptations. Many of them are reflected in the form of their shells, in the way the pores are arranged, the configuration of the pore apparatus itself, sutures, outgrowths, in the size of the cells and the thickness of the shell. In fact, the appearance of the shell of an individual diatom species reflects the result of a long evolutionary path of this species, and its adaptation to the spectrum of environmental conditions in which it is adapted to develop.

It is extremely important that in different groups of diatoms that evolved in their own ways, there are species with similar adaptations—and a similar shape and ornamentation of shells. In fact, with similar adaptive shapes. For example, there are diatoms from evolutionarily different groups that build colonies that branch out in the water column like trees. They often have cells that are similar in outline and in the way they release biopolymers. Moreover, the composition of such exopolymers can differ significantly for representatives of different evolutionary groups. But the shape of their colonies will be similar, as demonstrated in Figure 3.2.

Among the mobile diatoms, there are also representatives of different evolutionary groups that have convergent similarities. For example, large and mobile representatives of

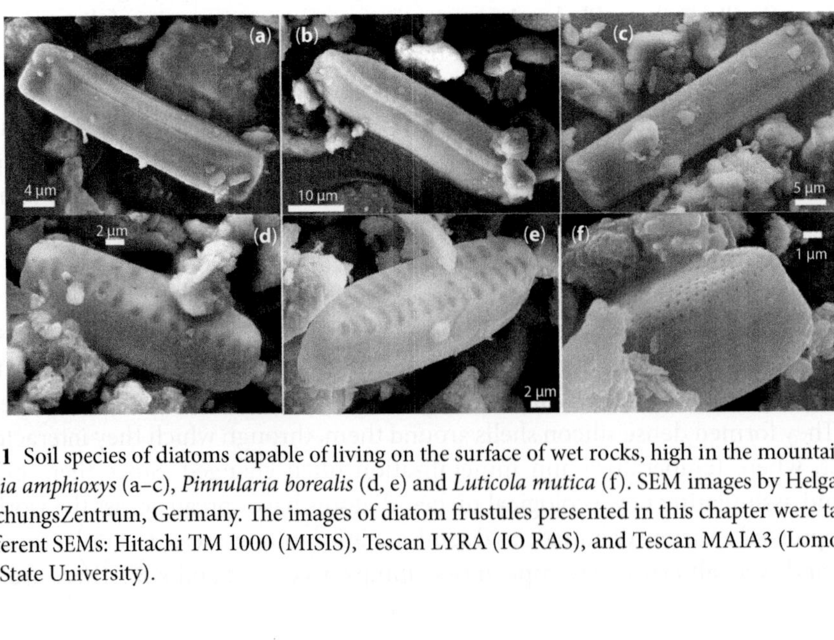

Figure 3.1 Soil species of diatoms capable of living on the surface of wet rocks, high in the mountains: *Hantzschia amphioxys* (a–c), *Pinnularia borealis* (d, e) and *Luticola mutica* (f). SEM images by Helga Kemnitz, GeoForschungsZentrum, Germany. The images of diatom frustules presented in this chapter were taken using three different SEMs: Hitachi TM 1000 (MISIS), Tescan LYRA (IO RAS), and Tescan MAIA3 (Lomonosov Moscow State University).

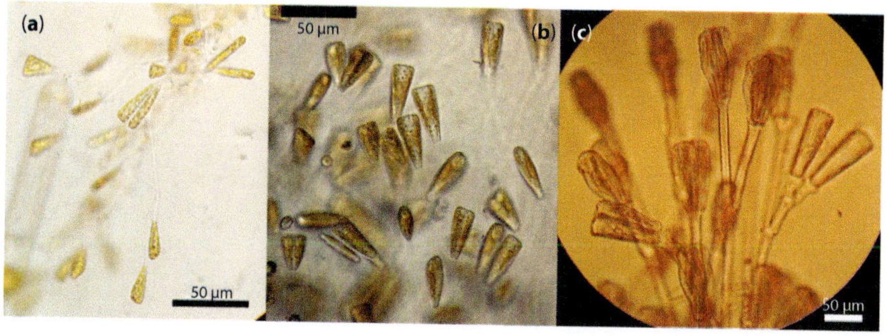

Figure 3.2 Treelike colonies of diatoms from different evolutionary groups: (a) *Licmophora gracilis* (West Antarctica, Powell Basin), (b) *Gomphonema olivaceum* (Moscow-River, Central Russia), (c) *Didymosphenia* sp. (Lake Baikal, East Siberia). (Light microscopy by Ph.V. Sapozhnikov, O.Y. Kalinina, SIO RAS).

the genera *Plagiotropis, Entomoneis, Donkinia, Auricula* and *Trachyneis*, as well as some species of *Nitzschia*, have habitually similar shells, as illustrated in Figure 3.3.

Planktonic diatoms are most often endowed with a radial symmetry of the shell, benthic and periphyton (feathery), or, in various forms, bilateral, although representatives of some genera look like a loaf of bread. They are shown in Figure 3.4 [3.8]. The diatom periphyton communities are the most diverse in terms of the set of adaptive forms. Like forests, they are layered. In turn, each tier gains an advantage in distributing their own sets of responsive forms.

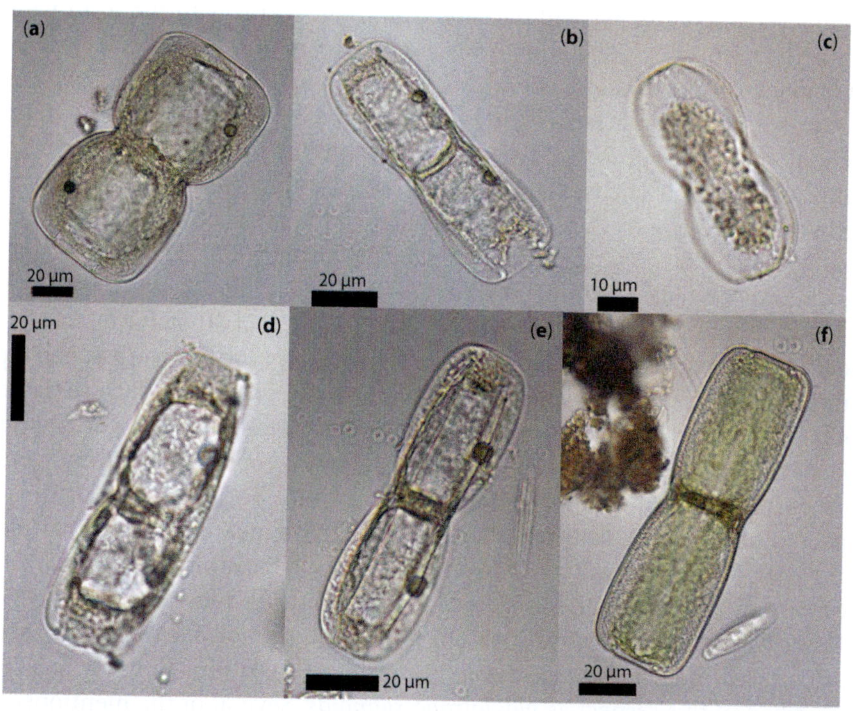

Figure 3.3 Habitual similarity of the shells of diatoms from different evolutionary groups: (a) *Donkinia carinata*, (b) *Plagiotropis longa*, (c) *Auricula intermedia*, (d) *Plagiotropis gibberula*, (e) *Pl. lepidoptera*, (f) *Trachyneis aspera*. (Light microscopy images by Ph. V. Sapozhnikov, O.Y. Kalinina, SIO RAS).

Figure 3.4 Diatoms with a radial symmetry of the frustule are mostly adapted to life conditions within planktonic populations (a–c, e–g); diatoms with a bilateral symmetry are more prevalent in benthos and periphyton (d, h–l). (SEM images by M.V. Kravchishina, Ph. V. Sapozhnikov, O.Y. Kalinina, SIO RAS).

If we talk about the main design options for diatom shells, then there will be quite a lot of such options. However, here we look at the interesting and beneficial properties of diatoms at the group level. Therefore, we will pay attention to life forms with the most widespread shell design. The point is that such life forms are most common in seas and fresh waters—and include many species of diatoms.

Let us start with representatives of the Fragillariophyceae class that are shown in Figure 3.5. These diatoms do not possess adaptations for rapid independent locomotion, but the very organization of their shells, as well as the colonies they form, allows them to be in streams of water that bring fresh nutrients and carry away metabolic products. This design of the organization of the carapace will be called fragilarioids [3.9]–[3.11].

Mobile diatoms capable of sliding at different speeds among thickets or sediment particles are grouped into the class Bacillariophyceae [3.12] [3.13]. They have a bilateral shell shape and are endowed with special devices for locomotion (variously organized raphes).

This large and diverse group includes particularly active wanderers such as naviculoids, nitzschioids and surirelloids, diploneoids, fallacioids, entomoneoids, gomphonemoids, cymbelloids and sigmoids. These names are based on a complex of macromorphological features of frustules. Based on their combination, it could be stated that diatoms having specific macromorphological features (shape and habitual appearance of the frustule, ornamentation, type of pore apparatus and raphes—and their location, as well as the method of displacement or substrate adhesion used) are probably the most evolutionarily similar to *Navicula* or *Nitzschia*, or *Diploneis*, etc. In relation to sigmoids, similarity to species of genera such as *Pleurosigma* and *Gyrosigma* is considered. Diatoms of the genus *Donkinia* may be described in this respect as *plagiotropo-sigmoids*. Several of the mentioned species are shown in Figure 3.6.

Figure 3.5 Fragilarioids are feathery diatoms without sutures. They are incapable of independent movement, but build colonies in the form of bundles, stars and ribbons. (SEM images by Ph. V. Sapozhnikov, O.Y. Kalinina, M.V. Kravchishina, SIO RAS).

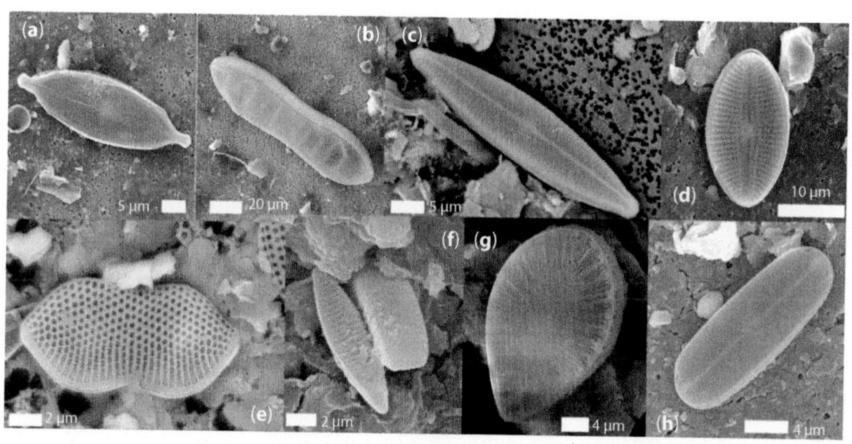

Figure 3.6 Active wanderers among diatoms: (a, c) naviculoids, (e, f) nitzschioids, (b, g) surirelloids, (d) diploneoids and (h) fallacioids. (SEM images by Ph. V. Sapozhnikov, O.Y. Kalinina, M.V. Kravchishina, SIO RAS).

They have specific paired facilities for easy mobility, located on different sides of the frustule (on opposite thecas)—the raphes or raphe cracks. There are some mobile species among single-raphe diatoms (monoraphids), having a raphe only on the bottom frustule (acanthoid, cocconeioids). However, displacement is often a forced measure for them and is used, as a rule, while searching for new places to adhere to a substrate. Diatoms of some life forms are capable of forming "shell coatings" (relatively dense colonial settlements) on the surface of solid substrates. These include amphoroids [3.14] [3.15] and cocconeoids [3.16]–[3.18], as well as some achnanthoids (*Karayevia* spp., *Planothidium* spp., *Psammothidium* spp.) [3.19]. Some of them are shown in Figure 3.7.

Such diatoms have either two raphes on one side of the shell (in amphoroids, flounders among diatoms), or one located on the side of the shell facing the substrate. Note that many achnantoids are capable of forming ribbons, in which cells use sutures to fasten them together. Despite the habitual similarity of the organization, achnanthoids are a very modular, polyphyletic group. Many genera among them do not build ribbons.

The forms with radial symmetry of the frustule are grouped into the *Coscinodiscophyceae* class. These diatoms have characteristic adaptations that allow them to float in the water column: singly or in colonies. For example, thalassiosiroids have cells in the form of tablets, forming chains during division using exopolymers [3.20]–[3.22]. They excrete these polymers through special pore structures (fultoportula and rimoportula). Many of them have a whole gallery of fultoportula on each face of the "tablet," releasing long organic threads through them, which also increase the ability of the colony to float in water.

The coscinodiscoids shown in Figure 3.8 [3.23] hover due to the large size of single cells, have thin-walled shells and contain large vacuoles in the cells, which balance their density with the surrounding water.

Chaetoceroids build colonies in the form of flat chains, often curling in a spiral, as shown in Figure 3.9 [3.24]. Their cells are united by means of long, hollow setae called chaetae. Most *Chaetoceros* species contain cytoplasm in the setae, and some also contain chloroplasts. Skeletonemoids float in the water column due to a combination of two valuable adaptive traits. First, their cells tend to contain a lot of oil, which lowers their density

Figure 3.7 Semi-mobile forms of the class Bacillariophyceae, which have a pronounced tendency to build widespread colonial settlements ("shell coverings") on the surface of solid substrates: (a–d) cocconeoids, amphoids (lying "belly up", e–h) and various achnantoids (i, j, l - from the dorsal side of the shell, k - from the ventral). (SEM and light microscopy images by Ph. V. Sapozhnikov, O.Y. Kalinina, M.V. Kravchishina, SIO RAS).

Figure 3.8 Different representatives of thalassiosiroids (a–f) and coscinodiscoids *Coscinodiscus radiatus* (g, h). (SEM images by M.V. Kravchishina, SIO RAS, taken through a light microscope by Ph. V. Sapozhnikov, SIO RAS).

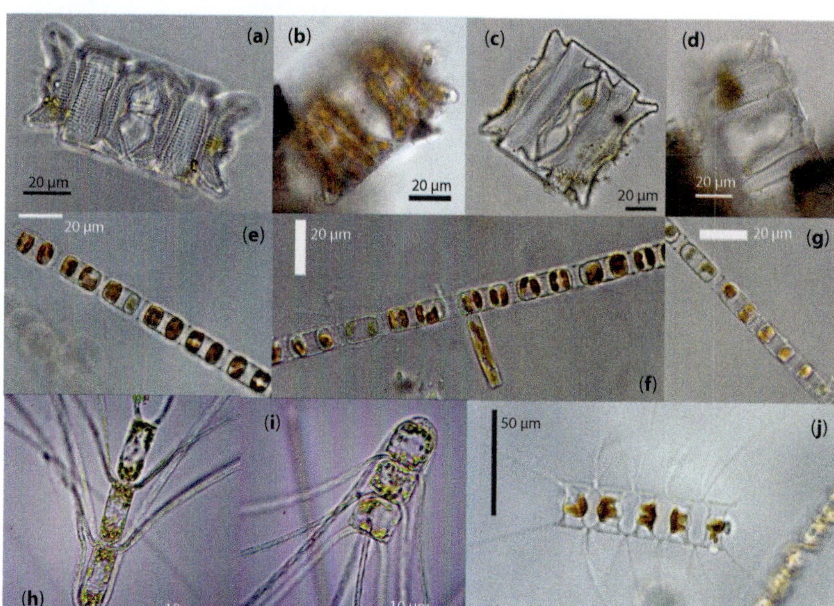

Figure 3.9 Widespread life forms of planktonic and periphyton-planktonic diatoms: (a–d) biddulphioids, (e–g) skeletonemoids and (h–j) chaetoceroids. (Images by Ph. V. Sapozhnikov, O.Y. Kalinina, SIO RAS).

relative to water. Second, the thin outgrowths of the frustules—due to which they are connected in straight chains—create delicate structures that are permeable to water but still perform the functions of sail elements. The total surface of the chains expands manifold due to such outgrowths. At the same time, the relative density of communities remains low. Thanks to such "translucent" compounds, communities of ovate-orbicular cells become like thin sails stretched in the water [3.25]. Finally, biddulphioids [3.26] form colonies of cells that resemble butterflies (from the side), but when viewed from above have 2-5-beam symmetry, depending on the species. These diatoms often live among fouling—it is from there that they come en masse to coastal plankton.

Of course, the question arises: Can, for example, representatives of naviculoids or nitzschioids build colonies in which cells have limited mobility or are completely immobile? Yes, among them there are separate genera (*Parlibellus, Fragilariopsis*) and some species from very mobile genera (some *Navicula* species) that form such colonies that are shown in Figure 3.10.

In the summaries on the analysis of the structure of communities, one can find references to the fact that benthic life forms of diatoms are found in the composition of plankton—and vice versa, plankton ones are found at the bottom.

However, to understand such images, one should take into account that benthic and periphyton diatoms, evolutionarily adapted to life among sediment particles or on surfaces, need to slide on something, stick to something, or grow, or rely on something, building their colonies (twine with ribbons around the supports, grow tree-like colonies from the substrate, form tubes on a solid foundation).

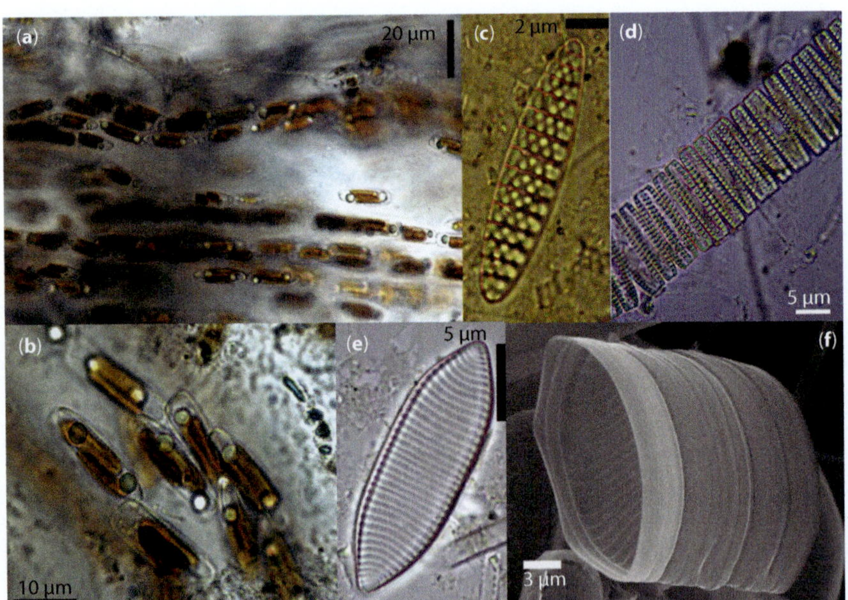

Figure 3.10 (a, b) Colonial form of naviculoids (*Navicula smithii*), a colony living in biopolymer tubes. (c–f) Colonial forms of nitzschioids: ribbon-like colonies (d, f) and separate valves (c, e) of different species from the genus *Fragilariopsis*. (Images by Ph. V. Sapozhnikov, O.Y. Kalinina, SIO RAS).

Of course, they are found in the water column on a permanent basis, and not after a storm temporarily lifts them from the bottom. However, in the pelagic zone they live in "aggregates"— clots of particles floating in water, built up with the participation of fragments of dead planktonic creatures (and/or microplastic particles), colonies of bacteria and microalgae, including these very diatoms.

On the contrary, it is much more convenient for planktonic life forms of diatoms to cling to water with their thin outgrowths, or to soar in streams in suspension. Their shells, as a rule, are much more fragile and delicate than those of the bottom forms. When they fall to the bottom, precipitating from the pelagic zone, they rarely live there for a long time; they are ground by the smallest particles of mineral sediment, or buried under themselves, deposited on them from above. If they get stuck among algae, where solid particles rush less, they can live much longer, successfully multiplying.

Ribbon-like colonies of *Fragilariopsis* species mainly inhabit the pores of sea ice, getting from there into sub-ice water, and also travel as part of "aggregates." Note that the shells of large planktonic diatoms that have fallen to the bottom often inhabit small benthic species, using them as a convenient porous substrate.

Diatoms often build monogroup communities, or taxocenes. In their composition, species with different life forms harmoniously divide the space of biotopes among themselves, since they are distributed over different ecological niches. The design of their shells also plays an important role in this.

3.4 Mechanical Performance of Diatom Frustules – Experimental Characterization

Diatoms are among the most diverse groups of microalgae detectable in all aquatic environments and moist terrestrial habitats. They play an immense role in keeping the planet's ecosystem working by producing 20% of the total oxygen and 40% of the annual carbon in the world's oceans [3.27] [3.28]. The number of extant species of diatoms is estimated to be between 30,000 and 100,000, which makes them the largest group of eukaryotic algae [3.29]. The environmental factors, including temperature, light, salinity, pH, nutrients, and other habitat characteristics, strongly affect the composition and diversity of diatom communities [3.30]–[3.32]. However, some studies showed that diatom community structure and diversity are also influenced by geographical variables (e.g., connection between lakes and the amount of isolation of lake districts) separately from other environmental conditions [3.33] [3.34].

According to the taxonomy used by Algaebase.org [Guiry, M.D., Guiry, G.M., 2021. AlgaeBase World-wide electronic publication, National University of Ireland, Galway searched on 06 October 2021], the diatoms (phylum Bacillariophyta) include the centric species with multiradial and bipolar symmetry (classes Coscinodiscophyceae and Mediophyceae), and pennate species with bilateral symmetry (class Bacillariophyceae). The latter includes the subclasses Fragilariophycidae and Urneidophycidae (without raphe on the valve) and Bacillariophycidae (with raphe). They appear as individual cells or colonies in different shapes (filaments, chains, ribbons, fans, zigzags, or stellate), either as free-floating (planktonic) or attached to the substrate (benthic or periphytic) [3.35]. These unicellular organisms can be distinguished by their hierarchically structured 3D porous cell wall called frustule, made of amorphous silica [3.36]. The structure and main characteristics of the cell wall of diatoms are shown in Figure 3.11 for the example of the pennate diatom *Achnanthidium sibiricum.*

The cytoplasmic content is enclosed between two thecae or valves: the upper epitheca and the lower hypotheca, connected by a series of siliceous rings called girdle bands [3.38] [3.39]. They are developed inside a membrane-bound compartment called silica deposition vesicle (SDV) [3.37] [3.38]. Among the structural elements of the frustule of *Coscinodiscus* sp. we can distinguish: foramen, cribrum, and cribellum [3.40]. Most pennate diatoms contain one or two longitudinal slits or fissures in the valve, called the raphe, which explains their ability to adhere to the substrate and move along the surface. A complex network of thickened ribs (costae) and striae, represented by a row of pores termed areolae, is formed by the precipitation of silica precursors generated from monosilicic acid $Si(OH)_4$ [3.41]. Frustules of various sizes (2-2000 microns) are penetrated with pores located more or less periodically (on a scale from microns to nanometers). The diameter of fine perforations ranges from 3 nm to 50 nm [3.42].

The pores facilitate mineral and organic nutrition of cells, their respiration, waste removal and mucus secretion, and thus ensure the access of diatom cells to the surrounding environment. Besides, the frustule protects the cytoplasm from parasites and DNA from UV radiation, increases nutrient diffusion and absorption through the porous systems, facilitates light-harvesting and focusing, as well as sinking [3.43]–[3.45]. Frustules are enveloped by large quantities of extracellular polymeric substances (EPS), mostly polysaccharides, which are essential for sessile adhesion, motility, protection against desiccation, and the formation of biofilms and colonies [3.42] [3.44]. Due to a large number of pores of different sizes, the frustules have a high specific surface area, high adsorption, and porosity, which allows them to be used in many areas. Other important features include biocompatibility, low toxicity

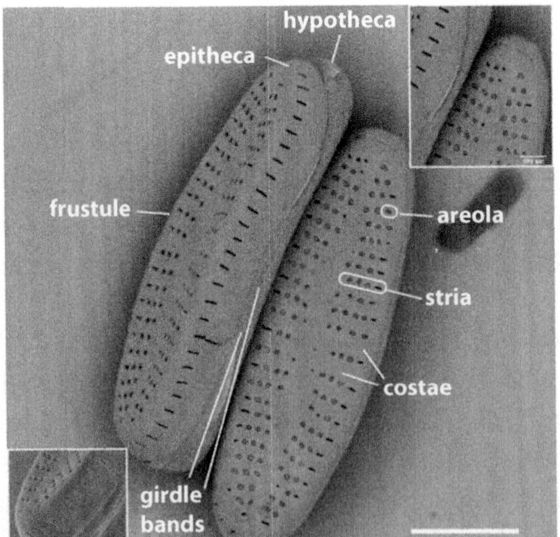

Figure 3.11 SEM images of *Achnanthidium sibiricum* diatoms showing the basic structure of the cell wall (frustules). Insets: close-up of individual sections of the frustules (scale bar: 500 nm). (Reproduced with permission from [3.37].)

of bare diatoms up to the level of 1000 μg/mL, easy modification of their frustules, their thermal and mechanical stability, chemical inertness, light focusing and luminescence, and low cost of cultivation [3.46]. These properties make diatoms a valuable resource for applications in various modern technologies for the production of biosensors, gas sensors, catalysts, adsorbents, filters, semiconductors, solar cells, a platform for smart nanofabrication, drug carriers, optoelectronic devices, etc. [3.47]–[3.54].

Given the fact that diatoms are not only a target for predators but are also exposed to abrasive particles in water, the possibility of using biosilica shells as protective armor has been discussed and analyzed in numerous scientific studies. The characterization of mechanical properties is critical to a better understanding of the organic and inorganic processes involved in silica biomineralization and the role of frustule geometry and its hierarchical 3D structure in the evolutionary design of diatoms. Different experimental techniques, such as nanoindentation [3.55], AFM probing [3.54] [3.56] [3.57], and bending by glass micro-needle [3.58], were implemented to investigate modulus, hardness, and failure strength of different species of diatoms. However, because of the challenges related to the application of force and detection at the appropriate scale, micromechanical properties of frustules were studied to a lesser extent.

The unique mechanical properties of diatoms open a wide range of their applications and attract the special attention of scientists in numerous investigations. In this section, the mechanical characteristics of biosilica of the most widely studied diatom species like *Coscinodiscus*, *Didymosphenia*, and *Synedra* (many of which are currently regarded as *Ulnaria*) will be briefly described.

Modern technological progress allows better understanding and studying laws of the surrounding environment at the nanoscale level. In addition to AFM and nanoindentation being the most widely used techniques, the cutting-edge equipment provides *in-situ* measurements that usually involves a micro/nanomechanical device inside a spectroscopy or imaging technique. Such modification possesses high accuracy data collection and provides synchronized high-resolution visualization during the experiment, while the post-processing allows performing multimodal correlations between the structure and measured characteristic.

3.4.1 Nanoindentation Testing of Diatom Frustules

The key characteristics of the indicated methods are Young's modulus and hardness. However, before conducting the mechanical test on diatom frustules, it is essential to select a controlled cleaning process that removes organic biomaterial inside the frustule and saves its sufficient strength with structural integrity. The removal of various compounds from the diatom cell must consider the nature of every substance starting from easily extractable carbohydrates to silaffins incorporated in biosilica. However, vigorous cleaning can modify the pore structure and thereby influence the structural properties. For this reason, the most common protocol to perform this procedure is based on the chemical treatments by using either oxidizing agents like hydrogen peroxide or the combination of surfactants with complexing substances.

Roman *et al.* [3.59] studied the effects of chemical treatment on organic material removal and mechanical properties in cleaned diatom frustules. They chose two various species of the diatom genus *Coscinodiscus* with the size ~150 μm in diameter collected from the

Trondheim Fjord in the Norwegian Sea and next cultivated monocultures using standard laboratory conditions in cell cultivation flasks. First, grown cells were washed multiple times and centrifuged between each washing and dried overnight at 60 °C. Afterwards, they were cleaned according to the two different methods. The first method, called "hard cleaning," involved stirring of dried material into H_2O_2 (10 mL, 30%) at 90 °C for 24 hours followed by the addition of HCl (1 mL, 37%) solution and centrifugation for 10 minutes. The second one, called "soft cleaning," included a 1-minute vortex of the dried material with the dispersion of sodium dodecyl sulfate (SDS)/ethylenediaminetetraacetic acid (EDTA) solution (50 g/L SDS in 100 mM Titriplex III) and their centrifugation for 10 minutes. After repeating six times, samples were stored in ethanol (96%) at −20 °C. The fluorescence microscopy and energy dispersive spectroscopy analysis showed a clear difference of residual organic into diatom frustule depending on the cleaning process, with little organic material left after the oxidizing method. Moreover, this approach induced a partial cracking of the frustules that was proved via the nanoindentation technique. To summarize, the used cleaning protocol correlates with the mechanics of diatom frustules, thereby showing the importance of paying special attention to sample preparation.

In their study, Subhash *et al.* [3.64] detected the strong dependence of mechanical properties on the test location. They used *Coscinodiscus* species with average dimensions of about 150×150 μm that were cleaned with the Brooks protocol before nanoindentation. A series of nanoindentation tests were carried out from the center to the edge of the frustule. The results postprocessing showed the variations of both elastic moduli from 0.591 to 2.768 GPa (center)/0.347 to 2.446 GPa (edge) and hardness from 0.033 and 0.116 GPa (center)/0.076 and 0.120 GPa (edge), respectively. To investigate the fracture modes, they increased the indentation depth to 400-500 nm and found that they were noticeably different in each orientation. Cracks initiate and propagate at approximately 120°, which corresponds to the angles between the three sharp edges of the Berkovich indenter. In a convex orientation, additional ring cracks are followed by the radial cracks, whereas in concave orientation, only cracks along the sharp edges could be observed. The obtained inconsistency was connected with the frustule porosity and curved structure. However, the nature of the nanoindenter made it impossible to determine the position of the indents accurately.

Another experiment was conducted on the *Coscinodiscus* sp. that were tested *in situ* by three-point bending. Aitken *et al.* [3.60] analyzed the bending of the beam with a square cross section and obtained a brittle fracture of the diatom frustule at the average stress of 1.1 GPa and strain 3.0% with the elastic modulus of 36.4 ± 8.3 GPa. Moreover, they estimated the relative frustule density near 30% and got the highest strength-to-density ratio, approximately 1700 N·m/kg among all known biologic materials, as provided by Ashby plot in Figure 3.12. The authors proved the given results via nanoindentation of the girdle band of the frustule with modulus from 21.0 ± 7.7 GPa at 28 nm to 39.7 ± 8.3 GPa at 217 nm, which corresponds with the mechanics of the bending test. To sum up, these results clearly showed how nature could build live organisms with hierarchical architecture simultaneously with the record lightweight and strength.

A summary of the information about nanoindentation and bending tests of different species of diatoms with the obtained values of Young's modulus and hardness is shown in Table 3.1.

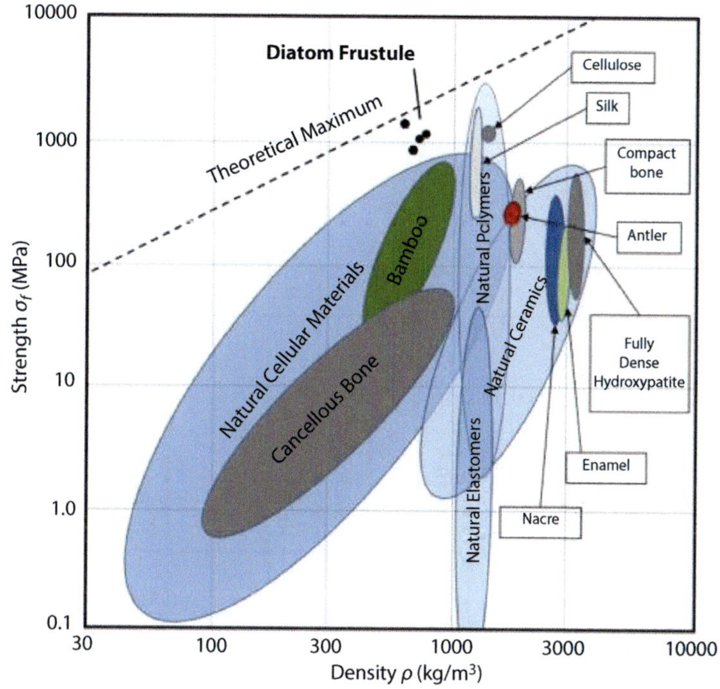

Figure 3.12 Ashby plot of strength vs. density for naturally occurring biological materials. (Reproduced with permission from [3.60].)

3.4.2 AFM Studies of Diatom Frustules

The first AFM studies related to the mechanical behavior of diatom silica nanostructure were done in 2001 on dried diatoms by Almqvist *et al.* [3.57]. They measured elasticity and hardness of the diatom *Navicula pelliculosa* semi-quantitatively with a resolution down to approximately 10 nm and found that Young's modulus varied from 7 GPa to hundreds of GPa, depending on the location. At the same time, hardness values ranged from 1 GPa to 12 GPa. The elastic modulus was calculated from the AFM force curves using indentation and deflection methods. Even though both methods revealed the same overall elastic pattern, differences in absolute elastic modulus observed on the hardest part of the frustules could be explained by the calibration of the AFM cantilever. Besides, it was assumed that the elastic response is due to the deformation of the shell material, which is valid only at the central nodule (CN) and central rib (CR) where the modulus is high, but not on the transapical rib (TR), where the shell gets elastically deformed.

On the other hand, hardness, which is directly related to the material's structure, was also higher on the CN and at the curved-end side of the raphe fissure (RF). These results confirmed that micromechanical characteristics are not restricted only to the biosilica itself, but are also influenced by the hierarchical three-dimensional morphology of the porous structures of the exoskeleton on scales from nanometers to micrometers. Also, the change in mechanical properties can potentially be explained by differences in silica content and phase of development of a frustule. This study firstly confirmed that the micromechanical properties could be studied using AFM method. However, the authors suggested that

Table 3.1 Summary of various experiments of diatom algae by nanoindentation and bending tests.

Genus	Test type	Specimen location	Stress or load	Strain or penetration depth	Young's Modulus, GPa	Hardness, GPa	Article
Coscinodiscus	A* B**	– girdle	1.1 ± 0.3 GPa – –	3.5 ± 0.7 % 28 nm 217 nm	36.4 ± 8.3 21.0 ± 7.7 39.7 ± 8.3	– – –	[3.60]
Coscinodiscus	B	center	0.01–0.03 mN	60–100 nm	1.1–10.6	0.10–1.03	[3.61]
Synedra	B	center	0.01 mN	20 nm	13.7–18.6	0.85–1.41	
Coscinodiscus	A	–	2.81 ± 0.15 mN	2 μm	36.4 ± 8.3	–	[3.62]
Coscinodiscus	B	center	0.02–0.07 mN	300–500 nm	0.591–2.768	0.033–0.116	[3.55]
	B	edge	0.02–0.07 mN	300–500 nm	0.347–2.446	0.076–0.120	
Didymosphenia	B	–	100–1400 μN	1–6 μm	31.8	–	[3.63]
Coscinodiscus	B	center	0.12 mN	200 nm	1.562	–	[3.64]

*A = three-point bending
**B = nanoindentation

in order to increase knowledge of silica biomineralization, the instrumentation should be improved along with the comparison of the properties of several species under physiological conditions and the study of the mucilaginous organic coating. The AFM technique was also readily used to examine the diatom cells in the native and hydrated state [3.65]–[3.68].

Later, Losic *et al.* [3.56] mapped the nanoscale mechanical properties of *Coscinodiscus* sp. at different locations along the frustule using an AFM nanoindenter and calculated the hardness and elastic modulus from force penetration curves. Figure 3.13 shows SEM and

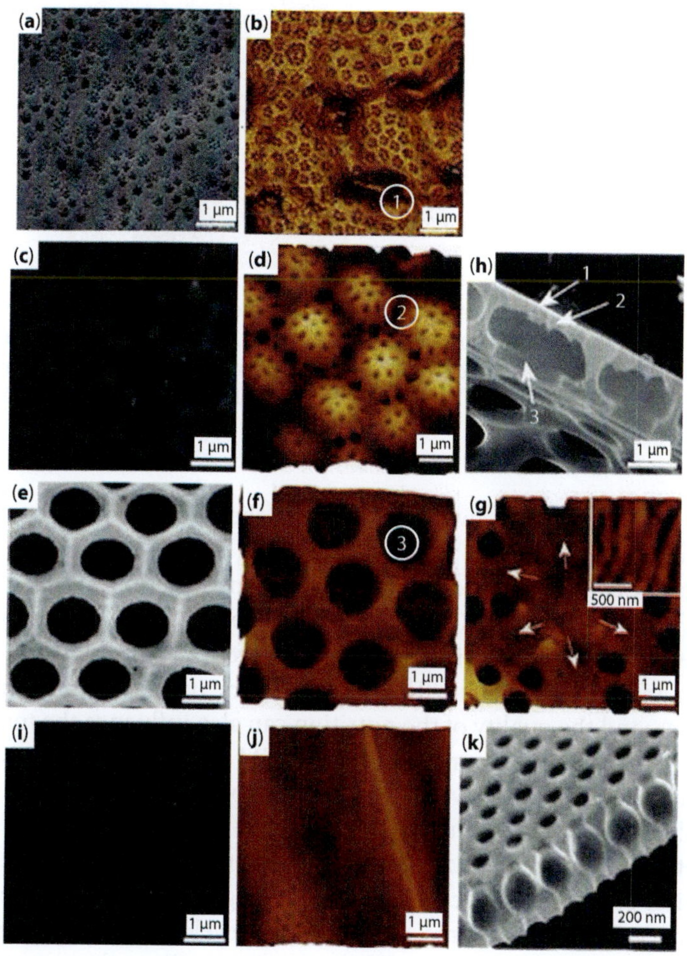

Figure 3.13 SEM and AFM structural characterization of porous layers of the silica frustule of *Coscinodiscus* sp. (a) SEM and (b) AFM images of the cribellum surface (z scale 500 nm). (c) SEM and (d) AFM images of the cribrum surface (z scale, 500 nm). (e) SEM and (f) AFM images of the internal plate (z scale, 700 nm). (g) AFM image of the central part of the internal surface (z scale, 700 nm) with inset showing channels and ribs. (h) SEM image of a cross section of the diatom frustule showing the internal shape of pores connecting cribellum and cribrum. (i) SEM and (j) AFM images of girdle band area of surfactant-cleaned diatom in side-on position (z scale, 200 nm) and (k) SEM image of a cross section of the girdle band. Numbers within white circles denote the three porous layers of the diatom frustule: (1) cribellum, (2) cribrum, and (3) internal plate. Cross-section image (h) is shown before images of the internal layer (e, f, g). (Reproduced with permission from [3.69].)

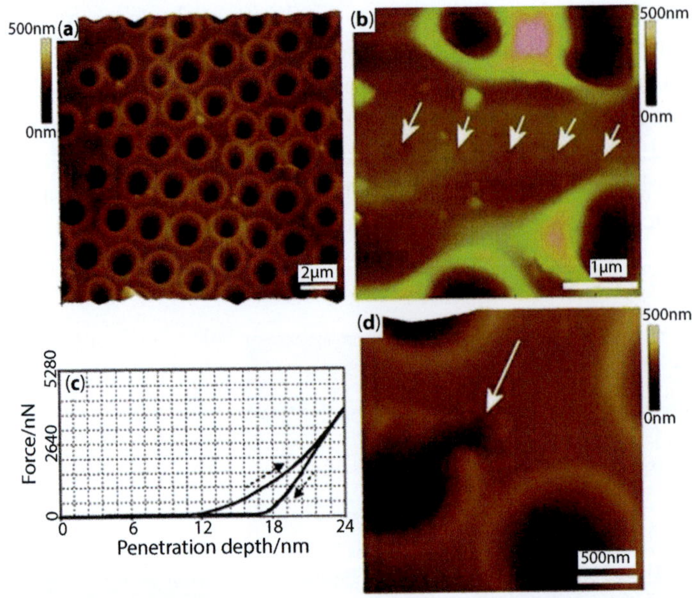

Figure 3.14 (a) Low-resolution AFM image of the internal plate after nanoindentation. (b) Series of indentation points (solid arrows) obtained using a force of 1.5-3 μN. (c) Typical force penetration curve showing loading and unloading parts (dashed arrows). (d) AFM image showing nanoindentation performed on an areola rib at higher applied load (12 μN). White scale bar is 10 μm. (Reproduced with permission from [3.69].)

AFM images of *Coscinodiscus* sp., which summarize the structure and organization of the porous silica layers.

The AFM nanoindentation tests confirmed significant variations of the elastic modulus *E* (1.7-15.6 GPa) and hardness *H* (0.076-0.53 GPa) depending on frustule morphological features as pore size, porosity, and distance between pores. The lowest values were obtained on the cribellum and the girdle band. The location of the indentation, different biomineralization processes, and the quality of the nutrition can also influence the micromechanical properties of diatom frustules. Figure 3.14 shows AFM images of the internal plate with a series of indentations performed at various positions from the central region to the periphery and a typical force penetration curve [3.69].

The contribution of the shape, size, pore distribution and geometry of the cell wall to general mechanical properties that can provide meaningful insights into the morphogenesis and evolutionary design of diatoms, as well as the development of biologically inspired hierarchically structured materials, are still being explored. Given the complexity, diversity, and range of sizes of biosilica shells, predicting the mechanics of diatom-inspired structures is quite challenging, and further research is needed to address all of these issues thoroughly.

3.5 Engineering Applications of Diatomaceous Earth

Behind their fascinating formation process and neatly formed nanoporous structure, diatoms are known to possess interesting mechanical properties [3.70]. The mechanical properties of diatom silica have been studied in several reports [3.71]–[3.74]. Indeed, such interesting mechanical properties observed are highly dependent on the diatom framework, with the valves and girdle bands reported to form a pill boxlike protection for the plasma membrane against the severe environmental conditions at their exposure [3.70]. In one such early study based on understanding the mechanical properties, Hamm *et al.* [3.58] observed an inverse relationship between the frustule size and the mechanical strength after measuring the force required to break a single cell. In their study [3.58], a range of allowable mechanical stresses (150 to 680 Nmm^{-2}) were observed, depending on the region of the frustule, and these variations were attributed to the frustule framework, in particular the presence of ribs and pores which are responsible for dispersing mechanical stresses. As far as the diatom framework is concerned, significant variations of mechanical properties are mostly observed, which is further evidenced in the reports in [3.55] [3.57]. In these reports, the elastic modulus, fracture mode and hardness of various diatom species were studied at the nanometer scale using atomic force microscopy (AFM). In both studies, a range of hardness (0.03-0.11 GPa) and elastic modulus (0.6-2.8 GPa) were measured across the frustule and these variations were attributed to the inhomogeneous porosity and silica content. In a similar study [3.69] performed on the diatom *Coscinodiscus* sp., lower limits of hardness (0.06 GPa) and elastic modulus (3.5 GPa) which are comparable to that of porous silicon (0.04 and 8 GPa) bearing similar pore size were observed.

Another interesting area of diatom studies is their self-lubricating response to the constant friction and wear taking place between various connecting parts of the frustule. This is highly significant in *Ellerbeckia arenaria*, where frequent changes in length by a factor of two occur via a spring-load mechanism [3.71]. From zipper-like features to bead-like features found at the edges of the girdle bands, which reportedly act as ball bearings and solid lubricants, diatoms indeed showcase interesting self-lubricating abilities. Even though such detailed features provide some ideas about this mastery, it remains an unknown area of diatom research due to unexplained roles of the other hundreds of nanostructures contributing to their lubrication strategy [3.68] [3.71]. Apart from their self-lubricating abilities, diatoms exhibit another wonder through their self-healing mechanism [3.72]. It is known that diatoms secrete strong adhesive exopolymers which prevents the dissolution of their silica content in fresh and sea water, as both systems are undersaturated with silica. This self-healing ability of diatom silica could inform steps in solving the problem of deterioration faced by glass-fiber-reinforced polymers in moisturous environments [3.72]. This mechanism was experimentally observed through an AFM study [3.73], where live *Navicula* sp. were brought into contact with hydrophobic Intersleek and hydrophilic mica surfaces. They observed comparable cell-adhesion strengths on the surfaces, showing diatoms secrete adhesives which are responsive to both surfaces. With continued attention towards biomimicry [3.74] [3.75], understanding how diatoms exhibit such fascinating mechanical properties could inspire a new generation of materials and solutions; lightweight structures with optimum mechanical properties, self-healing structural materials and self-lubricating materials. Additionally, the use of diatom silica as a mechanical support could require less energy input and materials unlike equally applicable biomaterials like calcite [3.76].

In some practical studies [3.77]–[3.84], the potential application of diatom silica as a mechanical support in composites and hydrogels and soil amendment have been reported. For some time now, diatom silica has been studied for its role in soil amendments

[3.81–3.83]. In the study by Aksakal *et al.* [3.77], the addition of diatom silica led to an improvement of the soil's mechanical strength by altering several soil parameters. In their work, they reported an increase in the liquid limit (LL), plastic limit (PL) and the optimum moisture content (OMC) after diatom silica addition, which was mainly attributed to its high surface area. On the contrary, the maximum dry bulk density (MBD) of the soil decreased after diatom silica addition. In effect, with the improved mechanical strength of the soil, it becomes increasingly easy to till it. Diatom silica has been studied as a reinforcement for resin composites for dental restoration [3.78]. Over the years, resin composites have attracted more attention over their counterparts like conventional dental amalgam for various reasons [3.78]. The resin composites boast of excellent mechanical properties, manipulative qualities, biocompatibility, and aesthetic appeal [3.84]–[3.86], unlike the poor aesthetic appeal, toxic, corrosive and pollution related problems often associated with the use of conventional dental amalgam [3.87] [3.88]. Although resin composites are superior to conventional dental amalgam, there is a currently existing challenge with the addition of the inorganic fillers due to the high viscosity of Bis-GMA (up to 1200 Pa·s) [3.78]. Without the inorganic filler, the resin composite exhibits poor mechanical properties coupled with a high polymerization shrinkage during curing, which can eventually result in secondary caries. This setback has recently been tackled with nano-inorganic particles [3.89]–[3.91]; nevertheless, there are still setbacks such as lower content of solid filler. Wang *et al.* [3.78] employed diatom silica as a porous filler in their studies following earlier studies and the suggestion that porous fillers could provide better results [3.92] [3.93]. In their work, they observed an increase in the mechanical properties of the diatom silica-based resin composite compared to the composite without diatom silica. In a recent study, Lv *et al.* [3.79] demonstrated the ability of diatom silica to improve the mechanical properties of poly(acrylamide) hydrogel. They reported an increase in the both tensile (twice) and compressive (13-fold) strengths of the hydrogel after diatom silica addition. They attributed the improved mechanical properties to the presence of more physical crosslinked joints created from the hydrogen bonding existing between –SiOH and $-CONH_2$. Recently, Mustafov *et al.* [3.80] showed the ability of diatom silica to improve the morphology and mechanical properties of a polyurethane foam for biomedical applications. In their study, they observed a significant increase in compressive strength after the addition of more than 1% diatom silica. These studies demonstrate the potentially critical role of diatom silica as a mechanical support for novel nanocomposite. Nevertheless, increasing the odds of success requires further understanding of how the mechanical properties of fresh diatom silica differ from pretreated diatom silica, currently unaddressed in existing reports.

The unique morphology of diatom frustules is considered a driving factor for the application of diatomaceous earth. Current studies showed the prospects of this material in agricultural, engineering, construction and biomedical areas. Due to the significant presence of silica starting from 80%, the diatoms are considered an abundant precursor for the production of silicon or silicate compounds. As a result of numerous rock deposits of fossilized frustules, the diatomite or diatomaceous earth is a widely available natural resource consisting of a vast diatom species. Being a silica-bearing material, the diatomite earth firstly focuses on microelectronic production that needs an accessible source for the synthesis of silicon and silicate formed zeolite. The role of diatomite is to decrease the cost of fabrication affecting different industries, such as Li-ion batteries, by suggesting the novel diatomite-derived nano-silicon anode [3.94]. In this approach, the author mentioned the technology of magnesiothermic reduction of diatomite earth as a lower thermal energy required method to convert the silica-based frustules to nano silicon (nanoSi) than the more high-temperature reduction with the help of carbon. Hence, it establishes the potential of diatomite earth as a low-cost precursor for the improved technology of anode materials [3.95].

The zeolite could potentially be obtained from the diatomite earth for the chemical production industry. One of the disadvantages is the presence of impurities in the diatomite ores that should be removed by chemical treatment and pre-calcination to improve physicochemical properties. The study indicated the efficiency of the rock diatomite mined from Egypt. The rock diatomite was used for the hydrothermal synthesis of mesoporous zeolite after pretreatment with aqua regia [3.96]. Moreover, the diatomite could serve as a precursor in the microwave heating synthesis to form sodalite zeolite as the catalyst for electrochemical studies [3.97]. The authors mentioned the advantages of the microwave method being eco-friendly with a shorter time required for synthesis than the hydrothermal one. The ordered porosity of zeolite is of interest for the production of ion filters, absorbent and catalysis. In the study presented, the diatomite frustules retain the pristine hierarchical mesoporosity after the various syntheses. Thus, diatomite is known to take the place of a source of silica for industrial synthesis.

Diatom fossils have captured attention for saving crops from insects [3.98]. The broken parts of diatomite frustules adhere to and induce critical dehydration of insect cuticle. This property of investigated diatomite has resulted in their application in the agricultural area. In the studies dedicated to agriculture, the diatomite caused the dehydration and damage of the insect body within a short period of treatment time, leading to the death of the weevils studied [3.99] [3.100]. SEM photomicrographs showed the particles of diatomite absorbed in the wax and the damage of cuticle morphology of the granary weevil [3.99]. Because the exoskeleton of insects facilitates the water absorbance from the ambient medium, the authors studied the effect that diatomite has in changing the lipids and water from the cuticle surface of the exoskeleton.

These researches showed the potential for the diatomite obtained from different regions as an insecticidal remedy for protective measures in agriculture. The results established the dependence of efficiency between such parameters of the morphology of diatoms as the mesoporous structure and high amorphous silica content. The high surface area of mesoporous frustules repeatedly characterized diatomite as a powerful absorbent that facilitates the removal of fatty acids and water from the insect cuticle [3.100].

Hence, diatomite is a highly porous material that uncovers the vast application for engineering solutions. There are strong reasons for using diatomaceous earth to produce filters, such as complex pore structure. In addition to the advantages of the material associated with a hierarchical pore system, the material is resistant to strong chemicals and has a large surface area. Besides, particularly due to its porosity, diatomite has good adsorption properties, especially for oils, is chemically inert for most substances, and has very low thermal conductivity.

Lo *et al.* [3.101] delivered the contribution of freeze-drying of diatomite-based suspensions to fabricate the separation membrane for oil and water. The resulting membrane was superhydrophilic and superoleophobic in the air due to the synergistic effect of diatomite. While submerged in water, the porous structure captured a lot of water in the intricate pores, forming a stable oil/water interface in the presence of oil. That proved the superoleophobicity of the membrane to various types of oils. Hence, the work results revealed the high selectivity for oil and water that provided more than 99.6% separation efficiency for free mixtures of oil and water driven solely by gravity. The water permeability of the obtained membrane was one to two orders of magnitude higher than the permeability for those synthesized by traditional methods.

Diatomite is considered as a material for filtration due to its high surface area of hierarchical porosity. The functions of filters include the separation of particles and substances, and absorption of the separated solutions. Diatomite was mentioned in the series of low-cost absorbent materials to clean wastewater. In the paper of Zhu *et al.* [3.102], an electrospun filter with superfine diatomite was suggested as a dye absorbent. The aim of dye removal was developed in the study dedicated to the synthesis of the porous composite derived from the diatomite. The removal of methylene blue (MB) was investigated for further application in the field of adsorption and photocatalysis [3.103].

Taking into account such properties of diatomite as thermal resistance and hierarchical porosity, this material is widely used to produce construction materials for thermal insulators. It is worth mentioning a study on the mechanical and physical properties of diatomite composites with different fraction character in which the authors tried to build a lightweight thermal insulative material for furnaces [3.104]. The study indicated the more suitable composition of diatomite up to 90% with sintering at 900 °C. The observed conditions produced a material with the lowest thermal conductivity coefficient among the studied composites of 0.32 W/mK. It was affirmed that with the increase of diatomite, the thermal conductivity and compressive strength decreased, which was caused by the increase of porosity and weight ratio of diatomite.

In the case of the rise of the firing temperatures up to 1100 °C, the compression strength started increasing due to the phase transitions of silica-bearing diatomite. The calcination of diatomite (1000 °C for 2 h) induces the high crystalline phase formation of silica polymorphs (cristobalite) [3.105]. The differential thermal analysis detected the phase transition of calcined diatomite to the cristobalite to demonstrate endogenous reaction after heating up to 980 °C. The specimens were obtained by replacing the Clinker with calcined diatomite at various rates (5%, 10% and 20%) by weight. The maximal value of compression strength showed the composite with 10% of diatomite addition after 28 days of the experiment (49.2 MPa).

Despite the use of diatomaceous earth for various syntheses, there are other potentials. Diatomite was applied with the aim to improve the mechanical properties of concrete. The proven compatibility of diatomite in the pozzolanic reaction made it applicable to build concrete compositions. In the following studies, the diatomite was a potential ingredient to partially substitute the cement [3.106] [3.107]. After the addition of diatomite (up to 10% to the weight of cement), the compressive strength of concrete was increased to 32 MPa. Due to the diatomite properties, such as microfilters, the reduced microporosity and water absorption of the modified concrete had an impact on the mechanical properties.

Moreover, diatomite is a scope of interest for biomedicine due to its proven bioactive properties. There is ongoing research of scaffolds for guided tissue regeneration [3.108] [3.109]. Diatomite serves as the bone-inducing component used in the bone tissue engineering approach [3.110]. As a source of silica, the diatomite could be mixed with the hydroxyapatite or the other biocompatible components of biomaterials. The addition of silica provides bone cell adhesion and bone tissue formation by increasing the bioactive properties of composites.

3.6 NEMS/MEMS Perspective

Successful fabrication of nanostructured silicon structures reduced from diatomaceous silica earth and firmly integrated into silicon wafers shows that neat nanostructured objects obtained in the framework of the BANSAM paradigm [3.53] [3.111] can be readily applied as elements in smart electronic chip technologies. Since diatoms are able to colonize silicon monocrystalline wafers [3.36], a simple technique of guided, highly precise deposition of single diatom cells might be developed for mass production of NEMS and MEMS containing diatom frustules (siliceous or silicon after reduction) as an integral part of a sensor, such as a microphone, or an energy harvesting element.

The description of vibration behavior for a single openwork architected element is a relatively simple task for modern computational math. Numerous FEA software packages are available on the market such as Ansys, COMSOL and many others. These software tools are capable of predicting main modes of vibrations and eigenfrequency spectra processing 3D-meshed geometrical models of diatom frustules fixed to the infinitely thick substrates in a manner (boundary conditions). Figure 3.15 exemplifies typical results of FEA analysis of diatom frustule vibrational patterns [3.112].

Calculated values of eigenfrequencies of *Ulnaria acus* and *Coscinodiscus* sp. are as high as 7...17 MHz, which is undoubtedly very attractive for NEMS and MEMS elements and microelectronics in general. It is important to note that elastic modulus of amorphous hydrated silica (22.4 GPa) was input for calculations. One can suggest that the reduction of silica to silicon will increase eigenfrequencies a few times due to the dimensional thinning

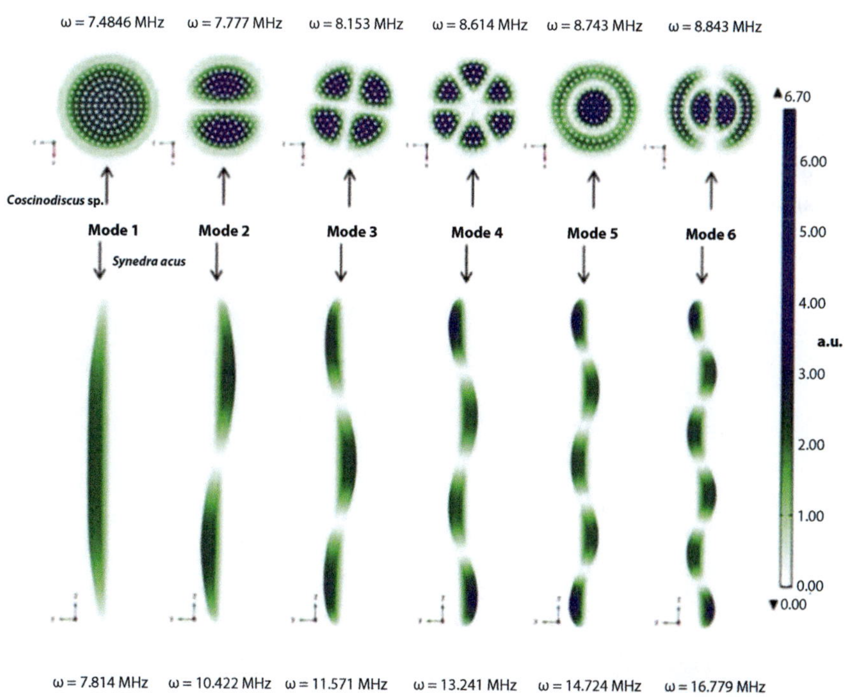

Figure 3.15 Mode shapes and natural frequencies of *Coscinodiscus* sp. (centric) and *Ulnaria acus* (pennate) observed in the simulation using COMSOL Multiphysics. (Reproduced with permission from [3.112].)

and material stiffening (Young's modulus of silicon is 140…180 GPa). The values of calculated eigenfrequencies for the geometrical models of diatoms are in good agreement with the data of photoacoustics measurements, which we recently conducted for several species of live diatoms [3.37] [3.53]. It is believed that live diatoms absorb the energy of laser pulses in chlorophyll molecules and convert some part of absorbed energy to a wide spectrum of vibrations resonating with frustules' vibrations at eigenfrequencies.

The drawback of this computational math applied to this aspect of diatom science is quite obvious—simple 3D geometrical models of frustules capture the rough details of frustules neat architecture only. Two approaches can be put forward to deal with this drawback:

- Parametrization of a simplified analytical solution adopted from continuum mechanics of thin shells, as demonstrated in the paper by Abdusatorov *et al.* [3.112].
- The development of destructive (through FIB-SEM milling – Figure 3.16) or non-destructive tomography (e.g., using ptychography with the help of synchrotron coherent beam – Figure 3.17).

Parameterization techniques are relatively simple since they require the calibration and fitting of a few parameters for the series of vibration modes, the i-th eigenfrequency of which obeys the formula:

$$\omega_i \propto \sqrt{\frac{E}{\rho}\frac{1}{D}\left[a_i\frac{t}{D}+b_i\right]P_i\left(\frac{d}{D}\right)} \tag{3.1}$$

where E, ρ, D, d and t are Young's modulus, density, diatom principal size (outer diameter for centric and longest size for pennate diatoms), pore diameter and frustule thickness, respectively.

Parameters a_i and b_i as well as morphology factor P_i are to be fitted and experimental measurements are vital in this case in order to retrieve the values of parameters from the correlation analysis.

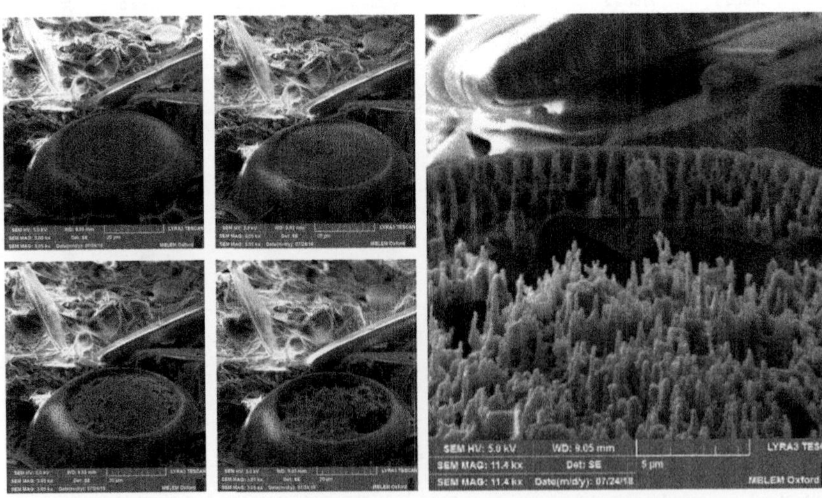

Figure 3.16 The example of FIB-SEM milling of diatom species (destructive method).

Figure 3.17 Ptychography of *Aulacoseira* diatom cells. (Images courtesy of T.E.J. Moxham, University of Oxford/Diamond Light Source, U.K.)

The 3D reconstruction of frustule architecture is a much more complex approach promising, on the other hand, a more valuable outcome—a robust 3D model of frustule for the computations of many particular properties such as mechanical (static and vibrational for MEMS), optical (light confinement, focusing and diffraction) and hydraulic (fine filtering).

3.7 On the Mathematical Description of Self-Organized Diatom Frustule Growth

Diatom algae display a diverse variety of morphologies. Taxonomists report more than 100000 different species distinguished by the frustule morphology [3.113].

Not only the shape of the whole frustule may vary between different species, ranging from circular to thinly elongated and razor-shaped, but also the morphology of the pores. Different species of diatoms exhibit different shapes and positions of micropores and pore occlusions (thin walls dividing microscale pores into nanoscale sections and channels) (Figure 3.18).

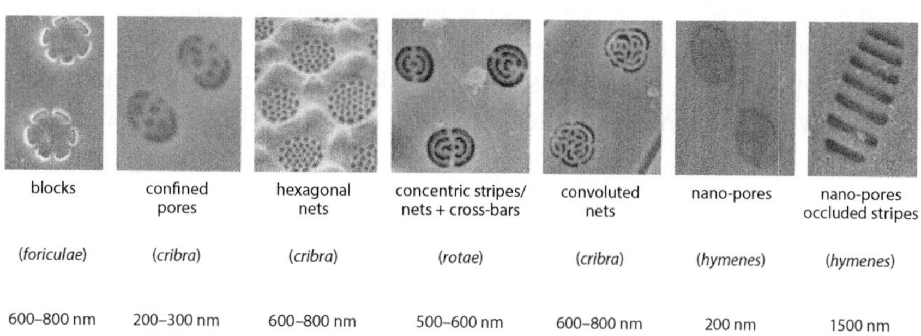

Figure 3.18 A classification scheme for pore occlusions, provided with Latin descriptive terminology. Listed at the bottom are lengths of the major axes of the pores. The pores in the first five panels from the left are characterized with wavelengths of 60–100 nm. For panel six, nanopores have a wavelength of 10 nm, and for panel seven one of 220 nm. (Reproduced from [3.114].)

The morphologies of pore occlusions remain consistent within the same species. The pattern of the occlusions is repeated when a cell undergoes mitosis, and daughter cells are left with a single inherited valve and rebuild a second matching valve within the silica deposition vesicle (SDV).

The genesis of valves along with the pore occlusion structure is governed by DNA-encoded mechanisms [3.38]. However, it is worth noting that very similar shapes for pore occlusions can be obtained via stochastic mathematical models. Willis *et al.* [3.42] used a stochastic, order-parameter model of interface movement that included a mean-field analysis. It could be seen as an extension of the threshold voter model [3.115] applied in social behavior and ecology [3.116].

The model, called laterally dependent stochastic aggregation (LDSA), is a continuous-time Markov chain that operates with white and black cells on a discrete square grid lattice. At any given moment, the cells that have at least one neighbor of the opposite color may flip or not flip their color depending on the parameters of the simulation and the contents of the short- and long-range environment. The probabilities of an x cell changing the color from black to white $(B \to W)$ and from white to black $(W \to B)$ at the time increment Δt are defined as

$$x \in \partial W : B \overset{U_x}{\longrightarrow} W \text{ @ probability } \mathrm{p} = \sqrt{K} e^{\lambda w_{L_x} + \sigma w S_x} \Delta t \tag{3.2}$$

$$x \in \partial B : W \overset{U_x}{\longrightarrow} B \text{ @ probability } \mathrm{p} = \frac{1}{\sqrt{K}} e^{\lambda b_{L_x} + \sigma b S_x} \Delta t \tag{3.3}$$

where parameters K, λ, σ w_L, w_S, b_L, b_S are related with the occupancy of long and short order spheres with the cells of certain color.

The LDSA with negative long-range feedback gives results (Figure 3.19) very similar to most of the bulk patterns that appear among diatom pore occlusions. Different patterns may be obtained by choosing the rate of precipitation (black cells flip to white) over the rate of dissolution (vice versa).

The agreement between the stochastic model and observation is due to the fact that the formation of the pore occlusion substructures occurs at a very small scale that approaches the macromolecular domain scale. Hence, the process is significantly affected by fluctuations, and a stochastic model may represent a suitable approach.

When the computational algorithms are considered, the ultimate emergent pattern of probabilistically self-organized matter depends on a number of model aspects and parameters:

- the initial distribution of black and white cells;
- the boundary conditions (fixed colors of particular cells along the boundary of mesh);
- the values of short and long order parameters and their factors.

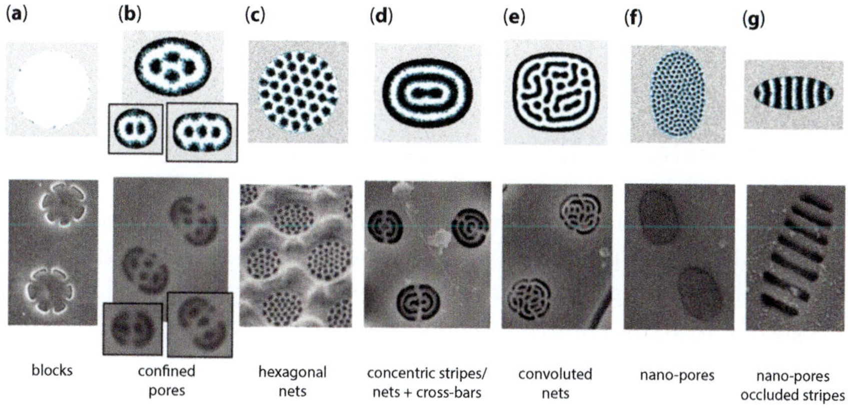

Figure 3.19 (a)-(g) Patterns produced by tuning negative long-range feedback of a lateral dependence stochastic algorithm (LDSA) compared to similar-looking SEM micrographs of pore occlusions of various diatom algae. (Reproduced from [3.114].)

In Figure 3.20 we exemplify the appearance of probabilistic patterns for some chosen values of the order parameters. This serves to illustrate that even in the absence of biological control, the complex genetic coding involved in the processes occurring within silica deposition vesicles that predetermine the overall architecture of diatom frustules, and the purely physicochemical and thermodynamical properties of the matter (i.e., the processes of silica globule precipitation from siliceous solutions to form frustules) may be sufficient to create ordered structures provided that the short and long order parameters

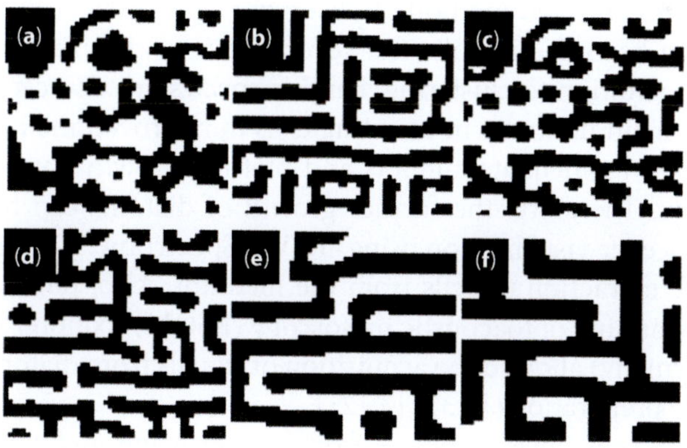

Figure 3.20 Probabilistic patterns of self-organized matter calculated for different long and short order parameters and factors: (a) Long-range parameter 4, long-range factor -1, short-range parameter 1, short-range factor 30; (b) Long-range parameter 4, long-range factor -2, short-range parameter 1, short-range factor 10; (c) Long-range parameter 4, long-range factor -1, short-range parameter 2, short-range factor 30; (d) Long-range parameter 4, long-range factor -2, short-range parameter 1, short-range factor 10; (e) Long-range parameter 5, long-range factor -1, short-range parameter 2, short-range factor 5; (f) Long-range parameter 5, long-range factor -1, short-range parameter 3, short-range factor 3.

have appropriate values (which may be determined and controlled by the biological activity of the cell).

3.8 On the Kinetics of Diatom Colony Growth

The growth kinetics of a culture of diatoms is very similar to the growth kinetics of prokaryotic organisms [3.117]. The life cycle of algae cultures consists of five different phases: lag, exponential growth, declining relative growth, stationary and dying phase [3.118]. The algal growth is regulated by light, temperature, nutrient quantity and quality, pH, turbulence, and salinity [3.119].

As a result of the work of Hill and Kolmogorov-Avrami [3.120] on classical kinetics and statistics of nucleation and growth, the first four phases of diatom colony evolution can be described reasonably well. The lag stage corresponds closely to the term *incubation time* used in materials science of phase change. It is perhaps not by chance that it has a biological connotation. The exponentially accelerating growth stage is associated with the constant division rate and the fact that growth is not yet moderated by the limitation of nutrient supply. Upon further proliferation, the growth is reduced due to the impossibility of sustaining a growing number of cells. The colony size then approaches steady state that corresponds to a dynamic equilibrium between nutrient supply and consumption that is reflected in the equal "birth" and death rates of diatoms. The dying phase represents a stage that is distinct from the phase change kinetics and is a hallmark of a biological system. Precise investigation of the processes that govern this stage is of particular interest in the context of creating and sustaining positive dynamics of diatom culture for applications in nanotechnology.

Cvjetinovic *et al.* [3.37] proposed novel approaches for monitoring the growth of diatoms using fluorescence and photoacoustic techniques. Although cell counting under a microscope is a well-established method of observing algal growth, due to the formation of cell clusters, it can become difficult after several days of culturing, especially for fast-growing cultures. In addition to clustering, benthic species are also capable of secreting a lot of mucus, so the main difficulty in counting cells in microvolume in a well plate is the impossibility of choosing adequate aliquots and plotting growth curves with sufficiently large technical errors. Therefore, the further growth of diatoms can be more accurately monitored by fluorescence visualization using an IVIS SpectrumCT imaging system, which does not require the extraction of cells from the medium. Also, photoacoustic measurements using raster-scanning optoacoustic mesoscopy (RSOM) Explorer P50 setup (iThera-Medical GmbH, Germany) allowed the acquisition of signals from diatom colonies without sophisticated sample preparations [3.37] [3.121].

We studied the fluorescence response from fast-growing *Achnanthidium sibiricum* diatoms using the IVIS SpectrumCT *in-vivo* imaging fluorescence tomography system (Xenogen Corp., CA, USA) [3.122]. *A. sibiricum* strain 256, isolated from the coastal bottom of Lake Baikal, was cultivated in a specially assembled incubator under optimal conditions (12 ± 1 °C, 12:12 day-night cycle) for 100 days. The fluorescence signal from cell culture flasks containing diatoms was obtained with a 465/680 nm excitation/emission filter. The results of fluorescence measurements of diatom algae obtained using the IVIS imaging system are shown in Figure 3.21. The total radiant efficiency increases with increasing incubation time up to 56 days of cultivation, after which a slight decrease in values can be observed (Figure 3.21b).

The results of fluorescence measurements were also compared with fluorescence spectra collected during 100 days of incubation. Figure 3.21c shows the fluorescence intensity spectra of *A. sibiricum* cells, obtained by excitation at 465 nm. An emission band with a maximum at about 680-682 nm can be attributed to chlorophyll a. Fluorescence intensity was normalized to one to compare the obtained results with the normalized average radiant efficiency depending on the sampling days, as shown in Figure 3.21d. The maximum values of fluorescence intensity were achieved after 30 days of cultivation, while the total radiant efficiency reached the maximum after 56 days. The difference in values can be explained by insufficient mixing of a medium with diatom cells prior to fluorescence spectroscopy measurements. The growth kinetics of diatoms is very similar to the growth kinetics of prokaryotic organisms [3.123]. Previously, we observed the lag phase followed by exponential growth by performing cell counting [3.37], while Figure 3.21b demonstrates the end of exponential growth of the culture followed by stationary phase, which is characteristic for all microorganisms. Therefore, in other to predict the composition and dynamics of phytoplankton communities or invasion of cyanobacteria, such as *Cylindrospermopsis raciborskii*, it is crucial to understand the nature of interaction between nutrients, light and temperature and properly monitor them [3.124].

This noninvasive, easy-to-use monitoring method, based on the chlorophyll fluorescence, allows monitoring the growth of diatoms very quickly, without preliminary sample preparation and introduction of contamination into the medium during the selection of aliquots. We believe that this method is promising for applications in aquaculture and bioreactor practice as it expands the possibilities of the cultivation of diatoms and their observation without affecting the integrity of the cells.

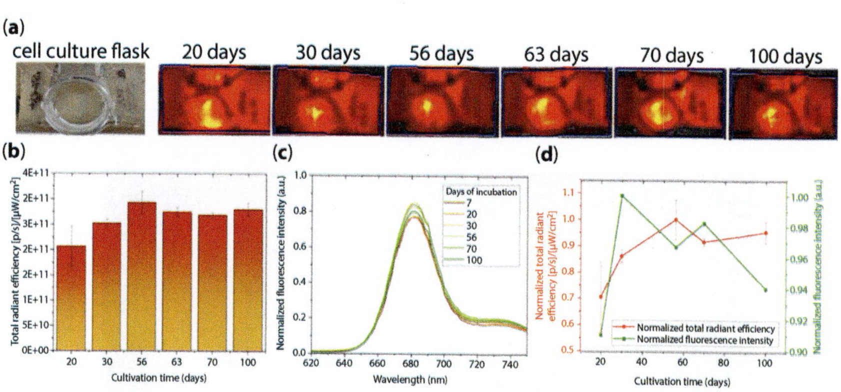

Figure 3.21 (a) IVIS fluorescence images of *A. sibiricum* in cell culture flasks; (b) the total radiant efficiency of diatoms measured during 100 days of cultivation; (c) fluorescence spectra of diatoms; (d) comparison of normalized fluorescence intensity and average radiant efficiency during the cultivation period. (Adapted from [3.122].)

3.9 Advanced Pattern Analysis of the Hierarchical Structure of Diatom Frustules

Figure 3.22 shows the electron micrograph of *Coscinodiscus oculus-iridis* diatom frustule that contains pores (areolae) arranged in clusters. In order to reveal the nature of imperfect, but readily discernible order that exists in the arrangement of the clusters and the pores within them, Fourier transform image analysis was carried out.

It is apparent that the arrangement of features displays a certain hierarchy across the length scales. Figure 3.23a presents the 2D Fourier transform of the image in Figure 3.22. Figure 3.23b illustrates the central region of that transform pattern.

In the 2D Fourier transform image in Figure 3.23, the rings located closer to the pattern center correspond to larger dimensions, whilst those with the larger radii represent shorter length scales. The wide bright outer rings (A) are associated with the arrangement of pores (areolae) within the clusters, whilst the smaller diameter rings closer to the pattern center (B) correspond to the arrangement of clusters. It is important to note the prominent hexagonal nature of rings B2 and B3. The smallest and most intense ring B1 appears to reveal an instance of icosahedral 12-fold order.

Figure 3.22 SEM image of the *Coscinodiscus oculus-iridis* frustule (Ehrenberg).

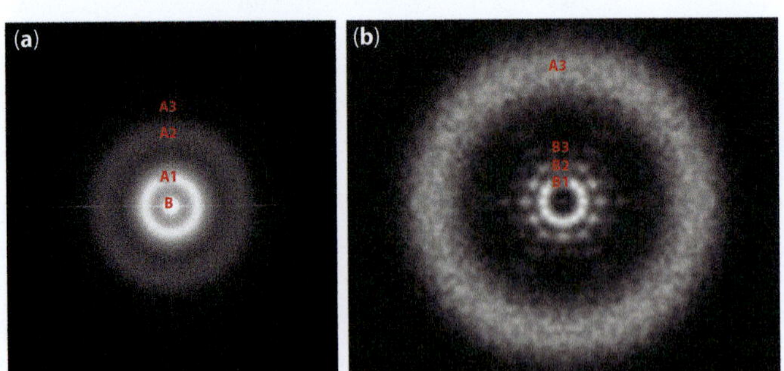

Figure 3.23 (a) Fourier transform of the selected circular area in Figure 3.22. (b) A close-up view of the central region of the Fourier transform in Figure 3.23a.

The radial intensity variation within the Fourier pattern in Figure 3.23 can be presented as a line profile in Figure 3.24.

The strongest 120 nm peak corresponds to the principal period of the alveolae pattern. The smaller 45 nm peak corresponds to the average diameter of an alveola. The major peak corresponding to the mean period of 840 nm is associated with the spacing between clusters, whilst the 460 nm peak corresponds to the cluster radius. Thus, the two most significant features of the 2D FFT pattern clearly reveal the two levels of hierarchical structural organization of *Coscinodiscus oculus-iridis* diatom frustule.

The features corresponding to specific rings can be revealed via Filtered Inverse Fourier Transform, as illustrated in Figure 3.25. Figure 3.25a shows a subset of the original image. Figure 3.25b contains the Inverse FFT that has been filtered down to and including the 120 nm peak. This operation removes the details of the underlying silica structure and prominently reveals the main features of the hierarchical organization: the strongest contrast is related to the pore arrangement within clusters, whilst the next apparent level of contrast concerns the pseudo-hexagonal arrangement of clusters. Figure 3.25c is the inverse FFT filtered down to and including the 460 nm peak, and reveals the smaller periodicity associated with the cluster radius. This is approximately equal to half of the distance between two adjacent clusters, as confirmed by the Filtered Inverse FFT down to and including the 860 nm period, as shown in Figure 3.25d. It is worth noting that the imperfection of the pseudo-hexagonal cluster arrangement is apparent, as is also reflected in the approximate azimuthal uniformity of ring A1 in Figure 3.23b.

Further insights into the pattern geometry of diatom frustules can be obtained using radial distribution function (RDF) analysis, along with its multi-dimensional equivalent, the spatial distribution function (SDF) [3.125]. Unlike the 2D FFT approach reported above, the RDF, also referred to as the pair correlation function (PCF), is obtained by calculating the correlation from all possible pairwise point selections within a given image. This is closely related to X-ray methods known as wide and small angle X-ray scattering (SAXS/WAXS). These experimental configurations represent nature's mode of computing the correlation function that occurs by the mechanism of interference between photons scattered from different parts of the object [3.126].

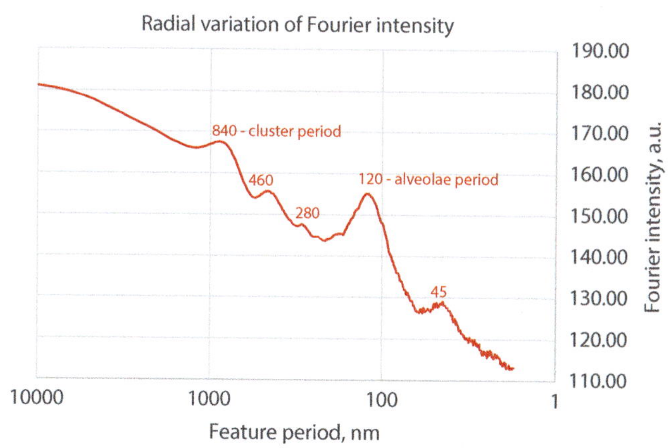

Figure 3.24 Prominent peaks in the radial variation of FFT corresponding to the features at different length scales (reverse logarithmic scale used to correspond to the peak order outward from pattern center).

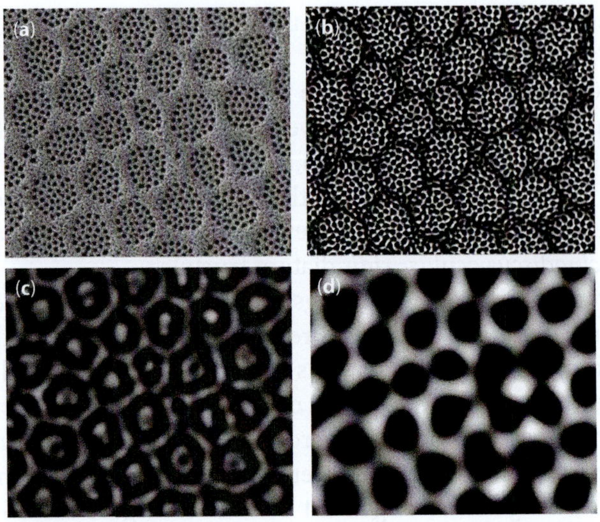

Figure 3.25 Reconstructed patterns corresponding to different peaks of Fourier Transform signal: (a) Initial picture provided for reference; (b) Features that correspond to 120 nm FFT peak; (c) Features that correspond to 460 nm FFT peak; (d) Features that correspond to 860 nm FFT peak.

Figure 3.26 illustrates the RDF $g(r)$ obtained from the same SEM image shown in Figure 3.22. The peaks corresponding to structural order are clearly and prominently seen across several orders of length scale.

To complete the illustration of this analytical approach, Figure 3.27 shows the 2D analogue of RDF, the SDF computed with account taken not only of the pairwise point distance, but also the point-to-point vector orientation. It is apparent that this approach brings out the fine detail of the spatial frequency and orientation of the complex arrangement of pores and clusters in diatom frustules.

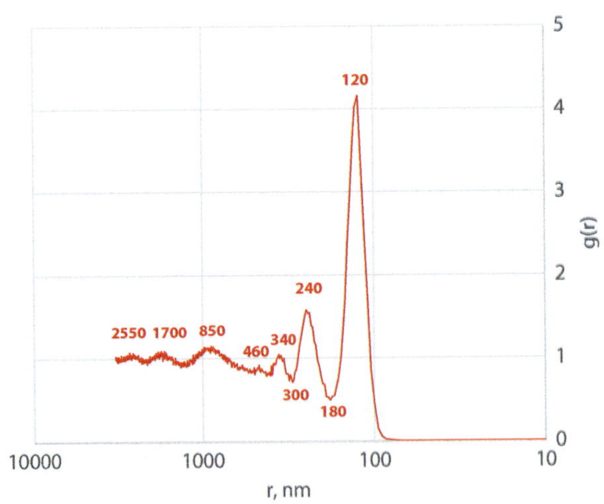

Figure 3.26 Radial distribution function (RDF) $g(r)$ plotted for convenience using the same inverted logarithmic distance scale used in Figure 3.24. Prominent peaks are observed corresponding to structural order across several orders of length scale.

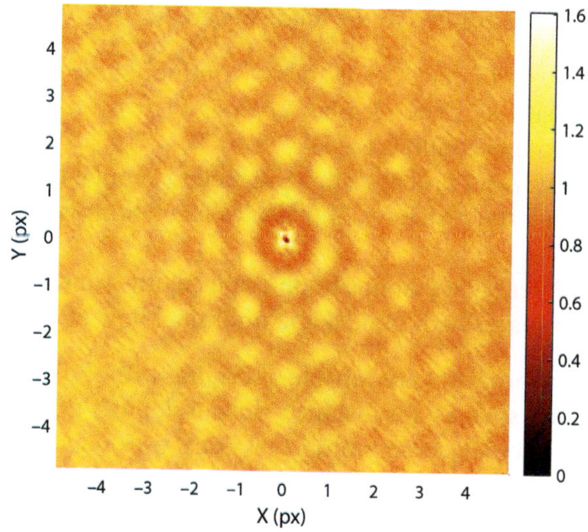

Figure 3.27 Spatial distribution function (SDF) illustration.

We draw the conclusion that mathematical analysis of diatom frustule structural patterns provides an excellent means of extracting order information in a situation when such order may be less than readily apparent, both in the angular and length scale sense. In fact, it is possible to view the "SDF signature" of diatom frustule structure as a key additional indicator of the biological species, and can be helpful in introducing rigor into distinguishing and classifying species in this vast animal class.

3.10 Concluding Remarks

The authors of this overview chapter are firmly convinced that modern diatom science is a growing multidisciplinary research domain integrating biology, molecular biology and genetics, ecology, materials science and engineering, chemistry, optics, etc. Ultimately, and inevitably, this also includes mathematics as the most powerful tool for rational theorization. As indicated in this report, mathematical approaches are needed to understand the processes of frustule pattern formation, as well as in the analysis of the resulting imperfectly ordered structures. Furthermore, mathematical descriptions of the growth and evolution of diatom colonies are needed to capture, quantify and rationalize the results of experimental observations, and possibly to obtain means of controlling the growth towards more or less rapid, and sustained vs. interrupted growth.

The authors expect that mathematics will support and perhaps at a certain stage in the future guide versatile studies of diatom species, reinforcing them with a number of efficient approaches, including statistics, topology, graph theory, systems theory, stability theory, machine learning and others. The present-day mathematical treatment of certain aspects of diatom science includes those overviewed above.

We anticipate the fastest advances in the following areas:

- basic differential and integral equations of continuum mechanics and related computation math as finite element analysis of frustule mechanics under static and vibrational conditions;
- combinatory math and probabilistic modeling to simulate self-ordering processes mimicking the processes of frustules formation within silica deposition vesicles;
- parametric differential equations to describe colony growth kinetics in confined and open spaces with limited and unlimited flow of nutrients.

These mathematical approaches are capable of dealing with the most promising applied aspects of diatom science, such as, for instance, the development of aquaculture technologies to produce diatom microalgae for multiple practical uses, such as

- CO_2 capture to improve air quality control in living rooms with potential implications for global climate;
- synthesis of diverse organic substances for pharmaceutics and drug delivery;
- production of nanostructured silica structural elements for industrial nanofabrication of miniature filters, NEMS/MEMS, gas sensors, micro-batteries and photosensitive elements.

Acknowledgement

The work was supported by Russian Science Foundation (RSF grant № 22-14-00209).

References

[3.1] Fratzl, P. & Weinkamer, R. (2007) Nature's hierarchical materials. *Prog. Mater. Sci. 52*: 1263-1334.

[3.2] Fabritius, H. O., Sachs, C., Triguero, P. R. & Raabe, D. (2009) Influence of structural principles on the mechanics of a biological fiber-based composite material with hierarchical organization: The exoskeleton of the lobster homarus americanus. *Adv. Mater. 21*: 391-400.

[3.3] Medlin, L.K., Kooistra, W.H.C.F., Gersonde, R., Sims, P.A. & Wellbrock, U. (1997) Is the origin of the diatoms related to the end Permian mass extinction? *Nova Hedwigia 65*: 1-11.

[3.4] Harwood, D.M. & Nikolaev, V.A. (1995) Cretaceous diatoms: morphology, taxonomy, biostratigraphy, in C.D. Blome, P.M. Whalen, and K.M. Reed (eds.), Siliceous Microfossils: Short Courses in Paleontology, 8: 81–106.

[3.5] Nemoto, T. (1956) On the diatoms of the skin film of whales in the northern Pacific. *Scientific Reports of the Whales Research Institute*, Tokyo, 11:97-132.

[3.6] Nemoto, T. (1958) Cocconeis diatoms infected on whales in the Antarctic. *Scientific Reports of the Whales Research Institute*, Tokyo, 13:185-191.

3.7] Nemoto, T., Best, P.B., Ishimaru, K., Takano, H. (1980) Diatom films on whales in South African waters. *Scientific Reports of the Whales Research Institute*, Tokyo, 32: 97-103.

[3.8] Kociolek, J.P. and Spaulding, S.A. (2003) Eunotioid and asymmetrical naviculoid diatoms. In: *Freshwater Algae of North America, London: Acad. Press*, 655-668.

[3.9] Morales, E.A., Edlund, M.B., Spaulding, S.A. (2010) Description and ultrastructure of araphid diatom species (Bacillariophyceae) morphologically similar to Pseudostaurosira elliptica (Schumann) Edlund *et al. Phycol. Res.*, 58:97-107.

[3.10] Kooistra, W., Forlani, G., De Stefano, M. (2009) Adaptations of araphid pennate diatoms to a planktonic existence. *Mar. Ecol.*, 30:1-15.
[3.11] Kociolek, J.P., Theriot, E.C., Williams, D.M., Julius, M., Stoermer, E.F., Kingston, J.C. (2015) Centric and araphid diatoms. In: *Freshwater Algae of North America: Ecology and Classification. Elsevier Inc.: Amsterdam, The Netherlands*, 653-708.
[3.12] Bruder, K., Medlin, L.K. (2008) Morphological and molecular investigations of naviculoid diatoms. II. Selected genera and families. *Diatom Res.*, 23:283-329.
[3.13 Wetzel, C.E., Ector, L., Van de Vijver, B., Compère, P. & Mann, D.G. (2015) Morphology, typification and critical analysis of some ecologically important small naviculoid species (Bacillariophyta). *Fottea*, 15(2):203-234.
[3.14] Stepanek, J.G., Kociolek, P.J. (2014) Molecular phylogeny of Amphora sensu lato (Bacillariophyta): an investigation into the monophyly and classification of the Amphoroid diatoms. *Protist*, 165:177-195.
[3.15] Stepanek, J.G., Kociolek, P.J. (2019) Molecular phylogeny of the diatom genera Amphora and Halamphora (Bacillariophyta) with a focus on morphological and ecological evolution. *Journal of Phycology*, 55:442-456.
[3.16] Romero, O.E. (2011) Morphological study of the genus Cocconeis Ehrenberg (Bacillariophyceae) collected during the 1897-1899 Belgian Antarctic Expedition. *Botanica Marina*, 54(2):179-188.
[3.17] De Stefano, M., Romero, O.E., Totti, C. (2008) A comparative study of Cocconeis scutellum Ehrenberg and its varieties (Bacillariophyta). *Botanica Marina*, 51:506-536.
[3.18] Sullivan, T. (2019) Cell Shape and Surface Colonisation in the Diatom Genus Cocconeis — An Opportunity to Explore Bio-Inspired Shape Packing? *Biomimetics*, 4, 29.
[3.19] Gogorev, R.M., Lange, E.K. (2015) Achnanthoid diatoms (Bacillariophyta) of relict Lake Mogilnoye (Kildin Island, Barents Sea). *Novosti Sist. Nizsh. Rast.*, 49:13-31.
[3.20] Hoppenrath, M., Beszteri, B., Drebes, G., Halliger, H., Van Beusekom, J.E.E., Janisch, S. & Wiltshire, K.H. (2007) Thalassiosira species (Bacillariophyceae, Thalassiosirales) in the North Sea at Helgoland (German Bight) and Sylt (North Frisian Wadden Sea) — a first approach to assessing diversity. *European Journal of Phycology*, 42(3):271-288.
[3.21] Urbani, R., Magaletti, E., Sist, P., and Cicero A.M. (2005) Extra-cellular carbohydrates released by the marine diatoms Cylindrotheca closterium, Thalassiosira pseudonana and Skeletonema costatum: Effect of P-depletion and growth status. *Sci. Total Environ.*, 353: 300-306.
[3.22] Alverson, A.J. (2006) Phylogeny and evolutionary ecology of thalassiosiroid diatoms. [*PhD dissertation*]. The University of Texas at Austin, 198 pp.
[3.23] Hasle, G.R., Lange, C.B. (1992) Morphology and distribution of Coscinodiscus species from the Oslofjord, Norway, and the Skagerrak, *North Atlantic. Diatom Res.*, 7:37-68.
[3.24] Bosak, S. (2013) Taxonomy and ecology of the planktonic diatom family Chaetocerotaceae (Bacillariophyta) from the Adriatic Sea. [*PhD dissertation*]. Department of Biology, Univ. Zagreb, 256 pp.
[3.25] Sarno, D., Kooistra, W.H.C.F.; Medlin, L.K., Percopo, I., Zingone, A. (2005) Diversity in the Genus Skeletonema (Bacillariophyceae). II. An Assessment of the Taxonomy of S. costatumlike species with the Description of four new species. *J. Phycol.*, 41 (1):151-176.
[3.26] Hoban, M.A. (1983) Biddulphioid diatoms. II. The morphology and systematics of the Pseudocellate species, Biddulphia biddulphiana (Smith) Boyer, B. alternans (Bailey) Van Heurck, and Trigonium arcticum (Brighywell) Cleve. *Botanica Marina*, 26:271-284.
[3.27] Sethi, D., Butler, T. O., Shuhaili, F. & Vaidyanathan, S. (2020) Diatoms for carbon sequestration and bio-based manufacturing. *Biology (Basel)*. 9, 1-29.
[3.28] L. Virta, J. Gammal, M. Järnström, G. Bernard, J. Soininen, J. Norkko, A. Norkko (2019) The diversity of benthic diatoms affects ecosystem productivity in heterogeneous coastal environments, *Ecology*. 100 1-11.

[3.29] Mann, D. G. & Vanormelingen, P. (2013) An inordinate fondness? the number, distributions, and origins of diatom species. *J. Eukaryot. Microbiol.* 60, 414-420.

[3.30] Soininen, J. (2007) Environmental and spatial control of freshwater diatoms—a review. *Diatom Res.* 22, 473-490.

[3.31] Finkel, Z. V., Vaillancourt, C. J., Irwin, A. J., Reavie, E. D. & Smol, J. P. (2009) Environmental control of diatom community size structure varies across aquatic ecosystems. *Proc. R. Soc. B Biol. Sci.* 276, 1627-1634.

[3.32] Svensson, F., Norberg, J. & Snoeijs, P. (2014) Diatom cell size, Coloniality and motility: Trade-Offs between temperature, Salinity and nutrient supply with climate change. *PLoS One* 9, e109993.

[3.33] Vanormelingen, P., Verleyen, E. & Vyverman, W. (2008) The diversity and distribution of diatoms: From cosmopolitanism to narrow endemism. *Biodivers. Conserv.* 17, 393-405.

[3.34] Vyverman, W., Verleyen, E., Sabbe, K., Vanhoutte, K., Sterken, M., Hodgson, D.A., Mann, D.G., Juggins, S., Van De Vijver, B., Jones, V., Flower, R., Roberts, D., Chepurnov, V.A., Kilroy, C., Vanormelingen, P., De Wever, A. (2007) Historical processes constrain patterns in global diatom diversity. *Ecology* 88, 1924-1931.

[3.35] Mishra, M., Arukha, A. P., Bashir, T., Yadav, D. & Prasad, G. B. K. S. (2017) All new faces of diatoms: Potential source of nanomaterials and beyond. *Front. Microbiol.* 8, 1-8.

[3.36] Korsunsky, A. M., Sapozhnikov, P. V., Everaerts, J. & Salimon, A. I. (2019) Nature's neat nanostructuration: The fascinating frustules of diatom algae. *Mater. Today* 22, 159-160.

[3.37] Cvjetinovic, J., Bedoshvili, Y., Nozdriukhin, D., Efimova, O., Salimon, A., Volokitina, N., Korsunsky, A., Gorin, D. (2021) In situ fluorescence/photoacoustic monitoring of diatom algae. *Proceedings Volume 11641, Dynamics and Fluctuations in Biomedical Photonics.* 8.

[3.38] Bedoshvili, Y., Gneusheva, K., Popova, M., Morozov, A. & Likhoshway, Y. (2018) Anomalies in the valve morphogenesis of the centric diatom alga Aulacoseira islandica caused by microtubule inhibitors. *Biol. Open* 7, 1-10.

[3.39] Grachev, M. A., Annenkov, V. V. & Likhoshway, Y. V. (2008) Silicon nanotechnologies of pigmented heterokonts. *BioEssays* 30, 328-337.

[3.40] Losic, D., Rosengarten, G., Mitchell, J. G. & Voelcker, N. H. (2006) Pore architecture of diatom frustules: Potential nanostructured membranes for molecular and particle separations. *J. Nanosci. Nanotechnol.* 6, 982-989.

[3.41] Soleimani, M., Rutten, L., Maddala, S.P., Wu, H., Eren, E.D., Mezari, B., Schreur-Piet, I., Friedrich, H., van Benthem, R. A.T.M. (2020) Modifying the thickness, pore size, and composition of diatom frustule in Pinnularia sp. with Al^{3+} ions. *Scientific Reports.* 10, 1-12.

[3.42] Willis, L., Page, K. M., Broomhead, D. S. & Cox, E. J. (2010) Discrete free-boundary reaction-diffusion model of diatom pore occlusions. *Plant Ecol. Evol.* 143, 297-306.

[3.43] De Tommasi, E., Gielis, J. & Rogato, A. (2017) Diatom Frustule Morphogenesis and Function: a Multidisciplinary Survey. *Marine Genomics* 35, 1-18.

[3.44] Finkel, Z. V. & Kotrc, B. (2010) Silica use through time: Macroevolutionary change in the morphology of the diatom fustule. *Geomicrobiology.* 27, 596-608.

[3.45] Pančić, M., Torres, R. R., Almeda, R. & Kiørboe, T. (2019) Silicified cell walls as a defensive trait in diatoms. *Proceedings of the Royal Society B: Biological Sciences.* 286.

[3.46] Uthappa, U. T., Brahmkhatri, V., Sriram, G., Jung, Ho-Young, Yu, J., Kurkuri, N., Aminabhavi, T.M., Altalhi, T., Neelgund, G.M., Kurkuri M.D. (2018) Nature engineered diatom biosilica as drug delivery systems. *Journal of Controlled Release* 281, 70-83.

[3.47] Parkinson, J. & Gordon, R. (1999) Beyond micromachining: The potential of diatoms. *Trends in Biotechnology.* 17, 190-196.

[3.48] Todd, T., Zhen, Z., Tang, W., Chen, H., Wang, G., Chuang, Y.J., Deaton, K., Pan, Z., Xie, J. (2014) Iron oxide nanoparticle encapsulated diatoms for magnetic delivery of small molecules to tumors. *Nanoscale* 6, 2073-2076.

[3.49] Ragni, R., Cicco, S., Vona, D., Leone, G. & Farinola, G. M. (2017) Biosilica from diatoms microalgae: smart materials from bio-medicine to photonics. *J. Mater. Res.* 32, 279-291.

[3.50] Rea, I., Terracciano, M. & De Stefano, L. (2017) Synthetic vs Natural: Diatoms Bioderived Porous Materials for the Next Generation of Healthcare Nanodevices. *[Progress Report]. Advanced Healthcare Materials.* Feb: 6(3).

[3.51] Jeffryes, C., Campbell, J., Li, H., Jiao, J. & Rorrer, G. (2011) The potential of diatom nanobiotechnology for applications in solar cells, batteries, and electroluminescent devices. *Energy and Environmental Science* 4, 3930-3941.

[3.52] Guo, X. L., Kuang, M., Li, F., Liu, X.Y., Zhang, Y.X., Dong, F., Losic, D. (2016) Engineering of three dimensional (3-D) diatom@TiO_2@MnO_2 composites with enhanced supercapacitor performance. *Electrochim. Acta* 190, 159-167.

[3.53] Korsunsky, A. M., Bedoshvili, Y.D., Cvjetinovic, J., Aggrey, P., Dragnevski, K.I., Gorin, D.A., Salimon, A.I., Likhoshway, Y. V. (2020) Siliceous diatom frustules - A smart nanotechnology platform. In *Materials Today: Proceedings* 33, 2032-2040.

[3.54] Svendsgaard, T. (2015) Diatoms Investigated by Atomic Force Microscopy. *Chemistry.*

[3.55] Subhash, G., Yao, S., Bellinger, B. & Gretz, M. R. (2005) Investigation of Mechanical Properties of Diatom Frustules Using Nanoindentation. *J Nanosci Nanotechnol,* 50-56

[3.56] Losic, D., Pillar, R. J., Dilger, T., Mitchell, J. G. & Voelcker, N. H. (2007) Atomic force microscopy (AFM) characterisation of the porous silica nanostructure of two centric diatoms. *J. Porous Mater.* 14, 61-69

[3.57] Almqvist, N., Delamo, Y., Smith, B.L., Thomson, N.H., Bartholdson, Å., Lal, R., Brzezinski, M., Hansma, P.K. (2001) Micromechanical and structural properties of a pennate diatom investigated by atomic force microscopy. *J. Microsc.* 202, 518-532.

[3.58] Hamm, C. E., Merkel, R., Springer, O., Jurkojc, P., Maiert, C., Prechtelt, K., Smetacek, V. (2003) Architecture and material properties of diatom shells provide effective mechanical protection. *Nature.*

[3.59] Romann, J., Chauton, M.S., Hanetho, S.M., Vebner, M., Heldal, M., Thaulow, C., Vadstein, O., Tranell, G., Einarsrud, M.A. (2016) Diatom frustules as a biomaterial: effects of chemical treatment on organic material removal and mechanical properties in cleaned frustules from two Coscinodiscus species. *J. Porous Mater.* 23, 905-910.

[3.60] Aitken, Z. H., Luo, S., Reynolds, S. N., Thaulow, C. & Greer, J. R. (2016) Microstructure provides insights into evolutionary design and resilience of Coscinodiscus sp. frustule. *Proc. Natl. Acad. Sci.* 113, 2017-2022.

[3.61] Moreno, M. D., Ma, K., Schoenung, J. & Dávila, L. P. (2015) An integrated approach for probing the structure and mechanical properties of diatoms: Toward engineered nanotemplates. *Acta Biomaterials.*

[3.62] Luo, S. & Greer, J. R. (2018) Bio-Mimicked Silica Architectures Capture Geometry, Microstructure, and Mechanical Properties of Marine Diatoms. *Adv. Eng. Mater.*

[3.63] Topal, E., Rajendran, H., Zgłobicka, I., Gluch, J., Liao, Z., Clausner, A., Kurzydłowski, K.J., Zschech, E. (2020) Numerical and experimental study of the mechanical response of diatom frustules. *Nanomaterials* 10, 1-14.

[3.64] Yao, S., Subhash, G. & Maiti, S. (2007) Analysis of nanoindentation response of diatom frustules. *J. Nanosci. Nanotechnol.* 7, 4465-4472.

[3.65] Crawford, S. A., Higgins, M. J., Mulvaney, P. & Wetherbee, R. (2001) Nanostructure of the diatom frustule as revealed by atomic force and scanning electron microscopy. *J. Phycol.* 37, 543-554.

[3.66] Higgins, M. J., Crawford, S. A., Mulvaney, P. & Wetherbee, R. (2002) Characterization of the adhesive mucilages secreted by live diatom cells using Atomic Force Microscopy. *Protist* 153, 25-38.

[3.67] Higgins, M. J., Sader, J. E., Mulvaney, P. & Wetherbee, R. (2003) Probing the surface of living diatoms with atomic force microscopy: The nanostructure and nanomechanical properties of the mucilage layer. *J. Phycol.* 39, 722-734.

[3.68] Gebeshuber, I.C., Kindt, J.H., Thompson, J.B., Del Amo, Y., Stachelberger, H., Brzezinski, M.A., Stucky, G.D., Morse, D.E., Hansma, P.K. (2003) Atomic force microscopy study of living diatoms in ambient conditions. *J. Microsc.* 212, 292-299.

[3.69] Losic, D., Short, K., Mitchell, J. G., Lal, R. & Voelcker, N. H. (2007) AFM nanoindentations of diatom biosilica surfaces. *Langmuir* 23, 5014-5021.

[3.70] Losic, D., Mitchell, J. G., & Voelcker, N. H. (2009). Diatomaceous lessons in nanotechnology and advanced materials. *Advanced Materials*, 21(29), 2947-2958.

[3.71] Gebeshuber, I., Stachelberger, H., & Drack, M. (2005) Diatom bionanotribology—biological surfaces in relative motion: their design, friction, adhesion, lubrication and wear. *J. Nanosci. Nanotechnol.* 5(1), 79-87.

[3.72] Wetherbee, R., Lind, J. L., Burke, J., & Quatrano, R. S. (1998) Minireview—the first kiss: establishment and control of initial adhesion by raphid diatoms. *Journal of Phycology*, 34(1), 9-15.

[3.73] Arce, F. T., Avci, R., Beech, I. B., Cooksey, K. E., & Wigglesworth-Cooksey, B. (2004) A live bioprobe for studying diatom-surface interactions. *Biophysical journal*, 87(6), 4284-4297.

[3.74] Aldersey-Williams, H. (2004) Towards biomimetic architecture. *Nature materials*, 3(5), 277279.

[3.75] Ball, P. (2001) Life's lessons in design. *Nature*, 409(6818), 413-416.

[3.76] Bäuerlein, E. (Ed.). (2004) *Biomineralization: Progress in biology, molecular biology and application.* John Wiley & Sons.

[3.77] Aksakal, E. L., Angin, I., & Oztas, T. (2013) Effects of diatomite on soil consistency limits and soil compactibility. *Catena*, 101, 157-163.

[3.78] Wang, H., Zhu, M., Li, Y., Zhang, Q., & Wang, H. (2011) Mechanical properties of dental resin composites by co-filling diatomite and nanosized silica particles. *Materials Science and Engineering: C*, 31(3), 600-605.

[3.79] Lv, J., Sun, B., Jin, J., & Jiang, W. (2019) Mechanical and slow-released property of poly (acrylamide) hydrogel reinforced by diatomite. *Materials Science and Engineering: C*, 99, 315321.

[3.80] Mustafov, S. D., Sen, F., & Seydibeyoglu, M. O. (2020) Preparation and characterization of diatomite and hydroxyapatite reinforced porous polyurethane foam biocomposites. *Scientific reports*, 10(1), 1-9.

[3.81] Aksakal, E. L., Angin, I., & Oztas, T. (2012) Effects of diatomite on soil physical properties. *Catena*, 88(1), 1-5.

[3.82] Dessalew, G., Beyene, A., Nebiyu, A., & Ruelle, M. L. (2017) Use of industrial diatomite wastes from beer production to improve soil fertility and cereal yields. *Journal of cleaner production*, 157, 22-29.

[3.83] Angin, I., Sari, S., & Aksakal, E. L. (2016). Effects of diatomite (DE) application on physical properties of soils subjected to freeze-thaw cycles. *Soil and Tillage Research*, 160, 34-41.

[3.84] Luo, J., Seghi, R., & Lannutti, J. (1997) Effect of silane coupling agents on the wear resistance of polymer-nanoporous silica gel dental composites. *Materials Science and Engineering: C*, 5(1), 15-22.

[3.85] Wilson, K. S., Zhang, K., & Antonucci, J. M. (2005) Systematic variation of interfacial phase reactivity in dental nanocomposites. *Biomaterials*, 26(25), 5095-5103.

[3.86] Xu, H. H., Quinn, J. B., Smith, D. T., Antonucci, J. M., Schumacher, G. E., & Eichmiller, F. C. (2002) Dental resin composites containing silica-fused whiskers—effects of whisker-to-silica ratio on fracture toughness and indentation properties. *Biomaterials*, 23(3), 735-742.

[3.87] Bindslev, D. A. (1992). Dental amalgam-environment, aspects. *Adv. Dent. Res.*, 6, 125-130.

[3.88] Clarkson, T. W., Magos, L., & Myers, G. J. (2003) The toxicology of mercury—current exposures and clinical manifestations. *New England Journal of Medicine*, 349(18), 1731-1737.

[3.89] Kim, J. W., Kim, L. U., & Kim, C. K. (2007) Size control of silica nanoparticles and their surface treatment for fabrication of dental nanocomposites. *Biomacromolecules*, 8(1), 215-222.

[3.90] Schulz, H., Schimmoeller, B., Pratsinis, S. E., Salz, U., & Bock, T. (2008) Radiopaque dental adhesives: dispersion of flame-made Ta_2O_5/SiO_2 nanoparticles in methacrylic matrices. *Journal of dentistry*, 36(8), 579-587.

[3.91] Khaled, S., Sui, R., Charpentier, P. A., & Rizkalla, A. S. (2007) Synthesis of TiO_2– PMMA nanocomposite: Using methacrylic acid as a coupling agent. *Langmuir*, 23(7), 3988-3995.

[3.92] Luo, J., Lannutti, J. J., & Seghi, R. R. (1998) Effect of filler porosity on the abrasion resistance of nanoporous silica gel/polymer composites. *Dental Materials*, 14(1), 29-36.

[3.93] Tan, Y., Liu, Y., Grover, L. M., & Huang, B. (2010) Wear behavior of light-cured dental composites filled with porous glass–ceramic particles. *Journal of the mechanical behavior of biomedical materials*, 3(1), 77-84.

[3.94] Campbell, B., Ionescu, R., Tolchin, M., Ahmed, K., Favors, Z., Bozhilov, K., Ozkan, C., Ozkan, M. (2016) Carbon-Coated, Diatomite-Derived Nanosilicon as a High Rate Capable Li-ion Battery Anode. Scientific Reports 6, 1-9.

[3.95] Entwistle, J., Rennie, A. & Patwardhan, S. (2018) A review of magnesiothermic reduction of silica to porous silicon for lithium-ion battery applications and beyond. *Journal of Materials Chemistry A* 6, 18344-18356.

[3.96] Ahmad Alyosef, H., Roggendorf, H., Schneider, D., Inayat, A., Welscher, J., Schwieger, W., Münster, T., Gert, K., Ibrahim, S., Dirk, E. E (2016) MFI-type zeolites from natural materials: a comparative study of MFI-type zeolites generated from different diatomite species (part I). *J. Porous Mater.* 23, 1609-1618.

[3.97] Zeng, S., Wang, R., Zhang, Z. & Qiu, S. (2016) Solventless green synthesis of sodalite zeolite using diatomite as silica source by a microwave heating technique. *Inorg. Chem. Commun.* 70, 168-171.

[3.98] Losic, D. & Korunic, Z. (2018) CHAPTER 10: Diatomaceous Earth, A Natural Insecticide for Stored Grain Protection: Recent Progress and Perspectives. In: *RSC Nanoscience and Nanotechnology 2018*-Janua, 219-247.

[3.99] Abd El-Aziz, S. E. & Abd El-Ghany, N. M. (2018) Impact of Diatomaceous Earth Modifications for Controlling the Granary Weevil, Sitophilus granarius (Linnaeus) (Coleoptera: Curculionidae). *J. Agr. Sci. Tech* 20.

[3.100] Prasantha, B. D. R., Reichmuth, C., Adler, C. & Felgentreu, D. (2015) Lipid adsorption of diatomaceous earths and increased water permeability in the epicuticle layer of the cowpea weevil Callosobruchus maculatus (F.) and the bean weevil Acanthoscelides obtectus (Say) (Chrysomelidae). *J. Stored Prod. Res.* 64, 36-41.

[3.101] Lo, Y. H., Yang, C. Y., Chang, H. K., Hung, W. C. & Chen, P. Y. (2017) Bioinspired Diatomite Membrane with Selective Superwettability for Oil/Water Separation. *Scientific Reports* 7.

[3.102] Zhu, H., Gao, W., Yan, S., Niu, M., Liu, G., Hao, H. (2019) Electrospun poly(vinyl alcohol) sub-micron fibers containing superfine diatomite for dye adsorption. *Mater. Chem. Phys.* 234, 354-360.

[3.103] Yuan, W., Yuan, P., Liu, D., Yu, W., Laipan, M., Deng, L.,Chen, F. (2016) In situ hydrothermal synthesis of a novel hierarchically porous TS-1/modified-diatomite composite for methylene blue (MB) removal by the synergistic effect of adsorption and photocatalysis. *J. Colloid Interface Sci.* 462, 191-199.

[3.104] Dal, S., Sutcu, M., Gok, M. S. & Gencel, O. (2020) Characteristics of lightweight diatomite-based insulating firebricks. *J. Korean Ceram. Soc.* 57, 184-191.

[3.105] Yilmaz, B. & Ediz, N. (2008) The use of raw and calcined diatomite in cement production. *Cem. Concr. Compos.* 30, 202-211.

[3.106] MacEdo, A. R. S., Silva, A. S., Da Luz, D. S., Ferreira, R. L. S., Lourenço, C. S., Gomes, U. U. (2020) Study of the effect of diatomite on physico-mechanical properties of concrete. *Ceramica* 66, 50-55.

[3.107] Ergün, A. (2011) Effects of the usage of diatomite and waste marble powder as partial replacement of cement on the mechanical properties of concrete. *Constr. Build. Mater.* 25, 806-812.

[3.108] Tamburaci, S. & Tihminlioglu, F. (2017) Diatomite reinforced chitosan composite membrane as potential scaffold for guided bone regeneration. *Mater. Sci. Eng. C* 80, 222-231.

[3.109] Maher, S., Kumeria, T., Aw, M. S. & Losic, D. (2018) Diatom Silica for Biomedical Applications: Recent Progress and Advances. *Adv. Healthc. Mater.* 7, 1800552.

[3.110] Tamburaci, S. & Tihminlioglu, F. (2018) Biosilica incorporated 3D porous scaffolds for bone tissue engineering applications. *Mater. Sci. Eng. C.* 91, 274-291.

[3.111] Aggrey, P., Abdusatorov, B., Kan, Y., Salimon, I.A., Lipovskikh, S.A., Luchkin, S., Zhigunov, D.M., Korsunsky, A.M. (2020) In situ formation of nanoporous silicon on a silicon wafer via the magnesiothermic reduction reaction (MRR) of diatomaceous earth, *Nanomaterials* 10, 601.

[3.112] Abdusatorov, B., Salimon, A. I., Bedoshvili, Y. D., Likhoshway, Y. V. & Korsunsky, A. M. (2020) FEM exploration of the potential of silica diatom frustules for vibrational MEMS applications. *Sensors and Actuators A: Physical* 315, 112270.

[3.113] Round, F. E., Crawford, R. A., Mann. D. G. (1990) *Diatoms: biology and morphology of the genera.* Cambridge, UK: Cambridge University Press.

[3.114] Willis, L, Cox, E.J., Duke, T. (2013) A simple probabilistic model of submicroscopic diatom morphogenesis. *J R Soc Interface* 10: 20130067.

[3.115] Cox, J. T., Durrett, R. (1991) *Nonlinear voter models. In Random walks, Brownian motion, and interacting particle systems*, editors: R Durrett, H Kesten. Boston, MA, USA.

[3.116] Ehrlich, P. R., Levin, S. A. (2005) The evolution of norms. *PLoS Biology 3(6)*: e194.

[3.117] Sullivan, C. W. (1977) Diatom mineralization of silicic acid. II. Regulation of $Si(OH)_4$ transport rates during the cell cycle of Navicula Pelliculosa. *J. Phycol.* 13, 86-91.

[3.118] Faraq, I. & Price, K. (2013) Resources Conservation in Microalgae Biodiesel Production. *International Journal of Engineering and Technical Research* 1, 49-56.

[3.119] Svensson F, Norberg J, Snoeijs P (2014) Diatom Cell Size, Coloniality and Motility: Trade-Offs between Temperature, Salinity and Nutrient Supply with Climate Change. *PLoS ONE* 9(10): e109993

[3.120] Korsunsky, A. M. (2021) *On the Hill and Kolmogorov statistics and the K-function representation of judiciously modified Avrami kinetics.*

[3.121] Cvjetinovic, J., Salimon, A.I., Novoselova, M.V., Sapozhnikov, P.V., Shirshin, E.A., Yashchenok, A.M., Kalinina, O. Yu., Korsunsky, A.M., Gorin, D.A. (2020) Photoacoustic and fluorescence lifetime imaging of diatoms. *Photoacoustics* 18,100171.

[3.122] Cvjetinovic J., Nozdriukhin, D.V., Bedoshvili, Y.D., Salimon, A.I., Korsunsky, A.M., Gorin, D.A. (2021) Assessment of diatom growth using fluorescence imaging. *Journal of Physics Conference Series* 1984(1):012017.

[3.123] Thompson, A. S., Rhodes, J. C., Pettman, I. (1998) *Culture Collection of Algae and Protozoa: Catalogue of Strains.* Amblesibe, UK.

[3.124] Mridul, T.K., Litchman, E. (2016) Effects of temperature and nitrogen availability on the growth of invasive and native cyanobacteria. *Hydrobiologia* 763(1), 357-369.

[3.125] Mac Fhionnlaoich, N., Qi, R., Guldin, S. (2019) Application of the Spatial Distribution Function to Colloidal Ordering. *Langmuir* 35, 16605-16611.

[3.126] Liu, H., Zwart, P.H. (2012) Determining pair distance distribution function from SAXS data using parametric functionals. *Journal of Structural Biology* 180, 226-234.

Part II

DIATOM DEVELOPMENT, GROWTH AND METABOLISM

4

Ring to the Linear: Valve Ontogeny Indicates Two Potential Evolutionary Pathways of Core Araphid Diatoms

Shigeki Mayama[1,2*] and Momoko Kushida[3,4]

[1]Advanced Support Center for Science Teachers, Tokyo Gakugei University, Koganei, Tokyo, Japan
[2]Tokyo Diatomology Lab, 2-3-2 Nukuikitamachi, Koganei, Tokyo, Japan
[3]Department of Biology, Tokyo Gakugei University, Koganei, Tokyo, Japan
[4]Takasago Elementary School, Urawa, Saitama, Japan

Abstract

The morphogenesis of diatom valves proceeds from the center, following one of two specific patterns. In centrics, there is a ring-shaped annulus, and pennates have a linear sternum. Pennate diatoms are evolutionarily derived from centric diatoms; however, the transition in valve morphogenesis pattern type from ring-shaped to linear remains unclear. Among pennates, raphid diatom valve formation has been examined by many researchers. However, investigation of araphid diatoms is limited. Herein, we observed sequential valve development in two araphid diatoms using electron microscopy. *Fragilaria mesolepta* has a narrow linear sternum in its mature valve. However, a single silicified chain comprising several rings appeared in earlier developmental stages. The rings comprised a series of spherules with an indistinct periphery ~55–75 nm in diameter with a core of ~30–40 nm. This chain was associated with laterally developing short ribs (virgae), which were aligned radially at one pole with bifurcation. The chain was short in the early stage but gradually elongated in the middle stage and fused to form a linear sternum. The mature valve of *Staurosira binodis* has a sternum expanded into a narrow spindle shape. In the early stage, two thin threads arranged as facing arcs were observed. They formed a continuous loop at one pole, were fused into a single linear structure at the other pole and had laterally developing virgae. As valve formation proceeded, the pair of threads were united and transformed into a sternum. The early structures of these diatoms suggest homology with the annulus of centrics and also indicate two potential pathways associated with the evolution of the sternum from annuli-possessing ancestors. Pattern center deformation during the ontogenetic process indicated that the siliceous structure in the early stages is not strictly fixed but retains plasticity within the silica deposition vesicle.

***Keywords*:** Araphid diatoms, annulus, evolution, *Fragilaria*, *Saurosira*, sternum, valve morphogenesis, valve ontogeny

**Corresponding author*: mayama@u-gakugei.ac.jp
Shigeki Mayama: https://researchmap.jp/simriver; https://scholar.google.com/citations?hl=ja&user=_HiomYkAAAAJ; https://www.researchgate.net/profile/Shigeki-Mayama

Janice L. Pappas (ed.) The Mathematical Biology of Diatoms, (105–120) © 2023 Scrivener Publishing LLC

4.1 Introduction

Over the past two decades, researchers have revealed the phylogeny of diatoms. Currently, the main scheme from the radial centrics to the raphid pennates has almost been finalized, although there are different views on the monophyly of the radial centrics and the polar centrics due to difference in the genes used, the handling of the secondary structure of SSU rDNA, and the number of outgroups used in phylogenetic trees [4.31] [4.33] [4.50] [4.51]. In the past, the araphid pennates comprised one major group in diatom taxonomy due to their distinctive morphology without raphe [4.39], but the genetic analyses have revealed their paraphyly [4.32] [4.41] [4.42] [4.50] [4.51]. Basically, the araphid pennates are distributed in two clades, the basal araphids (more ancestral) and the core araphids (a sister clade to raphid pennates), although the phylogenetic positions of several species remain unclear. In addition to molecular studies, combinations of characteristics shared by higher groups, such as sexual reproduction type, auxospore covering, and valvar areola structure, have been emphasized for understanding the evolution of diatoms [4.31] [4.33]. Valve ontogeny is useful for gaining systematic insights [4.9] [4.27] [4.30].

Previous research has indicated that a new valve begins to form from the center of the valve pattern. The typical pattern center is a ring-shaped annulus in centric diatoms and a linear sternum in pennates [4.27]. Although the annulus was originally termed for the ring of costa in *Bellerochea* Van Heurck [4.47], it has been observed in a variety of mature centric valves [4.39]. During valve morphogenesis, rings with short radiating ribs appear first in many diatoms, such as *Ditylum brightwellii* (T.West) Grunow [4.23] [4.24], *Chaetoceros rostulatus* Ralfs [4.25], *Stephanopyxis turris* (Greville) Ralfs [4.37], *Plagiogrammopsis vanheurckii* (Grunow) Hasle *et al.* [4.40], *Stephanodiscus meyeri* Genkal & Popovskaya [4.2], *Cyclotella cryptica* Reimann *et al.* [4.16] [4.48], *Thalassiosira pseudonana* Hasle & Heimdal [4.16], and *Hyalodiscus stelliger* Bailey [4.19].

As a variation of the annulus, a silicified central disk and an elongate ring have been reported in *Coscinodiscus weilesii* Gran & Angst [4.46] and two *Biddulphia* species, *B. biddulphiana* (J.E. Smith) Boyer [4.53] and *B. tridens* (Ehrenberg) Ehrenberg [4.19], respectively. Moreover, Li & Volcani [4.26] illustrated two elliptical rings with two parallel ribs extending from each of them to bridge each other in *Trieres sinensis* (Greville) Ashworth & E.C. Theriot (as *Odontella sinensis*). Kaczmarska *et al.* [4.19] reported a pseudo-annulus in *Ardissonea crystallina* (C.Agardh) Grunow, an extremely elongated centric diatom [4.34] that was formerly assigned to the araphid pennates.

In the pennates, sequential observations covering whole morphogenesis process have often been reported in various biraphid diatoms [4.1] [4.6] [4.7] [4.9] [4.17] [4.36] [4.37] [4.45] [4.54] [4.55]. Early valve formation had preceded in one side of the raphe slit (primary side), where a single longitudinal rib was formed first, followed by the formation of secondary ribs in the other side (secondary side), as reported first in *Mayamaea* sp. (as *Navicula pelliculosa* (Kützing) Hilse [4.5]), although the developmental procedures varied among genera in later stages. Moreover, the processes of losing the raphe slit were examined in the monoraphid *Achnanthes coarctata* (Brébisson ex W.Smith) Grunow [4.4] and *Achnanthidium saprophilum* (H. Kobayasi & Mayama) Round & Bukhtiyarova (as *Achnanthes minutissima* var. *saprophila* [4.29]), and the biraphid *Diadesmis gallica* W. Smith [4.8]. In *Eunotia* Ehrenberg, which emerged as the first branch in the monophyletic raphid diatoms, the short raphe extending from both poles to the valve mantle was developed after the formation of the main valve face, such as in araphid diatoms with a linear sternum [4.9] [4.30].

In contrast to the raphid pennates, ontogenetic information on the araphid pennates remains limited. Most observations that have been reported lacked information on early development and showed the virgae as already extending from the sternum to almost the complete valve width, as seen in *Gephyria media* Walker-Arnott [4.52], *Meridion circulare* (Greville) C. Agardh and *Synedra ulna* (Nitzsch) Ehrenberg [4.10], *Raphoneis amphiceros* (Ehrenberg) Ehrenberg [4.43], *Delphineis karstenii* (Boden) G. Fryxell and *Glyphodesmis acus* Mann [4.13], and *Synedra* sp. [4.10]. The formation process, including the early stage, has been reported only in *Synedra acus* subsp. *radians* (Kützing) Skabichevskii [4.18] [4.20] and *Ulnaria danica* (Kützing) Compère & Bukhtiyarova [4.4]. The initiation of both diatoms was a linear sternum with rudimentary virgae. However, there was nothing observed in either species to imply a relationship with the annulus of the centrics.

Mayama and Kuriyama [4.30] stated that during the valve formation of *Fragilaria*, a single chain of annulus-like rings appeared in the earliest stage that were associated with laterally developing virgae, and the chain fused into a linear sternum in a later stage. However, these authors did not provide a detailed description with photographs since the article was a comprehensive review. In the current paper, we describe the sequential valve development of *Fragilaria mesolepta* Rabenhorst and the newly observed *Staurosira binodis* (Ehrenberg) Lange-Bertalot from the early to complete stages and discuss the evolutionary relationship between the annulus and sternum.

4.2 Material and Methods

4.2.1 *Fragilaria mesolepta*

F. mesolepta was collected from the Tama River in Fuchu, Tokyo, in November 1988 by Y. Taira (Sample no. M-2630). The mature valves were linear in valve view, 30–31 μm long,

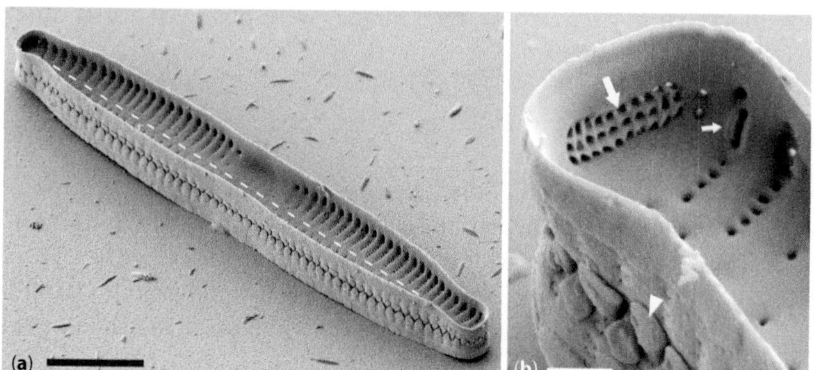

Figure 4.1 *Fragilaria mesolepta.* (a) A whole view of the two valves connected with linking spines. Upper valve showing the inside portion with the linear sternum (dashed line). (b) Enlarged valve end with the apical pore field (large arrow) and the linking spine (arrow head). Note the rimoportula (small arrow) placed in the border of the valve face and mantle. Scale bar = 5 μm (a), 0.5 μm (b).

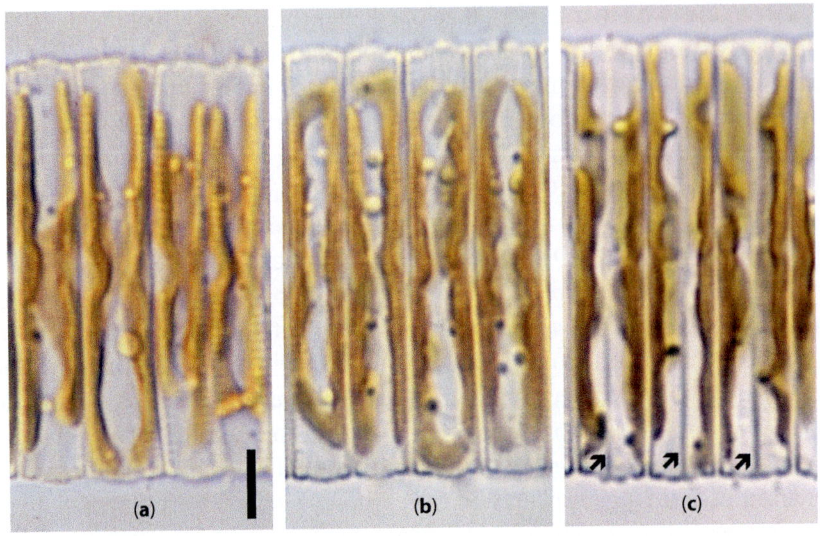

Figure 4.2 *Fragilaria mesolepta*. A part of a chain colony. (a) Each cell has one H-shaped chloroplast. (b) Divided chloroplast shortly before cell division. (c) After cell division. Note the newly forming valves (arrows). Scale bar = 5 μm.

2.5–3 mm wide, and had a narrow pattern center (sternum) between both apices (Figure 4.1a, dashed line). There were 18–20 striae in 10 mm. The valve apex had an apical pore field (Figure 4.1b, large arrow). Near one valve apex, a rimoportula was borne on the stria around the boundary between the valve face and mantle (Figure 4.1b, small arrow). The adjacent valves were connected to each other with wide-tipped linking spines (Figure 4.1a, arrowhead).

Cells had an H-shaped chloroplast, elongated parts underlaid each valve face, and the central part was appressed against one side of the girdle (Figure 4.2a). Prior to cell division, the chloroplast was divided into two (Figure 4.2b), and the formation of new valves began after cell division (Figure 4.2c, arrow). Live images were observed using an Optiphoto microscope (Nikon, Tokyo, Japan).

4.2.2 *Staurosira binodis*

S. binodis was collected from Shakujii Pond, Nerima, Tokyo in August 2017 (M-2343). The mature valves were lanceolate with two undulations on both sides, 23–24 mm long, 4.5 mm wide at the center, and 8–9 mm wide at the most inflated area (Figure 4.3). There were 12–14 striae in 10 mm. The sternum was linear (Figure 4.3a, dashed line) and expanded into a spindle-shaped axial area in the internal valve view (Figure 4.3b). Linking spines were aligned along the valve face and mantle juncture (Figure 4.3a, arrow).

Cells had an H-shaped chloroplast (Figure 4.4a, cells with a single asterisk) and divided into two just before cell division (Figure 4.4a, cell with a double asterisk). They were observed using an Axioskop microscope (Zeiss, Jena, Germany).

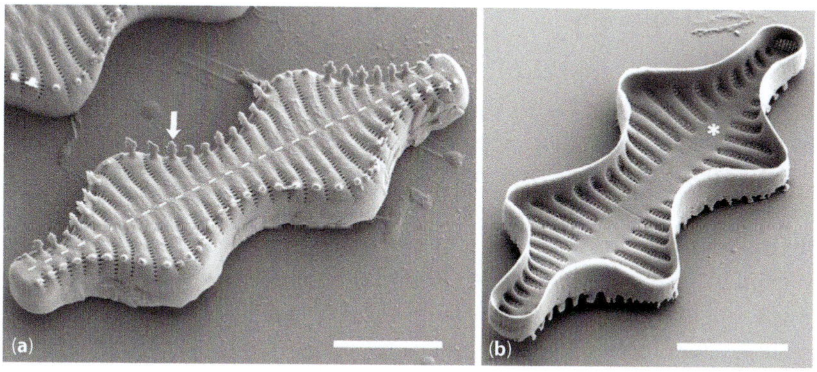

Figure 4.3 *Staurosira binodis*. (a) Valve exterior showing the narrow linear sternum (dashed line) and the linking spines (arrow) on the valve face and mantle juncture. (b) Valve interior with the spindle-shaped axial area (*). Scale bar = 5 μm.

4.2.3 Induction of Synchronous Division

A clone culture of *F. mesolepta* was established in 5-fold-diluted Bold's Basal Medium modified [4.28] under a condition of 20 °C, pH 7.8, and 12:12 (L:D) photoperiod. Synchronized cell division was induced by 48 h of continuous dark followed by light exposure on April 12, 1989. The culture at 2 h 40 min after lighting contained the highest number of cells with divided chloroplast (Figure 4.2b) and newly developing valves (Figure 4.2c), and we used this culture for observing valve morphogenesis.

In *S. binodis*, a clone was maintained under the same conditions as *F. mesolepta* and then placed in the dark for 36 h prior to light with N'-[2-(dimethylamino)ethyl]-2-[4-2-(4-pyridyl)-5-oxazolyl]phenoxy]acetohydrazide (PDMPO) (Molecular Probes, Eugene, OR, USA). The culture at 2–3 h after induction of synchronous division contained the highest number of cells with two chloroplasts (Figure 4.4a) and newly developing valves (Figure 4.4b). Therefore, they were used for electron microscopy.

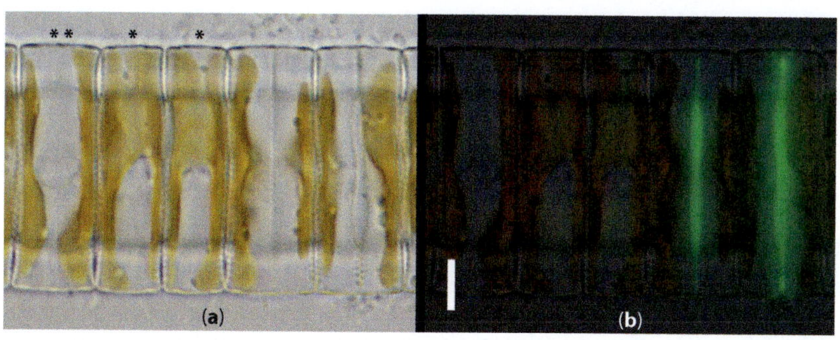

Figure 4.4 *Staurosira binodis*. A part of a chain colony. (a) Cells in three phases, with one chloroplast (*), two chloroplasts (**), and after cell division (no asterisk). (b) The same colony as (a). Green color indicates newly forming valves with PDMPO fluorescence. Scale bar = 5 μm.

4.2.4 Electron Microscopy

Live cells of both species were boiled in sulfuric acid with a small amount of potassium dichromate for 1 h and washed with water several times until the chemical was removed. The cleaned samples were dried on copper grids with a Formvar support membrane and observed using a 100CX-II transmission electron microscope (TEM) (JEOL, Tokyo, Japan) or S-8020 scanning electron microscope (SEM) equipped with a transmission device (STEM) (Hitachi High-Tech, Tokyo, Japan). For SEM, some portions of the samples were coated with osmium using an OPC40 osmium plasma coater (Filgen, Nagoya, Japan).

Voucher specimens cleaned with acid are available upon request from the author. The terminology used here for the valve structure followed that of Ross *et al.* [4.38] and Cox [4.9].

4.3 Results

In both species, the induced cellular synchronicity was not high in all cultures due to the absence of silica starvation pretreatment in the culture medium. However, it was sufficient for ontogenetic research, allowing us to observe prominently different valve formations in the early stages between both araphid diatoms.

4.3.1 *Fragilaria mesolepta*

Electron microscopy showed diverse forming valve lengths. The shorter the valve, the thinner and more delicate it was, while the longer the valve, the thicker and more developed it was (Figure 4.5), but the density of the virgae in all specimens was within the range of that in the mature valves (18–20 in 10 μm). In the sequential arrangement of these specimens, the forming valve of the earliest stage appeared like a chain of small rings or small annuli with extended unilateral short ribs, namely early virgae (Figures 4.5a, 4.6, arrow heads). The rings comprised a series of silica spherules with an indistinct periphery 55–75 nm in diameter with a core of ~30–40 nm. At one end of the chain, the developing virgae were bifurcated and arranged in semi-radial (Figure 4.6b). At the opposite end, multiple virgae were branched from ribs longer than the other virgae (Figure 4.6c). Subsequently the ribs were united into one, which had virgae bilaterally (Figures 4.5b–f). The difference between the two ends could be distinguished until the middle stages (Figures 4.5a–f, 4.7a, b). In the earliest stage, the growing edge of the virgae was finely fringed with silica with a size of 5–8 nm (Figures 4.6b, c).

As the valve developed, the chain of small rings became narrow and obscure (Figures 4.5b–f, 4.7a, b), and in the later stages, it was transformed into a single linear structure, i.e., the sternum (Figures 4.5g–i, 4.7c, d). This process proceeded toward both apices, so that the small holes remained until subsequent stages near the sternum ends (Figures 4.5d–h, 4.7c, d).

At the stage when the growing edge of the virgae reached the future mantle margin, the linking spines began to develop (Figure 4.7c, circled), followed by the formation of vimens to build up areolae, which started simultaneously throughout the whole valve (Figure 4.7d, arrows). The formation of the rimoportula had not yet begun at this stage. Thereafter, the

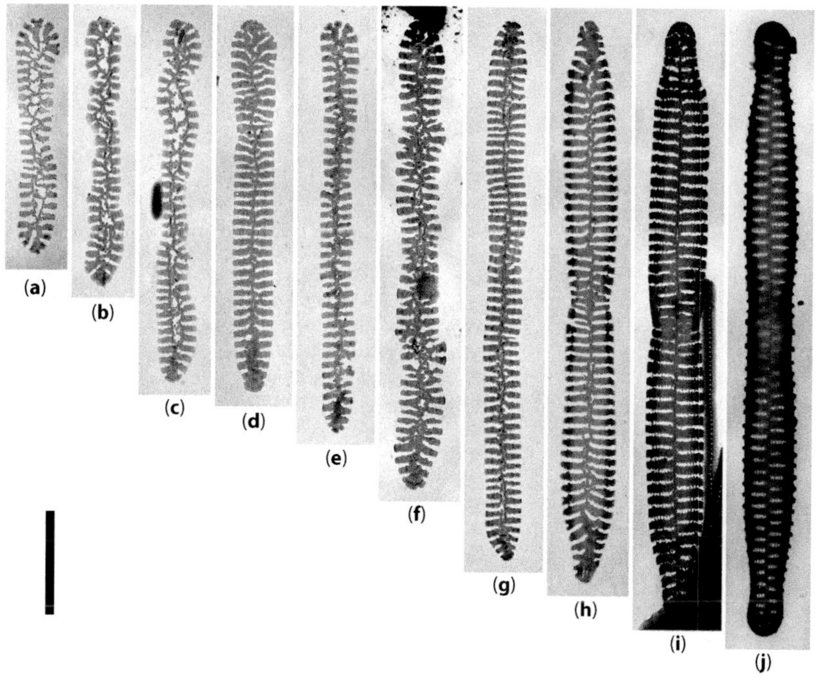

Figure 4.5 *Fragilaria mesolepta*. The series of forming valves arranged sequentially by the degree of development. As the valves get longer, the valve components become thicker and more developed. Scale bar = 5 μm.

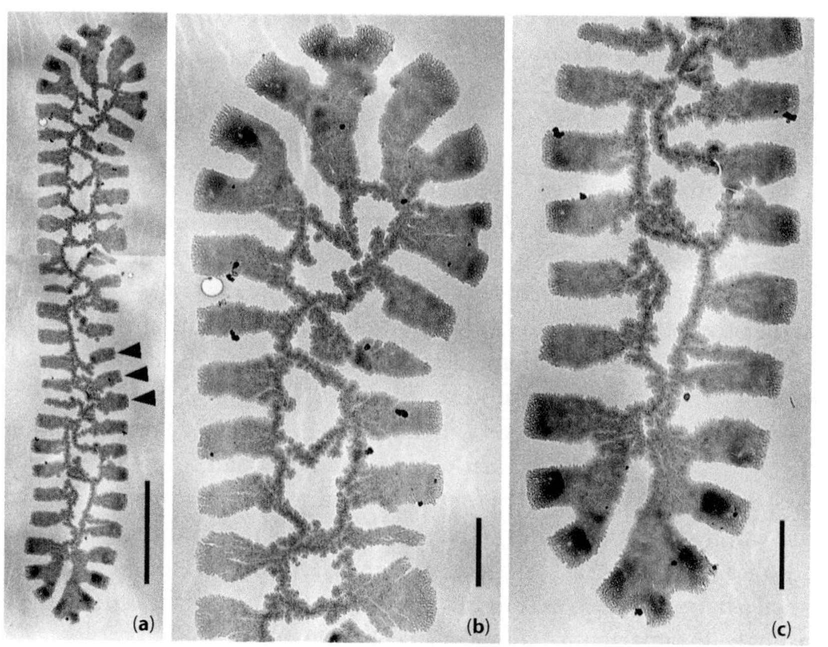

Figure 4.6 *Fragilaria mesolepta*. A forming valve in the earliest stage. (a) A whole view showing a chain of small rings with short virgae (arrow heads). (b) Enlargement of the upper half of (a). The ring chain is composed of silica spherules. The uppermost ring extends the branching virgae. (c) Enlargement of the lower half of (a). At the bottom end, two longer ribs extend multiple virgae. Scale bar = 2 μm (a), 0.5 μm (b, c).

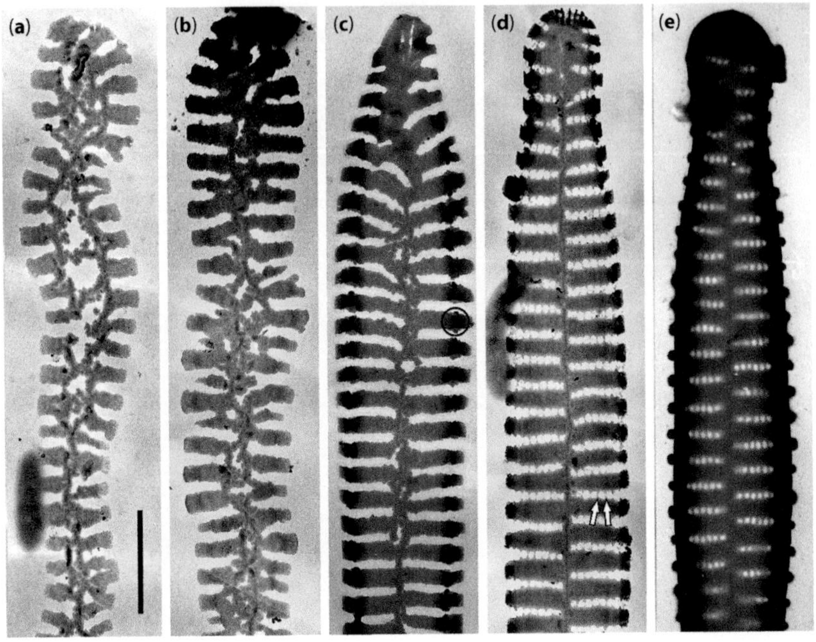

Figure 4.7 *Fragilaria mesolepta*. Sequential valve development after the earliest stage. The ring chain gradually becoming obscure and united into a single linear sternum. Note the initiation of the linking spine (c, circled) and the vimens (d, arrows). Scale bar = 2 μm.

rimoportula was created together with the areola occlusion and the silica-embedded central area before valve completion (Figures 4.1b, 4.7e).

4.3.2 *Staurosira binodis*

In the earliest stage that we observed, the forming valves were almost the same length as the complete valves (Figures 4.8a, b, d, 4.9c), and they were often observed as adjacent sibling valves sticking together (Figures 4.8d, 4.9a, b). The basic structure of the valves was a pair of siliceous threads arranged in arcs facing each other. These threads were ~80 nm wide, slightly thicker than the other structures and easily distinguishable. The young virgae grew outward from these threads, but their development varied between threads. In one thread (Figures 4.8c, e, 4.9a, TSV), the forming virgae were still short, small projection-like (Figures 4.8c, e, SV, 4.9b, arrows). In the other thread (Figures 4.8c, e, 4.9a, TLV), the developing virgae were longer (Figures 4.8c, e, LV, 4.9b, arrowheads) and part had a down feather-like structure (Figures 4.8e, circled, 4.9e). Except for this structure, the virgae density was within that of the mature valves (12–14 in 10 μm). The area between the two threads was occupied by unorganized silica (Figures 4.8a, b, d, 4.9a, b, asterisk). At one pole of the earliest valve, the threads formed a hairpin-like loop (Figures 4.8c, top, 4.9d), and at the other pole, they were fused into a single linear structure (Figures 4.8c, bottom, 4.9e).

In the following stages, two threads were united into a single linear sternum with bilaterally developing virgae (Figures 4.10a, b). However, the hairpin-like loop was still visible at one pole (Figure 4.10c). Even in the further development stage, in which the extended

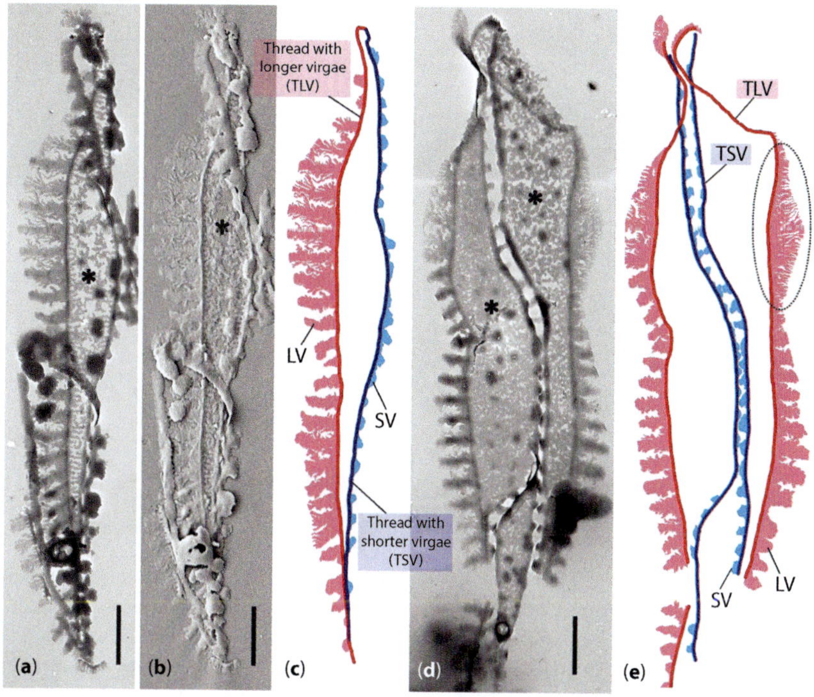

Figure 4.8 *Staurosira binodis*. Forming valves in the earliest stage. (a–c) Valve composed of a pair of threads arranged as arcs facing each other. Drawing (c) was prepared from multiple images observed with scanning electron microscope equipped with a transmission device (STEM) (a) and SEM (b). A couple of threads, one (TLV, red) with longer virgae (LV) and the other (TSV, blue) with shorter virgae (SV) make a loop at one end and fuse into a linear structure at the opposite end. (d, e) Sibling forming valves side by side. Two pairs of threads are partly broken and overlapped. Note the unorganized silica between two threads (*) and the down feather-like structure (circled). STEM image (d) and drawing (e). Scale bar = 2 μm (a, b, d).

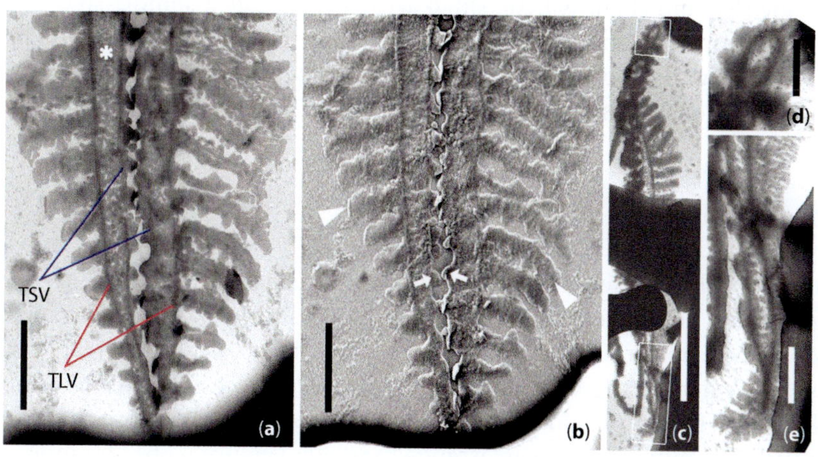

Figure 4.9 *Staurosira binodis*. Forming valves in the earliest stage. (a, b) Enlarged sibling valves side by side. Each valve comprises a thread with longer virgae (TLV) and a thread with shorter virgae (TSV). The densities of the longer virgae (arrow heads) and the shorter virgae (arrows) are the same. Note the unorganized silica (*). STEM (a) and SEM (b) images. (c–e) A pair of threads demonstrates both closed ends. Enlargement of top of (c) shows a rounded looping end (d). Enlargement of bottom of (c) shows a linearly fused end of two threads (e). Scale bar = 2 μm (a, b), 5 μm (c), 1 μm (d, e).

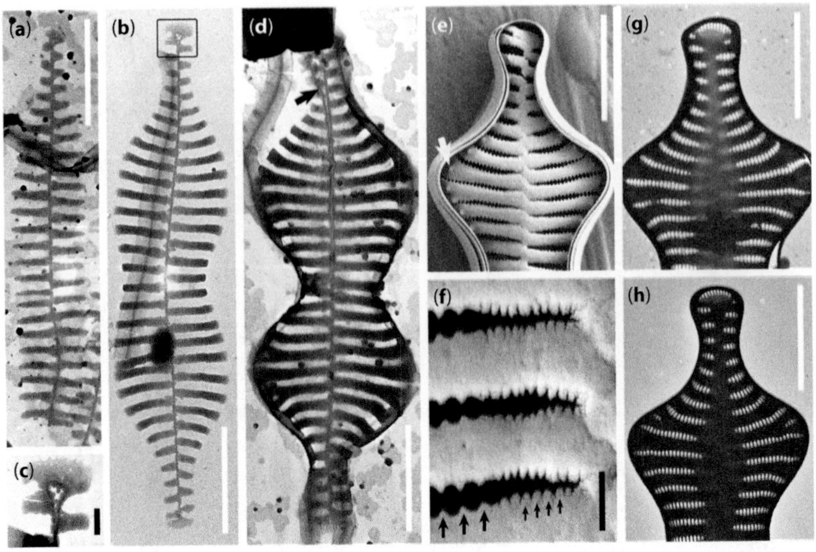

Figure 4.10 *Staurosira binodis.* Sequential valve development after the earliest stage. (a) An early forming valve with short virgae extending from a sternum. (c) Enlarged top of (b) showing the looping end. (d) A valve with the remaining of hairpin-like loop in the upper half (arrow). (e) A valve with early extending vimens. Note the initiation of linking spine (arrow). (f) Enlargement of (e). Note the virgae near the sternum bearing vimens (large arrows) and denser projections (small arrows). (g) A valve with forming axial area. (h) A valve with completed axial area. Scale bar = 5 μm (a, b, d, e, g, h), 0.5 μm (c, f).

virgae figured the valve outline with distinct double undulation, a hairpin-like loop that was half of the valve length was occasionally found (Figure 4.10d, arrow).

Thereafter, the vimens began to develop simultaneously throughout the valve (Figures 4.10e, f, large arrows). In addition, densely arranged projections also developed from the virgae near the sternum (Figure 4.10f, small arrows) to create a spindle-shaped axial area (Figure 4.10g). Moreover, the valves at the same stage began to form the linking spines (Figure 4.10e, arrow). In the last stage, the axial area became thicker, and the areola had occlusions to complete the mature valve (Figures 4.3, 4.10h).

4.4 Discussion

In the valve morphogenesis of pennate diatoms, the earliest valves show almost the same length as the complete valves [4.1] [4.5] [4.17] [4.29] except for a raphid diatom *Phaeodactylum*, which started valve formation from a 'π-like' structure [4.54]. *F. mesolepta* made its valve longer during valve development. However, it is debatable how new and additional chain elements are added. Both ends of the forming valve have characteristically different structures, bifurcated virgae arranged radially (Figure 4.6b) and longer rib with multiple virgae (Figure 4.6c). If new elements are added to either end, virgae with a specific form must appear in the middle of the valve. Therefore, it is unlikely that the new element is added in either end. However, it is not implausible that the additional element is inserted into the valve middle, because the alignment of silica spherules is highly plastic due to the small size of each binding surface and they are possibly disconnected easily, thus facilitating the addition of the new elements.

The chain of small rings formed in the early developmental stage suggests an evolutionary relationship with the annulus of the centrics. In particular, the appearance of the end of the round curving strand, which extends the branching virgae, may indicate homology with the annulus of the centrics.

A phenomenon similar to the extending pattern center in our specimens was observed in polar centric *Trieres* [4.26]. However, it differs from that of *F. mesolepta* in the absence of siliceous strings bridging between parallel ribs that extend from two elongate annuli. Moreover, due to lack of the whole view of forming valves, the details remain unclear. The SEM image of the mature valve of the same species did not show two elongated annuli (Figure 40e in [4.37]), but parallel ribs lay between two poles as an elongated annulus. Even though the annulus could not be identified in the complete valve, its formation in the early stage can be seen in *Stephanopyxis* [4.37] and *Hyalodiscus* [4.19]. This is due to the addition of a secondary structure on top of the basal siliceous layer or the initially formed siliceous structure [4.19], indicating the need for further studies on valve formation in *Trieres.*

The loop comprising a pair of threads in *S. binodis* resembles the elongated annulus observed during valve formation of the polar centric *B. biddulphiana* [4.52] and may be homologous to the annulus. The early development of annulus was observed in *D. brightwellii* [4.24] [4.25], *C. rostulatus* [4.25], and *T. sinensis* (as *Odontella* [4.26]) from the transverse section of the cells. Notably, the initiated silica deposition vesicle (SDV) was not a ring but a flat disk, and silicified annulus was generated around the margin of the SDV in a ring shape. If two threads of *S. binodis* were created in the margin of SDV like those diatoms, silica ion or silica sol remaining in the space surrounded by two threads could appear as unorganized silica after acid treatment (Figures 4.8a, b, d, 4.9a, b, asterisk), because strong acid promotes gelling of silica [4.14]. However, applying other sample preparation methods not using a strong acid, e.g., mechanical destruction of parental frustule by glass beads [4.16], will be necessary to substantiate this occurrence.

We observed the down feather-like structure several times in this study. The area of the structure (Figures 4.8d, c, 4.9e) is a space where virgae would normally occur. In other diatoms, virgae were generated in tube-like SDV, which expands separately like the fingers of a glove [4.12] [4.24] [4.44], while the SDV was continuous between virgae near to their junction with the central rib (axial area) in *Craticula cuspidata* (Kützing) D.G. Mann (as *Navicula* [4.12]). *C. cuspidata* is a large raphid diatom unlike *S. binodis*, but we suppose that the siliceous down feather was temporally generated in the continuous SDV and turning into virgae soon later, otherwise it was generated as an artifact in the vesicle.

Mann [4.27] illustrated a possible variant of the sternum but did not refer to the transformation of an annulus to the sternum. The morphological changes observed in the forming valves were drastic, and they indicate two potential pathways of how the sternum possibly evolves from annuli-possessing ancestors. Both *Fragilaria* and *Staurosira* are core araphid members and have their phylogenetic positions in different subclades [4.22] [4.33] [4.51]. Thus, our results possibly support their phylogeny, but *Synedra* and *Ulnaria* (formerly classified into *Synedra*), which start valve formation from a single linear structure [4.3] [4.18] [4.20], belong to the same subclade with *Fragilaria* as well. It seems contradictory that species within closely related lineages have different ontogenetic patters. However, the earliest developing valve in the reports of *S. arcus* was a fragment of 400 nm long, which formed a small rib unevenly on one side of the sternum with ~30 nm wide [4.20]. The earliest forming valve of *U. danica* showed a wide sternum (300–400 nm wide) [4.3], which was in contrast to the earliest sternum and annulus of other diatoms (measurements based on photographs of species from various reports and our observations, maximum 250 nm, many less than 140 nm). Moreover, the virgae densities of both species reduced to half or one-third as they developed, while the densities were stable through the valve formation process in the species in our study. Therefore, these specimens need to be further verified.

In contrast, a photograph of *Fragilaria* sp., which was shown as an example of the araphid diatom valve morphogenesis by Crawford and Schmid [4.11], is of interest as a linear sternum was shown to accompany a somewhat disconnected ring-like structure. However, the authors did not comment about this structure.

Among the basal araphids, *Raphoneis* was studied for valve ontogeny, but early forming valves were not covered [4.43]. Auxospore coverings of the basal araphid share both elements from core araphid+raphid and polar centric, the longitudinal perizonial bands and the properizonium, respectively [4.31]. In the valve ontogeny, the analysis of the early valve formation should be promoted further.

Moreover, the ontogenetic study of araphid pennate *Pseudostriatella oceanica* Sato, D.G. Mann et Medlin will also be important, since its mature valve has an annulus-like ring with several central small pores at one valve end (possibly at the other end as well) and a narrow rib runs to the other (possible) ring [4.41]. This is similar to the schematic drawing of the pattern center of the bipolar centric *T. sinensis* [4.23] [4.26]. *Pseudostriatella* has a close phylogenetic relationship with *Striatella*, but their phylogenetic positions have been indifferent depending on the analyses [4.41] [4.32] [4.42]. However, this species is one potential clue for pattern center evolution.

Most conventional studies of valve ontogeny have examined the development of the entire valve. However, TEM observation of silica precipitation in the SDV of the araphids as well as the polar centrics in cross-section will enhance our understanding of pattern development. In addition, recent advances in molecular biology have revealed proteins localized in the SDV and their functions [4.15] [4.21] [4.49], and have reported novel valve formation-related proteins [4.35]. It is highly expected that this field will lead to further research on valve evolution in the future.

4.5 Conclusion

Currently, there is no direct evidence documenting evolutionary transformation from the centric annulus to the pennate sternum. Initial valve development in two core araphid diatoms, *F. mesolepta* and *S. binodis*, exhibited dynamic changes in their silicified structure. In both species, a structure homologous to the annulus was transformed into the sternum during the ontogenetic process. Specifically, the annulus homologues are a chain of annulus-like rings with lateral virgae and the elongate loop with lateral virgae. The valves that formed first in these diatoms had high plasticity at the molecular level. The progression from the developmental origin of valves is not well documented in many araphid diatoms but is better understood in centric and biraphid species. Future efforts to investigate the early stages of valve formation in various diatoms, including basal araphids and polar centrics, are required for a deeper understanding of valve evolution for all diatoms.

References

[4.1] Bedoshvili, Y.D., Gneusheva, K.V., Popova, M.S., Avezova, T.N., Arsentyev, K.Y. and Likhoshway, Y.V. (2017) Frustule morphogenesis of raphid pennate diatom *Encyonema ventricosum* (Agardh) Grunow. *Protoplasma* 255, 911-921.

[4.2] Bedoshvili, Y.D. and Likhoshway, Y.V. (2019) Cellular mechanism of diatom valve morphologenesis. In: *Diatoms: Fundamentals and Applications*. J. Seckbach and R. Gordon (eds.) Scrivener Publishing LLC, Beverly: 97–112.

[4.3] Bedoshvili, Y.D., Volokitina, N.A. and Marchenkov, A.M. (2019) Valve morphogenesis and silicon dynamics in the synchronized culture of *Ulnaria danica. Limnology and Freshwater Biology* 5, 297–301.

[4.4] Boyle, J.A., Pickett-Heaps, J.D. and Czarnecki, D.B. (1984) Valve morphogenesis in the pennate diatom *Achnanthes coarctata. Journal of Phycology* 20, 563–573.

[4.5] Chiappino, M.L. and Volcani, B.E. (1977) Studies on the biochemistry and fine structure of silica shell formation in diatoms. VII. Sequential cell wall development in the pennate *Navicula pelliculosa. Protoplasma* 93, 205–221.

[4.6] Cox, E.J. (1999) Variation in patterns of valve morphogenesis between representatives of six biraphid diatom genera (Bacillariophyceae). *Journal of Phycology* 35, 1297–1312.

[4.7] Cox, E.J. (2001) What constitutes a stauros? A morphogenetic perspective. In: *Studies on diatoms. Festschrift for Horst Lange-Bertalot.* R. Jahn, J.P. Kociolek, A. Witkowski and P. Compère (eds.) A.R.G. Gantner Verlag K.G., Ruggell: 303–316.

[4.8] Cox, E.J. (2006) Raphe loss and spine formation in *Diadesmis gallica* (Bacillariophyta) – an intriguing example of phenotypic plasticity in a diatom. Nova Hedwigia, Beiheft 130, 163–176.

[4.9] Cox, E.J. (2012) Ontogeny, homology and terminology of pennate diatoms – wall morphogenesis as an aid to character state recognition and character state definition for pennate diatom systematics. *Journal of Phycology* 48, 1–31.

[4.10] Cox, E.J. and Kennaway, G.M. (2004) Studies of valve morphogenesis in pennate diatoms: investigating aspects of cell biology in a systematic context. In: *Proceedings of the 17th International Diatom Symposium.* M. Poulin (ed.) Biopress Ltd., Bristol: 35–48.

[4.11] Crawford, R. M. and Schmid, A.-M.M. (1986) Ultrastructure of silica deposition in diatoms. In: *Biomineralization in lower plants and animals.* B.S.C. Leadbeater and R. Ridig (eds.) Clarendon Press, Oxford:291–314.

[4.12] Edgar, L.A., and Pickett-Heaps, J.D. (1984) Valve morphogenesis in the pennate diatom *Navicula cuspidata. Journal of Phycology* 20, 47–61.

[4.13] Fryxell, G.A. and Miller III, W.I. (1978) Chain-forming diatoms: three araphid species. *Bacillaria* 1, 113–136.

[4.14] Gordon, R. and Drum, R.W. (1994) The chemical basis of diatom morphogenesis. *International Review of Cytology* 150, 243-241.

[4.15] Gorlich, S., Pawolski, D., Zlotnikov, I., Kröger, N. (2019) Control of biosilica morphology and mechanical performance by the conserved diatom gene Silicanin-1. *Communications Biology* 2, #245.

[4.16] Heintze, C., Formanek, P., Pohl, D., Hauptstein, J., Rellinghaus, B. and Kröger, N. (2020) An intimate view into the silica deposition vesicles of diatoms. BMC Materials 2, #11

[4.17] Idei, M., Sato, S., Nagumo, T. and Mann, D.G. (2018) Valve morphogenesis in *Diploneis smithii* (Bacillariophyta). *Journal of Phycology* 54, 171–186.

[4.18] Kaluzhnaya, O.V. and Likhoshway, Y.V. (2007) Valve morphogenesis in an araphid diatom *Synedra acus* subsp. *radians. Diatom Research* 22, 81–87.

[4.19] Kaczmarska, I., Ehrman, J.M., Davidovich, N.A., Davidovich, O.I. and Podunay, Y.A. (2020) Valve morphogenesis in selected centric diatoms. *Botany* 98, 725–733.

[4.20] Kharitonenko, K.V, Bedoshvili, Y.D. and Likhoshway, Y.V. (2015) Changes in the micro- and nanostructure of siliceous valves in the diatom *Synedra acus* under the effect of colchicine treatment at different stages of the cell cycle. *Journal of Structural Biology* 190, 73-80.

[4.21] Kotzsch, A., Groger, P., Pawolski, D., Bomans, P.H.H., Sommerdijk, N.A.J.M., Schlierf, M. and Kröger, N. (2017) Silicanin-1 is a conserved diatom membrane protein involved in silica biomineralization. *BMC Biology* 15, #65.

[4.22] Li, C.L., Ashworth, M.P., Witkowski, A., Lobban, C. S., Zgłobicka, I, Kurzydłowski, K. J. and Qin, S. (2016) Ultrastructural and molecular characterization of diversity among small araphid diatoms all lacking rimoportulae. I. Five new genera, eight new species. *Journal of Phycology* 52, 1018–1036.

[4.23] Li, C.-W. and Volcani, B.E. (1984) Aspects of silicification in wall morphogenesis of diatoms. *Phillosophical Transactions of the Royal Society, London. B* 304, 519–528.

[4.24] Li, C.-W. and Volcani, B.E. (1985a) Studies on the biochemistry and fine structure of silica shell formation in diatoms. VIII. Morphogenesis of the cell wall in a centric diatom, *Ditylum brightwellii. Protoplasma* 124, 10–29.

[4.25] Li, C.-W. and Volcani, B.E. (1985b) Studies on the biochemistry and fine structure of silica shell formation in diatoms. IX. Sequential valve formation in a centric diatom, *Chaetoceros rostratum. Protoplasma* 124, 30–42.

[4.26] Li, C.-W. and Volcani, B.E. (1985c) Studies on the biochemistry and fine structure of silica shell formation in diatoms. X. Morphogenesis of the labiate process in centric diatoms. *Protoplasma* 124, 147–157.

[4.27] Mann, D.G. (1984) An ontogenetic approach to diatom systematics. In: *Proceedings of the Seventh International Diatom Symposium.* D.G. Mann (ed.) Otto Koeltz, Koenigstein: 113–144.

[4.28] Mayama, S. (1992) Morphology of *Eunotia multiplastidica* sp. nov. (Bacillariophyceae) examined throughout the life cycle. *Korean Journal of Phycology* 7, 45–54.

[4.29] Mayama, S. and Kobayasi, H. (1989) Sequential valve development in the monoraphid diatom *Achnanthes minutissima* var. *saprophila. Diatom Research* 4, 111–117.

[4.30] Mayama, S. and Kuriyama, A. (2002) Diversity of mineral cell coverings and their formation processes: a review focused on the siliceous cell coverings. *Journal of Plant Research* 115, 289–295.

[4.31] Medlin, L.K. (2016) Evolution of the diatoms: major steps in their evolution and a review of the supporting molecular and morphological evidence. *Phycologia* 55, 79–103.

[4.32] Medlin, L.K. and Desdevises, Y. (2016) Phylogeny of 'araphid' diatoms inferred from SSU and LSU rDNA, rbcL and psbA sequences. *Vie et Milieu* 66, 129-154.

[4.33] Medlin, L.K. and Kaczmarska, I. (2004) Evolution of the diatoms: V. Morphological and cytological support for the major clades and a taxonomic revision. *Phycologia* 43, 245–270.

[4.34] Medlin, L.K., Sato, S., Mann, D.G. and Kooistra, W.H.C.F. (2009) Molecular evidence confirms sister relationship of *Ardissonea, Climacosphenia*, and *Toxarium* within the bipolar centric diatoms (Bachillariophyta, Mediophyceae), and cladistic analyses confirm that extremely elongated shape has arisen twice in the diatoms. *Journal of Phycology* 44: 1340–1348.

[4.35] Nemoto, M., Iwaki, S., Moriya, H., Monden, Y., Tamura, T., Inagaki, K., Mayama, S. and Obuse, K. (2020) Comparative gene analysis focused on silica cell wall formation: Identification of diatom-specific SET domain protein methyltransferases. *Marine Biotechnology* 22, 551–563.

[4.36] Pickett-Heaps, J.D., Cohn, S.A., Schmid, A.-M. and Tippit, D.H. (1988) Valve morphogenesis in *Surirella* (Bacillariophyceae). *Journal of Phycology* 24, 35–49.

[4.37] Pickett-Heaps, J., Schmid, A.-M.M. and Edgar, L.A. (1990) The cell biology of diatom valve formation. In: *Progress in phycological research. Vol. 7*. F.E. Round and D.J. Chapman (eds.) Biopress, Bristol: 1–168.

[4.38] Ross, R., Cox, E.J., Karayeva, N.I. Mann, D.G., Paddock, T.B.B., Simonsen, R. and Sims, P.A. (1979) An amended terminology for the siliceous components of the diatom cell. *Nova Hedwigia, Beiheft* 64, 513–533.

[4.39] Round, F.E., Crawford, R.M. and Mann, D.G. (1990) *Diatoms. Biology and morphology of the genera.* Cambridge University Press, Cambridge.

[4.40] Sato, S. (2010) Valve and girdle band morphogenesis in a bipolar centric diatom *Plagiogrammopsis vanheurckii* (Cymatosiraceae, Bacillariophyta). *European Journal of Phycology* 45, 167–176.

[4.41] Sato, S., Mann, D.G., Matsumoto, S. and Medlin, L. K. (2008) *Pseudostriatella* (Bacillariophyta): a description of a new araphid diatom genus based on observations of frustule and auxospore structure and 18S rDNA phylogeny. *Phycologia* 47, 371–391.

[4.42] Sato, S., Nanjappa, D., Dorrell, R.G., Vieira, F.R.J., Kazamia, E., Tirichine, L., Veluchamy, A., Heilig, R., Aury, J.-M., Jaillon, O., Wincker, P., Fussy, Z., Obornik, M., Muñoz-Gómez, S.A., Mann, D.G., Bowler, C. and Zingone, A. (2020) Genome-enabled phylogenetic and functional reconstruction of an araphid pennate diatom *Plagiostriata* sp. CCMP470, previously assigned as a radial centric diatom, and its bacterial commensal. *Scientific Report* 10:9449.

[4.43] Sato, S., Watanabe, T., Nagumo, T. and Tanaka, J. (2011) Valve morphogenesis in an araphid diatom *Rhaphoneis amphiceros* (Rhaphoneidaceae, Bacillariophyta), *Phycological Research* 59, 236–243.

[4.44] Schmid, A.-M.M. (1994) Aspects of morphogenesis and function of diatom cell walls with implications for taxonomy. Protoplasma 181, 43–60.

[4.45] Schmid, A.-M. (1979) The development of structure in the shells of diatoms. *Nova Hedwigia, Beiheft* 64, 219–237.

[4.46] Schmid, A.-M.M. and Volcani, B.E. (1983) Wall morphogenesis in *Coscinodiscus wailesii* Gran and Angst. I. Valve morphology and development of its architecture. *Journal of Phycology* 19, 387–402.

[4.47] Stosch, H.A. von (1977) Observations on *Bellerochea* and *Streptotheca*, including descriptions of three new planktonic diatom species. *Nova Hedwigia, Beiheft* 54, 113 –166.

[4.48] Tesson, B. and Heldebrand, M. (2010) Dynamics of silica cell wall morphogenesis in the diatom *Cyclotella cryptica*: Substructure formation and the role of microfilaments. *Journal of Structural Biology* 169, 62–74.

[4.49] Tesson, B., Lerch, S.J.L. and Hildebrand, M. (2017) Characterization of a new protein family associated with the silica deposition vesicle membrane enables genetic manipulation of diatom silica. *Scientific Report* 7, #13457.

[4.50] Theriot, E.C., Ashworth, M., Ruck, E., Nakov, T. and Jansen, R.K. (2010) A preliminary multigene phylogeny of the diatoms (Bacillariophyta): challenges for future research. *Plant Ecology and Evolution* 143, 278–296.

[4.51] Theriot, E.C., Ashworth, M.P., Nakov, T., Ruck, E. and Jansen, R.K. (2015) Dissecting signal and noise in diatom chloroplast protein encoding genes with phylogenetic information profiling. *Molecular Phylogenetics and Evolution* 89, 28–36.

[4.52] Tiffany, M.A. (2002) Valve morphogenesis in the marine araphid diatom *Gephyria media* (Bacillariophyceae). *Diatom Research* 17, 391–400.

[4.53] Tiffany, M.A. (2015) Valve and girdle band morphogenesis in the pseudocellate diatom species *Biddulphia biddulphiana* (J.E. Smith) Boyer and *Isthemia nervosa* Kütz. *Nova Hedwigia, Beiheft* 144, 61–95.

[4.54] Vartanian, M., Desclés, J., Quinet, M., Douady, S. and Lopez, P.J. (2009) Plasticity and robustness of pattern formation in the model diatom *Phaeodactylum tricornutum. New Phytologist* 182, 429–442.

[4.55] Watanabe, T., Mayama, S., and Idei, M. (2012) Overlooked heteropolarity in *Surirella* cf. *fastuosa* (Bacillariophyta) and relationships between valve morphogenesis and auxospore development. *Journal of Phycology* 48: 12265–1277.

5

Mathematical Basis for Diatom Growth Modeling

Dariush Sardari

Department of Medical Radiation Engineering, Science and Research Branch, Islamic Azad University, Tehran, Iran

Abstract

A physical-mathematical perspective on population growth of diatoms is presented. Depending on the conceptual needs, biomolecular interpretations are invoked. Beginning with the simple exponential growth of population, the mathematical approach is carried out on differential equations with time-varying coefficients. With decreasing growth rate coefficient, population reaches saturation at its maximum. There are various magnitudes of growth rate corresponding to different sizes of diatoms. Taking into consideration nutrient uptake and metabolism as well as internal structures and phenomena of diatoms, models with different degrees of complexity have been developed with their outlooks presented here. Monod and Michaelis-Menten models are mostly based on kinetics of growth and nutrient uptake. Droop, Aquaphy, and mechanistic models consider internal cell structure and go beyond to enzymatic and molecular reaction phenomena.

***Keywords*:** Marine biology, cell modeling, diatom growth, mathematical models, differential equations

5.1 Introduction

Diatoms are single cell algae that are surrounded by a cell wall made of silica (hydrated silicon dioxide). The diameter in various types of diatoms ranges from 1 to 200 μm [5.25] [5.46]. Diatoms can divide more rapidly than other phytoplanktons. During their physiologic process, they convert carbon dioxide into organic carbon. In this process, oxygen is generated. Hence, diatoms are known as the major contributor to global carbon fixation [5.32].

Living diatoms generate a significant proportion (30%) of the oxygen produced on the planet earth. Silicates are essential as a component of the cell wall as well as metabolic regulators. Silicon, constituting the major nutrients of diatoms, is the second most abundant element in the Earth's crust [5.26]. The existence of silicate macronutrient and selenium micronutrients in the culture medium is important in the environment.

Email: sardari@srbiau.ac.ir; dsardari@hotmail.com
Google Scholar: https://scholar.google.com/citations?user=8WvL1SwAAAAJ&hl=en&oi=ao
ResearchGate: https://www.researchgate.net/profile/Dariush-Sardari-2
ORCID: https://orcid.org/0000-0002-9400-8243

Janice L. Pappas (ed.) The Mathematical Biology of Diatoms, (121–140) © 2023 Scrivener Publishing LLC

Diatom growth maximization is becoming imperative for enhancing their biomass production. The biotechnological application of such photosynthetic microorganisms makes the study of their environmental characteristics in laboratory cultures an important matter. The numeric value of the growth rate is the combined effect of the cell size and environmental factors such as temperature, light, and nutrient's concentration. Besides, each kind of diatom has inherent differences from others in nutrient requirements and response to light and temperature [5.7] [5.14].

During the 1940s to 1960s, Monod [5.35] [5.36], Droop [5.8], and Michaelis-Menten [5.33] introduced the early mathematical concept for describing diatom physiology. The important part of their concept is the growth rate, which is a coefficient relating the increase of population in unit time to the existing population; in other words, cell concentration per volume of the environment.

In the simplest view, if only a few diatoms are located in a proper environment, their population begins to rise. Taking into account the limited amount of nutrients around cells, with increasing population, the competition between cells for taking up nutrients turns rather strict and population growth becomes slower [5.52] [5.19] [5.7] [5.37]. The role of light intensity has been investigated in the literature [5.51] [5.7] [5.27], as well as temperature [5.23] [5.3] [5.14].

The objective of this chapter is to pave the way towards a better understanding of the mechanisms and kinetics governing the growth or decline of a diatom population. In this regard, mathematical functions are invoked to describe cell concentration and its temporal variations with environmental factors such as temperature, light and nutrients. Furthermore, with the use of mathematical formulations one can put laboratory results in a predictable framework. In the present chapter the basic knowledge of mathematics for analyzing cell proliferation is demonstrated and a chronicle of mathematical models in diatom growth is reviewed.

5.2 General Physiology of Diatoms

Diatoms have gained interest as photosynthetic sources of nutraceutical lipids and biofuels. They need carbon (i.e., CO_2), light energy and elements, such as nitrogen, silicon and phosphorus, for growth. Carbon is assimilated through the process of photosynthesis where CO_2 is incorporated into organic form (carbohydrates of low molecular weight), using light as an energy source [5.9]. As an essential nutrient, nitrogen is necessary for synthesis of amino acids, nucleonic acids and lipids. It is taken up in inorganic form, usually as ammonium or nitrate, and enters into complex molecules such as proteins, that form the biosynthetic and photosynthetic apparatus of the algae. Another form of carbon compound, carbohydrates are vital for synthesizing proteins, as they act both as a source of carbon as well as energy [5.28].

Some physiological compounds that are important in cell activity are functional and structural metabolites. These are carbon monomers, early products of photosynthesis, internal carbon reserves in the form of carbohydrate and fatty acids plus intracellular soluble phosphate in polyphosphate reserves. Carbohydrates that are not channeled into protein production are converted into storage molecules (such as starch) that serve as a source of carbon and energy during times when photosynthesis stops (night).

In general, elemental density in diatoms is modulated by environmental conditions such as light intensity and ambient temperature. Cold temperature facilitates uptake of elements by the diatom while it grows. Hence, polar diatoms contain more lipids, and hence a higher amount of phosphorous. On the other hand, rates of consumption of phosphorus (P) and carbon (C) in diatoms are not similar. These elements are basic for making RNA and DNA in diatoms [5.31] [5.39]. Metabolic rate, the most important biological rate, varies with temperature, body size and light intensity. It is the rate at which organisms take up, transform and expend energy and materials. Geometric constraints affect packing, limiting diffusion and light absorption. Nutrient and light limitation can also undermine growth [5.10].

Total cellular carbon is divided into separate components on the basis of their function. These comprise synthetic and structural material (i.e., photosynthetic apparatus, ribosomes, genetic material, membranes, etc.), stored carbon (i.e., carbohydrates, lipids), and carbon monomers (i.e., photosynthetic products and precursors of complex molecules). The C:N ratio is proportional to diatom size [5.24].

Light intensity absorbed by a diatom cell modulates its activities through its photosynthesis process. For example, maximum specific growth rate for HL (high light) and LL (low light) is 0.74 and 0.56 day^{-1} [5.40]. Intracellular lipid is consumed overnight and dips to a minimum in early morning.

The most recent review for the role of nutrients in diatom physiology appears in Giri *et al.* [5.20]. The response of diatom growth to excess and lack of metals in the environment is discussed and the effect of heavy metals in mortality of diatoms is pointed out.

5.3 Mathematical View of Diatom Growth

Mathematical models are useful tools for anticipating the response of a diatom population to variations in nutrient concentration and other environmental factors. A realistic mathematical model should contain multiple inputs. In the wide sense of modeling, besides nutrients, such as concentration of nitrogen and silicon and carbon, the temperature must be taken into account. Hence, a mechanistic approach to modeling may be necessary [5.19]. A sufficient definition for a mechanistic model is presented in the literature [5.18]. Under the ideal condition, the population of dividing cells rise up in an exponential manner (Figure 5.1):

$$N(t) = N_0 e^{\mu t} \tag{5.1}$$

Where:
N_0 = cell population at t=0
$N(t)$ = cell population at time t
μ = growth rate
t = time lapsed

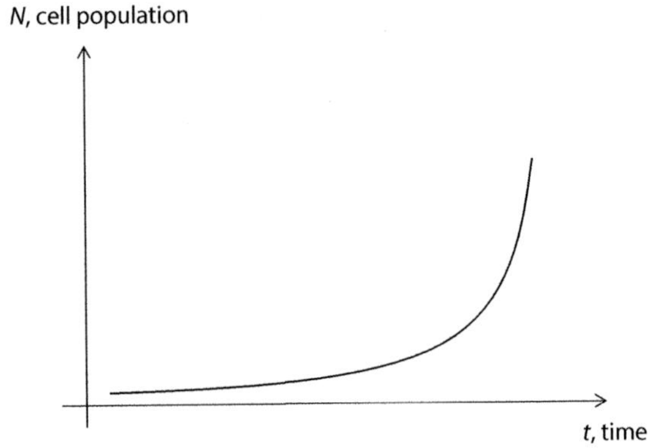

Figure 5.1 Exponential population growth.

Assuming that each cell undergoes one division per day, the cell population doubles in 24 hours. Thus, for t=1 day, $N(1)$ becomes $2N_0$. Putting this in Eq. (5.1), the growth rate is estimated as:

$$\ln(2) = \mu t;\ \mathrm{t} = 1;\ \text{and}\ \mu = 0.693\ \mathrm{d}^{-1}$$

which is in agreement with the usual data for specific growth rate of diatoms [5.52] [5.49] [5.39]. Using Eq. (5.1), the growth rate is obtained in Eq. (5.2) below:

$$\mu(d^{-1}) = \frac{\ln\left(\frac{C_n}{C_{n-1}}\right)}{t_n - t_{n-1}} \tag{5.2}$$

where C_n shows cell concentration at day n and t_n-t_{n-1} is actually one day. In a more mathematical way [5.1]:

$$\mu = \frac{1}{C} \cdot \frac{dC}{dt} \tag{5.3}$$

A similar parameter that implies the rate of rise in cell population is given [5.42]:

$$k\ (day^{-1}) = 3.322\ [\log(N_1 - N_0)]/(t_1 - t_0) \tag{5.4}$$

Where:
k = division/day
N_1 = cell density at time t_1
N_0 = cell density at the inoculation of culture

It is important to note that the exponential growth of cell concentration (or total population) happens for the case of constant growth rate:

$$\mu = \text{Const.} \Longrightarrow \text{pure exponential growth}$$

which leads to Eq. (5.1). In reality, this is not true and the growth rate varies from time to time. This is due to changes in the availability of nutrients in the environment and the higher demand of a dense cell population. In this case, one may consider the argument in the following paragraphs.

The basic equation of population growth appears as a first-order differential equation as follows [5.43]:

$$\frac{dP}{dt} = \mu_0 P\left(1 - \frac{P}{K}\right) \tag{5.5}$$

In which:
t = time
P = population or cell concentration
μ_0 = growth rate (constant)
K = carrying capacity (the maximum population that can be supported by the environment)

Hence, with increasing population P the effective growth rate $\mu_0\left(1-\frac{P}{K}\right)$ decreases. When $P=K$ the population growth stops (dP/dt=0) (Figure 5.2).

Besides other factors, such as temperature, light, type of diatom, etc., the growth rate is a function of diatom cell volume:

$$\mu_{max} = 3.4V^{-0.13} \tag{5.6}$$

where V is the cell volume and the equation is valid for: $13<V<7\text{x}10^5\ \mu\text{m}^3$ [5.44].

The above procedure is simplified because the growth rate is not constant over a long period of time and varies with nutrient concentration as well as population density. The population growth follows the differential equation:

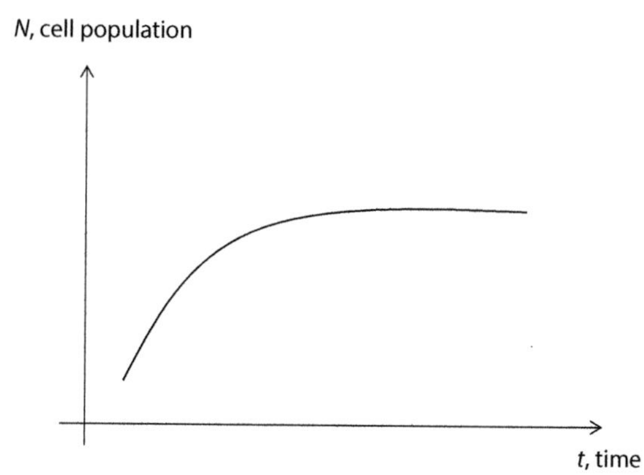

Figure 5.2 Simple growth curve depicting the saturable behavior of Eq. 5.5.

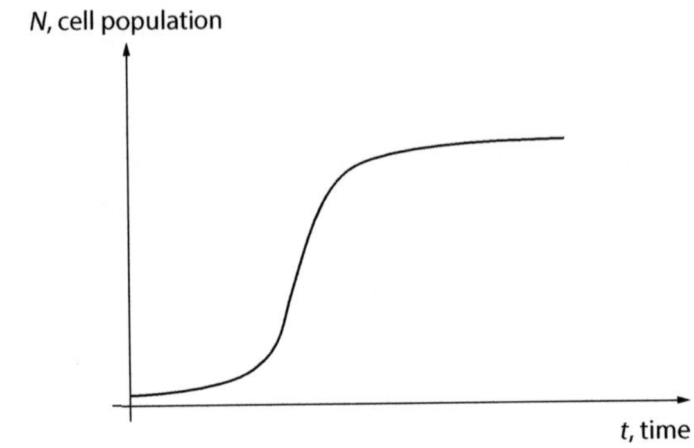

Figure 5.3 Gompertzian growth [5.40] [5.41] [5.29].

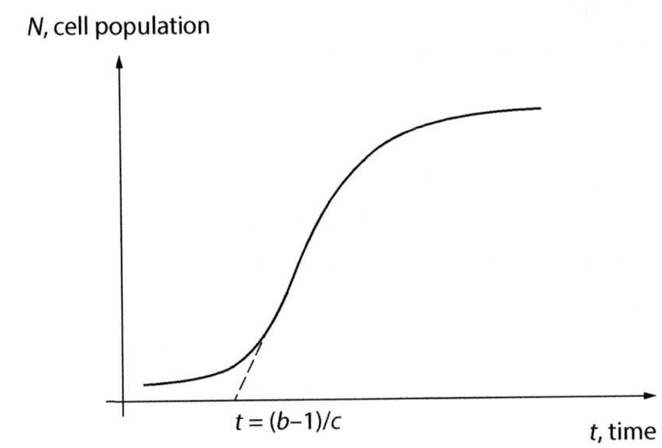

Figure 5.4 Gompertzian lag time. See Eq. (5.12) for parameters a, b, c. [5.54].

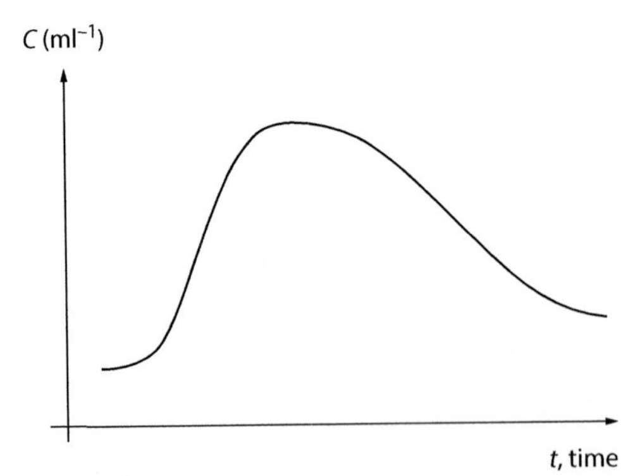

Figure 5.5 Cell population exceeds the limit that may be supported by nutrients leading to negative growth rate and reduction in population [5.37].

$$\frac{dN(t)}{dt} = \mu(t)N(t) \tag{5.7}$$

This is a simple differential equation, though its solution could be challenging if the function $\mu(t)$ becomes difficult to integrate.

In a more realistic case, the Gompertz law is valid for population growth. This is actually a sigmoid function (Figure 5.3). The detail of sigmoid shape (such as elongation in time scale) depends on maximum growth rate μ_{max} [5.50]. Lag time in the Gompertz law is described in Zwietering *et al.* [5.54] (Figure 5.4). Shortage of nutrients causes deviation from Gompertzian behavior (Figure 5.5).

5.4 Physical Basis for Diatom Modeling

In this section, a view of the physics of diatom cells and population growth is presented. All kinds of biological cell growth and reproduction takes place by absorption of nutrients that are transported across its wall (and membrane) from the surrounding environment. Hence, the surface area of the cell is a factor that governs the growth rate. On the other hand, as a rough estimate, the metabolism of the cell is proportional to its volume. We conclude that the surface to volume ratio is an important geometrical parameter in cell proliferation and physiological studies.

5.4.1 Diatom Dimensions

Assuming spherical shape for the cell with radius R, the surface to volume ratio is obtained as:

$$\frac{S}{V} = \frac{4\pi R^2}{(4/3)\pi R^3} \tag{5.8}$$

thus:

$$\frac{S}{V} = \frac{3}{R} \tag{5.9}$$

According to Eq. (5.9), one may conclude that bigger cells have smaller surface to volume ratio and hence grow slower. The size dependence of population growth rate takes effect by the balance of nutrient uptake and metabolism.

The uptake and growth rate of diatoms as a function of nutrient concentration is governed by the surface to volume ratio of the cell. For example, the term $(S/V)^{-3.66}$ appears in the half saturation constant of silicon and $(S/V)^{-1.77}$ for iron. Such dependency is universal among all types of diatoms with considerations about fitting errors [5.44]. When the cell grows, its average energy consumption per unit of its volume decreases [5.5].

$$R = aM^b \tag{5.10}$$

In which:
R = metabolic rate per unit mass
M = mass of the cell
'a' and 'b' = fitting parameters

Geider *et al.* [5.18] have stated that

$$-0.33<b<-0.25$$

Actually, R is equivalent to growth rate and reduces as the cell grows and becomes larger in volume.

Organismal metabolic rates influence many ecological processes, and the mass-specific metabolic rate of organisms decreases with increasing body mass according to a power law. The exponent in this equation is commonly thought to be the three-quarter power of total body mass, determined by fundamental physical laws that extend across various species. The value of 0.75 is an interspecies "average" of scaling exponents that varies naturally

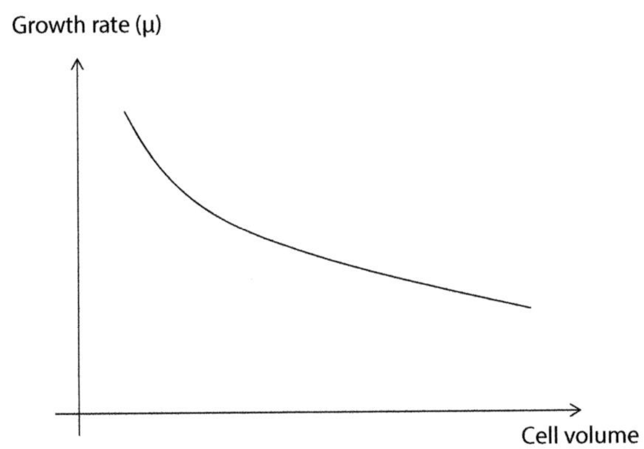

Figure 5.6 Small growth for large cells [5.18].

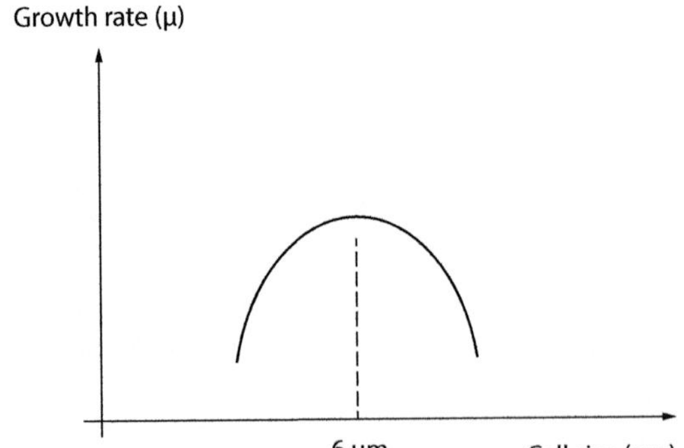

Figure 5.7 Growth vs. cell size variation is not monotonic [5.51].

between certain boundaries. Therefore, it is expected that among varieties of diatoms, those with larger size possess smaller growth rate (Figure 5.6). Later it was shown that the above trend is not monotonic and manifests itself with a maximum (Figure 5.7). This means that below a specific cell volume, the growth rate increases with the cell size [5.51]. For larger cells, the surface to volume ratio drops and less nutrients per mass can cross into the cell. On the other hand, too small cells absorb less light, leading to inadequate energy from photosynthesis. An interesting discussion appears in the literature [5.15].

5.4.2 Ambient Temperature

The effect of temperature on the biological process is expressed as Q10, which quantifies temperature dependence across a limited temperature [5.23] [5.14]. It is the ratio of increase in chemical reaction velocity when the temperature rises by 10 °C. The biological rate is approximately an exponential function of ambient temperature in degree Celsius with 15% error.

The effect of temperature on metabolic rate is described by the Hoff-Arrhenius relation [5.3]:

$$B = A.e^{-E/kT} \tag{5.11}$$

Where:
B = metabolic rate
T = ambient temperature in degrees Kelvin
E = activation energy
k = Boltzmann's constant
A = normalization factor

A more general equation taking the effect of organism mass in metabolic rate leads to:

$$B = A.M^{3/4}e^{-E_i/kT}$$

There have been interesting works on formulation of temperature effect on diatom growth rate. One example is:

$$\mu_{max} = 0.48e^{0.126T'}$$

in which T' is the environment temperature in degrees Celsius [5.52]. Although the above result is mostly experimental, the exponential behavior of the function agrees with that of theoretical work carried out decades later [5.23] [5.3]. Later, the temperature corrected growth rate was worked out [5.6]. The exponential rise in growth rate as a function of temperature (in Kelvin or Celsius) is called the Eppley curve [5.31] [5.14].

5.4.3 Light Intensity and Duration

The main source that provides energy for diatom growth is photosynthesis. Hence, it is expected to have a positive effect on growth rate with more light intensity and duration.

The effect of photon flux density (PFD) on the rate of metabolic process was investigated [5.18]. It was observed that diatoms undergoing a Si-limited experiment show higher growth rate in continuous light than in a light-dark cycle [5.12]. Concentration of light-sensitive pigments in a cell was considered in a composite model that included an allometric and bio-optic model [5.11].

The increase in growth rate does not have a linear relationship with light. Actually, light-limited growth rate has been observed in phytoplankton [5.28]. The photosynthesis rate varies with light intensity as:

$$P_h = P_{h-\max}\,(1 - e^{A.I})$$

in which P represents photosynthesis rate, A is light absorption coefficient and I stands for light intensity. The above equation does account for chlorophyll concentration inside the cell. Its concentration has an approximately linear effect on the growth rate assuming a constant light intensity.

5.5 Review of Existing Mathematical Models

In order to develop realistic and accurate mathematical models of diatom proliferation, the models should consider multiple variables as a function of nutrient concentration, at least N and Si [5.19]. Simplifying assumptions are made in the development of every model. One might assume that plenty of nutrients are available to diatoms that do not change remarkably as the cell population grows. At intracellular scale, an assumption is made that nutrients are homogeneously dispersed inside the cell volume. The parameters existing in differential equations are adjusted to make a proper fit of the curve to empirical data for specific growth rate, cellular N:C ratio, cellular ammonium uptake rate, and ammonium concentration measured.

5.5.1 Gompertz Model

The Gompertz model, which has long been used to describe cancer tumor growth, might be applied to the population growth (or cell concentration) of diatoms. When plotted, this mathematical function slowly increases at the beginning time followed by a sharp and fast growth. As it asymptotically approaches its maximum, the rate of growth becomes slow again. Gompertzian behavior has been used to describe the growth of cell population [5.40] [5.41] [5.29].

It has been observed that experimental data of diatom growth follows the Gompertz law [5.19]. In Pan *et al.* [5.40], a complete form of the Gompertz equation is shown and its specifications are cited in Zwietering *et al.* [5.54] such as the lag phase in the Gompertz equation, which is expressed by its parameters equalling $(b\text{-}1)/c$.

The proliferation of diatom cells is a function of nutrient's concentration around the cell. When there is a constant amount of nutrients available to the diatom population, their concentration as a function of time follows the Gompertz law. Actually, it has been applied to describe the rise of a diatom population [5.29].

Nutrient restriction leads to a decline in cell population, as shown in Figure 5.5 [5.37]. The population would rise to a maximum but due to high consumption of nutrients, the diatom concentration decreases with time. This behavior is not predicted by the Gompertz function. Diatom cell counts are plotted as a function of time (day) for different cases of nutrient ratio, such as nitrogen to carbon (N:C) [5.19]. The plot resembles the Gompertz function (Eq. 5.12).

$$y = a \exp(-\exp(b - ct)) \tag{5.12}$$

in which a, b, c are parameters describing the shape of the curve and are constant for each set of data. In Eq. (5.12), "y" is defined as:

$$y = \ln\left(\frac{N}{N_0}\right)$$

The maximum value of y is implied in parameter "a":

$$a = y_{\max} \ (t \to \infty)$$

Maximum growth rate is

$$\mu_{\max} = \frac{ac}{e} \tag{5.13}$$

where "e" is the renowned mathematical constant being approximately equal to "2.718".

Duration of lag phase is given by: (see Figure 5.4)

$$\Delta t = \frac{b-1}{c} \tag{5.14}$$

5.5.2 Monod Model

The simplest model that relates the concentration of nutrients to diatom growth rate is the Monod equation [5.35] [5.36]. The basic result of the model is losing the linear proportionality between growth rate and nutrient supply (S), especially for larger amounts of S. An example of this model is observed in experiments [5.52]. The relationship between growth rate and the external concentration of a limiting nutrient is generally expressed using the Monod expression as the following [5.35] [5.4] [5.34]:

$$\mu = \frac{\mu_m S}{K_s + S} \tag{5.15}$$

Where:
μ = specific growth rate
μ_m = maximum rate of growth [μ at $[S] \to \infty$]
K_s = half saturation constant (the substrate concentration at which $\mu=0.5\mu_m$)
S = the external concentration of the limiting nutrient.

From the above equation, $S=K_s$ when $\mu=0.5\mu_m$.

In the Monod model, unlike the Gompertz equation, there is no explicit dependency on time. It says the rate of growth is a nonlinear function of nutrient concentration, or, in other words, there is an upper limit for the growth coefficient although the population rises in an exponential manner with any positive value for μ. The basic view in the Monod model is the limiting role of nutrient concentration (S) in specific growth rate [5.47]. The number of cells increases rapidly with the concentration of silicon but tends to level off after a specific silicon concentration (see Figure 5.8).

The validity of Monod expression fades as the external nutrients become exhausted [5.2]. The model predicts that the cells cease to grow when the external nutrients downgrade to zero, while in reality diatom growth can proceed due to internal nutrient reserves. Another error happens in the transposition of multiple nutrients [5.13] that makes the Monod model unsatisfactory.

The Monod equation has been widely used to model relationships between specific growth rates and limiting substrate concentrations for a variety of nutrients, including nitrogen, carbon and silicon. The classic Monod might come up with coupled differential equations that predict time-dependent changes in concentrations of nutrients and total algal cell carbon (C_{cell}) in units of moles per liter of medium [5.47].

5.5.3 Michaelis-Menten Model

Though the subject of the Monod model is cell population growth rate, the Michaelis-Menten model describes nutrient uptake by a cell. It predicts the theoretical rate of uptake of nutrients into a cell that are transported across a membrane at a fixed concentration [5.12]. However, in reality the concentration of membrane transport proteins, and thus the V_{max} of the system (maximum uptake rate), is under negative feedback regulation, and consequently changes substantially with the intracellular nutrient status for virtually all nutrients, including ammonium, nitrate, phosphate and all trace metals examined to date.

In the Michaelis-Menten model it is assumed that nutrient uptake proceeds at a rate which only depends on external concentrations and enzymatic activity at the cell membrane as expressed in the original Michaelis-Menten function (Eq. 5.16).

Michaelis-Menten function for enzymatic reaction appears as:

$$V = \frac{V_m S}{K_s + S} \tag{5.16}$$

Where:
V = the specific rate of nutrient uptake
V_m = maximum specific uptake rate, i.e., V at $[S] \to \infty$

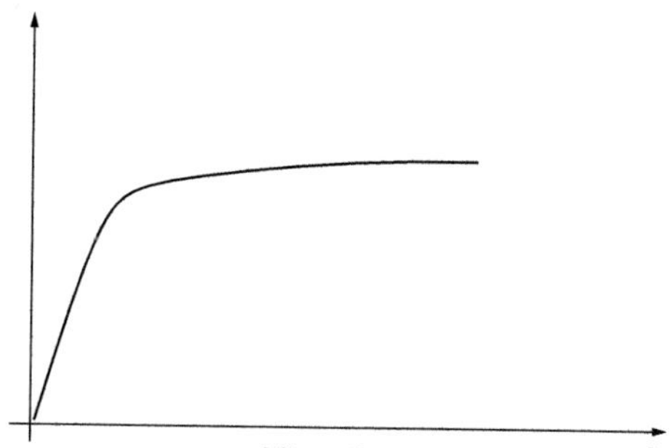

Figure 5.8 Silicon limited uptake rate in Michaelis-Menten model.

S = the external concentration of the limiting nutrient

Ks = the half-saturation constant defined as the substrate concentration at which $V = 0.5V_m$

It has been practically demonstrated that dependence of the specific uptake rate (V) on silicon concentration in silicic acid form agrees well with a Michaelis-Menten saturation model [5.38]. Maximum uptake rates (V_{max}) ranged from 0.0022 to 0.0028 h^{-1}, which corresponds to maximum growth rates of 0.08 to 0.10 d^{-1}. The standard Monod equation and the Michaelis-Menten uptake equation have the same mathematical form and both indicate zero growth or uptake rate at zero external nutrient concentrations. Unfortunately, there is usually little experimental data near zero growth rates to test this prediction, and in the case of ammonium limitation, there was essentially no published data for relationships among specific growth rates [5.47].

At steady state, the specific growth rate, the specific uptake rate and the rate of assimilation of the limiting nutrient are in balance and thus equivalent. An assumption that is implicitly made in development of the Michaelis-Menten model is that every cell in the population behaves identically and each cell takes up the limiting resource continuously and at a rate that is independent of cell cycle phases.

5.5.4 Droop Model

Michaelis-Menten and Monod equations, respectively, express nutrient uptake rate and growth rate as functions of the limiting nutrient concentration in the form of a saturation function. Diatoms can uptake nutrient in excess of current needs and continue to grow during nutrient-poor conditions. Hence, in reality, when the external nutrient is exhausted the growth would proceed. This is due to the ability of cells to consume the internal nutrient reserves. The Droop model takes into consideration that diatom cells store nutrients and that the growth rate depends on the stored nutrient. During a stable situation, the concentration of these nutrients is assumed to be constant. Time-dependent changes in nutrient concentration is predicted by the Droop modeling approach by a system of three differential equations [5.47].

A Quota model was presented by Droop [5.8] as an alternative to the Monod description. This class of models takes into consideration the ability of diatoms to accumulate nutrients in dedicated intracellular storage pools. The Quota model hypothesis is based on the fact that the primary rate of a molecular phenomena is controlled by the size of the internal pool of limiting nutrient and only indirectly by the medium concentration. The main factor that triggers the cell division is the reserved nutrients and energy inside the cell and external elements and compounds are transported to compensate the internal consumption. The nutrient quota Q is expressed relative to cellular carbon (C) [5.1] or in terms of nitrogen biomass per cell [5.50]. It is noteworthy that knowledge of one element can barely provide the ability to calculate the concentrations of other elements.

Cell quota is defined in terms of nitrogen biomass per cell. Net population growth requires a quota in excess of the basal cellular nutrient requirement (Q_{min}) and increases with increasing Q toward a theoretical maximum metabolic rate. This is effectively the maximum possible rate of cell division assuming an infinite (and unobtainable) internal nutrient reserve [5.8] [5.50].

5.5.5 Aquaphy Model

Aquaphy is a complex, physiological model that describes the growth and mortality of diatoms in response to nutrients (Si, P, and N), temperature and light [5.30]. Conceptually, it is based on the energy storage and conservation of biomass flowing between cell components. According to this, storage material is synthetized by diatoms through intracellular processes cells for the future requirements of growth. The Aquaphy model describes the cell as being composed of different pools and the cell growth is a function of Si, P, C, and N concentrations [5.48].

In the Aquaphy model, cellular biomass is categorized in three different state variables: low molecular weight carbohydrates (LMW), storage molecules (reserves), and the biosynthetic and photosynthetic proteins. All molecular concentrations are expressed in mmol/cm^3 [5.21]. The model structure includes as many as 12 state variables describing nutrient flows within a cell [5.22].

$$\frac{d(State\ Variable)}{dt} = sources - sinks$$

State variables include a concentration of carbon reserves (carbohydrates, fatty acids), functional and structural metabolites, dissolved inorganic nitrogen, intracellular soluble phosphate, etc.

5.5.6 Mechanistic Model

The three main modeling approaches in order of increasing complexity and matching the physical reality are the Monod, quota, and mechanistic models [5.12] [5.13]. This includes feedback processes and multiple internal pools. According to these models, synthesis of byproducts occurs when internal nutrient quota surpasses a threshold limit and increased acid production happens during silicon and phosphorus limitation. The mechanistic model, as implied by its name, deals with physical phenomena of molecular flow between intracellular pools and their transport across the cell membrane.

On the other hand, the use of the standard Michaelis-Menten model to drive the dynamics of the intracellular nutrient concentrations suggests that an enzymatic process controls the whole uptake/assimilation process, which can be the enzyme-mediated uptake through the cell membrane or the intracellular enzyme reactions [5.2].

5.6 Results

The need to anticipate variations in the diatom population is of utmost importance. The concentration of diatoms in laboratory experiments and the natural environment is a function of light, temperature, nutrients (elemental and small molecules), vitamins, etc. A mathematical model can provide researchers with a powerful tool that allows them to predict the population and/or concentration of diatom cells. Inputs to such a model determine its complexity and extent of applicability. The mathematical basis of models is made up of differential equations and state variables [5.17]. Models have been used to complement laboratory experiments and *in-situ* measurements, which are difficult and costly. Two general kinds of models are quantitative and qualitative [5.27].

The results obtained by the Droop model in predictions of diatom blooms are not in agreement with the Monod approach, especially when the dynamics of the ecosystem are not in equilibrium [5.8] [5.47]. Disagreements in the results of the Michaelis-Menten model observed experimentally are due to the assumption that Si uptake is continuous while it is actually restricted to specific sections of the cell cycle. The best fits to the Michaelis-Menten function produces better results when short uniform incubation periods are carried out.

Experimental data on diatom concentration as a function of time has produced curves very similar to the Gompertzian shape [5.42]. The information on cell limiting nutrient's quota appears in many models. For applicability of a model in a real situation, the concentration of limiting nutrient must be measured by separating the species from other components of the substrate and the environment. Hence, the observation through the experiment produces an inevitable error.

Typical values for specific Growth rate are:

$0.56 < \mu_{max} < 0.74$ day^{-1} [5.40]
1.7 div./day [5.53],
With division rates of the order of 1 doubling per day (0.69 day^{-1}) [5.49]
$0.4 < \mu_{max} < 3.3$ day^{-1} [5.44]
$0.2 < \mu_{max} < 1.4$ day^{-1} for $10^{-1} <$ cell volume $< 10^7\ \mu^3_{max}$ [5.50]
$0.66 < \mu_{max} < 1.04$ day^{-1} [5.29]

5.7 Conclusion

Models can be compared according to aspects of their complexity, precision, kinetic parameters, quantitative or qualitative approach, etc. In comparison of the models discussed in this chapter, Droop is more realistic than the Michaelis-Menten model. Comparing Monod and Michaelis-Menten models, the former expresses diatom growth as function of time while the latter does the same with nutrient uptake by the diatom cells. An assumption that is made in many existing simplistic models is that nutrients are homogeneously dispersed inside the cell volume. A realistic mathematical model should contain multiple inputs (concentration of nutrients such as N, Si, etc.). The precision with which a model can predict the specific growth rate is based on the physical and chemical specifications of the cell.

An assumption that is implicit in the formulation of equations is that every cell in the diatom population behaves identically. Statistical considerations in different responses among diatom cells of identical type might produce different results. External nutrient concentration influences the chemical composition of diatom structure. This is a matter of molecular transport across the membrane, which is governed by the diatom surface area. The size dependence of the maximum population growth rate varies among types of diatoms because of their differences in size that undermine the balance of nutrient uptake and metabolism. Significant differences have been observed between small fast-growing diatoms with 4.5 micrometer diameter and large slow-growing diatoms with 11 micrometer diameter.

5.8 Prospects

Mathematical models are described by a first-order differential equation. By definition, this kind of equation contains the variable (such as nutrient concentration) and its first derivative with respect to time. More advanced models take into account a higher number state variables, leading to a set of coupled differential equations. Solution of interwoven differential equations to forecast diatoms' population behavior is currently in progress as computational software [5.45]. Besides simulation software and model development, Flynn and Martin-Jézéquel [5.16] provide a source for bioreactor design that leads to biofuel production.

Application of radioisotopes, i.e., ^{14}C, ^{13}N and ^{32}P, is recommended to trace transport of molecules in the laboratory and movement of diatoms in the laboratory.

References

[5.1] Baklouti, M., Faure, V., Pawlowski, L. and Sciandra, A. (2006 a) Investigation and sensitivity analysis of a mechanistic phytoplankton model implemented in a new modular numerical tool (Eco3M) dedicated to biogeochemical modeling. Progress in Oceanography 71, 34–58.

[5.2] Baklouti, M., Diaz, F., Pinazo, C., Faure, V. and Queguiner C. (2006 b) Investigation of mechanistic formulations depicting phytoplankton dynamics for models of marine pelagic ecosystems. Progress in Oceanography. 71: 1-33.

[5.3] Brown J.H., Gillooly J.F., Allen A.P., Savage V.M. and West G.B. (2004) Toward a metabolic theory of ecology. Ecology 85(7):1771–1789.

[5.4] Brzezinski, M.A. (1992) Cell-cycle effects on the kinetics of silicic acid uptake and resource competition among diatoms. Journal of Plank ton Research 14, 1511–39.

[5.5] Carey, N., Sigwart, J.D. and Richards, J.G. (2013) Economies of scaling: More evidence that allometry of metabolism is linked to activity, metabolic rate and habitat, Journal of Experimental Marine Biology and Ecology 439, 7–14.

[5.6] Chen B. and Liu H. (2010) Relationships between phytoplankton growth and cell size in surface oceans: interactive effects of temperature, nutrients and grazing, Limnology and Oceanography 55, 965-972.

[5.7] Clark, J.R., Lenton, T.M., Williams, H.T.P. and Daines, S.J. (2013) Environmental selection and resource allocation determine spatial patterns in picophytoplankton cell size. Limnology and Oceanography 58, 1008–1022.

[5.8] Droop, M.R. (1968) Vitamin B-12 and marine ecology. IV. The kinetics of uptake, growth and inhibition in Monochrysis lutheri. Journal of Marine Biological Association of the United Kingdom 48, 689–733.

[5.9] Dutkiewicz, S., Cermeno, P., Jahn, O., Follows, M.J., Hickman, A.E., Taniguichi, D. A. A. and Ward, B.A. (2020). Dimensions of marine phytoplankton diversity. Biogeosciences. 17, 609-634.

[5.10] Li Q.P., Franks P. J.S., Landry M.R., Goericke R. and Taylor A.G. (2010) Modeling phytoplankton growth rates and chlorophyll to carbon ratios in California coastal and pelagic ecosystems. Journal of Geophysical Research: Biogeosciences 115(G4), #G04003.

[5.11] Finkel, Z.V. and Irwin, A.J. (2000) Modeling size-dependent photosynthesis: Light absorption and the allometric rule. Journal of Theoretical Biology 204, 361 – 369.

[5.12] Flynn, K.J. and Martin-Jézéquel, V. (2000) Modelling Si–N-limited growth of diatoms. Journal of Plankton Research 22(3), 447–472.

[5.13] Flynn, K.J. (2003) Modelling multi-nutrient interactions in phytoplankton; balancing simplicity and realism. Progress in Oceanography 56, 249–279.

[5.14] Flynn K.J. and Raven J.A. (2017) What is the limit on photoautotrophic plankton growth rates? Journal of Plankton Research. 39, 13–22.

[5.15] Flynn, K.J., Skibinski, D.O.F. and Lindemann, C. (2018) Effects of growth rate, cell size, motion, and elemental stoichiometry on nutrient transport kinetics. PLOS Computational Biology. 14: e1006118.

[5.16] Flynn, K.J. (2020) Enhancing microalgal production - constructing decision support tools using system dynamics modelling. Zenodo, http://dx.doi.org/10.5281/zenodo.50366.

[5.17] Follett, C.L., Dutkiewicz, S., Karl, D.M., Inomura, K. and Follows, M. J. (2018) Seasonal resource conditions favor a summertime increase in North Pacific diatom–diazotroph associations, The ISME Journal. 12:1543–1557.

[5.18] Geider, R.J., Platt, T. and Raven, J.A. (1986) Size dependence of growth and photosynthesis in diatoms: a synthesis. Marine Ecology – Progress Series 30, 93-104.

[5.19] Gilpin, L.C., Davidson, K. and Roberts, E. (2004) The influence of changes in nitrogen: silicon ratios on diatom growth dynamics, Journal of sea research 51, 21– 35.

[5.20] Giri, T., Goutam, U., Arya, A. and Gautam, S. (2022) Effect of nutrients on diatom growth: a review. Trends in Sciences 19(2), 1752.

[5.21] Ghyoot, C., Gypens, N., Flynn, K. J., and Lancelot, C. (2015) Modelling alkaline phosphatase activity in microalgae under orthophosphate limitation: the case of Phaeocystis globosa. Journal of Plankton Research, 37, 869-885. doi: 10.1093/plankt/fbv062.

[5.22] Ghyoot, C. Flynn, K.J., Mitra, A., Lancelot, C. and Gypens, N. (2017) Modeling plankton mixotrophy: A mechanistic model consistent with the shuter-type biochemical approach. Frontiers in Ecology and Evolution 5 (78).

[5.23] Gillooly, J.F., Brown, J.H., West, G.B., Savage, V.M. and Charnov, E.L. (2001) Effect of size and temperature on metabolic rate, Science, 293, 2248-2251.

[5.24] Halsey K.H., O'Malley RTO, Graff JR, Milligan A.J., and Behrenfeld, M.J. (2013) A common partitioning strategy for photosynthetic products in evolutionarily distinct phytoplankton species. New Phytologist 4, 1030–1038.

[5.25] Hasle, G. R., Erik E. and Syvertsen, E. E. (1996) "Marine diatoms". In Tomas C. R. (Ed.), Identifying Marine Diatoms and Dinoflagellates. Academic Press (pp. 5–385).

[5.26] Iler R. K. (1979) *The Chemistry of Silica: Solubility, Polymerization, Colloid and Surface Properties, and Biochemistry*. JohnWiley & Sons, New York.

[5.27] Inomura, K., Deutsch, C., Masuda, T., Prášil O., and Follows, M.J. (2020 a) Quantitative models of nitrogen-fixing organisms, Computational and Structural Biotechnology Journal, 18, 3905-3924.

[5.28] Inomura, K., Omta, A.W., Talmy, D., Bragg, J., Deutsch, C., and Follows, M.J. (2020 b). A mechanistic model of macromolecular allocation, elemental stoichiometry, and growth rate in phytoplankton. Frontiers in Microbiology 11, 86.

[5.29] Jauffrais, T., Agogué, H., Gemin, M.P., Beaugeard, L. and Martin-Jézéquel, V. (2017) Effect of bacteria on growth and biochemical composition of two benthic diatoms Halamphora coffeaeformis and Entomoneis paludosa. Journal of Experimental Marine Biology and Ecology 495, 65 - 74.

[5.30] Lancelot, C., Veth, C. and Mathot, S., (1991) Modeling ice-edge phytoplankton bloom in the Scotia-Weddell sea sector of the southern ocean during spring 1988. Journal of Marine Systems, 2, 333-346.

[5.31] Lomas, M. W., Baer, S.E., Acton, S. and Krause, J.W. (2019) Pumped up by the cold: Elemental quotas and stoichiometry of cold water diatoms. Frontiers in Marine Science, 6, 286.

[5.32] Martin-Jézéquel V., Hildebrand, M. and Brzezinski, M.A. (2000) Silicon metabolism in diatoms: implications for growth, Journal of Phycology 36, 821–840.

[5.33] Michaelis, L. and Menten, M.L. (1913) Die kinetik der invertinwirkung. Biochem ische Zeitschrift 49, 333–369.

[5.34] Moejes, F.W., Succurro, A., Popa, O., Maguire, J. and Ebenhöh, O. (2017) Dynamics of the bacterial community associated with phaeodactylum tricornutum cultures. Processes 5, 77.

[5.35] Monod, J. (1942) Recherche sur la croissance des cultures bactériennes: actualite´s scientifiques et industrielles. Annual Review of Microbiology. 3, 3 – 71.

[5.36] Monod, J. (1950) La technique de culture continue: theorie et applications. Annales de l'Institut Pasteur 79, 390–410.

[5.37] Moutier, W., Duforêt-Gaurier, L., Thyssen M. Loisel H, MeÂriaux X, Courcot L, *et al.* (2017) Evolution of the scattering properties of phytoplankton cells from flow cytometry measurements, PLoS One 12(7), #e0181180.

[5.38] Nelson, D.M. and Treguer, P. (1992) Role of silicon as a limiting nutrient to Antarctic diatoms: evidence from kinetic studies in the Ross sea ice-edge zone, Marine Ecology Progress Series 80, 255-264.

[5.39] Orefice, I., Musella, M., Smerilli A., Sansone, C., Chandrasekaran, R., Corato, F. and Brunet, C. (2019) Role of nutrient concentrations and water movement on diatom's productivity in culture, Scientific Reports, 9, 1479.

[5.40] Pan, Y., Subba Rao, D.V. and Warnock, R.E. (1991) Photosynthesis and growth of Nitzschia-pungens f. multiseries hasle, a neurotoxin producing diatom, Journal of Experimental Marine Biology and Ecology, 154, 77-96.

[5.41] Pan, Y., Subba Rao, D.V., Mann, K.H., Brown, R. and Pocklington, R. (1996) Effects of silicate limitation on production of domoic acid, a neurotoxin, by the diatom Pseudo-nitzschia multiseries. I. Batch culture studies. Marine Ecology Progress Series 131, 225–233.

[5.42] Ramirez, E.E., Gonzalez, M.A., Cifuente, A.S., Inostroza, I. and Urrutia, R.E. (2015) Culture and growth of two benthic diatoms species isolated from the Salar del Huasco (North of

Chile, 20° S) at different conditions of temperature, light and nutrient. Gayana Botanica 72(2), 165-176.

[5.43] Ryabov, A.B. and Blasius, B. (2008) Population growth and persistence in a heterogeneous environment: the role of diffusion and advection. Mathematical Modeling of Natural Phenomena 3, 42–86.

[5.44] Sarthou, G., Timmermans, K.R., Blain, S. and Tréguer, P. (2005) Growth physiology and fate of diatoms in the ocean: A review. Journal of Sea Research 53, pp. 25–42.

[5.45] Schneider, L.K., Gypens, N., Troost, T.A. and Stolte, W. (2021) Modeling mixoplankton along the biogeochemical gradient of the Southern North Sea, Ecological Modelling, 459, #109690.

[5.46] Sommer, U., Charalampous, E., Genitsaris, S. and Moustaka-Gouni, M. (2017) Benefits, costs and taxonomic distribution of marine phytoplankton body size, Journal of Plankton Research 39, 494–508.

[5.47] Sunda, W.G., Shertzer, K.W. and Hardison, D.R. (2009) Ammonium uptake and growth models in marine diatoms: Monod and Droop revisited. Marine Ecology Progress Series 386, 29–41.

[5.48] Terseleer, N., Gypens, N. and Lancelot, C. (2013) Factors controlling the production of domoic acid by Pseudo-nitzschia (Bacillariophyceae): A model study. Harmful Algae 24, 45–53.

[5.49] Vaulot, D. (1995) The Cell Cycle of Phytoplankton: Coupling Cell Growth to Population Growth. In Joint, I. [Ed.] Molecular Ecology of Aquatic Microbes. 38, 303–322. Springer. doi:10.1007/978-3-642-79923-5_17

[5.50] Ward, B.A., Maranon, E., Sauterey, B., Rault, J. and Claessen, D. (2017) The size dependence of phytoplankton growth rate: a trade-off between nutrient uptake and metabolism. American Naturalist,189, 170-177.

[5.51] Wirtz, K.W. (2011) Non-uniform scaling in phytoplankton growth rate due to intracellular light and CO_2 decline. Journal of Plankton Research 33, 1325–1341.

[5.52] Yoder, J.A. (1979) Effect of temperature on light limited growth and chemical composition of *Skeletonema costatum* (Bacillariophyceae). Journal of Phycology 15, 362–370.

[5.53] Yoder, J.A., Martin, J., and Nill, A. (1982) Cell division periodicity and the nitrate environment of a marine diatom. Limnology and Oceanography 27, 352–357.

[5.54] Zwietering, M.H., Jongenburger, I., Rombouts, F.M. and van 'tRiet, K. (1990) Modelling of the bacterial growth curve. Applied and Environmental Microbiology 56, 1875–1881.

6

Diatom Growth: How to Improve the Analysis of Noisy Data

Olga Kourtchenko[1*], Kai T. Lohbeck[2], Björn Andersson[1] and Tuomas Rajala[3]

[1]Department of Marine Sciences, University of Gothenburg, Gothenburg, Sweden
[2]Limnological Institute, University of Konstanz, Konstanz, Germany
[3]Applied Statistical Methods, Natural Resources Institute Finland, Helsinki, Finland

Abstract

Measuring cell culture growth is a standard phenotyping technique in microbial research. It allows the determination of key parameters, such as exponential growth rate, acclimation time, and yield, which have been used extensively to study microorganisms' gene function, physiology, and evolution. Diatoms, being unicellular eukaryotes, are commonly phenotyped in the same fashion as other microorganisms. However, the resulting growth data is distinct in many ways, making it challenging to apply standard parameter estimation methods used, for example, for heterotrophic bacteria. Diatom growth data analysis often translates to the manual, ad hoc, data fitting, typically involving linear regression for exponential growth rate. The rate value is subsequently regressed against the control factors in an experiment. The quality of rate estimate thus determines the meaningfulness of the following regression analysis, making the selection of the appropriate (most robust, accurate, and objective) method for rate calculation of utmost importance.

We explored a suite of mathematical methods for specific growth rate estimation using simulation-based trials. The experimental data set for the trials was generated *in silico* and included thousands of growth curves, featuring unique diatom-like characteristics combined with various challenges (e.g., culture crashes). The specific growth rate value was estimated for each of the simulated curves by parametric, nonparametric, and heuristic methods, and the performance of each of these methods was evaluated and ranked. We followed up with an empirical example of thermal responses in the chain-forming diatom *Skeletonema subsalsum*, demonstrating the possible consequences of choosing an inappropriate growth rate estimation method. We argue that the proposed approach is on the one hand scalable to high throughput studies, and on the other hand reduces bias and leads to higher reproducibility in smaller-scale studies. Finally, we provide a guide for selecting the most objective method for diatom growth data analysis on a case-by-case basis.

**Corresponding author*: olga.kourtchenko@marine.gu.se
Olga Kourtchenko: https://www.researchgate.net/profile/Olga-Kourtchenko, https://www.gu.se/en/about/find-staff/olgakourtchenko, https://scholar.google.se/citations?hl=en&user=dS5r-NgAAAAJ
Kai T. Lohbeck: https://www.researchgate.net/profile/Kai-Lohbeck, https://scholar.google.com/citations?user=0xm9MlgAAAAJ&hl=en, https://orcid.org/0000-0003-3421-7112
Björn Andersson: https://www.researchgate.net/profile/Bjoern-Andersson, https://scholar.google.com/citations?user=tLKwJTcAAAAJ&hl=en, https://orcid.org/0000-0003-4297-2927
Tuomas Rajala: https://www.luke.fi/en/experts/tuomas-rajala, https://scholar.google.com/citations?user=yTeGlJEAAAAJ, https://orcid.org/0000-0002-1343-8058

Janice L. Pappas (ed.) The Mathematical Biology of Diatoms, (141–164) © 2023 Scrivener Publishing LLC

***Keywords*:** Phenotyping, growth curve, growth rate, nonlinear estimation, functional data analysis

6.1 Introduction

6.1.1 What is a Growth Curve?

Microbiologists have long recognized that physiological/ecological consequences of microbial traits can be studied simply by measuring changes in biomass over time [6.40]. In laboratory studies, such growth characteristics are most commonly represented by a growth curve. A growth curve is defined as a graphical representation of a series of observations (or measurements) of a biomass proxy over a period of time (Figure 6.1a). Starting with the pioneering work of Jacques Monod [6.32], the growth of microbial cultures, primarily bacterial, have been mathematically formalized and shown to follow a predictable and quantifiable pattern, described by the growth function:

$$n = f(t) \tag{6.1}$$

where n is the cell number and t is time.

6.1.2 Why Measure Growth?

A growth curve reflects the intrinsic properties of an organism and thus can be used to compare, for example, different species, strains, acclimatization states, or treatment responses. Through their array of applications, growth curves and related parameters are widely used to help answer diverse biological questions. In diatom research, growth data acquisition regularly constitutes the starting point of many laboratory studies; and growth data analysis is often the primary focus.

A typical microbial growth curve is characterized by several phases: i) the initial adjustment phase or the lag phase; ii) the exponential growth phase, iii) the stationary phase, and, in some

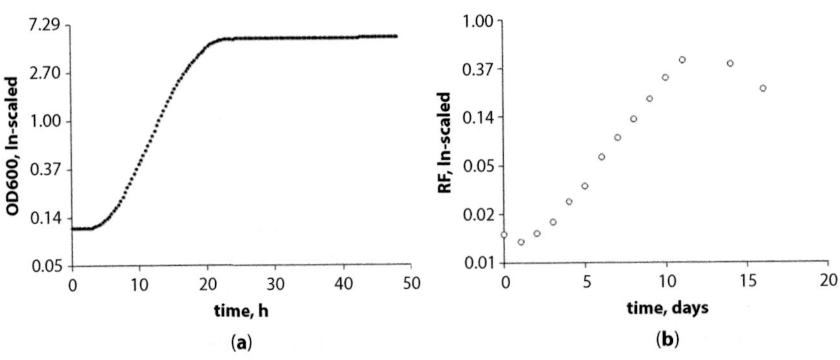

Figure 6.1 Growth curve examples: (a) A high data density yeast (*Saccharomyces cerevisiae*) growth curve; (b) a comparatively low data density diatom (*Skeletonema marinoi*) growth curve. Biomass is either recorded by optical density (light absorbance at 600 nm) or chlorophyll fluorescence (excitation and fluorescence emission wavelengths at 425 and 680 nm, respectively). h = hours; OD600 = optical density at 600 nm; RF = relative fluorescence.

cases, iv) the decline or death phase (Figure 6.1b). Each growth phase can be quantified or summarized in a growth parameter. For example, the length of a lag phase is indicative of the organism's physiological state at the start of the experiment, but also of its metabolic plasticity, while the maximum biomass at the stationary phase (maximum yield) would represent the efficiency of an organism in converting available resources into biomass [6.23] [6.35] [6.44].

The growth rate represents the organism's maximum capacity to utilize essential resources for growth in a given environment. The growth rate is therefore a function of the intrinsic properties of microbial culture and the environmental conditions. It is expected to decline as the environmental conditions are manipulated away from an organism's optimal conditions. This simple expectation makes it one of the most powerful, informative, and, thus, most commonly calculated parameters of growth curves. The "specific (or relative) growth rate" is the standard growth rate parameter used in microalgal research and is defined as the absolute growth rate $\left(\frac{dn}{dt}\right)$ per unit size (Eq. 6.2) [6.10].

$$Specific\ growth\ rate = \frac{d(ln\ n)}{dt} = \frac{1}{n} \cdot \frac{dn}{dt} \tag{6.2}$$

The specific growth rate is calculated as the slope of the growth curve in the exponential phase in a natural logarithm (ln) scale. Despite this simple definition, the determination of the exponential growth phase is not straightforward and requires detailed consideration, specifically in diatom growth studies. Let us then have a closer look at diatoms and what makes their growth distinct in the context of this chapter's focus subject.

6.1.3 Diatoms and Their Growth

Diatoms are ecologically successful, globally distributed, and the most species-rich group of microalgae on Earth today [6.41]. The global success of diatoms is attributed to a number of their unique features and metabolic peculiarities that have been reviewed elsewhere [6.22] [6.37] [6.43] and are beyond the scope of this study to describe. Perhaps, not surprisingly, diatoms have been the subject of intense research for many decades and are now entering the "molecular age" with the establishment of several model species and the development of molecular tools for genetic, genomic, and systems biology analyses [6.16]. Examples of such research include *in-situ* studies of natural diatom populations [6.1] [6.38] [6.55], laboratory studies of algal biomass production under various conditions [6.13] [6.31], and phenotypic screens of natural isolates or mutant strains [6.2] [6.4] [6.27]. However diverse the scientific questions we ask about these organisms, the sampling and the analysis of their growth remains one of the main tools in diatom research across the spectrum.

Diatom growth curves are likewise highly variable and often non-conforming to any idealized sigmoid growth model familiar to bacterial microbiology (Figure 6.1a). To illustrate this point, we compiled a number of diatom growth curves of various taxa, common laboratory model species, species with specific ecological relevance, and different study aims (Figure 6.2; Table 6.1). The selection in Figure 6.2 represents growth data typically reported by the diatom scientific community and exemplifies some of the common, "irregular" (i.e., deviating from the idealized sigmoid curve) features of diatom growth curves, caused by

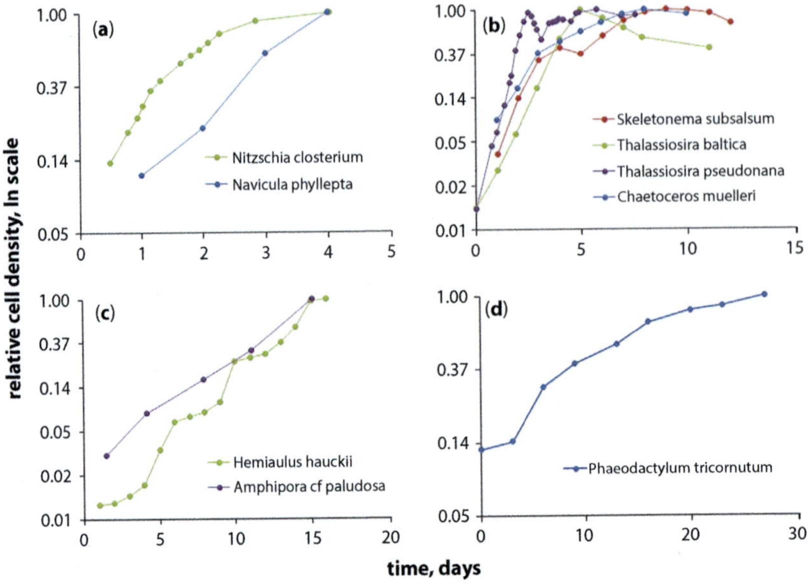

Figure 6.2 Examples of diatom growth curves: Data was retrieved from an arbitrary selection of published studies without considering data quality or research aims. The selection includes common model species and rarely cultured species, from both benthic and pelagic marine environments. Cell densities are based on cell counts, optical density, and chlorophyll fluorescence, and each species dataset has been scaled to the maximum observed density to enable a unit-free comparison (y-axis values, shown here in the natural logarithm scale). All curves represent an average of multiple replicates, and we have omitted error bars due to the inconsistency of variance reporting between studies. The data is presented on four panels based on the duration of a growth experiment (a) four days, (b) 12 days, (c) 16 days, (d) 27 days. Additional information about each species' growth curve and the respective study can be found in Table 6.1.

either their specific biology, or arising as a consequence of experimental design, or a combination of both.

Specifically, many diatom growth curves include a death, or culture senescence phase, often not captured in bacterial growth studies (e.g., Figure 6.2a: *S. subsalsum, T. baltica*). The culture's ultimate demise is expected for most cultivated microorganisms in a closed system, given enough time to exhaust vital resources and accumulate potentially harmful byproducts. Diatoms are no exception; moreover, many species are known to adhere to the "bloom-and-bust" life cycle mode, when the periods of rapid population growth during favorable conditions, or "blooms," are followed by their collapse as growth resources are being exhausted [6.26]. Programmed cell death may play a role in the "bust" process, as shown for laboratory cultures of the chain-forming centric diatom, *Skeletonema marinoi* [6.18] [6.49]. Biotic interactions between diatoms and their microbiome (heterotrophic bacteria) constitute another important driver of bloom dynamics [6.46] [6.42] (for a non-diatom, bloom-forming phytoplankton species) and could potentially be instrumental in sudden culture "crashes" in laboratory conditions (Figure 6.2a: *T. pseudonana*; Figure 6.3: *S. marinoi*).

Diatoms grow significantly slower than yeasts and many cultured bacteria, achieving at most two to three cell divisions per day (24-hour period) under optimized laboratory conditions. The resulting low data density growth curves (Figure 6.1b; Figure 6.2) introduce additional challenges for their interpretation and parametrization. From a technical perspective, the analysis

Table 6.1 Study details for each of the diatom species shown in Figure 6.2. The species are given in alphabetic order; the study details include species-specific information (symmetry type, natural habitat type), the growth experiment-specific metadata (type of treatment, biomass proxy, the length of the experiment, total number of observations, approximate number of observations in the exponential phase, light cycle conditions), and a reference. *Indicates unpublished growth data from the authors presented in Subsection 6.1.3.

Species	Symmetry type	Habitat	Treatment applied in a study	Biomass proxy	Length of the growth curve (days)	Total observations	Approx. observations in the log-phase	Light cycle	Reference
Amphiprora* cf. *paludosa	Pennate	Benthic	*In-situ* growth	Cell counts	15	5	2-4	Diel	[6.1]
Chaetoceros muelleri	Centric	Pelagic	UV light stress	Cell counts	10	9	3	Diel	[6.53]
Hemiaulus hauckii	Centric	Pelagic	Light intensity	Cell counts	16	15	8-10	Diel	[6.36]
Navicula phyllepta	Pennate	Benthic	Competing diatom species and bacteria	Cell counts	4	4	2-3	Diel	[6.27]
Nitzschia closterium	Pennate	Pelagic	Nutrient limitation and stress	Optical density	4	13	5-10	Constant	[6.44]
Phaeodactylum tricornutum	Pennate	Pelagic	Microplastic toxicity	Optical density	27	9	0-2	Diel	[6.13]
Thalassiosira baltica	Centric	Pelagic	Intraspecific diversity	Fluorescence	11	10	4-6	Diel	[6.4]
Thalassiosira pseudonana	Centric	Pelagic	Silicate vs. nitrate-limited stationary phase	Cell counts	8	20	7-10	Constant	[6.54]
Skeletonema subsalsum	Centric	Pelagic	Temperature	Fluorescence	12	12	3	Diel	[*]

of such data sets would be more sensitive to noise or unexplained measurement artifacts (e.g., Figure 6.2a: *S. subsalsum*). Typically, this issue is addressed in the experimental design using technical replicates, resulting in replicated data series. However, the number of experimental units, including technical replicates, in diatom growth studies is usually limited by the prevailing dependence on low throughput data acquisition methodology, such as manual cell counts (see Table 6.1 in Biomass proxy). Another consequence of slow growth is the "right-censoring" issue: a situation where the data acquisition has stopped before the culture has attained its maximum density (e.g., Figure 6.2b, c: *A.* cf. *paludosa*, *N. phyllepta*). "Right-censored," or truncated growth curves could be attributed to several technical reasons. However, a prolonged transition phase between the exponential and the stationary/decline phases is the likely culprit in many diatom studies (e.g., Figure 2b, d: *P. tricornutum*, *N. closterium*).

Given the right conditions, other state transitions are often evident in diatom growth data. These can be akin to the bacterial diauxic shifts [6.44] when distinct exponential growth phases are separated by intermittent lag or transition phases of metabolic shift (Figure 6.2c: *H. hauckii*). Finer scale transitions (within 24 h period) invariably happen in laboratory experiments where diatoms are exposed to diel changes in light conditions, or "day-night" transitions when a set period of constant light alternates with a set period of complete darkness. Sampling growth throughout the "light" period introduces a "stepwise" shape (Figure 6.3) to the diatom growth curve [6.6] [6.15] [6.39], complicating downstream analyses. The latter issue is sometimes circumvented by conducting diatom growth experiments in continuous light (no "night" phase). Alternatively, the sampling is confined to one measurement per day, ideally in the middle of the "light" phase and at the same hour each day for the duration of the experiment (for more details on diatom growth experimental setup we refer to [6.51]). The latter approach is quite common and inherently contributes to the low data density issue discussed above.

Diatom growth curves often appear to lack the lag phase, making them deviate further from the "expected" sigmoidal shape (Figure 6.2, all except *H. hauckii* and *P. tricornutum*). This phenomenon is unlikely unique to diatoms and generally arises for two reasons. First, effective subculturing from the inoculation culture already in the exponential phase can almost entirely remove the lag phase [6.30] [6.45]. Alternatively, the low sampling frequency may render the lag phase unresolved. It also appears to be a common practice to

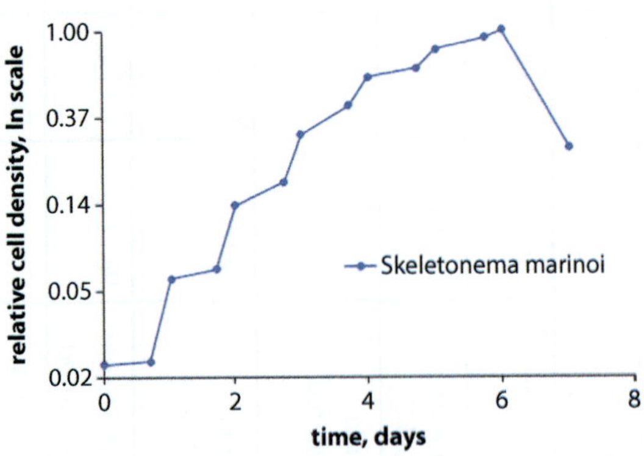

Figure 6.3 An example of a *Skeletonema marinoi* growth curve. Cells were counted twice a day, at the start and the end of the light "day" phase under laboratory conditions (authors own data, unpublished).

exclude a time zero measurement (left-censoring), which omits any lag phase shorter than 12-24 h. See, for instance, Figure 6.2 *C. muelleri, N. closterium, N. phyllepta, A.* cf. *paludosa, H. hauckii.*

The discussion of diatom growth is incomplete without mentioning their unique reductive cell cycle, where the average cell size of an asexually propagating culture decreases over time [6.11]. Though this feature in itself does not necessarily affect the shape of a growth curve, it is important to consider when comparing growth parameters of the same culture in long-term (over several months) studies or different genetic clones of the same species that each could be at different stages of the reductive cell cycle. Mean population cell size in the reproductive cycle of diatom cultures is relevant because numerous studies found a direct relationship between the diatom cell size and growth rate [6.5] [6.19] [6.48].

Clearly, diatom growth data, which may combine various deviant features simultaneously, provides an analytical challenge in terms of parametrization. In the following section, we will briefly review the suite of mathematical approaches available for this task.

6.1.4 Growth Data Analysis and Growth Parameter Estimation

The main goal of microbial growth data analysis is to gain an insight into a biological process under a given set of conditions. In technical terms, meeting this goal translates into a two-step analysis of the growth data: first, the growth curve is summarized in one value, or a growth parameter of interest (lag time, specific maximum growth rate, maximum growth level), meant to provide a mechanistic link to the investigated organism's biology in the context of the research question. The second step typically involves a statistical analysis of the growth parameters obtained during step one by regressing them to the control factors (experimental conditions in question). The latter is done, for instance, by the use of Analysis of Variance (ANOVA). In this kind of analysis, it is of utmost importance that reducing the complete information in the growth series into a single value, such as the maximum specific growth rate, is conducted in the most objective, accurate, and robust manner. If the uncertainty in that "response" variable is significant, the output of the following statistical analysis can be inefficient at best and misleading at worst.

Let us now consider the existing tools for the first, and most crucial, step in the two-step analysis: the growth data summary. We will focus the discussion on one growth parameter—the maximum specific growth rate—a central parameter in microbiology, including diatom research.

The mathematical tool kit for microbial growth data analysis includes a multitude of approaches, which can be generally classified as one of three types: i) parametric, ii) nonparametric, and iii) ad hoc.

The modeling approaches (both parametric and nonparametric) have been intensively developed and rigorously explored, in particular within the realm of food microbiology. We direct the readers to the following sources for a detailed overview of the state-of-the-art in microbial growth modeling: [6.3] [6.8] [6.25] [6.56].

Generally speaking, in a "parametric" approach the growth of a unit in an experiment, say a replicate culture i, is modeled as a function of time:

$$Y_i(t) = f(t;\theta_i), \quad t \in \{t_1, t_2 \ldots, t_{n_i}\} \tag{6.3}$$

Where $Y_i(t)$ is a measured quantity at time t, such as optical density, fluorescence, or cell count of the target unit i. An example unit is a given volume in a well of a multi-well plate, and Y_i is then the measured value for that well. f is a deterministic curve model and is parameterized for the target unit by θ_i. Popular parameterization is $\theta = (\mu, \lambda, A)$, where μ describes the maximum of the derivative of f or the maximum growth rate, λ is the initial lag in time before growth enters the exponential growth phase, and A is the upper asymptote, or maximum growth level, where the curve settles after the exponential growth phase.

Many mechanistic models for growth curves have been proposed and the popular types include, among others, logistic, Gompertz, and Richards [6.56] [6.57]. For each model the parameter μ is directly the maximum growth rate, λ controls the lag, and A sets the maximum growth level. The Richards model [6.57] has an additional parameter, v adjusting for the shape of the curve, and can model both logistic ($v = 1$) and Gompertz ($v \rightarrow 0$) variants.

When taking the parametric model approach to reduce a growth curve into a single value of interest, for example, a specific growth rate, the problem translates quantitatively to the problem of estimating the corresponding parameter μ either directly as a model parameter or indirectly as a combination or transformation of the model parameters. In statistical notation, we wish to find a good estimator for $\hat{\mu}$, which is both accurate (unbiased) and stable (low variance).

Parametric functions fulfill the condition of simplicity by including as few parameters as possible (often the three mentioned above), are easy to implement, and provide a fair fit to much of the reported experimental data. However, simplicity comes at the cost of rigidity. For example, many classical parametric models assume a continuous increase of the microbial biomass and are therefore unsuitable for the analysis of the experimental data that records culture crash or a decline phase, often observed in diatom studies.

Instead of formulating a curve model with a specific parametric form that may or may not fit the acquired growth data $Y_i(t)$, an alternative is to use "nonparametric" curve estimation. The model is syntactically identical to the formulation in Eq. (6.3) and its stochastic extensions. The only difference is that instead of imposing a parametric form for $f(t) = f(t; \theta)$ and then estimating $\hat{\theta}$, we use a nonparametric estimator $\hat{f}$ for the whole function directly.

A common approach to estimating f "freely" is using nonparametric techniques such as kernel estimators or splines. For example, let us consider the "cubic spline" model's assumptions: f is continuous; f is piecewise polynomial of order three; wiggliness of f should be limited. The curve's wiggliness is controlled by penalizing the magnitude of second derivatives, weighted by a smoothness parameter s. Generally, a continuous piecewise polynomial estimate $\hat{f}$ is called a spline. Figure 6.4 exemplifies three different spline fits to the example growth series and illustrates the role of the smoothing parameter.

The spline estimator of $\hat{f}$ avoids fixing a specific growth model. However, in some cases, it can be "too" flexible, as it does not encode the typical sigmoidal shape, is sensitive to the smoothness parameter, and does not guarantee continuous "growth." The latter provides an advantage of nonparametric approaches over most parametric ones, as it is equipped to handle the death phase, culture crashes, and diauxic growth.

One popular nonparametric method in growth data analysis, although rarely applied in the field of phycology to our knowledge, is local polynomial regression (loess), which is a local averaging technique [6.12]. Similar to cubic splines, the smoothness is controlled by

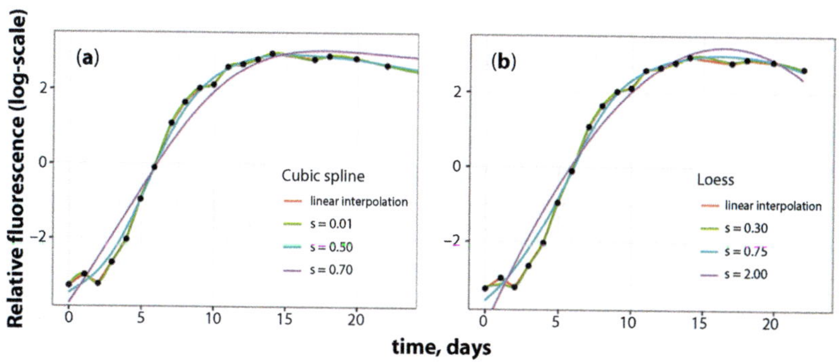

Figure 6.4 Examples of fitting a nonparametric curve to empirical *Skeletonema marinoi* data (black dots) using cubic splines (a) and local polynomial regression (b), with different smoothness parameter *s* values. The empirical data values represent relative *in-vivo* chlorophyll *a* fluorescence, log-transformed.

a parameter that in this case uses the span of the local neighborhood around a target time point as a "predictor" for the target value. Figure 6.4b illustrates the loess approach and highlights the influence of the smoothness parameter, which is essentially a user choice, on the shape of $\hat{f}$.

Whichever way we arrive at the nonparametric estimate $\hat{f}$, it enables interpolation in great detail to the exponential growth interval. The estimation of the maximum growth rate then simply becomes a matter of determining the maximum of the first derivative, $\hat{\mu} = max\ \hat{f}'$.

Despite the diversity of growth data modeling techniques, whether parametric or not, their application within basic microbiology is still very limited, with even fewer examples from diatom and other phytoplankton studies. The prevalent solutions are ad hoc methods, fine-tuned to a specific organism, individual researcher, or data type.

The simplest, and most intuitive method among the ad hoc, or heuristic solutions involves visual inspection of the growth curve in a semi-log scale for the purpose of defining the exponential growth phase timeframe and the corresponding sampled data points. Subsequently, the maximum specific growth rate is defined as the slope of the log-linear regression of the exponential growth phase. The described ad hoc method constitutes the recommended diatom growth data analysis tool widely employed within the diatom research community [6.51]. Although we did not perform a systematic literature review, the fact that ad hoc data analysis was used in all nine studies listed in Table 6.1, further highlights its popularity.

Though still a useful tool, assuming it can be conducted without the researcher's own bias, the ad hoc method becomes infeasible in the face of high-throughput data, with hundreds of often heterogeneous growth curves requiring an objective analysis.

Typical ad hoc solutions in such cases involve automated data processing tools, such as project-specific databases and R-packages (see examples in [6.9] [6.17]). In brief, the data is subjected to a smoothening step (typically a median filter) to address the technical noise and to remove or adjust outlier data points, which would otherwise bias the analysis. Next, the maximum growth rate is estimated in a robust manner by averaging a predetermined (three to seven) number of highest slopes (or first derivatives of the growth function), taken over the entire curve in a sliding window of three to five data points, but excluding the absolute maximum value [6.23] [6.50]. The automated ad hoc solutions depend on high sampling frequency data, and as such have limited use in the analysis of the diatom growth, given its peculiarities discussed in the previous section.

The question then arises, what options does a diatom researcher have in terms of objective and automated growth data analysis? The next section will present an *in silico* exploration on this subject and will provide a guide for evaluating and selecting appropriate analysis tools, following a two-step approach introduced above.

6.2 Simulation Trials

In any type of quantitative analysis, the researcher's main objective is to arrive at an estimate that accurately represents the studied phenomenon. When estimating a diatom's maximum specific growth rate, for example, we would want to find the model that fits perfectly to our observations, i.e., the correct model for our data. Clearly, this is no trivial task. However, statistics offer a solution in the form of simulation trials.

The principles of a simulation-based trial are widely used in statistics and are referred to as Monte Carlo methods [6.29]. Here, the true value of interest is known a priori: the underlying data is generated *in silico* following a user-defined set of parameters to mimic the experimental data as closely as possible. In the next step these synthetic data (e.g., growth curves) are subjected to parameter estimation (e.g., maximum specific growth rate) using estimation methods/models of choice. Repeating the "simulation-estimation" many times for an array of different growth series gives an idea of how every tested method performs on average, now and in the future, in terms of accuracy and reliability.

6.2.1 Methodology for the Simulation Trials

For the purpose of our simulation trials, we chose the Baranyi-Roberts function [6.7] as an archetypal sigmoid shape; it is flexible enough to produce a variety of growth curve shapes typically observed in diatom studies. Functions from the generalized logistic family, such as simple logistic, Richards, and Gompertz, would not have suited our purpose, as their relative growth is always monotonically decreasing and thus unable to exhibit log-scale lag-phases.

We generated a range of differently shaped growth series by randomly varying key parameters of the Baranyi-Roberts model. The maximum duration of a growth series was set to 14 days, reflecting the duration of typical diatom experiments. As the population responses vary in absolute magnitude, we fixed the asymptotic maximum of the function to 100 units, a convenient number, which allows easy scalability. Next, we varied the initial value of the growth series $Y_i(0)$ by uniformly sampling it from [0.01, 5] units, roughly the range observed in reported experiments. The intrinsic lag parameter was sampled uniformly from [0, 6], a range capturing both situations without lag and considerable lag (cf. Figure 6.2). The intrinsic growth rate parameter was fixed to 1 in order to reduce the degrees of freedom, due to its interconnectedness with the rest of the model's parameters: lag, $Y_i(0)$ value, and asymptotic maximum. The situation where the exponential phase is observed only briefly or missed altogether depends on the frequency of sampling the curve, which we included as a challenge.

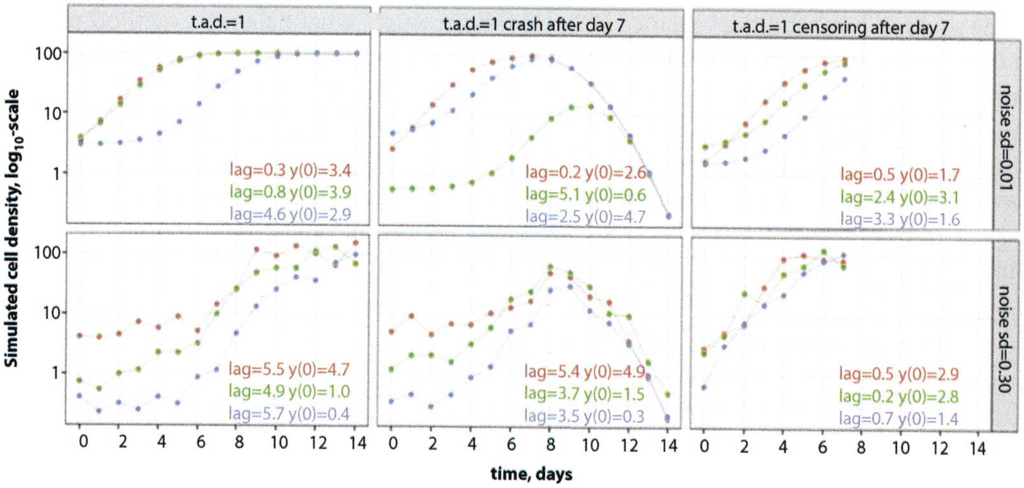

Figure 6.5 Three randomly chosen example growth series for 6 out of 27 challenge groups from the simulation trial. Each series was generated using the Baranyi-Roberts function with a random lag and initial value (shown), fixed maximum 100, and rate 1. Only the dots are taken as observations, the lines connecting the dots are merely part of the illustration and do not necessarily follow the true function. t.a.d. = times a day (sampling frequency).

Having generated a shape for the true growth function for each to-be-observed growth series, we then introduced challenges that mimic issues typically encountered in diatom growth experiments. First, the sampling frequency, with which the growth of the culture has been observed, was set to either twice a day, once a day, or once every two days ("times a day," abbreviated "t.a.d.", 2, 1 or 0.5). Then, we introduced the possibility for premature termination of the experiment, or "right-censoring," by either carrying measurements for the full 14 days or observing only 7 days. Typical to diatoms, we included the possibility of a death phase, mimicked by a multiplication by a negative rate logistic function after day 7 ("crash" after day 7, or no crash, i.e., perpetual growth towards the chosen asymptotic maximum). Finally, after being possibly crashed, censored or both, we added log-scale additive noise from the Gaussian distribution with zero mean, no temporal covariance, and three different levels ranging from practically no noise to considerable noise (standard deviation = 0.01, 0.15, 0.3). The noise encapsulates deviations from the ideal function, observational measurement errors, day-night growth cycles combined with small deviations in sampling time of day, or specific issues associated with various biomass proxies [6.30].

This two-step process of first generating an ideal sigmoid shape and then imperfectly "observing" the possibly disturbed growth series, was repeated 1000 times for each of the 27 combinations of challenges. Figure 6.5 illustrates three individual growth series from a selected set of challenges. For each such challenge group, we then computed the average (i.e., average over possible ideal shapes) error of estimation in terms of bias and variance. First, the estimation error per growth series is defined as the difference between the true specific growth rate, derived from the ideal function, and the estimated specific growth rate provided by the method. Then the bias of the method for the group is approximated by the mean error and the variance of the method by the variance of the errors. The overall quality of the method for a group can then be measured by the mean squared error (MSE), the sum of the squared bias, and the variance [6.24]. MSE measures what balance a method makes in the mathematically inescapable trade-off between bias and variance in statistical estimation. Smaller MSE is desirable, and sometimes a large reduction in variance at the cost of mild bias might be worth the price. And methods, which have very low variance but are systematically biased, i.e., are very certain of the wrong thing, will be penalized by the MSE metric. Box 1 provides a reminder of the quality metrics formulations used here.

Box 1. Quality metrics for parameter estimation

Variance: $\mathbb{E}\left[\breve{\mu}^2\right] - \mathbb{E}^2\left[\breve{\mu}\right]$

Systemataic bias: $\mathbb{E}\left[\breve{\mu} - \mu\right]$

Mean Squared Error (MSE): $MSE = \mathbb{E}\left[\left(\breve{\mu} - \mu\right)^2\right] = variance + bias^2$

6.2.2 Candidate Methods for Estimating the Specific Growth Rate

From each simulated growth series (varying in model parameters and applied challenges) we estimated the maximum specific growth rate using a selection of methods, some of which were introduced in the previous section.

The parametric curves included the logistic, Gompertz [6.57], Peleg [6.33], and Baranyi-Roberts [6.7] models (Box 2). The model parameters (lag, maximum rate, maximum asymptote) were estimated using nonlinear least-squares fitter (R function *nls*). To improve estimation, the target error minimization was carried out at the same scale as to where the noise was added, i.e., we minimized the sum of squared differences between the logarithm of the series values and the logarithm of the curve predictions. To further help the optimization, the parameter search space was confined to the following ranges: initial culture value ($Y_i(0)$) 0-100, lag 0-14, asymptote 50-150, rate 0-10; Peleg curve's additional death-lag 0-20 and death-rate -10-0.

Due to differences in the parameterization of these methods, the specific rate is not always one of the estimated parameters. In order to derive the specific growth rate, we compute the estimated curve values on a fine grid and then derive the target value numerically.

The nonparametric methods included fitting cubic splines and local polynomial regression [6.21]. Both methods provided a smooth curve through a growth series, from which the maximum specific growth rate was computed numerically. Cubic splines were estimated with the R-function 'smooth.spline' and the local regression with the function 'loess.as' from the package 'fANCOVA' [6.52]. The smoothing parameters for both methods were left to be chosen automatically by the algorithms based on generalized cross-validation (GCV).

The ad hoc methods included a naive "maximum 1-step change in log scale" ('max diff'), a slightly more robust "average 2nd and 3rd largest 1-step changes in log-scale" ('ave 2-3 diff'), and the "easylinear" method from the R-package "'GrowthRates," which implements the heuristic approach described in [6.23]. In addition, we tested versions of "max diff" and "ave 2-3 diff" after applying a running median filter with the window width 3 ('max diff med3' and 'ave 2-3 diff med3').

Box 2. Parametric curve models

General parametrization:

$(A, g, l, s, y_0) = (Asymptote, growth\ rate, lag, shape, starting\ value)$

and x = the time argument.

Logistic $A\{1 + \exp[4gA^{-1}(l - x) + 2]\}^{-1}$

Gompertz $A \exp\{-\exp[\exp(1)gA^{-1}(l - x) + 1]\}$

Richards $A\{1 + s \exp[1 + s + (1 + s)^{(1 + 1/s)}gA^{-1}(l - x)]\}^{-1/s}$

Peleg Logistic $(x; A, g_1, l_1) \cdot$ Logistic$(x; 1, g_2, l_2)$

with negative g_2

Baranyi - Roberts: $$y_0 \frac{\exp[gB(x)]}{1 + \frac{\exp[gB(x)] - 1}{A / y_0}}$$

with

$$B(x) = x + (1/g)\log[\exp(-gx) + \exp(gl) - \exp(-gx - gl)]$$

6.2.3 Simulation Trials Results

We will start by presenting the results of the simulation trials for every type of challenge (noise, crash, and censoring) individually, before looking at the overall performance ranking of all tested methods. For each type of challenge, we first provide a detailed overview of the error distribution for the best-performing method (loess GCV, panel A in the figures), and then contrast that to all other tested methods using variance, bias squared, and MSE metrics (panel B). The average ranking of each method, for each of the quality metrics, is summarized in Tables 6.2–6.4.

6.2.3.1 *Results with Only the Noise Challenge*

First, let us consider the quality of the performance estimators at the 9 challenge sets consisting of 3 noise levels and 3 sampling frequencies.

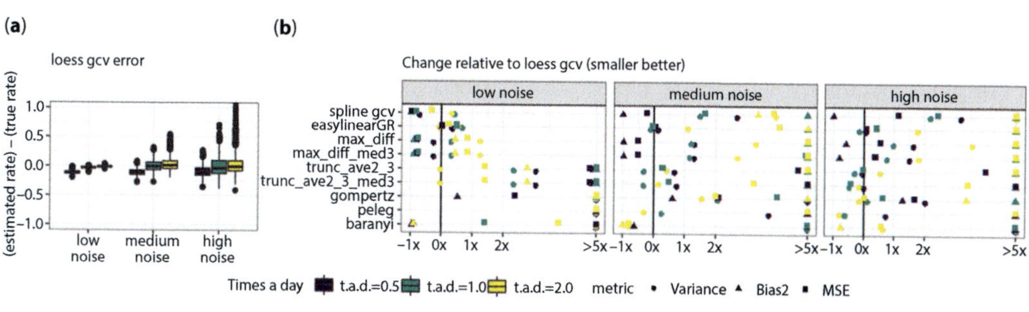

Figure 6.6 Quality of estimating the specific growth rate with only the noise challenge added, consisting of three levels of noise variance and three frequencies of sampling (t.a.d.). (a) The loess GCV error distributions. (b) Quality metrics relative to loess GCV (=0x; lower values indicate better performance).

Table 6.2 Average ranking of estimators in terms of quality metrics, averaged over challenge sets with varying noise and sampling frequency. Mean of ranks over challenge sets. Lower values signify better performance.

Estimator	Variance	Bias^2	MSE
loess gcv	3.22	3.11	2.56
spline gcv	7.00	2.00	3.67
easylinearGR	4.11	5.33	4.00
max diff	6.39	5.72	5.50
max diff med3	5.72	5.28	5.06
ave 2-3 diff	3.83	6.83	6.17
ave 2-3 diff med3	3.50	6.61	5.61
Gompertz	7.78	6.33	5.78
Peleg	5.33	9.67	9.22
Baranyi-Roberts	8.11	4.11	7.44

Figure 6.6a provides the distribution of the estimation error for the loess GCV method, stratified by noise and sampling levels (t.a.d., corresponding to the different colors in the figure). Starting with low noise, the errors are small, and they vary very little. With an increased level of noise, the average errors, i.e., biases, stay the same, but the variances increase as expected. Notice how increasing the sampling frequency, and hence the number of observations, reduces the bias, with once-a-day sampling frequency the bias being practically zero.

Figure 6.6b provides a comparison to the loess GCV in terms of relative variance, squared bias, and MSE. When considering MSE, no method is uniformly best. The bias of "max diff" is low, with good performance especially at the low sampling frequency, but the method is sensitive to noise. The "trunc ave" method behaves in the opposite way by having a large bias but low variance. Applying median filtering to these ad hoc methods changes the quality only marginally. All the parametric curve estimators have either a very high variance or bias, or both. These models seem to slightly improve their standing with increased noise.

The average ranking in terms of the means of the ranks of the estimators over the 9 "noise x sampling frequency" challenge sets is summarized in Table 6.2. In terms of the overall quality measured by the MSE metric, loess GCV performs best, followed by spline GCV and then by "easylinear." Spline GCV is slightly better than loess GCV when considering bias but has a higher variance. The Baranyi-Roberts model (the model that determines the shape of the underlying idealized curves in the simulation trials) does well in terms of bias but has the highest variance, making it unreliable. The Peleg model performed worst, due to having the wrong shape for our simulated data, which increases bias. In addition, it has a high parameter count, which increases variance.

6.2.3.2 *Results when Crashing Occurs*

Next, we consider the crash challenge and its effect on the estimators' quality. The crash is introduced into the data series after day 7. Note that the "noise x sampling frequency" challenge remains in effect in this set of simulations. Similar to the previous results, Figure 6.7 depicts the performance of the loess GCV and then contrasts the other estimators to that. Overall rankings are then given in Table 6.3.

Notably, loess GCV now has some positive bias at higher noise levels, an indication of possible undersmoothing (Figure 6.7a). The positive bias is also present at high sampling rates, indicating that the difficulty of automatic smoothing is increased by the crash event. The variances seem to be affected only slightly.

In the average ranking, loess GCV shares the top position with spline GCV, which has a slightly lower bias but higher variance. Similar to the noise-only challenge, all the parametric curves have the lowest ranks, the exception being Gompertz with a decent performance at high noise levels. The Peleg model performed poorly despite having the capability of

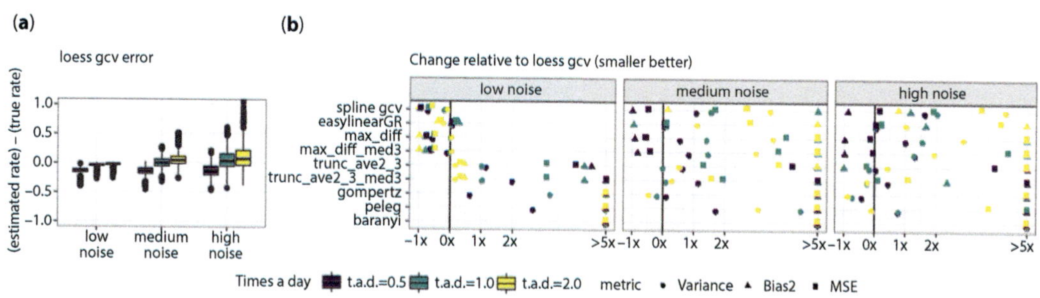

Figure 6.7 Quality of estimating the specific growth rate when the crash is introduced into the series after day 7. (a) The loess GCV error distributions. (b) Quality metrics relative to loess GCV (=0x).

Table 6.3 Average ranking of estimators in terms of quality metrics, averaged over challenge sets with a crash event on day 7 and varying noise and sampling frequency. Means of ranks over challenge sets. Lower values signify better performance.

Estimator	Variance	Bias^2	MSE
loess gcv	2.78	3.00	2.78
spline gcv	6.00	1.89	2.78
easylinearGR	4.11	4.11	4.22
max diff	5.78	4.67	4.44
max diff med3	5.78	4.22	4.33
ave 2-3 diff	4.72	5.17	5.17
ave 2-3 diff med3	5.06	5.61	5.17
Gompertz	3.33	8.78	8.67
Peleg	7.44	7.56	7.44
Baranyi-Roberts	10.00	10.00	10.00

modeling the crash event. The high parameter count of this model is a likely culprit, making its estimate unreliable even in the case of a high sampling rate with 29 available data points.

6.2.3.3 *Results when Censoring Occurs*

The last subset we studied includes cases where the experiment was prematurely stopped after day 7. Note that the "noise x sampling frequency" challenge remains in effect in this set of simulations too. Figure 6.8 depicts the performance of the loess GCV and then contrasts the other estimators to that, and the overall ranking in the censoring trial is given in Table 6.4.

Sampling only once every two days and stopping after seven days provides only four data points, which is clearly too little for loess GCV (Figure 6.8a), resulting in a large underestimation of the specific growth rate. In this scenario, almost all methods perform better, except Peleg and Baranyi-Roberts, due to their high parameter counts. In contrast, at a higher sampling rate there is little change in the quality of loess GCV compared to the results with the full 14-day series (for comparison, see Figure 6.6a).

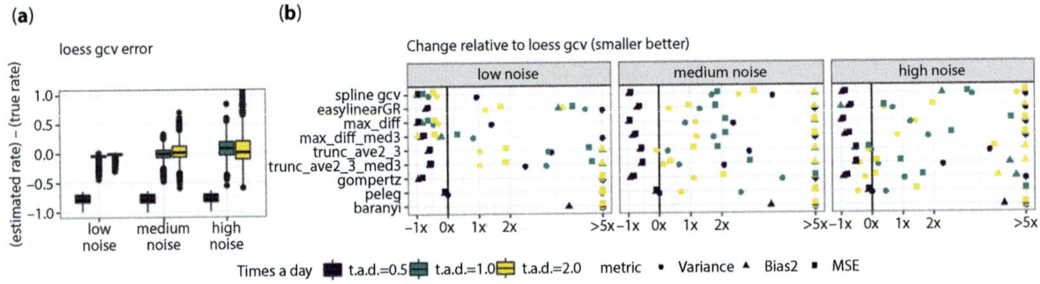

Figure 6.8 Quality of estimating the specific growth rate when the experiment is halted at day seven. (a) The loess GCV error distributions. (b) Quality metrics relative to loess GCV (=0x).

Table 6.4 Average ranking of estimators in terms of quality metrics, averaged over challenge sets with experiment stopped at day 7 and varying noise and sampling frequency. Means of ranks over challenge sets. Lower values signify better performance.

Estimator	Variance	Bias^2	MSE
loess gcv	2.33	4.56	4.33
spline gcv	5.67	2.22	3.33
easylinearGR	6.11	4.33	4.22
max diff	4.83	4.39	4.50
max diff med3	6.72	4.83	5.06
ave 2-3 diff	3.83	5.83	5.06
ave 2-3 diff med3	3.61	5.83	5.61
Gompertz	7.56	4.89	5.00
Peleg	4.33	8.11	7.89
Baranyi-Roberts	10.00	10.00	10.00

In the average ranking with censoring, spline GCV takes the lead, followed by "easylinear" and then closely by loess GCV and "max diff." The ad hoc methods perform much closer to the nonparametric approaches in this particular set, suggesting that if the number of observations (data points) is low, then simpler methods might be appropriate to use. The Gompertz model again performs quite well, while the high complexity models still rank the lowest.

6.2.3.4 *Overall Results and Ranking of the Methods*

To assess the overall quality of the estimators we provide the rankings in each of the 27 challenge sets in Figure 6.9, together with the average of those rankings. Based on the simulation trial we conclude that the nonparametric loess GCV and spline GCV perform best, in terms of the mean squared error metric. These methods are closely followed by "easylinear" and "max diff," then the truncated averaging, with the parametric curve models taking the last places.

Note how the rankings vary between the challenge sets: no single method appears to be uniformly best and even the on-average worst Baranyi-Roberts model works well under specific circumstances. However, without a priori knowledge of specific growth curve challenges, the nonparametric methods are more robust than the others. Particularly when the amount of simulated data (Figure 6.5) is reasonably close to what is commonly reported from the diatom literature (Figure 6.2), they outperform the other methods. The important difference between the two nonparametric methods is the way they balance variance and bias. Loess GCV has a lower variance, while spline GCV has lower bias, but their average MSE rankings are nearly identical.

The ad hoc robust methods of selective averaging have good variance properties, but the censoring challenge often introduces significant bias. If care is taken, they would still offer a reasonable alternative, even when only a few data points are available. The naive "max diff" performed well in the low noise cases and outperformed the truncated mean, but the quality, especially the variance, quickly deteriorated with increased noise.

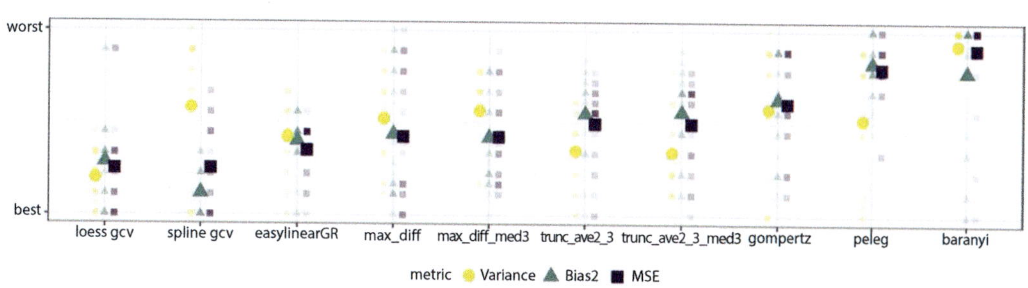

Figure 6.9 Ranking of the quality metrics over all 27 challenge sets. The small transparent symbols are the rankings per individual challenge set, while the big dark symbols represent the average score.

6.3 Empirical Example

The simulation trials allowed us to rank the performance of different methods in estimating specific growth rates on diatom-like growth data. Moreover, they could serve as a guide in model selection for future studies, supplying detailed information about which of the methods are more accurate, reliable, and robust, when faced with the particular challenges in the growth curve data.

The next logical step for us was to assess and quantify the consequence of the method's choice on the test statistic of interest in a follow-up analysis; for example, when comparing the specific growth rate estimates between treatment and control groups. To this end, we used empirical data from a growth experiment with the chain-forming diatom *Skeletonema subsalsum*. The experimental procedures for data acquisition are outlined in Box 3.

The data analyzed consisted of 48 growth series, recorded from 8 clones in 3 replicates, each at two temperature treatments (6 and 18 °C; Figure 6.10a). For each growth series, we estimated the specific growth rate using three different methods: the nonparametric spline (spline GCV), the Gompertz parametric curve model, and the ad hoc truncated average with median filter ('ave 2-3 diff med3').

Box 3. Experimental procedures

Skeletonema subsalsum cultures of 8 different strains were inoculated in triplicate with 10^4 cells ml^{-1} in 24-well culture plates and grown for 11 to 13 days at 6 and 18 °C under 100 $\mu mol^{-2}\ s^{-1}$ PPFD and a 12:12h light:dark cycle. Culture growth was measured daily at noon by chlorophyll fluorescence intensity using a Thermo Fisher Scientific Varioskan Flash plate reader machine at 425/680 nm excitation/emission wavelengths.

The rate estimates, presented in Figure 6.10b, clearly differ across methods and exhibit larger variation in the 18 °C group. The 6 °C group estimates showed relatively small variability, indicating a stable estimation, probably due to a clear lag and growth phases and the absence of a crash phase. In contrast, the growth curves in the 18 °C group have no discernible lag, while showing a clear crash or a decline phase. Given these challenges, it is

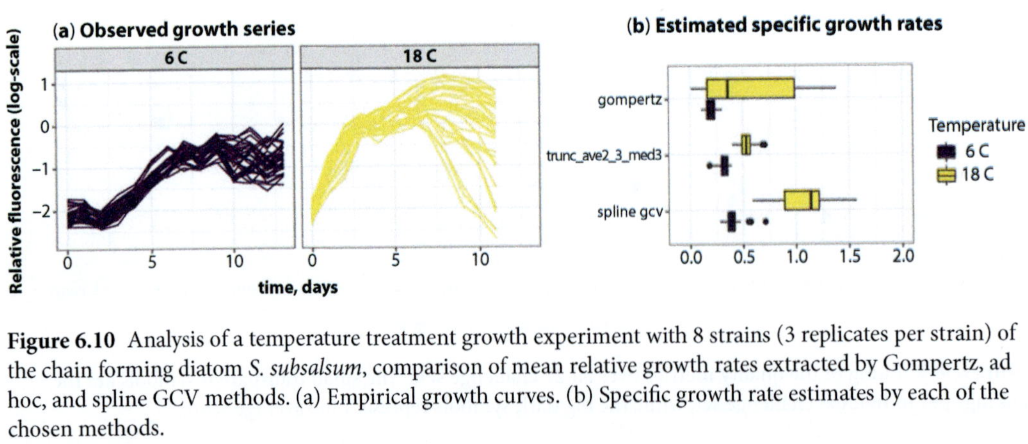

Figure 6.10 Analysis of a temperature treatment growth experiment with 8 strains (3 replicates per strain) of the chain forming diatom *S. subsalsum*, comparison of mean relative growth rates extracted by Gompertz, ad hoc, and spline GCV methods. (a) Empirical growth curves. (b) Specific growth rate estimates by each of the chosen methods.

Table 6.5 Estimated difference of mean growth rates in two temperatures. Growth rates are derived from different methods.

Method	Estimate	Std. Error	t value	Pr(>|t|)	2.5%	97.5%
ave 2-3 diff med3	0.22	0.02	13.62	0.00	0.18	0.25
Gompertz	0.42	0.11	3.20	0.00	0.16	0.68
spline gcv	0.68	0.04	15.26	0.00	0.59	0.77

not surprising that parametric curve model estimates varied the most in this group (Figure 6.10b, Gompertz, yellow bar). The spline method also demonstrated a noticeable variance (Figure 6.10b, spline GCV, yellow bar), likely due to splines being prone to artifacts near data series edges: the fastest growth is often found at the start of the growth series in examples with no lag, as was the case for many *S. subsalsum* curves in the 18 °C group.

After deriving the specific growth rates, we conducted a standard 1-way analysis of variance (ANOVA) to determine whether the average rate values were different in the two temperature groups. Table 6.5 provides the difference of group means and its 95% confidence intervals, along with variability and p-value statistics. Regardless of the method used to derive the "response," the results indicate a difference in the means. The effect sizes, however, varied greatly, such that the estimated group difference based on the spline estimates was more than 3 times the difference based on the ad hoc method (0.22 v 0.68). The parametric curve estimate of 0.42 settled in between the two but had many times higher uncertainty (se 0.11 vs. ca. 0.03). With a weaker signal, say 50% smaller difference in rate values between the treatment groups, the parametric curve estimate would no longer be statistically significant at the 5% level.

It might be clear to some that Gompertz is a poor model for the growth series with crashes. An example of a fix is an ad hoc truncation of any such series at the time of the crash event; see, for example, [6.28]. However, to avoid artifacts and biases in such ad hoc manipulations some form of manual quality control is required, and that can become very laborious for large experiments.

6.4 Conclusions and Recommendations

Diatom growth data present an analytical challenge due to heterogeneity and numerous types of irregularities, making it difficult to select the appropriate method for the analysis. We conducted a series of simulation trials in order to determine which of the available analytical methods provide the most accurate and robust estimate for specific growth rates in scenarios involving noise, crashes, premature termination of the experiment, and from a combined perspective. Furthermore, with an empirical example, we demonstrated the two-step data analysis workflow and how the choice of a rate parameter estimation method affects growth rate estimation and the statistical power of the study's conclusions. The presented example clearly showed that choosing an inappropriate method (in our example, choosing the wrong parametric model) may lead to the erroneous conclusion of no or non-significant (due to high uncertainty and possible bias) difference in a diatom growth response to temperature.

For the sake of full disclosure, we must add that the two-step approach does not represent the best analytical solution because the uncertainties in the estimates of the first step do not propagate properly to the second step. In an ideal scenario, both steps would be unified in a joint regression model, which would include all the available data, as well as uncertainties and sources of variance. A future challenge is to construct and generalize such a model to suit the variety of data types observed in diatoms. In the meantime, the two-step approach remains a viable alternative for a diatom researcher, being that it is straightforward in implementation and in fulfilling its purposes. Furthermore, "systematic" deviations from the idealized species mean curve by groups of replicates can be analyzed with mixed-effects regression models. However, in order for approaches based on growth summaries to succeed and produce meaningful results, several recommendations should be considered.

Our first advice concerns experimental design and pilot studies; both are invaluable in reducing technical noise and other artifacts that would obscure the real data. If the specific growth rate is the parameter of interest, obtaining a sufficient number of observations in the exponential phase can be equally important. Our results indicate that in the case of clean, high-frequency data, even the ad hoc solutions—the most common approach in diatom growth data analysis—could be sufficient and appropriate to obtain robust estimates. Note, however, that in none of these cases did the ad hoc methods surpass the nonparametric solutions.

Often, the diatom growth curve irregularities result from a biological phenomenon rather than technical noise and cannot be easily addressed by improving experimental design or performing pilot studies. Moreover, the treatment may impose a deviant shape to an otherwise "normal" growth curve observed in control conditions. Our simulation trials indicate that in such situations the nonparametric approaches (loess and splines) would provide the most objective alternative and should definitely receive higher attention from the phycology community. But even these methods were not fail-safe and occasionally produced an inaccurate or uncertain estimate (e.g., censored curves at low sampling frequency).

As a rule of thumb, we suggest reserving the parametric models for the analysis of data that is known or suspected to match that particular shape closely, such as a clear sigmoidal curve with high-density observations throughout. Even though the idealized curves in our study were based on a parametric model (Baranyi-Roberts), once the noise and other irregularities were introduced to the simulated data, this group of methods was unable to cope with it in a robust way. For methods, such as Peleg and Baranyi-Roberts, the larger number of in-built parameters might have partly been the reason for their poor performance. In the case of the simpler parametric models, the issue is likely explained by their numeric instability, especially at low data frequency. The numeric instability arises from a nonlinear dependence of a growth function on the model's parameters, leading to numerical optimization and identifiability issues. Nonparametric models have an advantage in this regard, as their internal parameters, such as local regression coefficients, relate linearly to the growth function, making them numerically more stable. Note, however, for extrapolation tasks, such as predicting future growth dynamics, parametric models, even with some biases, generally perform better than nonparametric models. This is because the latter rely heavily on interpolation and can behave unexpectedly outside the observed series.

All the parametric models explored here are derived from non-stochastic dynamic models, such as the logistic differential equation. These equations are solved under certain assumptions, which, as our results have shown, do not always fit the diatom-like growth data with its high levels of variability and artifacts. Recent years have seen the development of a new group of models that could be better suited for growth data analysis: the dynamic stochastic process model, e.g., a stochastic differential equation variant of the Gompertz model [6.14] [6.20] [6.47]. Though likely superior to the non stochastic variants we chose to study here, these models are still in early development in terms of the regression type analysis necessary for the broader statistical evaluation of experimental research questions. Moreover, these methods might require datasets larger than what current diatom studies typically produce.

In conclusion, we provide clear evidence that diatom growth data analysis would benefit from moving towards more robust and accurate types of solutions, facilitated by automated nonparametric approaches. Since no single method (of those explored here) fits all data variations, we would recommend first identifying what specific types of issues are characteristic to the data set that will be analyzed. Can these issues be addressed by improving experimental design? If not, the next recommended step would be to consult the detailed results of the here-presented simulation trials for the relevant data challenges. The full set of results for each challenge set and method, as well as the R-script for the related figures, can be found at https://github.com/antiphon/diatomtrials. Once the best-performing method for the given data is identified, the final step would be implementing it and obtaining the specific growth rate estimates. It could be useful to additionally test other (e.g., second best and one of the worst) methods to get an idea of possible bias and variance ranges associated with the method's choice. In line with this study, we encourage the reader to implement their chosen models in R (for more details refer to the candidate methods for estimating specific growth rate in Subsection 6.2.2 of this chapter), and take advantage of the ready-made and regularly updated R-packages, such as GrowthRates [6.34].

References

[6.1] Admiraal, W. and Peletier, H. (1980) Influence of Seasonal Variations of Temperature and Light on the Growth Rate of Cultures and Natural Populations of Intertidal Diatoms. Mar. Ecol. Prog. Ser. 2, 35–43.

[6.2] Ajani, P. A., Petrou, K., Larsson, M. E., Nielsen, D. A., Burke, J., Murray, S. A. (2021) Phenotypic trait variability as an indication of adaptive capacity in a cosmopolitan marine diatom. Environ Microbiol. 23(1), 207-223.

[6.3] Altman, N. S. and Casella, G. (1995) Nonparametric Empirical Bayes Growth Curve Analysis. J. Am. Stat. Assoc. 90, 508–515.

[6.4] Andersson, B., Godhe, A., Filipsson, H. L., Rengefors, K. and Berglund, O. (2020) Differences in metal tolerance among strains, populations, and species of marine diatoms – Importance of exponential growth for quantification. Aquat. Toxicol. 226, https://doi.org/10.1016/j.aquatox.2020.105551.

[6.5] Ando, Y. and Katano, T. (2018) Effect of Cell Size on Growth Rates of two Skeletonema Species and their Competitive Interactions. J. Phycol. 54, 850–859.

[6.6] Ashworth, J. *et al.* (2013) Genome-wide diel growth state transitions in the diatom Thalassiosira pseudonana. PNAS 110, 7518–7523.

[6.7] Baranyi, J. and Roberts, T. A. (1994) A dynamic approach to predicting bacterial growth in food. Int. J. Food Microbiol. 23, 277–294.

[6.8] Beckon, W. N., Parkins, C., Maximovich, A. and Beckon, A. V. A (2008) General Approach to Modeling Biphasic Relationships. Environ. Sci. Technol. 42, 1308–1314.

[6.9] Bukhman, Y. V. *et al.* (2015) Modeling Microbial Growth Curves with GCAT. Bioenergy Res. 8, 1022–1030.

[6.10] Causton, D. R. (1983) The exponential and related functions of biological importance. In: A biologist's basic mathematics. A. J. Willis and M. A. Sleigh (eds.) Edward Arnold, London: 148-168

[6.11] Chepurnov, V., Mann, D. G., Sabbe, K. and Vyverman, V. (2004) Experimental studies on sexual reproduction in diatoms. International Review of Cytology: a Survey of Cell Biology, 237, 91–154.

[6.12] Cleveland, W. S., Grosse, E. and Shyu, W. M. (1992). Local regression models. In: Statistical Models in S. J. Chambers and T. Hastie, (eds.) Wadsworth & Brooks/Cole.

[6.13] Cunha, C. *et al.* (2020) The effect of microplastics pollution in microalgal biomass production: A biochemical study. Water Res. 186, https://doi.org/10.1016/j.watres.2020.116370.

[6.14] Donnet, S., Foulley, J.-L. and Samson, A. (2010) Bayesian Analysis of Growth Curves Using Mixed Models Defined by Stochastic Differential Equations. Biometrics 66, 733–741.

[6.15] Eppley, R. W., Rogers, J. N., McCarthy, J. J. and Sournia, A. (1971) Light/dark periodicity in nitrogen assimilation of the marine phytoplankters Skeletonema costatum and Coccolithus huxleyi in N-limited chemostat culture. J. Phycol. 7, 150–154.

[6.16] Falciatore, A., Jaubert, M., Bouly, J. P., Bailleul, B. and Mock, T. (2020) Diatom molecular research comes of age: Model species for studying phytoplankton biology and diversity[open]. Plant Cell 32, 547–572.

[6.17] Fernandez-Ricaud, L., Kourtchenko, O., Zackrisson, M., Warringer, J. and Blomberg, A. (2016) PRECOG: a tool for automated extraction and visualization of fitness components in microbial growth phenomics. BMC Bioinformatics 17, 1–15.

[6.18] Gallo, C., Ippolito, G., Nuzzo, G., Sardo, A. and Fontana, A. (2017) Autoinhibitory sterol sulfates mediate programmed cell death in a bloom-forming marine diatom. Nat. Commun. 8, 1–11.

[6.19] Geider, R. J., Platt, T. and Raven, J. A. (1986) Size dependence of growth and photosynthesis in diatoms: a synthesis. Mar. Ecol. Prog. Ser. 30, 93–104.

[6.20] Ghosh, H. and Prajneshu. (2017) Gompertz growth model in random environment with time-dependent diffusion. J. Stat. Theory Pract. 11, 746–758.

[6.21] Green, P.J. and Silverman, B.W. (1994) Nonparametric Regression and Generalized Linear Models: A Roughness Penalty Approach. Chapman and Hall

[6.22] Gross, M. (2012) The mysteries of the diatoms. Curr. Biol. 22, R581–R585.

[6.23] Hall, B. G., Acar, H., Nandipati, A. and Barlow, M. (2014) Growth rates made easy. Mol. Biol. Evol. 31, 232–238.

[6.24] Hastie, T., Tibshirani, R. and Friedman, J. (2009) Overview of Supervised Learning. In: The Elements of Statistical Learning. T. Hastie, R. Tibshirani, and J. Friedman, (eds.) Springer, New York, 2nd edition: 9-42.

[6.25] Horowitz, J., Normand, M. D., Corradini, M. G. and Peleg, M. (2010) Probabilistic model of microbial cell growth, division, and mortality. Appl. Environ. Microbiol. 76, 230–242.

[6.26] Huysman, M. J. J., Vyverman, W. and Veylder, L. De. (2014) Molecular regulation of the diatom cell cycle. J. Exp. Bot. 65, 2573–2584.

[6.27] Koedooder, C., Stock, W., Willems, A., Mangelinckx, S., De Troch, M., Vyverman, W. and Sabbe, K. (2019). Diatom-bacteria interactions modulate the composition and productivity of benthic diatom biofilms. Frontiers in microbiology, 10, #1255.

[6.28] Kourtchenko, O., Rajala, T., & Godhe, A. (2018). Growth of a common planktonic diatom quantified using solid medium culturing. Scientific reports, 8(1), 1-10

[6.29] Kroese, D. P., Brereton, T., Taimre, T. and Botev, Z. I. (2014) Why the Monte Carlo method is so important today. WIREs Comput. Stat. 6, 386–392.

[6.30] MacIntyre, H. L., Cullen, J. J. (2005) Using cultures to investigate the physiological ecology of microalgae. In: Algal culturing techniques. R. A. Andersen (ed) Elsevier, Amsterdam: 287-326.

[6.31] Megía-Hervás, I. *et al.* (2020) Scale-up cultivation of Phaeodactylum tricornutum to produce biocrude by hydrothermal liquefaction. Processes 8, https://doi.org/10.3390/pr8091072.

[6.32] Monod, J. (1949) THE GROWTH OF BACTERIAL CULTURES. Annu. Rev. Microbiol. 3, 371–394.

[6.33] Peleg, M. A (1996) Model of Microbial Growth and Decay in a Closed Habitat Based on Combined Fermi's and the Logistic Equations. J. Sci. Food Agric. 71, 225–230.

[6.34] Petzoldt, T. (2020). Estimation of Growth Rates with Package growthrates. https://cran.r-project.org/web/packages/growthrates/vignettes/Introduction.html

[6.35] Pirt, S. J. (1967) A Kinetic Study of the Mode of Growth of Surface Colonies of Bacteria and Fungi. J. gen. Microbiol. 47, 181–197.

[6.36] Pyle, A. E., Johnson, A. M. and Villareal, T. A. (2020) Isolation, growth, and nitrogen fixation rates of the Hemiaulus-Richelia (diatom-cyanobacterium) symbiosis in culture. PeerJ 8, 1–25.

[6.37] Sabater, S. (2009) Diatoms. In: Encyclopedia of Inland Waters. G. Likens, (ed.) Academic Press, Oxford: 149–156.

[6.38] Sakshaug, E. (1977) Limiting nutrients and maximum growth rates for diatoms in narragansett bay. J. Exp. Mar. Bio. Ecol. 28, 109–123.

[6.39] Sakshaug, E. and Andresen, K. (1986) Effect of light regime upon growth rate and chemical composition of a clone of Skeletonema costatum from the Trondheimsfjord, Norway. J. Plankton Res. 8, 619–637.

[6.40] Schaechter, M. (2015) A brief history of bacterial growth physiology. Front. Microbiol. 6, 1–5.

[6.41] Seckbach, J. and Kociolek, J. P. (eds.) (2011) The Diatom world [Volume 19 in the series: Cellular origin, life in extreme habitats and astrobiology, series editor: Joseph Seckbach]. Springer Dordrecht Heidelberg London New York.

[6.42] Seyedsayamdost, M. R., Case, R. J., Kolter, R. and Clardy, J. (2011) The Jekyll-and-Hyde chemistry of Phaeobacter gallaeciensis. Nat. Chem. 3, 331–335.

[6.43] Sims, P. A., Mann, D. G. and Medlin, L. K. (2006) Evolution of the diatoms: Insights from fossil, biological and molecular data. Phycologia 45, 361–402.

[6.44] Solopova, A. *et al.* (2014) Bet-hedging during bacterial diauxic shift. Proc. Natl. Acad. Sci. U. S. A. 111, 7427–32.

[6.45] Spencer, C. P. (1954) Studies on the Culture of a Marine Diatom. J. Mar. biol.Ass. U.K. 33, 265–290.

[6.46] Tol, H. M. Van, Amin, S. A. and Armbrust, E. V. (2017) Ubiquitous marine bacterium inhibits diatom cell division. ISME J. 11, 31–42.

[6.47] Tolvanen, A. *et al.* (2011) A new approach for estimating a nonlinear growth component in multilevel modeling. Int. J. Behav. Dev. 35, 370–379.

[6.48] Von Dassow, P., Chepurnov, V. A. and Armbrust, E. V. (2006) Relationships between Growth Rate, Cell Size, and Induction of Spermatogenesis in the Centric Diatom Thalassiosira weissflogii (Bacillariophyta). J. Phycol. 42, 887–899.

[6.49] Wang, H. *et al.* (2017) Metacaspases and programmed cell death in Skeletonema marinoi in response to silicate limitation. J. Plankton Res. 39(4), 729–743.

[6.50] Warringer, J. and Blomberg, A. (2003) Automated screening in environmental arrays allows analysis of quantitative phenotypic profiles in Saccharomyces cerevisiae. Yeast 20, 53–67.

[6.51] Wood, A. M., Everroad, R. C., Wingard, L. M. (2005) Measuring growth rates in microalgal cultures. In: Algal culturing techniques. R. A. Andersen (ed) Elsevier, Amsterdam: 269–285.

[6.52] Xiaofeng Wang (2020). fANCOVA: Nonparametric Analysis of Covariance. R package version 0.6-1. https://CRAN.R-project.org/package=fANCOVA

[6.53] Ying, L., Beardall, J. and Heraud, P. (2006) Effect of UV radiation on growth, chlorophyll fluorescence and fatty acid composition of Phaeodactylum tricornutum and Chaetoceros muelleri (Bacillariophyceae). Phycologia 45(6), 605-615.

[6.54] Yu, E. T., Zendejas, F. J., Lane, P. D. *et al.* (2009) Triacylglycerol accumulation and profiling in the model diatoms Thalassiosira pseudonana and Phaeodactylum tricornutum (Baccilariophyceae) during starvation. J Appl Phycol 21, 669-681.

[6.55] Zhu, Z. *et al.* (2020) Effects of Microscopic Hydrodynamic Conditions and Exposure Time on the Biomass of Benthic Diatoms. Research Square, https://doi.org/10.21203/rs.3.rs-63435/v1.

[6.56] Zimmerman, D. L. and Nunez-Anton, V. (2001) Parametric modelling of growth curve data: An overview. Soc. Estad. e Investig. Oper. 10 (1), 1–73.

[6.57] Zwietering, M. H., Jongenburger, I., Rombouts, F. M. and van 't Riet, K. (1990) Modeling of the bacterial growth curve. Appl. Environ. Microbiol. 56, 1875–1881.

7

Integrating Metabolic Modeling and High-Throughput Data to Characterize Diatoms Metabolism

Juan D. Tibocha-Bonilla[1†], Manish Kumar[2†], Karsten Zengler[2,3,4] and Cristal Zuniga[2,5*]

[1]*Bioinformatics and Systems Biology Graduate Program, University of California, San Diego, La Jolla, California, USA*
[2]*Department of Pediatrics, University of California, San Diego, La Jolla, California, USA*
[3]*Department of Bioengineering, University of California, San Diego, California, USA*
[4]*Center for Microbiome Innovation, University of California, San Diego, La Jolla, California, USA*
[5]*Department of Biology, San Diego State University, San Diego, California, USA*

Abstract

Metabolic modeling is a quantitative tool that uses genome-scale networks to predict growth phenotypes with high accuracy. During the reconstruction of metabolic networks of diatoms the genome annotation is manually improved by including new gene-protein-reaction associations of organelle-specific pathways, as well as intra- and extracellular transporters. Metabolic modeling combined with omic data of diatoms has been applied to identify suitable culture conditions, unravel interwoven regulatory mechanisms, explore genetic enhancement strategies for bioproduction, and quantify changes in resource allocation under nutrients limitation.

Computational models offer insights into the function of pathways critical for diatom physiology, from central carbon metabolism to lipid, pigment, and pathways involved in photon uptake. Here, we review the state-of-the-art of metabolic modeling of diatoms and how recently developed control biosystems can be applied to optimize rational culture cultivation conditions. We also discuss how metabolic models of diatoms and heterotrophs can be linked in a community framework to better understand the interactions of diatoms in natural environments.

***Keywords*:** Genome-scale metabolic modeling, constraint-based models, reconstruction of diatoms, metabolomics, transcriptomics, proteomics, diatom-heterotroph interactions

**Corresponding author*: czuniga2@sdsu.edu
†These authors contributed equally.
Juan D. Tibocha-Bonilla: https://scholar.google.com/citations?user=yGcA-F8AAAAJ&hl=en, https://www.researchgate.net/profile/Juan-Tibocha-Bonilla, https://www.linkedin.com/in/juan-d-tibocha-bonilla-2a999567/
Manish Kumar: https://profiles.ucsd.edu/manish.kumar, https://scholar.google.com/citations?user=beuzxi8AAAAJ&hl=en, https://orcid.org/0000-0001-8035-3399
Karsten Zengler: https://www.researchgate.net/profile/Karsten-Zengler, https://profiles.ucsd.edu/karsten.zengler, https://scholar.google.com/citations?user=qZjgIt0AAAAJ&hl=en
Cristal Zuniga: https://orcid.org/0000-0002-0135-7429, https://scholar.google.com/citations?user=wgM6f0YAAAAJ&hl=es, https://www.researchgate.net/profile/Cristal-Zuniga

Janice L. Pappas (ed.) The Mathematical Biology of Diatoms, (165–192) © 2023 Scrivener Publishing LLC

7.1 Introduction

Diatoms are intriguing organisms with an intertwined evolutionary past and unique metabolism. These organisms contain powerful photosynthetic machinery, which, coupled with unique response mechanisms to stress and low nutrient levels, allow them to be ubiquitous in oceans and freshwater ecosystems [7.1]. In addition to their metabolic traits, diatoms can accumulate valuable compounds with industrial value, such as lipids for the production of fuel alternatives [7.2]. As a result, both academic and biotechnological research has been focused on these organisms. Genome-scale metabolic models (GEMs) have been widely employed for the study and optimization of microbial metabolism [7.2]–[7.7]. In the last decade, reconstruction of diatom GEMs has become feasible after the genome sequencing of *Thalassiosira pseudonana* [7.8], *Phaeodactylum tricornutum* [7.9], and more recently, the polar diatom *Fragilariopsis cylindrus* [7.10]. The early efforts in metabolic modeling of diatoms described the central metabolism of diatoms [7.11], including enzyme localization, cross-talk of organelles, and unique metabolic pathways. Newer diatom GEMs were deployed to analyze diatom bioenergetics, thus discovering response mechanisms to excess reducing power [7.12] [7.13]. However, several questions remain unanswered, which will fuel extensive metabolic modeling research on diatoms in the following years. Such questions include the metabolic role of diatoms in communities using community models (CM models) [7.14] and the prediction of proteome allocation in diatoms using models of metabolism and gene expression (ME models) [7.15]. This chapter reviews the state-of-the-art of diatom GEMs and their outcomes, tools for the reconstruction and omics data integration of these models, and an overview of the future research potential in the field of metabolic modeling of diatoms.

7.2 Characterization of Diatom Genomes

7.2.1 Available Genomics Data

Genomics of diatoms has enabled the sequencing of several genomes, which are available in public domains, such as the Joint Genome Institute (JGI) genome portal [7.16], and the National Center for Biotechnology Information (NCBI) [17] (Figure 7.1). Table 7.1 contains a summary of different tools used in annotating the diatom genomes. The centric diatom *Thalassiosira pseudonana* was the first diatom whose complete genome was sequenced and annotated in 2004 [7.8]. This work reported 11,242 protein-coding genes, and 47% overall GC content. Later on, the genome of the pennate diatom *Phaeodactylum tricornutum* was sequenced and since then it has become the most studied diatom genome to date. Since its publication in 2008, the annotations of this genome have been improved a few times by using transcriptome and proteome data and updated gene models [7.18]–[7.20]. In the first report, nearly 10,402 protein-coding genes were predicted in the *P. tricornutum* genome (34.5 Mb) [7.9]. During the second and third updates an additional 1,489 and 606 novel genes were identified based on improved gene models and omics data in two separate studies [7.19] [7.20]. In 2012, Lommer *et al.* [7.21] sequenced the genome of the centric diatom *Thalassiosira oceanica*. The authors predicted 10,109 protein gene models based on the AUGUSTUS gene finder tool [7.22] and best BLAST hits

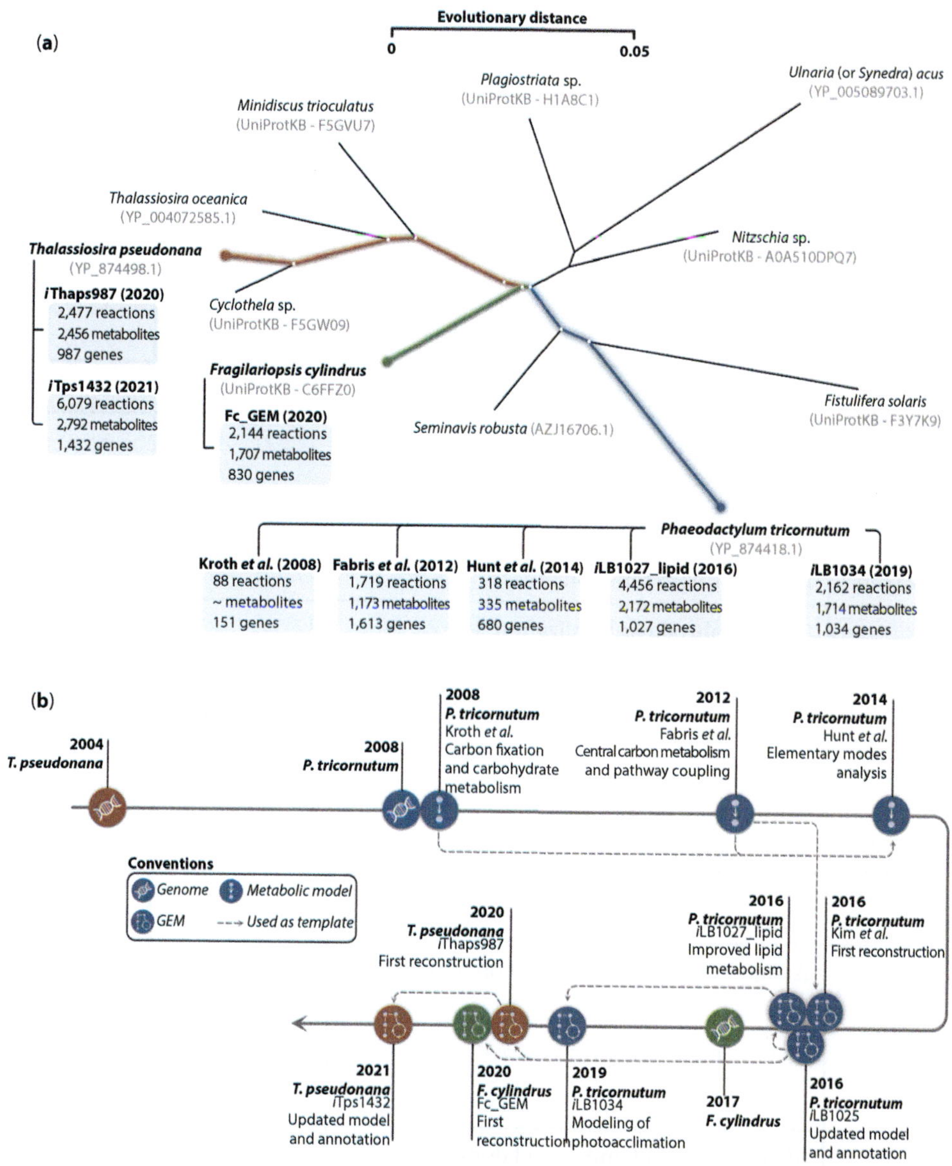

Figure 7.1 State-of-the-art genomics and metabolic modeling of diatoms: (a) Neighbor-joining tree based on full-length of the ribulose-1,5-bisphosphate carboxylase/oxygenase large subunit (chloroplast) gene sequences, showing phylogenetic relationships between diatoms with complete genome sequence available. NCBI or UniProt accession numbers are given in parentheses. GEMs properties (i.e., number of reactions, metabolites, and genes) are given under the diatom name for those that have an available genome-scale metabolic model). (b) Timeline of first genome annotations and published metabolic reconstructions of *T. pseudonana, P. tricornutum,* and *F. cylindrus*. The straight line represents the timeline from the first diatom sequenced until the present. Dotted lines connect models that were used as templates (line start) for other models (line end).

with a conservative E-value cutoff of 1.0E-30. The size and GC content of this genome were 81.6 Mb and 53.3%, respectively [7.21].

The complete nucleotide sequence of an araphid pennate diatom *Synedra acus* subsp. *radians* was recently determined. This genome (98.4 Mb) contained 11,184 protein-coding

Table 7.1 Comparison of annotation methods used on diatom genomes.

Annotation method	Reported use	Source*	References**
JAZZ	Assemble whole genome shotgun sequences	[7.28]	[7.8] [7.9]
Phred	Call bases from DNA sequence traces	[7.29]	[7.8] [7.9]
Phrap	Assemble DNA sequences	[7.29]	[7.8] [7.9]
Consed	Predict Inverted Repeat in the chloroplast genome	[7.30]	[7.8] [7.9]
Genewise	Map protein sequences to DNA sequences	[7.31]	[7.8]–[7.10]
GrailEXP	Predict genome regions, e.g., exons, genes, promoters, polyAs, etc.	[7.32]	[7.8]
Fgenesh	Predict protein-coding genes	[7.33]	[7.9] [7.10]
InterProScan	Predict protein function	[7.34]	[7.9] [7.10]
CELLO	Predict subcellular localization	[7.35]	[7.20]
OrthoMCL	Predict orthologous groups in eukaryotes	[7.36]	[7.37]
SignalP	Predict subcellular localization	[7.38]	[7.10] [7.23]
TMHMM	Predict protein topology and transmembrane domains	[7.39]	[7.10] [7.23] [7.40]
tRNAscan-SE	Predict tRNA sequences	[7.41]	[7.10] [7.23]
RNAmmer	Annotate rRNA genes	[7.42]	[7.43]
Hmmer	Assign EC annotations based on homology	http://hmmer.org/	[7.43]
MAKER	Identify protein-coding genes	[7.44]	[7.23]
PANTHER	Predict function from protein sequences	[7.30]	[7.23]
SMART	Identify protein families and domains	[7.45]	[7.23]
Infernal	Predict non-coding RNA sequences	[7.46]	[7.23]

* The column "Source" compiles the publications that first described the method.
** The column "References" compiles the studies that have employed each method.

genes out of 27,337 gene models, which were filtered based on the annotation edit distance (AED) cutoff of < 1. EC numbers were found to be associated with 2,729 genes. In 2015, Tanaka *et al.* sequenced the genome of a pennate diatom *Fistulifera solaris* and predicted 11,448 protein-coding genes out of 20,455 gene models by removing redundant genes from AUGUSTUS predictions [7.21]. This genome (24.9 Mb) included 46% GC content. The genome of a centric diatom *Cyclotella cryptica* (161.7 Mb) was sequenced and annotated by two separate studies between 2016 and 2020 [7.21] [7.23]. The latest version comprises 20,506 protein-coding genes out of 21,250 gene models based on AED cutoff of <0.5 [23]. In 2017, the genome *Fragilariopsis cylindrus* (61.1 Mbp), a cold-adapted pennate diatom, was annotated [7.10] using JGI's annotation pipeline [9]. This study predicted 21,066 protein-coding genes and reported 39.8% GC content. The genome (117 Mbp) of *Seminavis robusta* (a biofilm-forming pennate diatom) was sequenced and annotated in 2020 [7.24]. In this study, the authors sequenced one reference strain (D6) and 48 other strains of *S. robusta*. A total of 36,254 protein-coding genes were predicted based on integrated sequencing data of 49 strains and gene expression data under various environmental conditions [7.24]. The complete genome sequence of a few more diatoms (*Pseudo-nitzschia multistriata* [7.25], *Plagiostriata* sp. CCMP470 [7.26], *Minidiscus trioculatus* CCMP495 [7.16] (https://genome.jgi.doe.gov/portal/pages/projectStatus.jsf?db=Mintr2), and *Nitzschia* sp. [7.27]) were also made available between 2017 and 2020.

7.2.2 Computational Tools to Allocate Gene Functions by Subcellular Localization

Diatom metabolism is compartmentalized in various organelles such as cytoplasm, chloroplast, mitochondria, nucleus, peroxisome, and thylakoid. Since experimental data for compartmentalization of diatom proteins is limited, several computational tools, such as ASAFind [7.47], HECTAR [7.48], MitoProt [7.49], predictNLS [7.50], SignalP [7.51], TargetP [7.52], TMHMM [7.39], and WoLF PSORT [7.53], are employed to predict subcellular localization of each protein in a given genome. This step is critical to contextualize genomics information and to further understand the metabolism of diatoms.

Table 7.2 shows the main characteristics of different subcellular localization tools. In general, these computational tools use amino acid sequences of N-terminal peptides to predict the subcellular localization of proteins. The typical examples of such peptides are signal peptides (SPs), mitochondria targeting peptides (mTPs), and chloroplast targeting peptides (cTPs).

Some of these tools have been customized for diatoms. For example, ASAFind [7.47] was developed to identify plastid-localized proteins and tested on the genomes of diatoms such as *T. pseudonana* and *P. tricornutum*. Predictions were based on the presence of a conserved "ASAFAP" motif with plastid-localized peptides. This workflow included SignalP 3.0 [7.54] to filter out the proteins without signal peptides. Next, the output of SignalP 3.0 was used to identify cleavage sites. If the next amino acid from the cleavage site or first amino acid of ASAFind predicted transit peptide was 'F', 'W', 'Y', or 'L' then the sequence was classified as a potential plastid-targeted peptide. Furthermore, if the ASAFind predicted cleavage site is the same as the SignalP predicted cleavage site, proteins were considered plastid-targeted proteins with low confidence. But if the ASAFind cleavage site score of transit peptide was also more than 2, the proteins were classified as plastid-targeted proteins with high confidence. The cleavage site score was calculated for 25-position sequence windows surrounding the SignalP predicted cleavage site.

Another multi-algorithm tool is HEterokont subCellular localization TARgeting (HECTAR) [7.48], which includes various localization tools such as HMMTOP [7.55], iPsort [7.56], MitoProt2 [7.49], Phobius [7.57], PrediSi [7.58], Predotar [7.59], PredSL

Table 7.2 Tools for predicting subcellular localization of the proteome of diatoms. This table contains information about the tools that have previously been applied to diatoms' genomes.

	Tool	Prediction	Organisms	Accuracy	Sensitivity	References
1	ASAFind	Plastids-targeted proteins	Diatoms	-	-	[7.47]
2	HECTAR	Signal peptides, type II signal anchors, chloroplast-targeted proteins, mitochondrion-targeted proteins, and proteins without targeting peptides	Heterokonts	-	-	[7.48]
3	MitoProt	Mitochondrion-targeted proteins	Any	-	-	[7.49]
4	predictNLS	Nuclear-targeted proteins	Any	100%	-	[7.50]
5	SignalP	Secretory signal peptides in any organism and distinguishing three kinds of signal peptides (Sec substrates cleaved by SPase I, Sec substrates cleaved by SPase II, and Tat substrates cleaved by SPase I) in prokaryotes	Any	-	-	[7.52]
6	TargetP	Chloroplast-targeted, mitochondrion-targeted, and secretory signal proteins	Phototrophs and non-phototrophs	85% (plants) and 90% (non-plants)	-	[7.52]
7	TMHMM	Transmembrane proteins	Any	97-98%	99%	[7.39]
8	WoLF PSORT	Nuclear, mitochondrion, cytosol, plasma membrane, extracellular, and chloroplast-targeted proteins	Phototrophs and non-phototrophs	50% (mouse proteins)	70%	[7.53]

[7.60], SignalP [7.52], and TargetP [7.52] to predict five kinds of protein-like signal peptides (e.g., type II signal anchors, chloroplast-targeted proteins, mitochondrion-targeted proteins, and proteins without targeting peptides). Initially, this tool was examined on heterokonts [7.48], but eventually was also deployed for other diatoms (*P. tricornutum* [7.13] and *T. pseudonana* [7.61]).

Another tool, MitoProt [7.49], was developed to identify the mitochondrion-targeted proteins. This tool uses various parameters, such as targeting sequences, profiles, amino acid compositions, amphiphilicity, cleavage site, and hydrophobic characteristics, to classify proteins as mitochondrial.

One of the most accurate tools is predictNLS [7.50], which distinguishes nuclear and non-nuclear proteins. The predictNLS functions are based on an expert database of experimentally identified and potential nuclear targeting peptides. This tool searches for nuclear targeting peptides in amino acid sequence and DNA-binding regions within nuclear targeting peptides to classify a protein as nuclear targeted. The known DNA-binding regions were considered a parameter in this framework because these regions were found to overlap with nuclear targeting peptides in 90% of the proteins. The predictNLS has been used to compartmentalize nuclear proteins in the genomes of many organisms, such as human, *Drosophila*, *Caenorhabditis elegans*, *Saccharomyces cerevisiae*, and *Escherichia coli*, but also in diatoms (*P. tricornutum* [7.13] and *T. pseudonana* [7.61]).

A transport-proteins bioinformatics tool, namely SignalP [7.52], was developed and improved a few times to predict signal peptides [7.38] [7.51] [7.54] [7.62] [7.63]. In general, signal peptides are small peptides of a few amino acid sequences that target proteins across or into membranes. The first four versions of SignalP [7.38] [7.54] [7.62] [7.63] were based on artificial neural networks and hidden Markov models, and they were partially capable of distinguishing different types of signal peptides. However, the fifth version (SignalP 5.0) [7.51], which is based on deep neural networks, performed well for predicting signal peptides in any organism and distinguishing between three types of signal peptides (Sec substrates cleaved by SPase I, Sec substrates cleaved by SPase II, and Tat substrates cleaved by SPase I) in prokaryotes.

Another neural-network-based tool, TargetP [7.52], was found 85% (plants) and 90% (non-plants) accurate for predicting chloroplast- and mitochondrion-targeted proteins and secretory signal peptides. This tool also predicted approximately 40-50% correct cleavage sites for chloroplast and mitochondrion targeting peptides and higher (70%) for signal peptides. Another hidden Markov model-based tool, TMHMM [7.39], was explicitly developed to predict membrane proteins. WoLF PSORT [7.53] was designed to classify nuclear, mitochondrion, cytosol, plasma membrane, extracellular, and chloroplast-targeted proteins. This tool converts amino acid sequences into numerical values based on amino acid sequences and functional regions like DNA binding regions. Next, based on the k-nearest neighbor classifier, each protein is assigned a subcellular location.

Besides significant development in generating bioinformatics tools for predicting subcellular localization of proteins, to date, there is no universal tool available to localize diatom proteins of all known compartments. A few pipelines were published to address this issue systematically, including the above tools, and to predict accurate subcellular locations of diatom proteins based on their sequences. Recently, such pipelines were employed to compartmentalize the proteins of diatoms such as *Fistulifera solaris* [7.64], *P. tricornutum* [7.13], and *T. pseudonana* [7.61].

7.3 Metabolic Modeling of Diatoms: Data and Outcomes

7.3.1 Using Genomic Information to Build Genome-Scale Metabolic Models

Recently, diatoms have been attracting the attention of many research groups in academia and industry due to their capability to produce various valuable products [7.65]–[7.67]. Several experimental efforts have been conducted to explore the metabolic routes and apply metabolic engineering to improve the synthesis of desired products in diatoms (e.g., *F. cylindrus*). To accelerate these efforts, genome-scale metabolic modeling has been applied to several diatoms to predict their metabolic capabilities [7.13] [7.61] [7.68] [7.69]. Genome-scale metabolic models (GEMs) have been proven as an excellent tool to reveal uncharacterized mechanisms in the metabolism of plenty of organisms from all domains of life [7.70] [7.71]. The fundamentals of GEMs have been well described in previous publications [7.5] [7.72]; here, we will focus on diatom GEMs and their applications.

The metabolism of diatoms is still poorly characterized, rendering their GEM reconstruction resource- and time-intensive endeavors, which could extend to months or years of manual curation. Several tools to automatically reconstruct a metabolic model from an annotated genome currently exist and have been discussed in previous reports [7.73] [7.74]. Automated tools have been extensively used in bacteria and other well-described eukaryotes, but not diatoms. Diatom models have been reconstructed *de novo* following standard protocols [7.69] or from manual adaptation and curation of template models.

P. tricornutum's small genome and ease of cultivation quickly rendered it the model diatom for metabolism. For almost a decade after the sequencing of *P. tricornutum*'s genome, diatom core metabolism was studied extensively through the metabolic reconstructions of this diatom. Early efforts focused on characterizing carbon fixation and central carbon metabolism through the topological analysis of the metabolic network of *P. tricornutum* with reconstructions of 88 [7.11] and later 1,719 reactions [7.70], then followed by an elementary modes (EM) analysis with 318 reactions [7.72]. These first approaches further led to the reconstruction of three *P. tricornutum* GEMs: Kim *et al.* [7.75], *i*LB1025 [7.13], and *i*LB1027_lipid [7.13]. In recent years, these *P. tricornutum* models served as templates for the reconstruction of: 1) one updated *P. tricornutum* model *i*LB1034 [7.12] with a special focus on photoacclimation; 2) a *T. pseudonana* model *i*Thaps987 [59], including its specific biomass composition and gap-filling related to new precursor requirements; 3) an updated *T. pseudonana* model *i*Tps1432 [7.61] with a drastic improvement in genome coverage and updated pathways; 4) the single available *F. cylindrus* model [7.76]. The authors have not given any standard name to the latter model in the original publication, so we are using a generic name "Fc_GEM" in this chapter to keep this consistent with other diatom GEMs. Figure 7.1 shows the properties of the *P. tricornutum* models over the course of time. The general properties of these models, such as the number of genes, metabolites, and reactions, can be seen in Figure 7.1a. So far, *i*Tps1432 [7.61] is the most detailed diatom model published in terms of genes, metabolites, and reactions.

Figure 7.1b shows that *i*LB1025 (and *i*LB1027_lipid) served as templates for the latest models of *F. cylindrus* (FC_GEM [59]) and *T. pseudonana* (*i*Tps1432 [7.61]) (dotted lines). In metabolic modeling, it is an important step to assign at least one compartment to each reaction/metabolite in the network that helps to run compartment-specific simulations. As a result, reactions can often be inherited from template models, though they may be localized

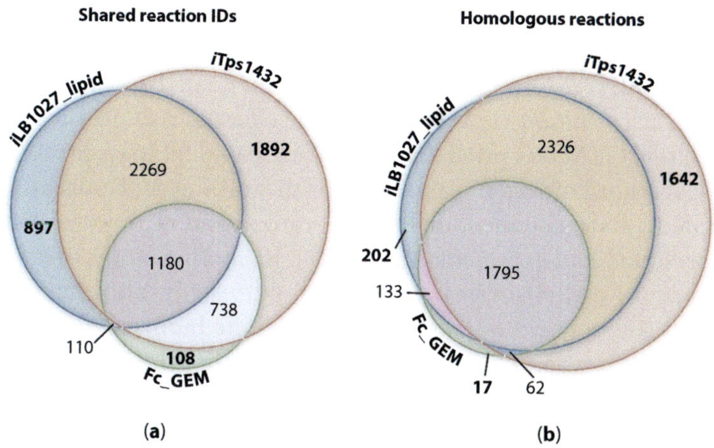

Figure 7.2 Shared reactions across the latest *T. pseudonana* and *F. cylindrus* GEMs and their template *i*LB1027_lipid. (a) Shared reactions with the same ID across models *i*LB1027_lipid, *i*Tps1432, and Fc_GEM. (b) Shared homologous reactions including: 1) same reaction with a different ID, 2) the same reaction at a different compartment, 3) the same reaction with a different associated enzyme.

in different compartments. Figure 7.2 illustrates the number of shared reactions across the three models, which include reactions with the same ID (Figure 7.2a) or reactions with different enzymes or localization, even though they represent the same chemical transformation (Figure 7.2b). More specifically, a reaction was deemed shared by two models in Figure 7.2b if at least one reaction in both models involved the same reactants and products, with the same stoichiometry. It is evident that *i*LB1027_lipid was critical for the reconstruction of Fc_GEM and *i*Tps1432, as 4,254 out of 4,456 of its reactions are found in the two models either with the same ID, or in a different compartment or enzyme (Figure 7.2b), and 1180 out of 4456 reactions of *i*LB1027_lipid were kept unchanged in both models (Figure 7.2a). *i*Tps1432 is currently the most comprehensive diatom model available, which largely resembles the highly curated *i*LB1027_lipid (4,121 homologous reactions), and includes 1,642 unique reactions.

7.3.2 Comprehensive Diatom Omic Datasets Are Useful to Constrain Metabolic Models

The GEMs employ a constraint-based framework, in which the metabolic network of the model is constrained to reflect certain phenotypic traits under defined environmental conditions (e.g., light and dark cycles, nutrients limitations, and growth in isolation or as part of a community). Without such constraints, GEMs often illustrate many possibilities that drastically decrease the prediction power of a model. One vital way to implement such constraints to the model is to integrate omics data (e.g., metabolomics or transcriptomics) into the reaction boundaries. With increasingly existing omics data for diatoms, it is promising to enhance the predictive power of diatom GEMs.

Plenty of multi-omics data (genomics, transcriptomics, proteomics, and metabolomics) have been generated to explore the biology of diatoms in terms of novel metabolic pathways, their stress responsiveness, and applications in biofuel production, the pharmaceutical industry, or wastewater remediation (Table 7.3). Integration of omics data with a GEM can aid in revealing the uncharacterized metabolic capabilities of diatoms. For example, Levering *et al.* [7.77] integrated transcriptomics data based on 179 experimental samples with the *P. tricornutum* GEM to explore the effect of rising atmospheric CO_2 levels. The *P. tricornutum* GEM was predicted to have increased biomass production based on increasing CO_2 levels until nitrogen reached a growth-limiting concentration. Overall, this study demonstrated a strong connection between carbon and nitrogen metabolisms based on the model prediction and expression data [7.77]. In another study, metabolomics data, including amino acids, carbohydrates, nucleotides, fatty acids, phospholipids, and other lipids, was used to compute time- and condition-specific biomass reactions. In general, biomass reaction is used as an objective function to predict the growth of the organism [7.78]. As a result, integrating metabolomics data into the *P. tricornutum* GEM helped to understand the metabolic variations in different organelles and pathways under stress conditions [7.79]. Omics data can also be used to validate the GEM predictions. For example, Broddrick *et al.* [7.12] used transcriptomics data to validate the glycolate pathway-related predictions from *P. tricornutum* GEM. Glycolate oxidase is found in mitochondria and peroxisome in *P. tricornutum*; however, flux predictions suggested that mitochondrial glycolate oxidase carried a much higher flux compared to peroxisomal glycolate oxidase. Similarly, the transcript abundance of mitochondrial glycolate oxidase was found to be 40 times higher than peroxisomal glycolate oxidase, confirming the flux predictions. During this study, upregulation of branched-chain amino acid synthesis and catabolism pathways in the dark and in response to iron limitation were found consistent between GEM predictions and transcriptomics data.

In several research efforts, multi-omic datasets were obtained based on monoculture and communities of diatoms with other microorganisms. Ashworth *et al.* [7.80] published the previously reported monoculture-based transcriptomics datasets of seven different diatoms (*P. tricornutum*, *T. pseudonana*, *C. cryptica*, *F. cylindrus*, *T. weissfogii*, *T. oceanica*, and *P. multiseries*) from 1,075 samples under different environmental conditions. Most of the diatom transcriptomics datasets in this public compendium were obtained from *P. tricornutum* (540 samples) and *T. pseudonana* (380 samples) monocultures, which are the most studied diatoms to date [7.80]. Moreover, several *P. tricornutum* monoculture-based studies reported multi-omics data for studying the production of triacylglycerol production metabolism under nitrogen limitation during light and dark cycles [7.81], to reannotate the diatom genome and identify novel transcripts [7.19], and for transcriptional co-regulation of genes involved in the metabolism of chloroplast and mitochondrial [82]. Apart from *P. tricornutum*, the biology of other diatoms, such as *Guinardia flaccida*, *Haslea* sp., *Minidiscus* sp., *Navicula* sp., *Pleurosigma* sp., *Shionodiscus bioculatus*, and *Synedra* sp., was studied based on multi-omics data to explore the evolutionary success of diatoms through horizontal gene transfer from bacteria and eukaryotes [7.83]. Metabolism of some more diatoms (*Pseudo-nitzschia multistriata* [7.84], *Chaetoceros calcitrans* [7.84], and *Stephanodiscus hantzschii* [7.85]) have been explored using metabolomic profiling to uncover the role of 7-dehydrodesmosterol during lipid accumulation in sexual reproduction, extract and identify novel metabolites, and compare endometabolome under different growth temperatures.

In another set of studies, multi-omics data were generated based on diatoms communities with other microorganisms. For example, the metabolomics and transcriptomics data based on a community of *T. pseudonana* and *Ruegeria pomeroyi* revealed metabolite exchanges

Table 7.3 Omics datasets that will facilitate the metabolic modeling of diatoms.

Omics technique	Microorganism(s)	Culture condition*	Outcome	Ref.
Isotope labeling experiment	*P. tricornutum*	M	Elucidation of the overexpression of diacylglycerolacyl-transferase under lipids production	[7.74]
Metabolomics and transcriptomics	*T. pseudonana, R. pomeroyi*	M and C	Omics results showed changes in expression of vitamin B12 and f 2,3-dihydroxypropane-1-sulfonate metabolism and metabolite exchange between marine plankton.	[7.86]
Metabolomics and transcriptomics	*P. multiseries, Sulfitobacter* sp.	M and C	Experimental evidence about indole-3-acetate and tryptophan exchange.	[7.87]
Metabolomics and transcriptomics	*T. pseudonana, R. pomeroyi, Stenotrophomonas* sp., *P. dokdonensis*	M and C	Characterization of molecules secreted by diatoms and uptaken by three heterotrophic bacteria	[7.88]
Metabolomics, Proteomics, Transcriptomics	*P. tricornutum*	M	Time-course data set taken over 24 h divided in 4 time points during the first 16 h of cultivation under light conditions and a final point at 24 h under dark conditions.	[7.81]
Transcriptomics	*P. tricornutum, T. pseudonana, Cyclotella cryptica, T. weissflogii, F. cylindrus, T. oceanica, P. multiseries*	M	Compendium of 1,075 samples of seven diatoms under various conditions.	[7.80]
Transcriptomics	*P. tricornutum*	M	Collection of 90 datasets from different growth conditions to unravel new genes functions.	[7.19]

(*Continued*)

Table 7.3 Omics datasets that will facilitate the metabolic modeling of diatoms (*Continued*).

Omics technique	Microorganism(s)	Culture condition*	Outcome	Ref.
Transcriptomics	*P. tricornutum*	M	Unraveling of transcriptional co-regulation of transcription factors and metabolic genes in chloroplast and mitochondrial metabolism using a compendium of 187 datasets	[7.82]
Transcriptomics	*Guinardia flaccida*, *Haslea* sp., *Minidiscus* sp., *Navicula* sp., *Pleurosigma* sp., *Shionodiscus bioculatus*, *Synedra* sp.	M	Characterization of horizontal-gene-transfer based on functional homology with *P. tricornutum* genome.	[7.83]
Transcriptomics	*P. tricornutum*	M	Integrated regulatory and metabolic network linking carbon and nitrogen metabolism in response to changes in carbon dioxide using 179 datasets.	[7.77]
Metabolomics and transcriptomics	*Asterionellopsis glacialis*	C	Study about the role of rosmarinic and azelaic acids to benefit some bacterial populations, four time-course dataset under axenic and reseeded conditions.	[7.89]
Metabolomics	*Pseudo-nitzschia multistriata*	M	Characterization of the role of 7-dehydrodesmosterol during lipids accumulation in sexual reproduction	[7.84]

(*Continued*)

Table 7.3 Omics datasets that will facilitate the metabolic modeling of diatoms (*Continued*).

Omics technique	Microorganism(s)	Culture condition*	Outcome	Ref.
Metabolomics	*Chaetoceros calcitrans*	M	Identification of fucoxanthinol, chlorophyll c2, and 3-hexadecenoic acid such as possible anti-inflammatory metabolites	[7.91]
Metabolomics	*Phaeocystis pouchetii*	M and C	Three time course dataset taken over batch culture and from open space to characterize biomass changes	[7.90]
Metabolomics	*Stephanodiscus hantzschii*	M	Comparison of biomass composition under replete nutrients at 5° and 15° C	[7.85]
Meta-transcriptomics	Mostly *Nitzschia*	C	Collection of 100 datasets of metatranscriptomics data from ocean samples	[7.73]

*Culture conditions are classified in either monoculture (M) or co-culture (C).

between diatom and bacterium. In co-culture, *R. pomeroyi* provided a growth-limiting vitamin B12 to *T. pseudonana* [7.86]. Similarly, the metabolomics and transcriptomics data were used to study the communities of *P. multiseries* and *Sulfitobacter* sp. [87], and *T. pseudonana* and *R. pomeroyi*, *Stenotrophomonas* sp., and *Porphyrobacter dokdonensis* [7.88] to uncover the growth interdependencies via metabolite exchanges between diatoms and bacteria. Shibl *et al.* [7.89] obtained transcriptomics and metabolomic datasets of *Asterionellopsis glacialis* in the presence of its naturally associated bacteria. This study revealed that *A. glacialis* reprogrammed its transcriptomic and metabolomic profiles to provide growth beneficial metabolites (e.g., rosmarinic acid and azelaic acid) to bacteria [7.89].

In a community-level analysis, metabolomic profiles were obtained to understand the link between the growth physiology of *Phaeocystis pouchetii* and natural plankton blooms [7.90]. Similarly, Lambert *et al.* [7.73] used a machine learning approach based on hundreds of metatranscriptomics datasets of aquatic protists, including diatoms, to study the potential drivers involved in mixotroph ecosystem functions, such as the ocean's carbon cycle [7.73]. Collectively, various multi-omics datasets have been generated, but only a very few of them have been integrated with diatom GEMs to characterize unknown metabolic traits and provide an opportunity for more research in this direction.

7.3.3 Unraveling New Knowledge About Central Carbon Metabolism of Diatoms

Before 2017, *P. tricornutum* models were used to understand diatom central carbon metabolism. Studies focused on identifying differences of diatom metabolism with plants and eukaryotic microalgae, considering the vast available knowledge on plant metabolism and available photosynthetic models of *Chlamydomonas reinhardtii* [7.92] and *Synechocystis* sp. [7.93] [7.94]. In this section, we summarize the main diatom knowledge produced by the early *P. tricornutum* metabolic models and their impact in subsequent diatom models.

Carbon metabolism in photosynthetic organisms begins with carbon fixation through the Calvin-Benson-Bassham cycle (CBB) in the chloroplast [7.95] [7.96]. Part of this incoming carbon flux is broken down to obtain energy in the oxidative branch of the pentose phosphate pathway (PPP), which in higher plants can occur in the chloroplast. In diatoms, co-localized CBB and PPP in the chloroplast would render a futile and energetically detrimental cycle, in which the oxidative branch of the PPP undoes carbon fixation in the CBB with a global cost of three moles of ATP [7.97]. As a result, *P. tricornutum* physically separates the two pathways: CBB is localized in the chloroplast and the oxidative branch of the PPP in the cytosol [7.11] [7.13] [7.75]. Physical separation of CBB and PPP is also present in *T. pseudonana* [7.61] [7.67] and *F. cylindrus* [7.76]. Interestingly, no diatom metabolic model has found this branch active under any growth conditions [7.13] [7.75].

After carbon fixation, most of the carbon flux is directed through glycolysis. In most organisms, including microalgae, such as *C. reinhardtii* [7.92], *Synechocystis* sp. [7.93] [7.94], and *Chlorella vulgaris* [7.98] [7.99], all glycolytic enzymes are localized in the cytosol. As one of the unique traits of diatom metabolism, enolase was found to be localized in the mitochondria for *P. tricornutum* (PHATRDRAFT_draft1572), *T. pseudonana* (Thaps3a_40391), and *F. cylindrus* (184892). This bisection relocates the glycolysis payoff phase to the mitochondria [7.100]–[7.102], which has been theorized to facilitate the feeding of pyruvate to the tricarboxylic acid cycle (TCA) [7.13] [7.70]. Interestingly, the TCA has been found to operate incompletely and mainly toward producing amino acid and lipid precursors in both *P. tricornutum* [7.13] and *T. pseudonana* [7.67].

Two other fates of assimilated carbon have been identified in *P. tricornutum*: the phosphoketolase pathway (PKP) and the Entner-Doudoroff pathway (EDP). While some studies have found a preference for PKP to feed carbon to the TCA [7.13] [7.70], phosphoketolase has only been identified in *P. tricornutum* (PHATRDRAFT_36257), and there is no other report of its activity [7.75]. It is worth noting that phosphoketolase was included in the *F. cylindrus* model, but no gene has been identified for this reaction. Similarly, although EDP enzymes have been identified for *P. tricornutum*, *T. pseudonana*, and *F. cylindrus*, there is no report of its activity in any diatom [7.13] [7.75] [7.103]. Further integration of proteomics and metabolomics in diatom GEMs will be needed to tackle the activity of the EDP and the PKP.

7.3.4 Light-Driven Metabolism that Enables Acclimation to High Light Intensities

Recently, diatom metabolic modeling has shifted the attention toward the response mechanisms to excess reducing power. As with many other photosynthetic organisms, diatoms have been shown to adjust pigment content to light levels [7.104], dissipate light energy through non-photochemical quenching (NPQ) [7.105], and direct electron flow through alternative pathways (cyclic electron flow) [7.11]. Interestingly, photorespiration was initially proposed in diatoms [7.106] [7.107], though no metabolic model has found this pathway to be active [7.70]. It has been suggested that photorespiratory glycolate is reassimilated as serine [7.12] [7.108]. Moreover, diatom-specific mechanisms have been discovered via metabolic modeling of *P. tricornutum*, such as branched-chain amino acid (BCAA) metabolism [7.12], and the ornithine-glutamine shunt [7.13].

Diatoms have been shown to use alternative electron pathways to respond to excess reducing power. Linear electron flow (LEF) is included in current diatom models in five major steps (complexes): 1) photosystem II (PSII), 2) cytochrome complex, 3) photosystem I (PSI), 4) NADP reductase, and 5) ATP synthase. LEF results in both NADPH and ATP production by NADP reductase and ATP synthase, respectively. On the other hand, under excess reducing power conditions (i.e., high light intensities), continued activity of NADP reductase can damage the photosystems. Under these conditions, green algae and higher plants process excited electrons through the cyclic electron flow (CEF), which only utilizes the cytochrome complex, PSII, and ATP synthase. Through CEF around the PSI, only ATP is produced, with no conversion of NADP to NADPH or water photolysis [7.109].

The CEF cannot be attained in diatoms as easily, as PSII abundance is typically more than twice the abundance of PSI [7.75]. However, CEF has been proposed to account for under 5% of the electron flow in diatoms under high light intensities [7.110]. Fabris *et al.* [7.70] proposed that CEF activity is dependent not only on NADPH production but also on ATP maintenance requirements. Their model predicted that while growth and non-growth associated ATP maintenance stays below 37.3, CEF is negligible. More recently, the *T. pseudonana* model *i*Tps1432 [7.61] and the *P. tricornutum* model *i*LB1025 [7.13] predicted that CEF contribution to the total electron flow increases at low light levels, which was previously hypothesized as a consequence of a rise in ATP production efficiency [7.111]. The authors of *i*Tps1432 highlighted that the model might be overestimating the contribution of CEF, which was predicted above 50%, due to a lack of better ATP maintenance constraining. On the other hand, *i*LB1025 predicted a CEF contribution of 30% at high light intensities and 5% at low light intensities.

Apart from decreasing NADPH production by NADP reductase through CEF, an ornithine-glutamine shunt was found using *i*LB1025 [7.13] to pump reductants in the chloroplast to the mitochondria in *P. tricornutum*. In this shunt, plastidic glutamine is converted to ornithine consuming oxidizing ferredoxin and consuming ATP through the acetyl cycle. Glutamine is then transported to the mitochondria, converted back to ornithine while releasing NADH, and is transported to the chloroplast. In a later study by Broddrick *et al.*, the updated model *i*LB1034 found this shunt to operate 100% cyclically at high light intensities, meaning that it was functioning solely to decrease excess reducing power in the chloroplast.

Adjustment in pigment content and composition as a response to high light intensities was recently incorporated in the *P. tricornutum* GEM *i*LB1034 [7.12]. In this study, Broddrick *et al.* obtained biomass compositions of *P. tricornutum* acclimated to different light levels throughout a diurnal sinusoidal light regime, which allowed the prediction of circadian cycle oscillations and metabolic effects of changing light intensities. Moreover, *i*LB1034 photon uptake rates were constrained by light intensity-specific chlorophyll *a* absorption coefficients. This improved model confirmed the activity of the ornithine-glutamine shunt and discovered another energy pump draining energy from the chloroplast and transporting it to the mitochondria. According to the simulations of *i*LB1034, branched-chain amino acid (BCAA) synthesis and transport to the chloroplast increased with higher light intensities, followed by higher BCAA degradation rates in mitochondrial TCA, resulting in pyruvate, which is then transported back to the chloroplast. The interplay of plastidic BCAA synthesis and mitochondrial TCA is supported by the observed co-regulation of both pathways in reported transcriptomics data [7.12] [7.13] [7.108].

7.4 Modeling Applications to Study Bioproduction and Genome Changes in Diatoms

7.4.1 Predicting Diatom-Heterotroph Interactions and Horizontal Gene Transfer Using Community Metabolic Models

Interactions between phototrophic and heterotrophic microbes form the base of different ecosystems on Earth. Diatoms play critical roles in oceanic ecosystems since they transform light energy and inorganic nutrients into biomass and organic molecules. Diatoms actively exchange metabolites and genetic material with heterotrophic bacteria dominated by Rhodobacterales, Flavobacteriales, and families within the Gammaproteobacteria [7.88] [7.112]. These heterotrophic organisms are often attracted to photosynthetically-derived metabolites leaked from these living cells. For example, experimental studies of the metabolic exchange of diatoms in open environments have identified the presence of 36 organic molecules with evidence of bacterial uptake; 53% contained nitrogen (including nucleosides and amino acids), 11% were organic sulfur compounds (including dihydroxy propanesulfonate and dimethyl sulfoniopropionate), and 28% were components of polysaccharides (including chrysolaminarin, chitin, and alginate) [7.88].

Metabolic exchange can enable syntrophic interactions; for example, when *P. tricornutum* and the bacterium *Donghicola* sp. were co-cultured on monomethylamine as the sole source of nitrogen, *P. tricornutum* consumed the NH_4 secreted by *Donghicola* sp. and in exchange, the diatom secreted organic carbon to sustain the bacterium [7.113] [7.114]. Other syntrophic interactions have been observed between the diatom *T. pseudonana* and the bacterium *R. pomeroyi* that had an obligate dependency on the organic carbon fixed by the diatom while providing vitamin B12 in exchange [7.86]. However, the same *T. pseudonana* diatom interacts competitively with the bacterium *Croceibacter atlanticus* [7.115].

Comparative genomics analyses have shown that diatoms are highly adaptable to different environments by changing their genome. For example, they thrive under gene loss, horizontal gene transfer that involved organic carbon degradation, and acquisition of cell wall transporters that facilitate the lifestyle of a nonparasitic, free-living secondary heterotroph [7.116].

Recently, metabolic models have been integrated with metagenomics, metatranscriptomics, metabolomics, and phenotyping data to decipher key metabolic functions in complex microbial communities. In addition, models have identified environmental conditions triggering syntrophic or competitive interactions between members [7.117], highlighting the complex interdependency of community members. These community metabolic models

(CM models) included cyanobacterium-bacterium [7.14], diatom-bacterium [7.118], and alga-fungus [7.117] pairs to delineate specific interactions between individual members.

All CM models were reconstructed using the Constraint-Based Reconstruction and Analysis Toolbox [7.119]. These CM models contain a shared metabolite pool compartment, which streamlines the connectivity of the member models based on each microorganism's capabilities. Model simulations are based on experimental constraints such as sucrose secretion by an engineered *Se-cscB+* strain and metabolic exchange predictions for the cyanobacteria-bacterium or ions exchange for the diatom-bacterium model. Furthermore, CM models were validated using growth rates in monoculture and co-cultures, predicting that the overall biomass growth rate can increase between 20-45% when comparing co-cultures to monocultures because of metabolic interactions. CM models predictions indicate that co-cultivation can help circumvent nutrient limitations [7.14], bypass member-specific metabolic bottlenecks (e.g., histidine- and transport-related reactions), and compensate for lethal genetic traits, achieving up to 27% recovery from lethal knockouts [7.14].

The diatom-bacterium model of *P. tricornutum* and *Pseudoalteromonas haloplanktis* was expanded to dynamic flux balance analysis by incorporating a simplified set of ordinary differential equations. These equations qualitatively describe metabolic exchanges in the phototroph-heterotroph system by including mutualism and competition for inorganic ions (i.e., phosphate and sulfate). Dynamic simulations were also validated using growth rates for both the diatom and the bacterium. Under steady-state, diatom and bacterial concentrations matched those determined in experimental co-cultures [7.118]. Interestingly, the steady-state reached an oscillatory trend, a behavior typically observed in the presence of metabolic co-dependencies. These mechanistic modeling frameworks provide insights into the physiology and metabolic response to environmental and genetic perturbation of these ubiquitous microbial associations. Additionally, it lays the foundation for robust predictive approaches for the rational design of synthetic communities with optimized growth sustainability using phototrophic partners.

7.4.2 Optimization and Scale-Up of the Production of Valuable Metabolites

The development of metabolic models of different algae has resulted in novel biosystems that control growth and lipid production by optimizing nutrient supply while characterizing algal metabolic pathways under different conditions. Other studies used optimization methods to predict the best light and nutrients levels [7.120]. Specifically, GEMs are constrained using different biomass compositions, typically set as the objective function to accurately drive the optimization of the model and predict cell growth rates [7.79] [7.98]. However, constraint-based optimization allows for alternate optimal solutions, which directly questions the biological relevance of any single given solution [7.121]. In previous studies, several solutions have been proposed to increase the robustness of the predicted flux distribution, namely flux variability analysis (FVA) [7.121] or sampling [7.122]. To date, metabolic modeling of diatoms has not accounted for the effect of alternate optima, which will likely shed light on new mechanisms and might disprove various predictions.

A second concern in the metabolic modeling of diatoms is the dynamicity of their metabolism and biomass compositions. While standard metabolic modeling does not capture changes in biomass composition and growth through the culture timespan, the metabolic model control (MMC) approach integrates time-course growth and biomass composition data to capture dynamic growth and lipid productivity. This biosystem approach was validated under autotrophic growth of the algae *C. vulgaris* [7.123]. The concentrations of nitrate required were lowered as much as possible while retaining robust algal growth. MMC used optical density measurements of *C. vulgaris* to predict subsequent growth rates during the exponential phase. Then, MMC predicted sustained nitrogen environments by controlling inputs of nutrients (e.g., nitrate and glucose). As a result, the authors demonstrated the capacity to utilize this biosystem to effectively optimize algae cultivation strategies that include nutrients and light supplementation. The MMC approach can be broadly applicable for optimizing nutrient inputs and biomanufacturing across a wide spectrum of organisms, including diatoms, to produce various biotechnology products in the coming decades.

7.4.3 Potential for the Study of Proteome Allocation in Diatoms

The new generation of genome-scale metabolic models of metabolism and gene expression (ME models) couple gene expression and protein synthesis to predicted metabolic requirements [7.15]. As a result, ME models can predict the optimal proteome allocation for a specific phenotype [7.15] and proteomic responses to changes in environmental conditions [7.124]–[7.126]. Recently, the *E. coli* ME model *i*JL1678b-ME was successfully employed to predict metabolic and proteomic responses to phenomena that are critical in algae (e.g., green algae and diatoms) growth, such as temperature shifts [7.124], oxidative stress [7.126], and pH stress [7.125]. For example, changes in temperature alter cell size, resting stage formation, and germination of diatoms [7.127], which ultimately affects bloom formation as in other phytoplankton [7.128]. Temperature dependence of metabolic activity in diatoms can be directly assessed by ME models, further informing about cell-cycle gene expression programs, organelle-specific proteomic shifts, and energy efficiency strategies.

The ME model, *i*JL1678b-ME, effectively captured the effect of elevated levels of reactive oxygen species (ROS) on iron-sulfur cluster damage and repair, and demetallation and mismetallation of metalloproteins in *E. coli* [7.126]. In diatoms, UV light induces ROS production, triggering defense mechanisms such as lipid peroxidation and expression of superoxide dismutase and ascorbate peroxidase [7.18]. ME models can be used to understand the underlying mechanism of the relative tolerance of diatoms to UV exposure. Furthermore, while silicon metabolism is of great interest, the disconnection between ions and metabolic pathways in GEMs has not allowed a thorough study of the role of silicon in diatom metabolism. On the other hand, ME models directly couple ion uptake and transport with the post-translational modification of metal-ion activated enzymes. The *Clostridium ljungdahlii* ME model captured metabolic responses of this bacterium to varying nutrient levels [7.129], including ions, thus opening the door for a similar analysis of silicon metabolism of diatoms.

Redox imbalance has been a focus of metabolic modeling in diatoms. GEMs have been extensively used to understand diatom adaptation to oxidative stress resulting from high light intensities. While GEMs were able to describe the activity of the ornithine-glutamine shunt, BCAA synthesis, and photorespiration in diatoms, ME models can be employed to assess the biosynthetic cost of these pathways, their impact on biomass composition, and potentially discover new response mechanisms.

Additionally, *i*JL1678b-ME accurately predicted the metabolic and gene expression response of *E. coli* to acid stress [7.125]. Changes in pH producing acid stress are of particular interest in diatom metabolism since their biomass composition is susceptible to pH levels [7.130]. It has been shown that high pH levels relative to intracellular pH trigger lipid accumulation in microalgae useful for biofuel production [7.131]. Both GEMs [7.2] and ME models can be further employed to optimize lipid production and accumulation as a response to not only pH but also nitrogen and other micronutrient limitations.

7.5 Conclusions

Diatoms are organisms with unique metabolic traits and intriguing evolutionary background. Their interesting metabolism encoded in their genomes was revealed by sequencing of several diatoms. Genome sequences paved the way for GEM reconstructions of *T. pseudonana, P. tricornutum*, and *F. cylindrus*. Extensive omics data integration and numerous enzyme localization prediction databases and tools were deployed to reconstruct these models. The first metabolic models focused on the central metabolism of *P. tricornutum*. More recently, new models of *T. pseudonana* and *F. cylindrus* have allowed us to understand critical similarities and differences across diatoms. More attention has been paid to cell bioenergetics and homeostasis, with a particular interest in response mechanisms to excess reducing power. While GEMs have discovered new metabolic traits of diatom metabolism, there is plenty of work to be done to fully understand the most intriguing aspects of diatoms. Future work will include silicon metabolism, proteome allocation responses on top of the already-discovered metabolic mechanisms, and the role of diatoms as members of microbial communities.

References

[7.1] Bowler C, Vardi A, Allen AE. Oceanographic and biogeochemical insights from diatom genomes. Annual Review of Marine Science. 2010;2: 333–365.

[7.2] Tibocha-Bonilla JD, Zuñiga C, Godoy-Silva RD, Zengler K. Advances in metabolic modeling of oleaginous microalgae. Biotechnology for Biofuels. 2018;11: 241.

[7.3] Campos DT, Zuñiga C, Passi A, del Toro J, Tibocha-Bonilla JD, Zepeda A, *et al.* Modeling of nitrogen fixation and polymer production in the heterotrophic diazotroph *Azotobacter vinelandii* DJ. Metabolic Engineering Communications. 2020;11: e00132.

[7.4] Seif Y, Monk JM, Mih N, Tsunemoto H, Poudel S, Zuniga C, *et al.* A computational knowledge-base elucidates the response of Staphylococcus aureus to different media types. PLoS Computational Biology. 2019;15: 1–27.

[7.5] Kumar M, Ji B, Zengler K, Nielsen J. Modelling approaches for studying the microbiome. Nature Microbiology. 2019;4: 1253–1267.

[7.6] Lewis NE, Nagarajan H, Palsson BO. Constraining the metabolic genotype-phenotype relationship using a phylogeny of in silico methods. Nature Reviews Microbiology. 2012;10: 291–305.

[7.7] O'Brien EJ, Monk JM, Palsson BO. Using genome-scale models to predict biological capabilities. Cell. 2015;161: 971–987.

[7.8] Armbrust EV, Berges JA, Bowler C, Green BR, Martinez D, Putnam NH, *et al.* The genome of the diatom *Thalassiosira Pseudonana*: Ecology, evolution, and metabolism. Science. 2004;306: 79–86.

[7.9] Bowler C, Allen AE, Badger JH, Grimwood J, Jabbari K, Kuo A, *et al.* The *Phaeodactylum* genome reveals the evolutionary history of diatom genomes. Nature. 2008;456: 239–244.

[7.10] Mock T, Otillar RP, Strauss J, McMullan M, Paajanen P, Schmutz J, *et al.* Evolutionary genomics of the cold-adapted diatom *Fragilariopsis cylindrus*. Nature. 2017;541: 536–540.

[7.11] Kroth PG, Chiovitti A, Gruber A, Martin-Jezequel V, Mock T, Parker MS, *et al.* A model for carbohydrate metabolism in the diatom *Phaeodactylum tricornutum* deduced from comparative whole genome analysis. PLoS One. 2008;3: e1426.

[7.12] Broddrick JT, Du N, Smith SR, Tsuji Y, Jallet D, Ware MA, *et al.* Cross-compartment metabolic coupling enables flexible photoprotective mechanisms in the diatom *Phaeodactylum tricornutum*. New Phytologist. 2019;222: 1364–1379.

[7.13] Levering J, Broddrick J, Dupont CL, Peers G, Beeri K, Mayers J, *et al.* Genome-scale model reveals metabolic basis of biomass partitioning in a model diatom. PLoS ONE. 2016;11: 1–22.

[7.14] Zuñiga C, Li T, Guarnieri MT, Jenkins JP, Li CT, Bingol K, *et al.* Synthetic microbial communities of heterotrophs and phototrophs facilitate sustainable growth. Nature Communications. 2020;11: 3803.

[7.15] Lloyd CJ, Ebrahim A, Yang L, King ZA, Catoiu E, O'Brien EJ, *et al.* COBRAme: A computational framework for building and manipulating models of metabolism and gene expression. PLoS Computational Biology. 2018;14: 106559.

[7.16] Nordberg H, Cantor M, Dusheyko S, Hua S, Poliakov A, Shabalov I, *et al.* The genome portal of the Department of Energy Joint Genome Institute: 2014 updates. Nucleic Acids Research. 2014;42: D26–D31.

[7.17] Agarwala R, Barrett T, Beck J, Benson DA, Bollin C, Bolton E, *et al.* Database resources of the National Center for Biotechnology Information. Nucleic Acids Research. 2018;46: D8–D13.

[7.18] Rijstenbil JW. Assessment of oxidative stress in the planktonic diatom *Thalassiosira pseudonana* in response to UVA and UVB radiation. Journal of Plankton Research. 2002;24: 1277–1288.

[7.19] Rastogi A, Maheswari U, Dorrell RG, Vieira FRJ, Maumus F, Kustka A, *et al.* Integrative analysis of large scale transcriptome data draws a comprehensive landscape of *Phaeodactylum tricornutum* genome and evolutionary origin of diatoms. Scientific Reports. 2018;8: 1–14.

[7.20] Yang M, Lin X, Liu X, Zhang J, Ge F. Genome annotation of a model diatom *Phaeodactylum tricornutum* using an integrated proteogenomic pipeline. Molecular Plant. 2018;11: 1292–1307.

[7.21] Lommer M, Specht M, Roy AS, Kraemer L, Andreson R, Gutowska MA, *et al.* Genome and low-iron response of an oceanic diatom adapted to chronic iron limitation. Genome Biology. 2012;13.

[7.22] Hoff KJ, Stanke M. Predicting genes in single genomes with AUGUSTUS. Current Protocols in Bioinformatics. 2019;65: e57.

[7.23] Roberts WR, Downey KM, Ruck EC, Traller JC, Alverson AJ. Improved reference genome for *Cyclotella cryptica* CCMP332, a model for cell wall morphogenesis, salinity adaptation, and lipid production in diatoms *(Bacillariophyta)*. G3: Genes, Genomes, Genetics. 2020;10: 2965–2974.

[7.24] Osuna-Cruz CM, Bilcke G, Vancaester E, de Decker S, Bones AM, Winge P, *et al.* The Seminavis robusta genome provides insights into the evolutionary adaptations of benthic diatoms. Nature Communications. 2020;11.

[7.25] Basu S, Patil S, Mapleson D, Russo MT, Vitale L, Fevola C, *et al.* Finding a partner in the ocean: molecular and evolutionary bases of the response to sexual cues in a planktonic diatom. New Phytologist. 2017;215: 140–156.

[7.26] Sato S, Nanjappa D, Dorrell RG, Vieira FRJ, Kazamia E, Tirichine L, *et al.* Genome-enabled phylogenetic and functional reconstruction of an araphid pennate diatom *Plagiostriata* sp. CCMP470, previously assigned as a radial centric diatom, and its bacterial commensal. Scientific Reports. 2020;10: 1–12.

[7.27] Pendergrass A, Roberts WR, Ruck EC, Lewis JA, Alverson AJ. The genome of a nonphotosynthetic diatom provides insights into the metabolic shift to heterotrophy and constraints on the loss of photosynthesis. bioRxiv. bioRxiv; 2020. p. 2020.05.28.115543.

[7.28] Chapman J, Putnam N, Ho IY, Rokhsar DS. JAZZ: A whole genome shotgun assembler Publication Date. 2002.

[7.29] Ewing B, Green P. Base-Calling of Automated Sequencer Traces Using Phred. II. Error Probabilities. Available: www.genome.org

[7.30] Thomas PD, Campbell MJ, Kejariwal A, Mi H, Karlak B, Daverman R, *et al.* PANTHER: A Library of Protein Families and Subfamilies Indexed by Function. Genome Research. 2003;13: 2129–2141.

[7.31] Birney E, Clamp M, Durbin R. GeneWise and Genomewise. Genome Research. 2004;14: 988–995.

[7.32] Uberbacher EC, Hyatt D, Shah M. GrailEXP and Genome Analysis Pipeline for genome annotation. Current protocols in bioinformatics. 2004;Chapter 4.

[7.33] Salamov AA, Solovyev V v. Ab initio Gene Finding in Drosophila Genomic DNA. Genome Research. 2000;10: 516–522.

[7.34] Jones P, Binns D, Chang HY, Fraser M, Li W, McAnulla C, *et al.* InterProScan 5: genome-scale protein function classification. Bioinformatics (Oxford, England). 2014;30: 1236–1240.

[7.35] Yu C-S, Lin C-J, Hwang J-K. Predicting subcellular localization of proteins for Gram-negative bacteria by support vector machines based on n-peptide compositions. Protein Science. 2004;13: 1402–1406.

[7.36] Li L, Stoeckert CJ, Roos DS. OrthoMCL: Identification of Ortholog Groups for Eukaryotic Genomes. Genome Research. 2003;13: 2178–2189.

[7.37] Galachyants YP, Zakharova YR, Petrova DP, Morozov AA, Sidorov IA, Marchenkov AM, *et al.* Sequencing of the complete genome of an araphid pennate diatom Synedra acus subsp. radians from Lake Baikal. Doklady Biochemistry and Biophysics. 2015;461: 84–88.

[7.38] Petersen TN, Brunak S, von Heijne G, Nielsen H. SignalP 4.0: Discriminating signal peptides from transmembrane regions. Nature Methods. Nature Publishing Group; 2011. pp. 785–786.

[7.39] Krogh A, Larsson B, von Heijne G, Sonnhammer ELL. Predicting transmembrane protein topology with a hidden Markov model: Application to complete genomes. Journal of Molecular Biology. 2001;305: 567–580.

[7.40] Traller JC, Cokus SJ, Lopez DA, Gaidarenko O, Smith SR, McCrow JP, *et al.* Genome and methylome of the oleaginous diatom Cyclotella cryptica reveal genetic flexibility toward a high lipid phenotype. Biotechnology for Biofuels. 2016;9: 1–20.

[7.41] Chan PP, Lowe TM. tRNAscan-SE: Searching for tRNA Genes in Genomic Sequences. Methods in molecular biology (Clifton, NJ). 2019;1962: 1–14.

[7.42] Lagesen K, Hallin P, Rødland EA, Stærfeldt HH, Rognes T, Ussery DW. RNAmmer: consistent and rapid annotation of ribosomal RNA genes. Nucleic Acids Research. 2007;35: 3100–3108.

[7.43] Nelson DR, Chaiboonchoe A, Fu W, Hazzouri KM, Huang Z, Jaiswal A, *et al.* Potential for Heightened Sulfur-Metabolic Capacity in Coastal Subtropical Microalgae. iScience. 2019;11: 450–465.

[7.44] Holt C, Yandell M. MAKER2: An annotation pipeline and genome-database management tool for second-generation genome projects. BMC Bioinformatics. 2011;12.

[7.45] Letunic I, Doerks T, Bork P. SMART 7: recent updates to the protein domain annotation resource. Nucleic Acids Research. 2012;40: D302–D305.

[7.46] Nawrocki EP, Eddy SR. Infernal 1.1: 100-fold faster RNA homology searches. Bioinformatics. 2013;29: 2933–2935.

[7.47] Gruber A, Rocap G, Kroth PG, Armbrust EV, Mock T. Plastid proteome prediction for diatoms and other algae with secondary plastids of the red lineage. Plant Journal. 2015;81: 519–528.

[7.48] Gschloessl B, Guermeur Y, Cock JM. HECTAR: A method to predict subcellular targeting in heterokonts. BMC Bioinformatics. 2008;9.

[7.49] Claros MG. Mitoprot, a macintosh application for studying mitochondrial proteins. Bioinformatics. 1995;11: 441–447.

[7.50] Cokol M, Nair R, Rost B. Finding nuclear localization signals. EMBO Reports. 2000;1: 411–415.

[7.51] Almagro Armenteros JJ, Tsirigos KD, Sønderby CK, Petersen TN, Winther O, Brunak S, *et al.* SignalP 5.0 improves signal peptide predictions using deep neural networks. Nature Biotechnology. 2019;37: 420–423.

[7.52] Emanuelsson O, Nielsen H, Brunak S, von Heijne G. Predicting subcellular localization of proteins based on their N-terminal amino acid sequence. Journal of Molecular Biology. 2000;300: 1005–1016.

[7.53] Horton P, Park KJ, Obayashi T, Fujita N, Harada H, Adams-Collier CJ, *et al.* WoLF PSORT: Protein localization predictor. Nucleic Acids Research. 2007;35.

[7.54] Bendtsen JD, Nielsen H, von Heijne G, Brunak S. Improved prediction of signal peptides: SignalP 3.0. Journal of Molecular Biology. 2004;340: 783–795.

[7.55] Tusnády GE, Simon I. The HMMTOP transmembrane topology prediction server. Bioinformatics. 2001;17: 849–850.

[7.56] Bannai H, Tamada Y, Maruyama O, Nakai K, Miyano S. Extensive feature detection of N-terminal protein sorting signals. Bioinformatics. 2002;18: 298–305.

[7.57] Käll L, Krogh A, Sonnhammer ELL. A combined transmembrane topology and signal peptide prediction method. Journal of Molecular Biology. 2004;338: 1027–1036.

[7.58] Hiller K, Grote A, Scheer M, Münch R, Jahn D. PrediSi: Prediction of signal peptides and their cleavage positions. Nucleic Acids Research. 2004;32.

[7.59] Small I, Peeters N, Legeai F, Lurin C. Predotar: A tool for rapidly screening proteomes for N-terminal targeting sequences. Proteomics. 2004;4: 1581–1590.

[7.60] Petsalaki EI, Bagos PG, Litou ZI, Hamodrakas SJ. PredSL: A tool for the N-terminal sequence-based prediction of protein subcellular localization. Genomics, Proteomics and Bioinformatics. 2006;4: 48–55.

[7.61] van Tol HM, Virginia Armbrust E. Genome-scale metabolic model of the diatom *Thalassiosira pseudonana* highlights the importance of nitrogen and sulfur metabolism in redox balance. PLoS ONE. 2020.

[7.62] Choo KH, Tan TW, Ranganathan S. A comprehensive assessment of N-terminal signal peptides prediction methods. BMC Bioinformatics. 2009;10.

[7.63] Emanuelsson O, Brunak S, von Heijne G, Nielsen H. Locating proteins in the cell using TargetP, SignalP and related tools. Nature Protocols. 2007;2: 953–971.

[7.64] Yi Z, Xu M, Di X, Brynjolfsson S, Fu W. Exploring valuable lipids in diatoms. Frontiers in Marine Science. Frontiers Media S. A; 2017. p. 17.

[7.65] Hussein HA, Abdullah MA. Anticancer compounds derived from marine diatoms. Marine Drugs. MDPI AG; 2020. p. 356.

[7.66] Huang W, Daboussi F. Genetic and metabolic engineering in diatoms. Philosophical Transactions of the Royal Society B: Biological Sciences. 2017.

[7.67] Ahmad A, Tiwari A, Srivastava S. A genome-scale metabolic model of *Thalassiosira pseudonana* CCMP 1335 for a systems-level understanding of its metabolism and biotechnological potential. Microorganisms. 2020;8: 1–20.

[7.68] Gu C, Kim GB, Kim WJ, Kim HU, Lee SY. Current status and applications of genome-scale metabolic models. Genome Biology. 2019;20: 1–18.

[7.69] Thiele I, Palsson BØ. A protocol for generating a high-quality genome-scale metabolic reconstruction. Nature Protocols. 2010;5: 93–121.

[7.70] Fabris M, Matthijs M, Rombauts S, Vyverman W, Goossens A, Baart GJE. The metabolic blueprint of *Phaeodactylum tricornutum* reveals a eukaryotic Entner-Doudoroff glycolytic pathway. Plant Journal. 2012;70: 1004–1014.

[7.71] Moseley B, Passi A, Tibocha-Bonilla JD, Kumar M, Tec-Campos D, Zengler K, *et al.* Genome-Scale Metabolic Modeling Enables In-Depth Understanding of Big Data. Metabolites 2022, Vol 12, Page 14. 2021;12: 14.

[7.72] Hunt KA, Folsom JP, Taffs RL, Carlson RP. Complete enumeration of elementary flux modes through scalable demand-based subnetwork definition. Bioinformatics. 2014;30: 1569–1578.

[7.73] Lambert S, Groussman RD, Schatz MJ, Coesel SN, Durham BP, Alverson AJ, *et al.* The dynamic trophic architecture of open-ocean protist communities revealed through machine-guided metatranscriptomics. bioRxiv. 2021; 2021.01.15.426851.

[7.74] Cheah YE, Young JD. Isotopically nonstationary metabolic flux analysis (INST-MFA): putting theory into practice. Current Opinion in Biotechnology. Elsevier Ltd; 2018. pp. 80–87.

[7.75] Kim J, Fabris M, Baart G, Kim MK, Goossens A, Vyverman W, *et al.* Flux balance analysis of primary metabolism in the diatom *Phaeodactylum tricornutum*. The Plant Journal. 2016;85: 161–176.

[7.76] Lavoie M, Saint-Béat B, Strauss J, Guérin S, Allard A, Hardy S v., *et al.* Genome-scale metabolic reconstruction and *in silico* perturbation analysis of the polar diatom *Fragilariopsis cylindrus* predicts high metabolic robustness. Biology. 2020;9: 30.

[7.77] Levering J, Dupont CL, Allen AE, Palsson BO, Zengler K. Integrated regulatory and metabolic networks of the marine diatom *Phaeodactylum tricornutum* predict the response to rising CO_2 levels. mSystems. 2017;2: e00142-16.

[7.78] Feist AM, Palsson BO. The biomass objective function. Current Opinion in Microbiology. 2010;13: 344–349.

[7.79] Tibocha-Bonilla JD, Kumar M, Richelle A, Godoy-silva RD, Zengler K, Zuñiga C. Dynamic resource allocation drives growth under nitrogen starvation in eukaryotes. npj Systems Biology and Applications. 2020; 1–9.

[7.80] Ashworth J, Ralph PJ. An explorable public transcriptomics compendium for eukaryotic microalgae. bioRxiv. bioRxiv; 2018. p. 403063.
[7.81] Remmers IM, D'Adamo S, Martens DE, de Vos RCH, Mumm R, America AHP, *et al.* Orchestration of transcriptome, proteome and metabolome in the diatom Phaeodactylum tricornutum during nitrogen limitation. Algal Research. 2018;35: 33–49.
[7.82] Ait-Mohamed O, Novák Vanclová AMG, Joli N, Liang Y, Zhao X, Genovesio A, *et al.* PhaeoNet: A Holistic RNAseq-Based Portrait of Transcriptional Coordination in the Model Diatom Phaeodactylum tricornutum. Frontiers in Plant Science. 2020;11.
[7.83] Dorrell RG, Villain A, Perez-Lamarque B, Audren de Kerdrel G, McCallum G, Watson AK, *et al.* Phylogenomic fingerprinting of tempo and functions of horizontal gene transfer within ochrophytes. Proceedings of the National Academy of Sciences of the United States of America. 2021;118.
[7.84] Fiorini F, Borgonuovo C, Ferrante MI, Brönstrup M. A Metabolomics Exploration of the Sexual Phase in the Marine Diatom Pseudo-nitzschia multistriata. Marine Drugs. 2020;18: 313.
[7.85] Jeong K-S, Jeong K-Y, Hong Y-S, Kim D-K, Oh H-J, Chang K-H. Application of nuclear magnetic resonance for analyzing metabolic characteristics of winter diatom blooms. Beisner BE, editor. Journal of Plankton Research. 2020;42: 31–39.
[7.86] Durham BP, Sharma S, Luo H, Smith CB, Amin SA, Bender SJ, *et al.* Cryptic carbon and sulfur cycling between surface ocean plankton. Proceedings of the National Academy of Sciences of the United States of America. 2015;112: 453–457.
[7.87] Amin SA, Hmelo LR, van Tol HM, Durham BP, Carlson LT, Heal KR, *et al.* Interaction and signalling between a cosmopolitan phytoplankton and associated bacteria. Nature. 2015;522: 98–101.
[7.88] Ferrer-González FX, Widner B, Holderman NR, Glushka J, Edison AS, Kujawinski EB, *et al.* Resource partitioning of phytoplankton metabolites that support bacterial heterotrophy. ISME Journal. 2021;15: 762–773.
[7.89] Shibl AA, Isaac A, Ochsenkühn MA, Cárdenas A, Fei C, Behringer G, *et al.* Diatom modulation of select bacteria through use of two unique secondary metabolites. Proceedings of the National Academy of Sciences of the United States of America. 2020;117: 27445–27455.
[7.90] Kuhlisch C, Althammer J, Sazhin AF, Jakobsen HH, Nejstgaard JC, Pohnert G. Metabolomics-derived marker metabolites to characterize *Phaeocystis pouchetii* physiology in natural plankton communities. Scientific Reports. 2020;10.
[7.91] Azizan A, Maulidiani M, Rudiyanto R, Shaari K, Ismail IS, Nagao N, *et al.* Mass spectrometry-based metabolomics combined with quantitative analysis of the microalgal diatom (*Chaetoceros calcitrans*). Marine Drugs. 2020;18: 403.
[7.92] Chang RL, Ghamsari L, Manichaikul A, Hom EFY, Balaji S, Fu W, *et al.* Metabolic network reconstruction of *Chlamydomonas* offers insight into light-driven algal metabolism. Molecular Systems Biology. 2011;7.
[7.93] Nogales J, Gudmundsson S, Knight EM, Palsson BO, Thiele I. Detailing the optimality of photosynthesis in cyanobacteria through systems biology analysis. Proceedings of the National Academy of Sciences. 2012;109: 2678–2683.
[7.94] Knoop H, Zilliges Y, Lockau W, Steuer R. The metabolic network of *Synechocystis sp.* PCC 6803: Systemic properties of autotrophic growth. Plant Physiology. 2010;154: 410–422.
[7.95] Michels AK, Wedel N, Kroth PG. Diatom plastids possess a phosphoribulokinase with an altered regulation and no oxidative pentose phosphate pathway. Plant Physiology. 2005;137: 911–920.

[7.96] Gruber N, Gloor M, Mikaloff Fletcher SE, Doney SC, Dutkiewicz S, Follows MJ, *et al.* Oceanic sources, sinks, and transport of atmospheric CO2. Global Biogeochemical Cycles. 2009;23: 1–21.

[7.97] Waldbauer JR, Rodrigue S, Coleman ML, Chisholm SW. Transcriptome and proteome dynamics of a light-dark synchronized bacterial cell cycle. PLoS ONE. 2012;7.

[7.98] Zuñiga C, Levering J, Antoniewicz MR, Guarnieri MT, Betenbaugh MJ, Zengler K, *et al.* Predicting dynamic metabolic demands in the photosynthetic eukaryote *Chlorella vulgaris.* Plant Physiology. 2018;176: 450–462.

[7.99] Zuñiga C, Li C-T, Huelsman T, Levering J, Zielinski DC, McConnell BO, *et al.* Genome-scale metabolic model for the green alga *Chlorella vulgaris* UTEX 395 accurately predicts phenotypes under autotrophic, heterotrophic, and mixotrophic growth conditions. Plant physiology. 2016;172: 589–602.

[7.100] Liaud M-F, Lichtl C, Apt K, Martin W, Cerff R. Compartment-Specific Isoforms of TPI and GAPDH are Imported into Diatom Mitochondria as a Fusion Protein: Evidence in Favor of a Mitochondrial Origin of the Eukaryotic Glycolytic Pathway. Molecular Biology and Evolution. 2000;17: 213–223.

[7.101] Bártulos CR, Rogers MB, Williams TA, Gentekaki E, Brinkmann H, Cerff R, *et al.* Mitochondrial glycolysis in a major lineage of eukaryotes. Genome Biology and Evolution. 2018;10: 2310–2325.

[7.102] Smith SR, Abbriano RM, Hildebrand M. Comparative analysis of diatom genomes reveals substantial differences in the organization of carbon partitioning pathways. Algal Research. 2012;1: 2–16.

[7.103] Singh D, Carlson R, Fell D, Poolman M. Modelling metabolism of the diatom *Phaeodactylum tricornutum*. Biochemical Society Transactions. 2015;43: 1182–1186.

[7.104] Lepetit B, Goss R, Jakob T, Wilhelm C. Molecular dynamics of the diatom thylakoid membrane under different light conditions. Photosynthesis Research. Springer; 2012. pp. 245–257.

[7.105] Su W, Jakob T, Wilhelm C. The impact of nonphotochemical quenching of fluorescence on the photon balance in diatoms under dynamic light conditions. Journal of Phycology. 2012;48: 336–346.

[7.106] Schnitzler Parker M, Armbrust EV, Piovia-Scott J, Keil RG. Induction of photorespiration by light in the centric diatom *Thalassiosira weissflogii* (*Bacillariophyceae*): molecular characterization and physiological consequences. Journal of Phycology. 2004;40: 557–567.

[7.107] Leboulanger C, Oriol L, Jupin H, Desolas-gros C. Diel variability of glycolate in the eastern tropical Atlantic Ocean. Deep Sea Research Part I: Oceanographic Research Papers. 1997;44: 2131–2139.

[7.108] Smith SR, Gillard JTF, Kustka AB, McCrow JP, Badger JH, Zheng H, *et al.* Transcriptional orchestration of the global cellular response of a model pennate diatom to diel light cycling under iron limitation. PLoS Genetics. 2016;12: 1–39.

[7.109] Huang W, Yang Y-J, Hu H, Zhang S-B. Different roles of cyclic electron flow around photosystem I under sub-saturating and saturating light intensities in tobacco leaves. Frontiers in Plant Science. 2015;6: 923.

[7.110] Bailleul B, Berne N, Murik O, Petroutsos D, Prihoda J, Tanaka A, *et al.* Energetic coupling between plastids and mitochondria drives CO_2 assimilation in diatoms. Nature. 2015;524: 366–369.

[7.111] Fisher NL, Halsey KH. Mechanisms that increase the growth efficiency of diatoms in low light. Photosynthesis Research. 2016;129: 183–197.

[7.112] Amin SA, Parker MS, Armbrust E v. Interactions between Diatoms and Bacteria. Microbiology and Molecular Biology Reviews. 2012;76: 667–684.
[7.113] Suleiman M, Zecher K, Yücel O, Jagmann N, Philipp B. Interkingdom cross-feeding of ammonium from marine methylamine-degrading bacteria to the diatom Phaeodactylum tricornutum. Applied and Environmental Microbiology. 2016;82: 7113–7122.
[7.114] Zecher K, Hayes KR, Philipp B. Evidence of interdomain ammonium cross-feeding from methylamine- and glycine betaine-degrading *Rhodobacteraceae* to diatoms as a widespread interaction in the marine phycosphere. Frontiers in Microbiology. 2020;11: 2431.
[7.115] van Tol HM, Amin SA, Virginia Armbrust E. Ubiquitous marine bacterium inhibits diatom cell division. ISME Journal. 2017;11: 31–42.
[7.116] Kamikawa R, Mochizuki T, Sakamoto M, Tanizawa Y, Nakayama T, Onuma R, *et al.* Genome evolution of a non-parasitic secondary heterotroph, the diatom *Nitzschia putrida.* bioRxiv. 2021; 2021.01.24.427197.
[7.117] Zuñiga C, Li CT, Yu G, Al-Bassam MM, Li TT, Jiang L, *et al.* Environmental stimuli drive a transition from cooperation to competition in synthetic phototrophic communities. Nature Microbiology. 2019;4: 2184–2191.
[7.118] Fondi M, di Patti F. A synthetic ecosystem for the multi-level modelling of heterotroph-phototroph metabolic interactions. Ecological Modelling. 2019;399: 13–22.
[7.119] Heirendt L, Arreckx S, Pfau T, Mendoza SN, Richelle A, Heinken A, *et al.* Creation and analysis of biochemical constraint-based models using the COBRA Toolbox v.3.0. Nature Protocols. 2019;14: 639–702.
[7.120] Tibocha-Bonilla JD, Zuniga C, Broddrick JT, Zengler K. A multiscale model predicts the sensitivity of *Chlorella vulgaris* to light and nitrogen levels in photobioreactors. bioRxiv. 2021; 2021.04.14.439858.
[7.121] Mahadevan R, Schilling CH. The effects of alternate optimal solutions in constraint-based genome-scale metabolic models. Metabolic Engineering. 2003;5: 264–276.
[7.122] Bordel S, Agren R, Nielsen J. Sampling the Solution Space in Genome-Scale Metabolic Networks Reveals Transcriptional Regulation in Key Enzymes. PLOS Computational Biology. 2010;6: e1000859.
[7.123] Li CT, Yelsky J, Chen Y, Zuñiga C, Eng R, Jiang L, *et al.* Utilizing genome-scale models to optimize nutrient supply for sustained algal growth and lipid productivity. npj Systems Biology and Applications. 2019;5: 1–11.
[7.124] Chen K, Gao Y, Mih N, O'Brien EJ, Yang L, Palsson BO. Thermosensitivity of growth is determined by chaperone-mediated proteome reallocation. Proceedings of the National Academy of Sciences. 2017;114: 11548–11553.
[7.125] Du B, Yang L, Lloyd CJ, Fang X, Palsson BO. Genome-scale model of metabolism and gene expression provides a multi-scale description of acid stress responses in Escherichia coli. PLoS Computational Biology. 2019;15: 1–21.
[7.126] Yang L, Mih N, Anand A, Park JH, Tan J, Yurkovich JT, *et al.* Cellular responses to reactive oxygen species are predicted from molecular mechanisms. Proceedings of the National Academy of Sciences of the United States of America. 2019;116: 14368–14373.
[7.127] Montagnes DJS, Franklin DJ. Effect of temperature on diatom volume, growth rate, and carbon and nitrogen content: Reconsidering some paradigms. Limnology and Oceanography. 2002;47: 1272.
[7.128] Bravo I, Anderson DM. The effects of temperature, growth medium and darkness on excystment and growth of the toxic dinoflagellate *Gymnodinium catenatum* from northwest Spain. Journal of Plankton Research. 1994;16: 513–525.

[7.129] Liu JK, Lloyd C, Al-Bassam MM, Ebrahim A, Kim JN, Olson C, *et al.* Predicting proteome allocation, overflow metabolism, and metal requirements in a model acetogen. PLoS Computational Biology. 2019;15: 1–16.
[7.130] Mus F, Toussaint JP, Cooksey KE, Fields MW, Gerlach R, Peyton BM, *et al.* Physiological and molecular analysis of carbon source supplementation and pH stress-induced lipid accumulation in the marine diatom *Phaeodactylum tricornutum.* Applied Microbiology and Biotechnology. 2013;97: 3625–3642.
[7.131] Guckert JB, Cooksey KE. Triglyceride accumulation and fatty acid profile changes in *Chlorella* (*Chlorophyta*) during high pH-induced cell cycle inhibition. Journal of Phycology. 1990;26: 72–79.

Part III

DIATOM MOTILITY

8

Modeling the Synchronization of the Movement of *Bacillaria paxillifer* by a Kuramoto Model with Time Delay

Thomas Harbich

Independent Researcher, Am Brüdenrain, Weissach im Tal, Germany

Abstract

Bacillaria paxillifer (also known as *Bacillaria paradoxa*) forms band-shaped colonies in which the diatoms can actively move relative to each other. Often a synchronization of sections or the whole colony is observed. According to the tracking data of adjacent diatoms, an easily calculable function is proposed for its relative positions as a function of time. Observations made on small colonies and individual diatoms indicate that the propulsion systems of neighboring diatoms interact in such a way that these systems can be regarded as one oscillator. The interaction includes the control of the own movement as well as the recognition of the position of the adjacent diatom. The correlated motion of the colony is assumed to be due to the interaction of neighboring oscillators, which presupposes communication between them. The hypothesis is made that the coupling of the oscillators according to the Kuramoto model is suitable for describing the movement of *Bacillaria* colonies. The phase shifts always observable in colonies require the assumption of a time delay in the coupling. This reproduces the typical observed dynamic movement patterns such as linear, bowed and S-shaped colonies. The Kuramoto model predicts that for oscillator frequencies which are near to each other, the movement from resting states in extreme positions starts from the ends of the colony, as is often observed. An analysis of the equations and numerical simulations allow statements to be made about the critical coupling above which the chain of oscillators oscillates synchronously and about the phenomenon of partial synchronization. An extended Kuramoto model, which additionally considers a periodic illumination whose frequency is close to that of the oscillators, suggests that the diatoms synchronize their motion with it. This could be confirmed experimentally.

Keywords: *Bacillaria paxillifer*, synchronization, Kuramoto model, swarm behavior, simulation

8.1 Introduction

Bacillaria paxillifer (O. F. Müller) Hendey is a pennate motile diatom that forms chain-like colonies. Diatoms are adjacent to each other at the valve faces and can slide in apical

Email: thomas.harbich@diatoms.de

Thomas Harbich: https://orcid.org/0000-0002-9151-4052, https://www.researchgate.net/profile/Thomas_Harbich, https://diatoms.de/en/

Janice L. Pappas (ed.) The Mathematical Biology of Diatoms, (195–228) © 2023 Scrivener Publishing LLC

direction relative to the neighboring diatom (see, e.g. [8.23]). In particular, in the case of small colonies, a synchronous back and forth movement of the entire colony can often be observed. The colony remains in an extended state for a few seconds. Thereafter, the diatoms begin to move relative to their neighbors, whereby these movements typically do not begin simultaneously. Once the colony has reached the opposite elongated position, a pause in the movement occurs before it moves back to the initial position. In this sequence, the diatoms either move in the same direction relative to the neighboring diatoms or they rest. Larger colonies with several tens of diatoms show more complex patterns. Typical is the occurrence of S-shaped and multiple S-shaped structures, which change dynamically. Here, different relative directions of movement occur simultaneously. The motion sequences were described in more detail by Kapinga [8.22] [8.24] and Ussing *et al.* [8.38]. The first description of the movement and drawings of typical forms of colonies (Figure 8.1) were made by the discoverer of the species, the Danish zoologist Otto Friedrich Müller (Müller [8.29] translated in [8.38]).

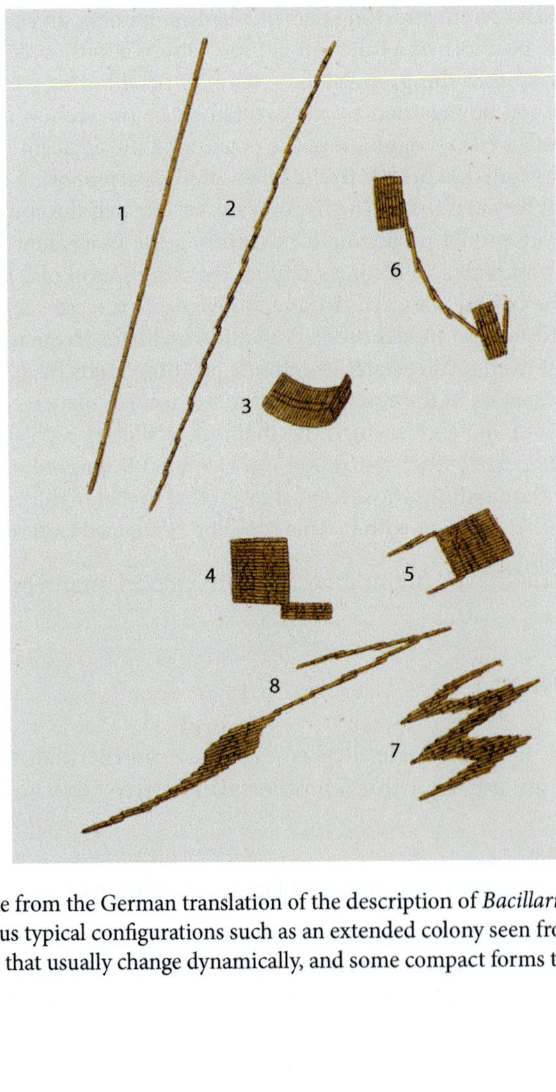

Figure 8.1 Colored plate from the German translation of the description of *Bacillaria* colonies by Müller [8.28]). Shown are various typical configurations such as an extended colony seen from two perspectives, S-shaped configurations that usually change dynamically, and some compact forms that are found statically or with little movement.

The stretched colonies nos. 1 and 2 in Figure 8.1 correspond to the mentioned short resting states within a synchronous motion, but they can also occur statically. Typical views of dynamic patterns are shown in nos. 7 and 8. Colonies 3, 4, and 5 in Figure 8.1 illustrate characteristic static arrangements in which most of the apices of the diatoms are next to each other. Kapinga and Gordon [8.24] have shown that for cultures in an alternating light-dark cycle, the majority of colonies are in motion in the light phase, while in the dark phases most colonies remained in such a non-mobile aligned state. During the transition from a static to a dynamic pattern, there is initially no fixed phase relationship between the movements of the diatoms (transient oscillation). In addition, small amplitudes are typically observed in the beginning. The response to light evidences a strong dependence of behavioral patterns on environmental parameters. Remarkably, Funk [8.12] reports on colonies kept in flowing water that took on a stationary elongated state during daylight hours.

With all movements, the diatoms do not slide completely over one another, but an overlap of at least 15% of their length remains. No silicate structures are apparent on the valves that prevent the diatoms from sliding completely over each other [8.23]. Neither are there any structures on the valve that cause the cells to stick together. However, elastic fibrils were found that could hold the frustules together and an irregular flange structure that may support parallel gliding [8.4]. Own observations show that in chains under mechanical tension, the parallel alignment can be locally destroyed, so that neighboring diatoms adhere only at one point and form a large angle. When appropriate external forces are applied or tension is released, the diatoms successfully close back into a parallel configuration and continue with regular motion (see video at https://diatoms.de/en/restoration-of-the-structure). Kapinga and Gordon suggest that the mucilage excreted by the raphes may be the means for cell attachment. This would also explain the elastic properties of the colonies [8.19]. In addition to the importance of mucilage for movement and adhesion, it could cause lubrication of the sliding movement [8.14] [8.15]. The raphes of *Bacillaria* are essential structures for motility, and their cross section in *B. paxillifer* exhibits a tongue-and-groove structure [8.8] (see also [8.34]). The various theories regarding the motility of many pennate diatoms will not be addressed here, but reference will be made to a review of suggested mechanisms [8.17].

Under the modeling assumption of constant velocities and the neglect of the rest state in the extended configuration, Ussing *et al.* [8.38] have pointed out that a chaotic motion can appear, which shows a transient synchronism. For simplicity, the authors assume the hypothesis that no cell-to-cell communication occurs. Essential, however, is that different velocities and cell lengths are present.

An alternative approach is given by Drum *et al.* [8.9]. When individual cells are separated by a laser beam by destruction of their nearest neighbor, a periodic oscillation is reproducibly observed, the frequency of which is close to that of any oscillating colony. They postulate that the partial synchronicity of the colony is due to the periodic motion of individual diatoms and a weak coupling between them. Kapinga [8.21] proposes a cell-to-cell communication to explain synchronous motion. She refers to her own investigations on the temperature dependence of the synchronicity. Gordon [8.16] gives suggestions for the investigation of this approach. These include the simulation of different kinds of interactions and the derivation of coupling parameters from observed motion. This paves the way for a comparison of theory and observation and a principal examination of whether the concept of coupled oscillators is a suitable explanation model. However, this presupposes a corresponding mathematical modeling. Synchronization processes of biological oscillators have been investigated extensively. An introduction to biological synchronization is given by Strogatz and Stewart [8.36]. Various approaches and application examples for the concept of coupled oscillators and their synchronization in nonlinear systems are given by Pikovsky *et al.* [8.32]. The synchronously flashing fireflies described by Engelbert Kaempfer [8.20] for the first time are a well-known example of coupled biological oscillators. Buck *et al.* [8.3] have described and measured the pulses and their synchronicity. An adaptive model for synchrony in the firefly *Pteroptyx malaccae* can be found in [8.10].

The aim of this work is to quantitatively describe the relative motion of adjacent diatoms when moving synchronously. Building on this, a modeling of synchronization based on coupled oscillators shall be developed. For this purpose, the Kuramoto model is adapted to the structure of *Bacillaria* colonies.

8.2 Materials and Methods

Freshwater diatoms of the species *B. paxillifer* were harvested from the Neckar river in Germany (49°04'41.8"N 9°09'17.9"E). For each sampling, several clonal strains were cultivated in polystyrene Petri dishes, grown in medium according to Walne [8.39] and kept at light intensities ranging from 300 lx to 500 lx, with the light phase lasting 11 hours and the dark phase 13 hours. It should be noted that cultivation in an f/2 medium according to Guillard [8.18] was also successful, but proved to be less favorable in terms of growth rate and quality of the cultures.

Most observations (particularly long-term observations) were performed on cultures in Petri dishes by means of an inverted microscope (Zeiss Axiovert A1). For higher resolution imaging, upright microscopes (Zeiss Standard and Zeiss Axioplan) were used, with suitable colonies pipetted onto slides and covered with a coverslip.

For tracking video recordings made by light microscopy, methods are particularly suitable in which the tracked object does not have to differ significantly from a sufficiently homogeneous background. The approach of choosing a template within the moving object and following this pattern in the frames of the video proved to be successful [8.41]. The video analysis and modeling tool Tracker 5.1.3 (https://www.softpedia.com/get/Science-CAD/Douglas-Tracker.shtml) was used for this purpose [8.2].

For light exposure with periodic light source either an interval switch or a signal generator with a power amplifier was used (self-constructed). The illumination was done with a white LED (cool-white, diameter approx. 30 mm, 7W, 12V AC/DC) optionally in combination with a blue filter (B+W filter 081). The flat LED was mounted on the microscope stage of the inverted microscope (Zeiss Axiovert A1) so that only a small part of the light came into the light path of the microscope. The microscope illumination was turned on at a very low intensity. VirtualDub (www.virtualdub.org) was used to generate videos from a sequence of individual shots, to change the time-lapse factor before tracking and to enhance image properties (e.g., contrast enhancement). All simulation programs were coded in Visual Basic for Applications (VBA).

8.3 Time Dependence of the Relative Motion of Adjacent Diatoms

A quantitative analysis of the movement of a colony of *B. paxillifer* requires consideration of the movement coordinates of all its cells, whereby only the relative positions of the cells in the apical direction with respect to neighboring cells are relevant to the dynamics considered here. Such information can be obtained from microscopy videos by classical tracking methods or deep learning analysis [8.1]. Sufficiently well analyzable recordings usually can be achieved only with colonies of a few diatoms. In many cases, they lie in the focal plane of the microscope for a long time. *Bacillaria* colonies are not exactly flat because their pervalvarous axis form a small angle. Larger colonies are trough-shaped bent in the compact resting configuration (Figure 8.1, no. 3), very large ones often curled up, which makes it difficult or impossible to observe the onset of movement of all diatoms at the same time. Figure 8.2 shows a colony consisting of four diatoms which could be recorded over 17 minutes with good spatial resolution.

Tracking with the "Tracker" tool [8.2] resulted in the difficulty that diatom #4 was not always completely visible in the observation area. Therefore, its trajectory was composed of several individual time ranges with different tracked templates. The displacement related to the neighboring diatom is obtained by calculating the difference between the coordinates of the diatoms in the direction of movement, where the zero point is defined in the manner that the apices of the diatoms are next to each other. These displacements are shown exemplarily in Figure 8.3 over a period of 400 seconds.

An analysis of two of the three trajectories for the same observation can be found in [8.1]. *B. paxillifer* is a remarkable exception among the motile diatoms regarding the recognizable strict periodicity. In other species, the time intervals between motion reversals have a strong random component.

One further recognizes a uniform sequence of the phases at the zeros, whereby the phase shift is not exactly kept, but is subject to fluctuations. However, a reference oscillation with the period matching the observation can be well adapted to the observed curves over several oscillation periods. The position of diatom #3 relative to #2 (green line) follows the position of diatom #4 relative to #3 (red line) by about 3.2 seconds, which is about 5% of the period of oscillation. At a distance of about 9.7 seconds, the position of diatom #2 relative to #1 (blue line) follows the relative position of diatom #3 to #2 (green line) by about

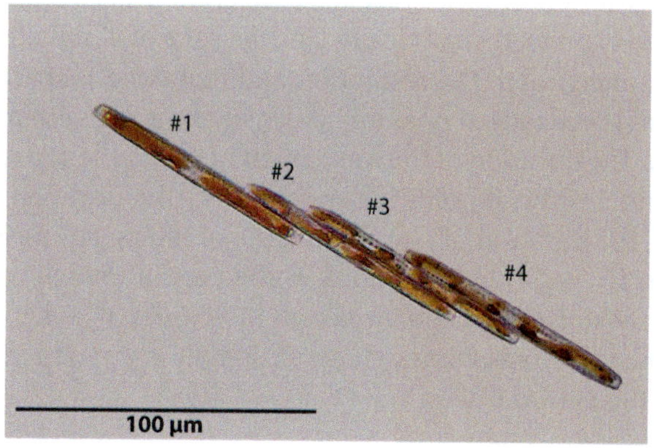

Figure 8.2 Image of a moving small colony on a microscope slide with coverslip. Diatom #1 was permanently attached to the slide, so only the paths of the other diatoms had to be tracked. Since the colony did not change its orientation in the plane of observation, it was sufficient to determine the coordinates of the diatoms with respect to a coordinate axis in apical orientation.

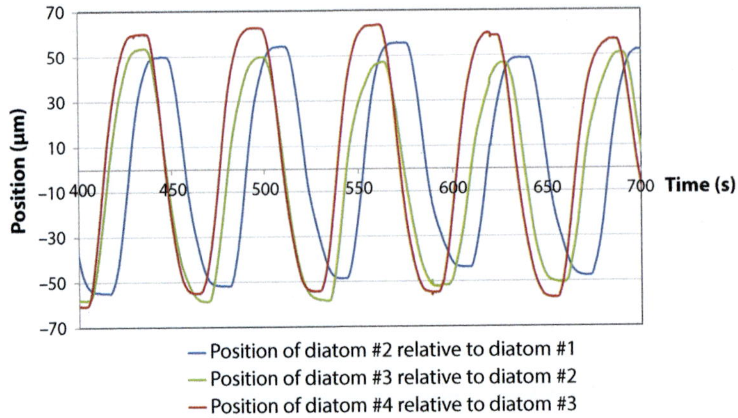

Figure 8.3 Time curve of the differences of the positions of neighboring diatoms, where the numbering corresponds to the identifiers from Figure 8.2. The duration of one cycle is 62.5 seconds.

16% of the oscillation period. A fixed phase relationship for all phases implies synchronous movement of the colony.

Figure 8.4a shows a short time segment of the relative motion of diatom #2 to diatom #1. and the corresponding relative velocity. A similar diagram can be found in [8.40].

One can see that the movement between neighboring diatoms in the extremal positions comes to a standstill for a certain time (vertical blue bars). In the present case, these rest periods are not exactly identical at the two reversal points. When the movement starts, the velocity increases and reaches its maximum when the diatoms are next to each other, i.e., when they have the largest common overlap. After that, the velocity decreases until the reversal point, after which a rest phase begins again. In Figure 8.4b, the rest phases were removed from the curves and the curve of the velocity was approximated by straight lines. This illustrates that in the time ranges shown there are approximately linearly increasing or decreasing velocities and thus constant accelerations $+a$ or $-a$. In this approximation, the position curves do not represent sinusoidal curves outside the rest periods, but parabolas.

Under the model assumption of a constant acceleration between the rest phases, the relative position can be expressed analytically. For the sake of simplicity, it is assumed that the amplitudes are symmetrical to the zero point and that there is strict periodicity. Figure 8.4c shows one cycle of the idealized position and velocity curve with an arbitrarily chosen origin of the time axis. The duration of the movement between the zero point (both diatoms next to each other) and a reversal point is denoted by t_0, the duration of the rest phase at a reversal point by t_b. The duration of a cycle consists of four sections with movement and two rest periods, thus $T = 4t_0 + 2t_b$. If we look at the part of the curve labeled 1, here the speed decreases linearly starting with the maximum velocity v_{max}, i.e., $v(t) = v_{max} - at$. At the reversal point t_0 the velocity becomes $v(t_0) = 0$, so that $v_{max} = a\, t_0$. Integration over time yields the position in the period $0 \leq t < t_0$

$$B(t) = a\left(t_0 t - \frac{1}{2}t^2\right) = A\left(2\frac{t}{t_0} - \left(\frac{t}{t_0}\right)^2\right). \tag{8.1}$$

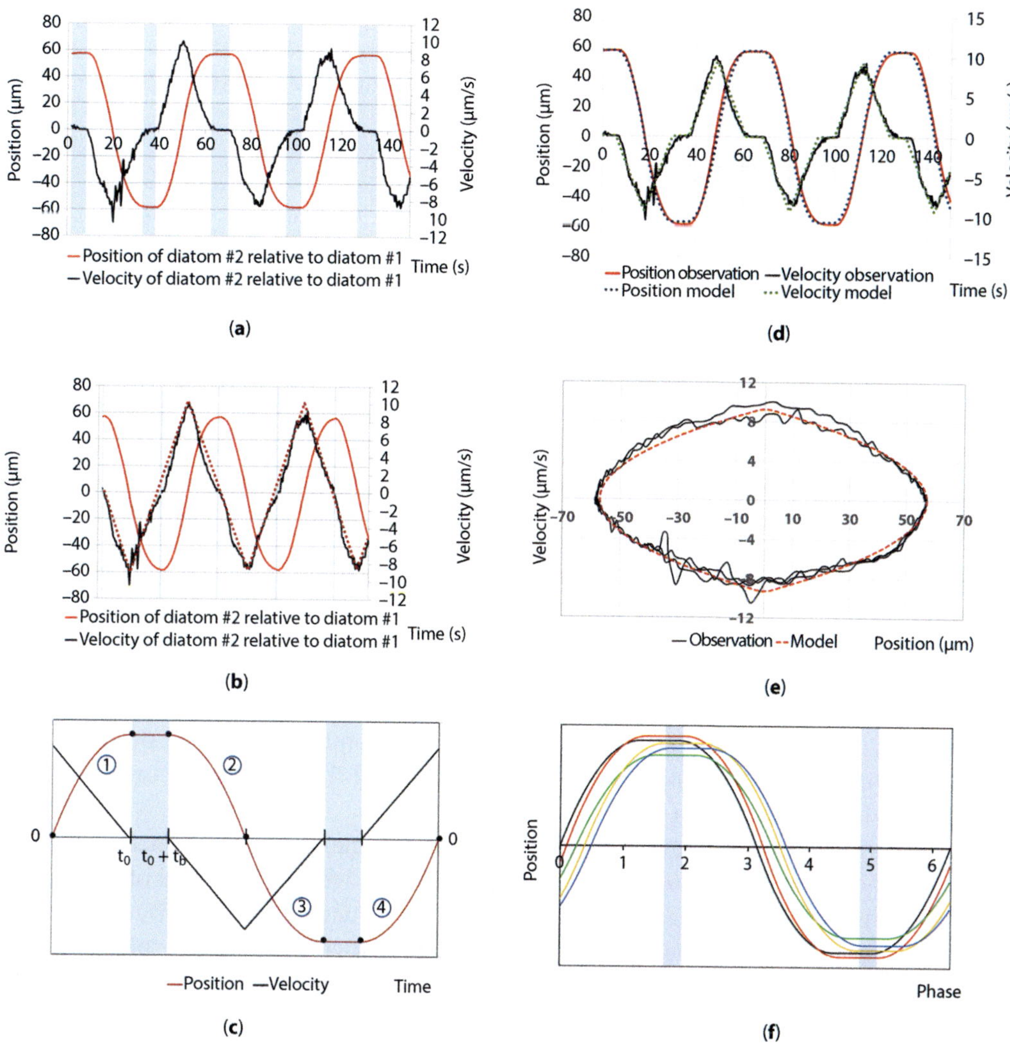

Figure 8.4 Path-time diagrams, velocity-time diagrams, and a phase diagram of the relative motion of diatoms covering about 2 periods. (a) Position and velocity from a measurement, where the velocity was obtained by numerical differentiation. The inaccuracies in the tracking of positions contribute to the noise of the velocity curve. (b) Figure obtained by removing the rest periods. The adjustment of the velocity curve by straight lines was done manually. (c) Idealization of the curves assuming linearly increasing or decreasing velocity and exactly maintained rest periods. For clarity of presentation, the parameters were not taken from the measured data shown. (d) In addition to the two curves from Figure 8.1a, fitted curves have been plotted based on the above assumptions. Such a good fit based on one set of parameters usually succeeds only over a few periods, presumably because environmental parameters and internal parameters of the diatoms do not remain constant over a long time. (e) Velocity as a function of displacement according to observation and model. (f) Exemplary plot of five displacements (six diatoms) versus phase, with identical parameters chosen except for amplitudes and phase shifts ($t_b/T = 0.1$).

Here we used that at time $t = t_0$ the displacement is equal to the amplitude A and thus $A = at_0^2/2$. Analogously, one could now calculate the parabolas denoted by 2 ... 4 from the given accelerations. However, one can quickly obtain a concise representation, which is also well suited for the implementation, if one takes advantage of the fact that these result from the first branch by mirroring and translations, because there is always a linear increase or decrease of the velocity between equidistant positions and the resulting trajectory must be continuous. Denoting the relative position for all segments of the illustrated cycle by $X(t)$, it follows

$$X(t)=\begin{cases} B(t) & \text{for } 0\le t<t_0 \\ A & \text{for } t_0\le t<t_0+t_b \\ B(2t_0+t_b-t) & \text{for } t_0+t_b\le t<2t_0+t_b \\ -B(t-2t_0-t_b) & \text{for } 2t_0+t_b\le t<3t_0+t_b \\ -A & \text{for } 3t_0+t_b\le t<3t_0+2t_b \\ -B(4t_0+2t_b-t) & \text{for } 3t_0+2t_b\le t<4t_0+2t_b. \end{cases} \tag{8.2}$$

For $t > T$, $X(t)$ is continued periodically. In Figure 8.4d, the dotted lines represent a fit according to (8.2) to the observation. Alternatively, the relative positions can be described by a function of the phase Θ, which has the period 2π

$$\bar{X}(\theta)=X(T\theta/(2\pi))=X(\theta/\omega). \tag{8.3}$$

A motion is often illustrated by a diagram in which the speed is plotted against the displacement. Here, the rest phases are mapped to two points. Figure 8.4e shows in this representation the considered period of the movement and the theoretical curve with the parameters, which were used in Figure 8.4d.

In strictly synchronously moving colonies, any displacement between adjacent diatoms is described by a function according to Eqs. (8.2) and (8.3), where the period of oscillation is identical, and the time or phase is shifted by a constant magnitude. Amplitudes and the rest time t_0 may differ, as can be seen in Figure 8.3. This provides a simple explanation for the resting states often observed in smaller colonies when the chain is elongated. Figure 8.4f shows an example of the displacements in a small colony of 6 diatoms (5 displacements). Since the phases of the curves are sufficiently close together in this case, periodic time intervals occur (marked with a blue bar) in which all displacements take on the stationary state of maximum amplitude. Considering a colony or a section of it where the phases (as in Figure 8.3) correspond to the order of the diatoms, the sum of the phase differences of neighboring diatoms which are involved must be less than t_b for a common resting state to occur. Therefore, from the observation of such resting states, one can conclude that the phase differences are small compared to the whole period.

If there were no phase differences between the displacements, curved or S-shaped colonies would not appear. If the phases grow successively along the colony, a half-arc is reached when the phase differences add up to π. In the case of colony 7 in Figure 8.1, about 8 diatoms are involved in such a half-arc. In the mathematical treatment of synchronously oscillating colonies, mainly solutions with phase differences considerably smaller than π will be considered.

Under close-to-nature constant observation conditions, the parameters of this modeling are regarded to be sufficiently constant, but they obviously depend on environmental parameters such as light intensity and temperature. To a high degree, the duration of the rest periods varied, ranging from 0 seconds to about 15 seconds in our cultures. In comparison, the oscillation period T of synchronized colonies is closely distributed around the typical oscillation period of 60 seconds. The oscillation durations of about 20 small colonies of a young culture in daylight were in a range of ±3 seconds around their mean value. In [8.19], however, a period duration of 80 seconds is reported. The reason for this difference is not known. It is worth mentioning that large colonies with tens of cells move synchronously only in certain spatial sections. Frequently, one also observes shorter chains that seem to consist of two or more apparently independent synchronous chains. A general requirement of the given modeling is that the colony should be attached to the substrate at no more than one point, so that the degrees of freedom of movement are not restricted. It should be noted that although the loosening of such attachments during movement was observed several times, systematic locomotion by this mechanism did not occur in the cultures.

8.4 Modeling Interacting Oscillators of a *Bacillaria* Colony

8.4.1 Observation of the Movement Activity at Uncovered Rhaphes

The activity at completely uncovered raphes of a *Bacillaria* colony, i.e., the raphes at the two outer diatoms of the colony, can be detected similarly to many other motile diatoms by marking with small particles, such as pigment particles of Indian ink. According to our own observations, such particles move along the raphe at approximately the same speed as the relative movement of the diatoms. Another method of detection is to observe the movement of individual cells on a substrate [8.9]. Single diatoms are frequently found in *Bacillaria* cultures, especially when the nutrients of the culture medium are becoming exhausted. These single cells often show a movement corresponding to that of artificially separated diatoms. However, there are large irregularities. An example of motion in one direction on smooth substrate together with a fitted curve according to (1) and (2) is shown in Figure 8.5 (see video at https://diatoms.de/en/movement-of-a-single-diatom).

The periodic time is slightly less than 60 seconds and thus within the typical range of synchronous movement in colonies. This was also the case for all other observations made on individual diatoms. Obviously, the trajectory is not at all as smooth as in the relative

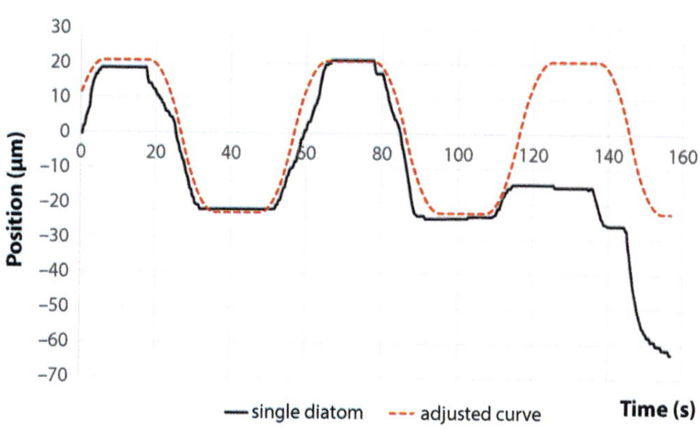

Figure 8.5 Trajectory of a single diatom relative to the substrate. As the motion had a constant orientation during the observation period, only one coordinate had to be considered. The period of the fitted theoretical curve (dashed) is 59.5 seconds.

motion of connected diatoms. Possibly, the mechanical coupling of the single diatom to a flat substrate is not sufficient. Particularly striking in Figure 8.5 is the termination of the periodic motion. Such an alternation between more or less exactly periodic movements and longer straight paths occurs frequently. In one observed case, 860 μm were covered without reversal of direction and without a stop. There seems to be an inherent periodicity already in a raphe system; on the other hand, very regular periodic trajectories rarely occurred for several minutes in individual diatoms. Nevertheless, the described observations basically show that activities with periodic change of driving directions already occur at a single raphe system and are not the result of the interaction of two diatoms.

8.4.2 Interaction of Neighboring Diatoms

In view of the fact that the raphes of the mechanically coupled *B. paxillifer* are laterally displaced [8.8], it seems plausible to assume that each drive system exerts forces on the opposite valve. The observed motion is the result of the interaction of the forces of both diatoms. However, since the frequencies, phases and amplitudes of the diatoms involved should not be exactly the same, but the interaction of the forces results in a regular periodic movement and the movement is limited up to a certain overlap, no simple control mechanism is sufficient for explanation. Therefore, the hypothesis is proposed that the diatoms possess information about their relative position. The raphe is understood as a functional component of a sensomotoric system. In addition, a control loop is required that controls the own raphe activity depending on the relative position. The phase of the movement is likely to be particularly relevant. As each diatom has its own sensor, actuator and controlled system, two control loops work together. A system-theoretical modeling is illustrated in Figure 8.6.

It should be emphasized that there need not be any correspondence between the functional blocks shown and biological entities, that this is merely an abstract modeling. The dotted line separates the two diatoms, with the controlled system containing the relative position as a controlled variable. Each diatom contains a controlled oscillator for its two

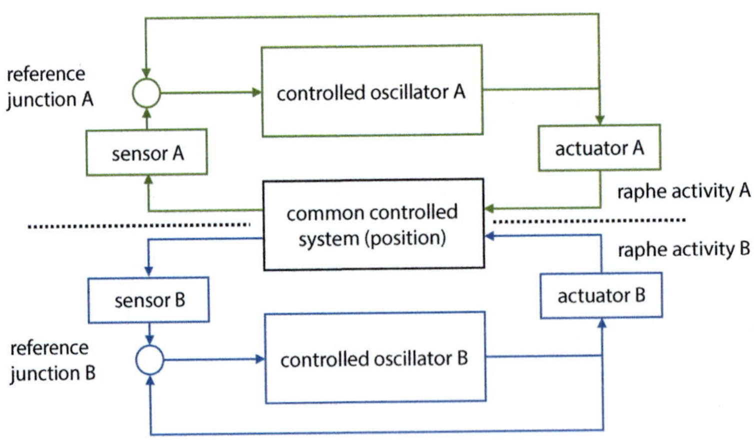

Figure 8.6 Control-oriented modeling of the interaction of two adjacent diatoms. The synchronization of two incomplete oscillators by separate controllers is assumed. The function of the controller is included in the controlled oscillators.

propulsion systems, similar to a voltage controlled oscillator (VCO), which generates time patterns according to the discussed trajectory. Without correcting action, the oscillators will have different parameters, i.e., different frequencies. Using a formally assumed actuator, they generate a raphe activity from it, which leads to a movement. At different raphe activities of the involved diatoms, mechanical tensions arise, resulting in an observable motion, which deviates from the motion as given by the oscillators. A sensor on each side measures the relative position of the diatoms. If it is in accordance with the respective oscillator, there is no correction of its frequency. Otherwise, the system tries to synchronize the oscillator with the movement. The correction function is the actual controller and is modeled in Figure 8.6 as part of the oscillator for simplicity. The oscillators are coupled and synchronized via the controlled system. The frequency of this system should be between the frequencies of the two oscillators with which they oscillate without corrections.

Without accurate knowledge of the mechanical interaction, one can only speculate about the behavior of the controlled system. Equally unknown is the reaction of the controlled oscillators on control deviations and thus the properties of the controller. It must also ensure that the diatoms do not exceed a minimum overlap of their valves.

In the case of a single diatom moving on a substrate, Figure 8.6 reduces to the logical blocks on one side of the dotted line and the controlled system. Information about the position of a neighboring diatom does not exist or is random depending on the contact with the substrate. This could explain deviations from continuously periodic trajectories. Although in the synchronized state the individual oscillators and the motion between the diatoms seem to be synchronous, it is significant that this synchronicity has to emerge when the light exposure begins, and the motion starts.

In the following, only the higher-level view will be taken. The entire structure in Figure 8.6 represents an oscillator which is not assigned to a specific diatom, but to the contact of neighboring diatoms.

8.4.3 Coupled Oscillators

In the presented formal modeling of a contact between diatoms, synchronization effects on the other diatoms are not considered. For this purpose, an interaction between the oscillators must be assumed. It is postulated that during regular observable autonomous motion, communication occurs between neighboring oscillators, hence through the cells. This is supported by the constant phase difference between neighboring oscillators in synchronism. If communication were mediated by light or mechanical tensions in the colony, the interaction would not exist only between neighboring oscillators and therefore a nearly simultaneous stimulation of motion would have been expected. In fact, this case occurs in a context other than autonomous synchronous motion, namely, mechanical stimulation of an entire colony. Funk [8.11] reports about the reaction when touched with the tip of a thin brush hair. When the colony was elongated and in a resting and stimuli-receptive condition, this stimulus triggered a contraction of the colony, the number of diatoms involved depending on the strength of the stimulus (see also [8.22]). He also describes the stimulation of the whole colony. Funk distinguishes between an autonomous and a stimulus movement. This could be reproduced in our own observations, whereby the stimulus reaction is particularly striking when a colony adheres to the substrate at both ends, is obviously under tension and is mechanically separated.

Usually both remaining parts of the chain contract (see videos at https://diatoms.de/en/stimulation-by-glass-fiber). By visual observation, no phase shift can be recognized at the beginning of the movement of the diatoms. This observation indicates that at the contact surface the tensile stress is detected by the diatoms. An alternative would be an extremely fast transfer of information about the stimulus from cell to cell.

A colony of 4 diatoms with oscillators and the proposed communication links is illustrated in Figure 8.7.

The information of the sending oscillator must contain at least its phase and could have an impact on the frequency of the receiving oscillator. A difficult question to answer is whether the incomplete oscillators at the two ends of the colony are involved in the coupled chain of oscillators. At the present state of investigation, this question cannot be decided with certainty. In an elementary chain composed of two vital diatoms under adequate light conditions, periodic oscillation is observed. Evidence for a transient response at the onset of light is not clearly evident. This supports the assumption that the oscillators at the ends are not involved. In the following, we will assume this. It is worth noting that this question affects the mathematical description of a chain, because it influences the assignment of oscillators to diatoms. In the example of Figure 8.7, a chain of four diatoms is described by either 3 or 5 oscillators.

The presented modeling of a coupling by information exchange within a cell has consequences for its behavior during cell division. Gordon [8.16] proposes to observe the motility of dividing cells in comparison to non-dividing cells. When a cell division is at a late stage of division, so that new valves are formed but the diatoms have not yet separated, hence show no relative motion, then in our observations they appear to split the chain into independently moving sections (see video at https://diatoms.de/en/cell-division-within-a-colony). This can be interpreted in the context of the assumptions as a break in communication between oscillators involved in the movement.

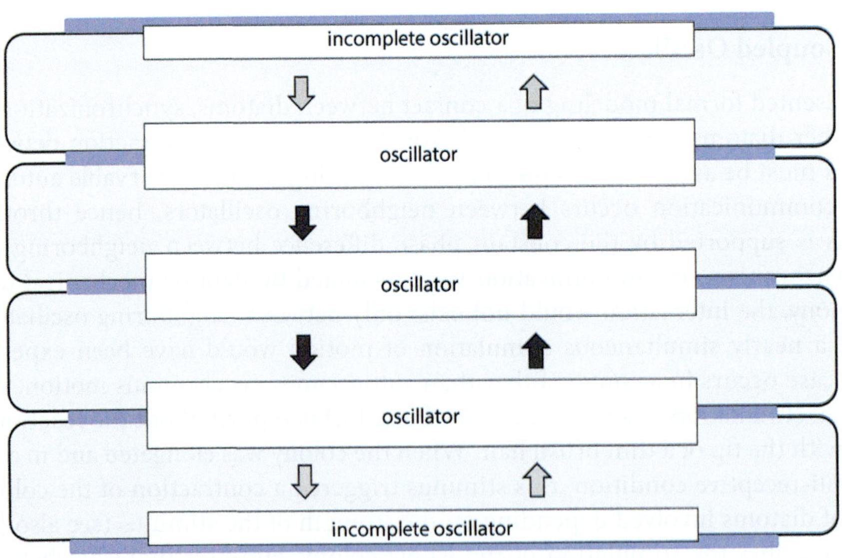

Figure 8.7 Modeling of a small colony with oscillators and their communication links, as far as it is necessary for the emergence of synchronous motion. The communication within the diatoms, is shown with dark arrows, the hypothetical coupling to the outer incomplete oscillators with light gray arrows.

8.5 Kuramoto Model

8.5.1 Adaptation of the Kuramoto Model for a *Bacillaria* Colony

Since the observations do not immediately yield an approach to the mathematical treatment of the coupling of the oscillators, it seems reasonable to use a simple model that does not require a formal representation of the oscillators. Such a model was presented by Kuramoto [8.25] in 1975. It describes the dynamics of the phases of the oscillators in the form

$$\frac{d\Theta_j(t)}{dt} = \omega_j - \frac{k}{n}\sum_{\nu=1}^{n} sin(\Theta_j(t) - \Theta_\nu(t)) \quad for\ j = 1,2,\ldots,n \tag{8.4}$$

The n ($n \geq 2$) oscillators have frequencies ω_j with which they oscillate periodically when they are not coupled to other oscillators, their natural frequencies, so to speak. The sum runs over the oscillators, whereby only summands with $\nu \neq j$ contribute. The phase Θ_j of an oscillator is coupled with the phases of all other oscillators, where only phase differences play a role. The oscillators "try" to match their phases to the phases of all other oscillators. The strength of this matching is described by the coupling strength k ($k \geq 0$). In (8.4) it is related to the number of oscillators, which has the consequence that the correcting effects on a considered oscillator do not increase when the number of oscillators is enlarged. Apart from the sine function, piecewise-linear periodic functions (sawtooth functions) have also been studied [8.13].

Based on this model, different variants adapted to the systems under consideration were developed (for a survey see [8.7]). If one allows pairs of oscillators to be assigned specific coupling strengths $k_{j\nu}$, one obtains a generalization with which different connectivities can be expressed

$$\frac{d\Theta_j(t)}{dt} = \omega_j - \sum_{\nu=1}^{n} k_{j\nu} sin(\Theta_j(t) - \Theta_\nu(t)) \quad for\ j = 1,2,\ldots,n. \tag{8.5}$$

The weights $k_{j\nu}$ can be understood as an edge-weighted adjacency matrix for a graph connecting the oscillators ($k_{jj} = 0$). Dekker and Taylor [8.6] described tree-like connected structures of oscillators. A special case of this is a finite linear chain. Cyclically closed linear chains were investigated by Ochab [8.30].

In case the oscillators are spatially distributed, it may be useful to introduce a distance-dependent coupling by appropriate weights. Furthermore, Eq. (8.5) can be extended by a retardation which considers finite propagation times [8.42], where the propagation time can be chosen according to the spatial distance of the oscillators [8.37].

The hypothesis will be put forward that the dynamics of the phases of the oscillators of a *B. paxillifer* colony under constant environmental conditions can be described by a Kuramoto model. Each oscillator is thereby coupled only with the adjacent one, as shown in Figure 8.7. It tends to reduce the distance between its phase and the phases of the coupled oscillators, whereas the two outer oscillators only have to align with one oscillator. This is described by the following set of equations:

$$\begin{aligned}
&\frac{d\Theta_1(t)}{dt} = \omega_1 - k\sin(\Theta_1(t) - \Theta_2(t-\tau)) \\
&\frac{d\Theta_n(t)}{dt} = \omega_n - k\sin(\Theta_n(t) - \Theta_{n-1}(t-\tau)) \\
&\frac{d\Theta_j(t)}{dt} = \omega_j - k[\sin(\Theta_j(t) - \Theta_{j-1}(t-\tau)) + \sin(\Theta_j(t) - \Theta_{j+1}(t-\tau))] \\
&\text{for } 2 < j < n-1.
\end{aligned} \tag{8.6}$$

This is a special case of (8.5) tailored to the problem, where additionally a retardation is considered, i.e., each oscillator perceives the neighboring ones as they were at time $t - \tau$. The question is not about the correctness of a retardation in principle, but about its relevance. As will be discussed later, without retardation there are no phase shifts between the oscillators in case all of them have the same natural frequency. With retardation, however, a phase shift occurs even in this idealization. However, since one typically observes phase shifts for colonies of all lengths, as illustrated in Figure 8.3, it is assumed that one must allow at least a small delay τ. Furthermore, in (8.6) it is assumed that the retardation and the coupling constant are homogeneous in the chain.

The relative positions of the diatoms follow curves according to Eqs. (8.1), (8.2) and (8.3), where the amplitudes can vary within a certain range, as can be seen in Figure 8.3. If the incomplete oscillators at the ends do not couple to the adjacent ones, the n oscillators correspond to $n + 1$ diatoms, otherwise $n - 1$ diatoms.

It should be noted that some mathematical methods which are successfully applied to Kuramoto systems cannot be used here. There is no global coupling in (8.6), the limit of a very large number of oscillators is not applicable and a decoupling of the equations by means of the usually introduced order parameter does not prove successful.

8.5.2 Analyses and Approximations

For oscillators oscillating exactly synchronously, we expect solutions of (8.6) to be of the form

$$\theta_j = \omega t + \vartheta_j \quad \text{for } j = 1, 2, \ldots, n, \tag{8.7}$$

where ϑ_j are constant in time and all oscillators oscillate with the same frequency ω. One of the constants (e.g. ϑ_1) can be chosen arbitrarily, thus defining origin of the time axis. Substituting (8.7) into (8.6) and applying the addition theorem for the sine, we obtain

$$\omega_1 - \omega = k[-sin(\varphi_2)cos(\omega\tau) + cos(\varphi_2)sin(\omega\tau)]$$

$$\omega_n - \omega = k[sin(\varphi_n)cos(\omega\tau) + cos(\varphi_n)sin(\omega\tau)]$$

$$\omega_j - \omega = k[sin(\varphi_j)cos(\omega\tau) + cos(\varphi_j)sin(\omega\tau) - sin(\varphi_{j+1})cos(\omega\tau) + cos(\varphi_{j+1})sin(\omega\tau)]$$

$$\text{for } 2 < j < n-1. \tag{8.8}$$

Therein the phase differences of adjacent oscillators were introduced

$$\varphi_j = \vartheta_j - \vartheta_{j-1} \quad \text{for } j = 2, 3, \ldots, n \tag{8.9}$$

The set of equations (8.8) consists of n equations for the unknowns ω and the $n-1$ phase differences. We restrict ourselves to small retardations, i.e. $\omega\tau \ll \pi$, so that the cosine of $\omega\tau$ has been replaced by 1 and the sine of $\omega\tau$ by its argument. With the mean of the frequencies

$$\bar{\omega} = \frac{1}{n}\sum_{j=1}^{n} \omega_j \tag{8.10}$$

and the abbreviation

$$\overline{cos(\varphi)} = \frac{1}{n-1}\sum_{j=2}^{n} cos(\varphi_j) \tag{8.11}$$

an expression for ω is obtained by summing all equations (8.8)

$$\omega = \frac{\bar{\omega}}{1 + 2k\tau \frac{n-1}{n}\overline{cos(\varphi)}}. \tag{8.12}$$

Without retardation ($\tau = 0$), the colony oscillates synchronously with the mean of the individual frequencies ($\omega = \bar{\omega}$) in agreement with the result of Dekker and Taylor [8.6]. The term $\overline{cos(\varphi)}$ can be interpreted as an order parameter. If all phase differences are small, i.e., $|\varphi_j| \ll \pi$, then $\overline{cos(\varphi)}$ is only slightly less than its maximum value of 1. For small phase differences, the following then holds approximately

$$\omega \approx \frac{\bar{\omega}}{1 + 2k\tau \frac{n-1}{n}}. \tag{8.13}$$

This is also a lower limit for ω. The consideration of a retardation leads to a decrease of the oscillation frequency ($\omega < \bar{\omega}$). As the coupling strength becomes smaller, the phase differences are expected to grow and $\overline{cos(\varphi)}$ gets smaller. The frequency ω then tends to the mean value of the frequencies of the oscillators. In the approximation of small delays, it follows from (8.8) using the abbreviations b_j

$$b_1 = sin(\varphi_2) = \frac{1}{k}(\omega - \omega_1) + \omega\tau\ cos(\varphi_2)$$

$$b_n = -sin(\varphi_n) = \frac{1}{k}(\omega - \omega_n) + \omega\tau\ cos(\varphi_n) \tag{8.14}$$

$$b_j = sin(\varphi_{j+1}) - sin(\varphi_j) = \frac{1}{k}(\omega - \omega_j) + \omega\tau(cos(\varphi_{j+1}) + cos(\varphi_j))$$

for $2 < j < n-1$.

Apparently, the sine of the phase differences is obtained by summation

$$sin(\varphi_j) = \sum_{i=1}^{j-1} b_i. \tag{8.15}$$

Carrying out this summation for $j > 2$, we finally yield

$$sin(\varphi_j) = \frac{1}{k}\sum_{i=1}^{j-1}(\omega - \omega_i) + \tau\omega\left\{2\sum_{i=2}^{j-1} cos(\varphi_i) + cos(\varphi_j)\right\}, \tag{8.16}$$

where ω is given by (8.12). Obviously, the absolute value of the right-hand side can become larger than 1, so that no solution for the phase differences results. Therefore, let us assume that the coupling constant is so large and $\tau\omega$ is so small that this does not happen. Numerically, the solutions of this equation can be found by iteration substituting a chosen set of phase differences (e.g., $\varphi_j = 0$ for all j) into the right-hand side of (16) and obtaining an improved set from the result, which is again used as the initial value (contracting mapping). For parameters that lead to the typical small phase differences, the insertion of $\varphi_j = 0$ already gives a good result.

Analytically, we consider two specific cases. With a negligible delay $\tau = 0$, (8.16) simplifies according to (8.12) under consideration of $\omega = \bar{\omega}$ to

$$sin(\varphi_j) = \frac{1}{k}\sum_{i=1}^{j-1}(\bar{\omega} - \omega_i). \tag{8.17}$$

Eq. (8.16) yields small phase differences if the frequency deviations from the mean are small compared to the coupling constant and if $\omega\tau$ is sufficiently small. Under these conditions, (8.16) is approximated using (8.13) to give:

$$sin(\varphi_j) = \frac{j-1}{k}\left[\bar{\omega}\frac{1 + k\tau(2j-3)/(j-1)}{1 + 2k\tau(n-1)/n} - \frac{1}{j-1}\sum_{i=1}^{j-1}\omega_i\right]. \tag{8.18}$$

This corresponds to the first step of the iteration mentioned above. If the frequencies of the oscillators are identical, i.e., $\omega_j = \omega_0$ for all j, then (8.18) simplifies to

$$sin(\varphi_j) = \omega_0 \tau \frac{2j-n-2}{n+2k\tau(n-1)}. \tag{8.19}$$

Without retardation, there would be no phase differences in this idealization. Since we assume small phase differences, one can approximate the sine by its argument and then calculate the phases analytically. To do this, one uses the definition of phase differences and obtains for $j > 1$

$$\vartheta_j = \sum_{i=1}^{j} \varphi_i = \omega_0 \tau \frac{j^2-(n+1)j+n}{n+2k\tau(n-1)}. \tag{8.20}$$

Here, $\vartheta_1 = 0$ was chosen without loss of generality (see above). In Figure 8.8a, the phases ϑ_j are shown for the exemplary case of 12 oscillators. According to the symmetry of the

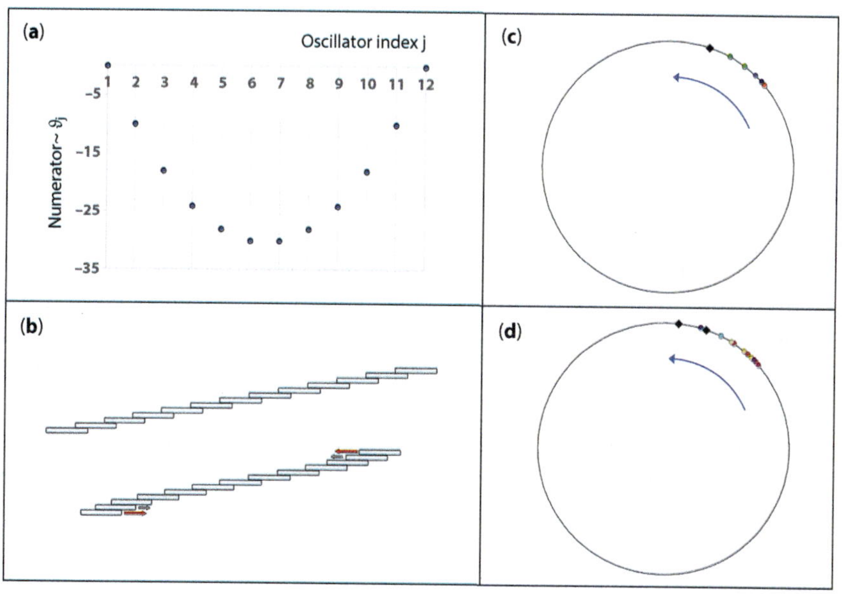

Figure 8.8 Illustration of phases and chronological sequence for identical and nearly identical oscillator frequencies in the case of 12 oscillators. (a) Numerator of Eq. (8.20), which is proportional to the phases ϑ_j. For an even number of oscillators there are two identical minimal phases of equal size, for an odd number there is exactly one minimal phase. (b) elongated colony at rest and appearance of the colony shortly after the onset of motion. The relative displacements are indicated by arrows. It does not matter whether the colony is fixed at one end. (c) Circulating phases from a simulation with 12 oscillators and identical oscillator frequencies. Only 6 marks for the phases are seen, as each mark covers another. In units related to $\bar{\omega}$, a delay of $0.04T'(T' = 2\pi / \bar{\omega})$ and a coupling constant $0.5\bar{\omega}$ of were used. (d) Circulating phases from a simulation with different oscillator frequencies. The parameters correspond to those of c, but with an empirical standard deviation of the angular frequencies of $0.1\bar{\omega}$.

chain, the phases are also symmetrical to the center of the chain, with the largest values of the phases belonging to the oscillators at the ends of the chain.

This is in accordance with the observations. The colony in Figure 8.8b is initially in the resting phase, i.e., in an elongated position (top). The following contraction starts in many cases with the outermost diatoms and continues successively with a phase shift from both sides towards the center of the colony. This motion sequence refers to the relative motions of the diatoms. Often one of the diatoms at the end of the chain sticks to the substrate and therefore does not move with respect to it. The movement with the contraction from both ends is described in [8.22] (see video at https://diatoms.de/en/typical-movement-of-a-colony). Often it is also other diatoms in the vicinity of the ends of the colony with which the movement starts. In the context of coupled oscillators, one often illustrates their phases in the complex plane. In Figure 8.8c, for each phase the values $e^{i\theta_j} = e^{i(\omega t+\vartheta_j)}$ are indicated by a mark on the unit circle. Since the chain is in a synchronous state, the set of marks moves in a mathematic positive direction without relative movements. The phases of the two oscillators on the outer side are marked by a rhombus. For reasons of symmetry only 6 marks are recognizable. The two rhombuses lie on top of each other. This diagram was obtained by numerical integration of the initial equations (8.6) and describes a state after the transient oscillation (see videos at https://diatoms.de/en/identical-natural-frequencies). The good agreement with the results (8.19) and (8.20) proves their applicability in case of small delays and small phase shifts. Figure 8.8d shows the result of a simulation with different angular frequencies ω_j, which were selected according to a normal distribution. The symmetry of the phases no longer occurs and the phase of an oscillator in the second place from the outside leads the group of phases. Therefore, on one side of the chain, the contraction starts with the second oscillator. Unequal frequencies ω_j lead to deviations from the "ideal picture," as they are also found in observations. In long chains, the phases on the circle are no longer in a close group, even if the phase difference of neighboring oscillators is small. Oscillators can then also oscillate in opposite phase. The aforementioned bowed, S-shaped or multiple S-shaped chains can result, as observed in nature.

The strength of the coupling k is a dimensioned variable. It is convenient to relate it to a suitably chosen angular frequency. For a single colony under consideration, the mean value $\bar{\omega}$ provides an appropriate reference, so that the coupling strength is characterized by the dimensionless quantity $k/\bar{\omega}$. Accordingly, the delay τ is expressed in the dimensionless unit $T' = 2\pi/\bar{\omega}$. These units are used in Figure 8.8. If a large number of colonies with statistically distributed natural frequencies are considered, their expected value μ is analogously suitable as reference.

8.5.3 Critical Coupling

8.5.3.1 Uncoupled Oscillators

In the limiting case $k = 0$, all oscillators oscillate independently at their natural frequency ω_j. The motion of a colony is then periodic or quasi-periodic, depending on whether the trajectory of the entire system $(\bar{X}(\theta_1), \bar{X}(\theta_2), \ldots \bar{X}(\theta_n))$, which is the vector consisting of all relative positions, is closed. This distinction is not relevant for real systems, since neither the frequencies are exactly constant, nor the measurement accuracy is unlimited. Apart from the modeled waveform, which also describes the rest positions, this modeling is equivalent to the "null hypothesis" of Ussing *et al.* [8.38]. For sufficiently large values of the coupling strength, on the other hand, we expect a periodic motion of all oscillators with uniform frequency ω. Therefore, there should be a critical coupling k_c above which the oscillators are synchronized. Three simple cases will be discussed.

8.5.3.2 Two Oscillators

As the simplest possible case we consider two coupled oscillators with small delay. This case was not explicitly considered in (8.16). The equations according to (8.14) for two oscillators are with $n = 2$ and the simplified notation $\varphi = \varphi_2$ read

$$b_1 = sin(\varphi) = \frac{1}{k}(\omega - \omega_1) + \omega\tau\ cos(\varphi)$$

$$b_2 = -sin(\varphi) = \frac{1}{k}(\omega - \omega_2) + \omega\tau\ cos(\varphi)\,. \tag{8.21}$$

Adding and subtracting these equations yields

$$\omega = \frac{\bar{\omega}}{1 + k\tau\ cos(\varphi)} \tag{8.22}$$

and

$$sin(\varphi) = \frac{1}{k}\frac{\omega_2 - \omega_1}{2}\,. \tag{8.23}$$

Hence, for (8.23) to have a solution, it must hold that

$$k \geq \left|\frac{\omega_2 - \omega_1}{2}\right|. \tag{8.24}$$

Therefore, the critical coupling is as in the case without retardation (see [8.6])

$$k_c = \left|\frac{\omega_2 - \omega_1}{2}\right| = |\bar{\omega} - \omega_1| = |\bar{\omega} - \omega_2|\,. \tag{8.25}$$

As k approaches this value from above, the phase difference φ approaches $\pi/2$ or $-\pi/2$ and $\omega = \bar{\omega}$ is obtained in this limiting case. The approximation of small phase differences is valid only if k is sufficiently large compared to k_c. Using $sin^2(\varphi) + cos^2(\varphi) = 1$, we finally obtain for $k \geq k_c$ from (8.22), (8.23) and (8.25)

$$\omega = \bar{\omega}\frac{1-\tau\sqrt{k^2-k_c^2}}{1-\tau^2\left(k^2-k_c^2\right)}. \tag{8.26}$$

8.5.3.3 *Chains without Retardation*

For chains without retardation ($\tau = 0$), one can immediately give an expression for the critical value of the coupling. According to (8.17), for each j must hold:

$$\frac{1}{k}\left|\sum_{i=1}^{j-1}(\bar{\omega}-\omega_i)\right| \leq 1 \tag{8.27}$$

so that

$$k_c = \max_j \left|\sum_{i=1}^{j-1}(\bar{\omega}-\omega_i)\right|. \tag{8.28}$$

This result can also be found in [8.6].

8.5.3.4 *Identical Oscillator Frequencies and Sufficiently Small Delay*

In the case of identical frequencies ω_j and small but finite delays, one sees from Eq. (8.19) that synchronization already occurs at arbitrarily small but finite couplings.

8.5.3.5 *Remarks on the General Case*

Nontrivial is the determination of the critical coupling for the general case of different natural frequencies at finite delay according to (8.16). For a synchronously oscillating chain, $|sin(\varphi_j)| \leq 1$ must again hold for all oscillators (j = 1, 3, … n). Therefore, when approaching the critical coupling from above, the phase differences do not all tend towards $\pm\pi/2$, but scatter, as can also be seen in the simulation. For the $\omega\tau \ll \pi$ assumption, which is often justified for *Bacillaria* colonies, calculating k_c using $sin(\varphi_j)$ according to (8.18) (neglecting the increasing of phase differences as the critical point is approached) provides an approximate value that can be easily calculated. More accurate results are obtained numerically by finding the zero of $|sin(\varphi_j)| - 1$ with respect to k according to (8.16), starting with a sufficiently large k and successively decreasing its value, determining the phases of each value by iteration, as described above. It is convenient to start each iteration for a new value of the coupling constant with the determined phases of the previous one. If one $|sin(\varphi_j)|$ reaches the value 1, one has found the critical value. Acceptable matching for graphical representations is obtained by finding zeros using (8.18), assuming that one restricts oneself to small delays (up to 0.16 T in the test), small deviations of the frequencies from their mean (tests with normally distributed frequencies and standard deviations up to $0.1\,\bar{\omega}$), and small chain lengths (tests up to 50 oscillators).

8.5.4 Statistical Considerations and Monte Carlo Simulations

8.5.4.1 Expected Value and Standard Deviation

In Kuramoto models with global coupling, the limiting case of very many oscillators is often investigated. The frequencies of the oscillators can then be statistically described by a time-varying density function satisfying a Fokker-Planck equation. A Kuramoto model with retardation studied with this method can be found in [8.5]. The situation in the synchronization of *Bacillaria* colonies does not allow this approach, because there are not enough oscillators to describe their frequencies by a density function and there is only a nearest-neighbor interaction. However, a statistical treatment is relevant if one does not want to consider only chosen individual cases whose parameters are difficult to extract from observations.

The key assumption is that the frequencies of the uncoupled oscillators ω_j can be described by a distribution function. Then the frequencies of a chain are a sample to this distribution. With the expected value μ of this distribution, the expected value of ω is obtained from (8.13) under the condition $|\varphi_j| \ll \pi$

$$E(\omega) = \frac{\mu}{1+2k\tau\dfrac{n-1}{n}}. \tag{8.29}$$

From (8.18), the expected value of φ_j is obtained for small phase differences

$$E(\varphi_j) \approx E(sin(\varphi_j)) = \mu\tau\frac{2j-n-2}{n+2k\tau(n-1)}. \tag{8.30}$$

The expected value (8.30) has the form of Eq. (8.19) which was derived for identical circular frequencies, except that instead of ω_j in (8.19) the expected value μ appears. Accordingly, ω_j can be replaced by μ in (8.20). This provides an extended applicability of (8.19) and (8.20), because even at statistically distributed oscillator frequencies, on average, the motion starts at the outer diatoms.

In the following, statistically independent frequencies of the oscillators are assumed. The covariances between adjacent frequencies then do not have to be considered. With the standard deviation σ for the distribution of the frequencies, we obtain the variance of ω in the approximation (8.13)

$$Var(\omega) = \frac{\sigma^2}{n\left(1+2k\tau\dfrac{n-1}{n}\right)^2}. \tag{8.31}$$

The variance of the phase differences according to (8.18) becomes

$$Var(\varphi_j) \approx Var(sin(\varphi_j)) = \left(\frac{\sigma}{k}\right)^2 (j[1-2d]+2d+nd^2-1), \tag{8.32}$$

where the abbreviation

$$d = \frac{j-1+k\tau(2j-3)}{n+2k\tau(n-1)} \tag{8.33}$$

was introduced. Note that d depends on the strength of the coupling. For a small delay $k\tau \ll 1$ the standard deviation of the phase differences is approximately proportional to the dimensionless quantity σ/k.

8.5.4.2 Distribution of Critical Coupling

Information about the distribution of critical coupling strength can be provided by a Monte Carlo simulation. Samples of oscillator frequencies selected by pseudo-random numbers for many oscillator chains simulate an ensemble of colonies and show how far their critical

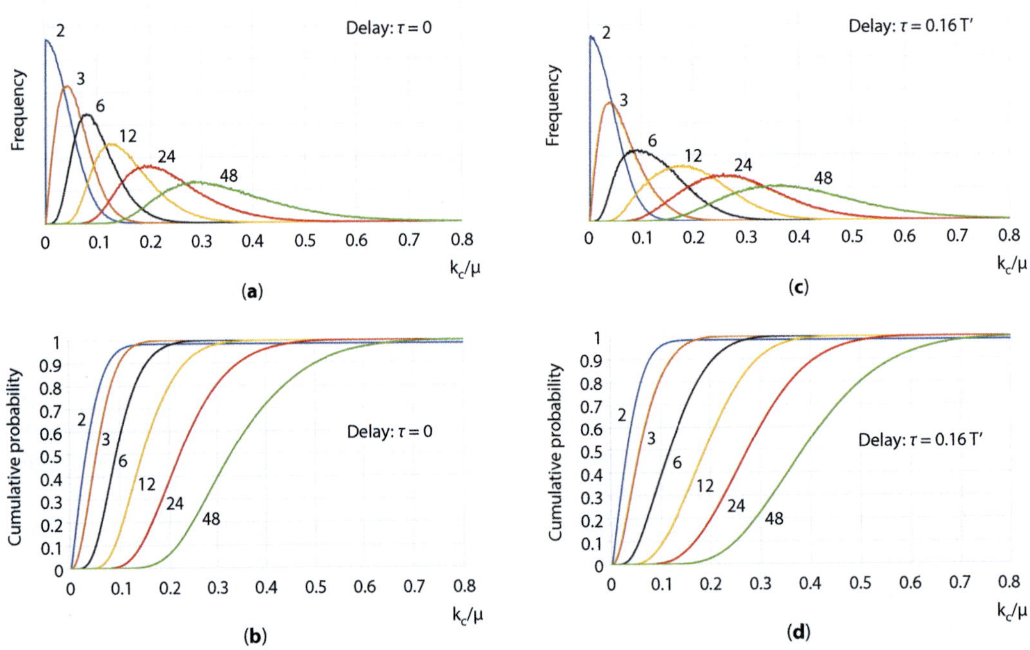

Figure 8.9 Histograms as approximation of the probability density function and corresponding cumulative distribution from a Monte Carlo simulation for different numbers of coupled oscillators. Their counts are noted next to the curves. The units are related to the standard deviation μ. Here, $\sigma = 0.064\ \mu$ was always chosen. (a) Histograms without delay. (b) Cumulative probability without delay. (c) Histograms with delay $\tau = 0.16\ T'$ with ($T' = 2\pi/\mu$). If μ corresponds to an oscillation period of 60 seconds, then the delay is 9.6 s. (d) Cumulative probability with delay.

coupling strength scatters for fixed parameters (number of oscillators, delay, standard deviation). A statistical independence of the frequencies of the oscillators ω_j is assumed again. Furthermore, a normal distribution with the expected value μ and the standard deviation σ shall be presumed. The difficulty of calculating the critical coupling based on the presented relationships has already been discussed. In the simulations presented below, as an approximation to their determination, the zero search of $|sin(\varphi_j)| - 1$ is carried out using (8.18). The critical value k_c is then the maximum over j of the determined zeros. As an approximation for the probability density function the representation of the frequencies in the form of a normalized histogram over the values, which k_c takes on, is used. From it, an approximation of the cumulative distribution function is obtained by partial summation. In terms of Monte Carlo simulation, it is an empirical distribution. To reproduce a large ensemble of colonies, 10^6 samples, in other words 10^6 chains, were chosen for each set of parameters.

Figure 8.9a shows the density function of the critical coupling approximated by a histogram without delay and with a finite standard deviation of the angular frequencies for a different number of oscillators. With increasing chain length, the density function broadens and its mean shifts to higher values. The corresponding cumulative distribution function can be seen in Figure 8.9b. For example, assuming that the coupling strength in a *Bacillaria* colony is 0.2 μ, about 40% of the chains consisting of 24 oscillators, 80% of the chains consisting of 12 oscillators, and almost all chains consisting of 6 oscillators will oscillate synchronously. Only about 7% of the chains of 48 oscillators show synchronous motion. If a delay is considered, as in Figure 8.9c for the example of $\tau = 0.16\ T'$, the respective curves

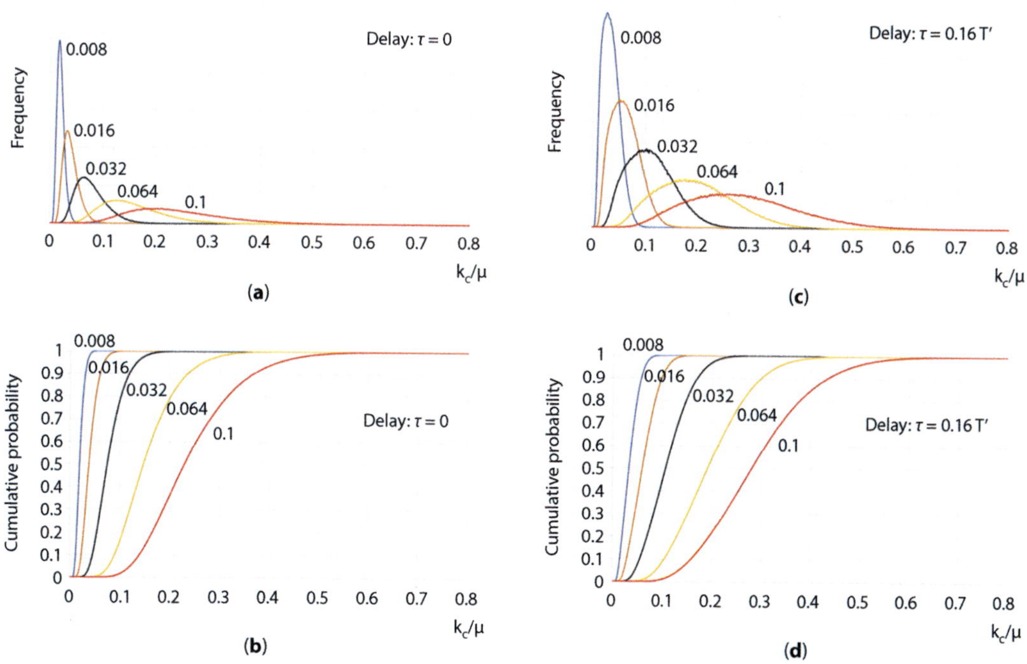

Figure 8.10 Illustration analogous to Figure 8.9, but for different values of the standard deviation. Their values in units of μ are noted next to the curves. Here, a chain consisting of 12 oscillators and $\sigma = 0.064\ \mu$ was always chosen. (a) Histograms without delay. (b) Cumulative probability without delay. (c) Histograms with delay $\tau = 0.16\ T'$ ($T' = 2\pi/\mu$). (d) Cumulative probability with delay $\tau = 0.16\ T'$.

shift to higher values and broaden. Therefore, the fraction of synchronized chains at a given coupling strength decreases (Figure 8.9d).

The influence of the standard deviation at constant chain length on the density of the critical coupling without delay is shown in Figure 8.10a. As expected, the curves shift to higher coupling values and broaden with increasing standard deviation. The effect of a finite delay is shown in Figures 8.10c and 8.10d.

8.5.5 Simulation of Non-Synchronous States

8.5.5.1 *Numerical Integration*

The previous discussion of the solution was limited to a synchronous motion of the entire colony, where the phase differences of all oscillators are constant. This does not describe the solutions for a coupling strength below the critical coupling nor the transient behavior. A numerical integration of Eqs. (8.6) is free from this limitation and also from the assumption of small delays. To calculate the phases at a time t, one must know the phases at time $t - \tau$. Therefore, the phases from the time window between the start of the calculation at time $t = 0$ and $-\tau$ in the time grid of the integration must be specified as initial values.

8.5.5.2 *Visualization of the Transient*

When one simulates the transient into the synchronous state with randomly chosen initial conditions, the marks for the phases in a phase portrait (see snapshots in Figure 8.8c and Figure 8.8d) initially run at different angular velocities and eventually stabilize in an arrangement with fixed phase spacings. This is primarily appropriate for an animated representation. For illustration purposes, it is useful to plot the phase differences φ_j or their sine versus time. Figure 8.11a shows the sine of the phase differences as an example for a transient with 12 oscillators (11 phase differences). These values converge to the value according to Eq. (8.16) within the approximation. Alternatively, the coupling term on the right side in Eq. (8.6) can be shown as a function of time for each oscillator (see Figure 8.11b). It describes the correction to frequency ω_j and tends towards $\omega - \omega_j$ in case of convergence.

An intuitively comprehensible example of a simulation will be visualized by means of the coupling term. Consider a chain with 24 oscillators, where the first 12 oscillators

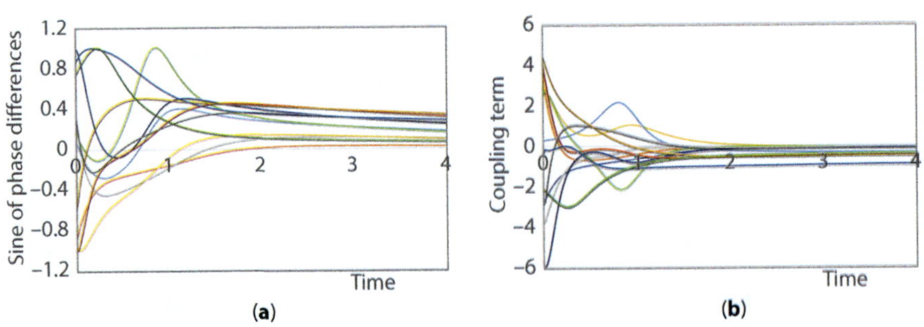

Figure 8.11 Illustration of the transient of 12 oscillators with normally distributed frequencies. Here the delay is $0.01T'(T' = 2\pi / \bar{\omega})$, the coupling constant is $0.48\bar{\omega}$ and the standard deviation of the frequencies ω_j is $0.034\,\bar{\omega}$. (a) Sine of the phase differences versus time measured in units of T'. (b) Coupling term for each phase.

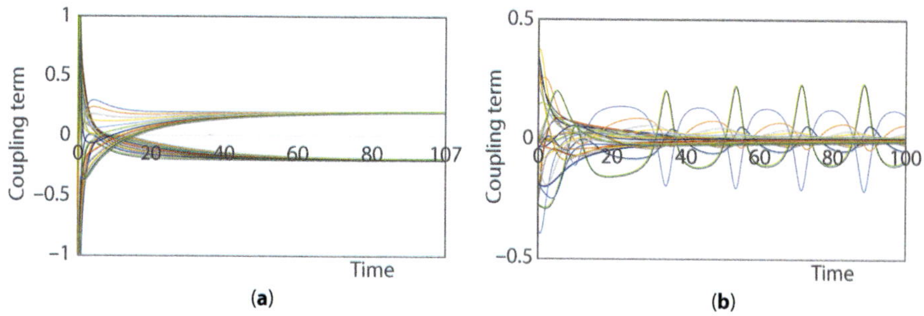

Figure 8.12 Illustration of the transient oscillation of a colony of 24 diatoms consisting of two groups of 12 oscillators with the same frequency. The distance of the frequencies from their mean is $\pm 0.032\,\bar{\omega}$. (a) Coupling constant $0.5\bar{\omega}$. (b) Coupling constant $0.032\bar{\omega}$.

($j = 1, 2, \ldots, 12$) all have frequency ω_a and the subsequent oscillators ($j = 13, 14, \ldots, 24$) all have frequency ω_b. Let $\omega_a > \omega_b$ be assumed. A delay is not present to be able to determine the critical coupling simply. In Figure 8.12a a coupling constant $k > k_c$ is chosen. With frequency spacing $\Delta\omega = \omega_a - \omega_b$ the correction terms converge to $\pm\Delta\omega/2$.

For the calculation of Figure 8.12b, $k < k_c$ was chosen so that the chain does not oscillate synchronously. The largest oscillations are found in the correction terms of those oscillators which are adjacent to the other half. To the outside these oscillations decrease rapidly, but a small oscillation remains, because by the coupling each oscillator is subject to the influence of the other half of the chain. The correction terms of the outer diatoms oscillate in a band that is narrow compared to $\Delta\omega$. These oscillators vibrate mainly at their natural frequency. Although there is no constant phase difference in time, the motion in the outer diatoms differs only slightly from that of separated chains. Simplified, one can use the term of partial synchronization, which is common in *Bacillaria* colonies. It seems to me that the term "local synchronization" is more specific because it expresses that certain sections of the chain oscillate synchronously. In the sense of this naming this a contrast to "global synchronization" for a synchronously oscillating colony.

In the discussed case, the sum in (8.28) increases linearly up to the center of the chain and then decreases linearly. If both parts of the chain are lengthened or shortened, k_c increases or decreases linearly.

8.5.5.3 *Discrete Fourier Transform*

To analyze a partial synchronization for a given chain, one can look at the Fourier spectrum of the relative motion. This does not provide any information about phase locking, but how the frequencies converge with increasing coupling is well shown. A spectrum for a signal according to (8.2) contains harmonics which are outside the investigated frequency range and are not relevant for the question of synchronicity. Therefore, only the sine of the phases $sin\,(\Theta_j)$ is considered. The result of a simulation of the equations of motion (8.6) are phases at equidistant time points, where the time interval is small compared to the occurring oscillation periods. Therefore, the study of the spectrum can be limited to phases in a coarser temporal grid. From this reduced number sequence, N values are subjected to a discrete Fourier transform (DFT), according to (see, e.g. [8.35]):

$$F_j(m) = \sum_{l=0}^{N-1} f_{j,l} e^{-i\frac{2\pi}{N}ml} \tag{8.34}$$

with

$$f_{j,l} = \sin(\theta_{j,l}), \tag{8.35}$$

where $\Theta_{j,l}$ denotes the l-th number of the sequence of the phase to the oscillator with index j. The calculated Fourier coefficients $F_j(m)$ represent a discrete spectrum for this oscillator. Denoting the time interval between two successive values of the time signal by Δt, the frequency at index m is given by

$$\omega'_m = \frac{2\pi m}{N\Delta t}. \tag{8.36}$$

The quantity $N\Delta t$ is the length of the examined time window. As we are not interested in the phase of the Fourier transform here, only the modulus of $F_j(m)$, i.e., its amplitude, is needed. For analysis, a time window after the transient is chosen.

In each plot of Figure 8.13, the amplitudes of the Fourier coefficients are plotted versus the frequency and the index of the oscillators for a chain of 24 oscillators. The oscillator

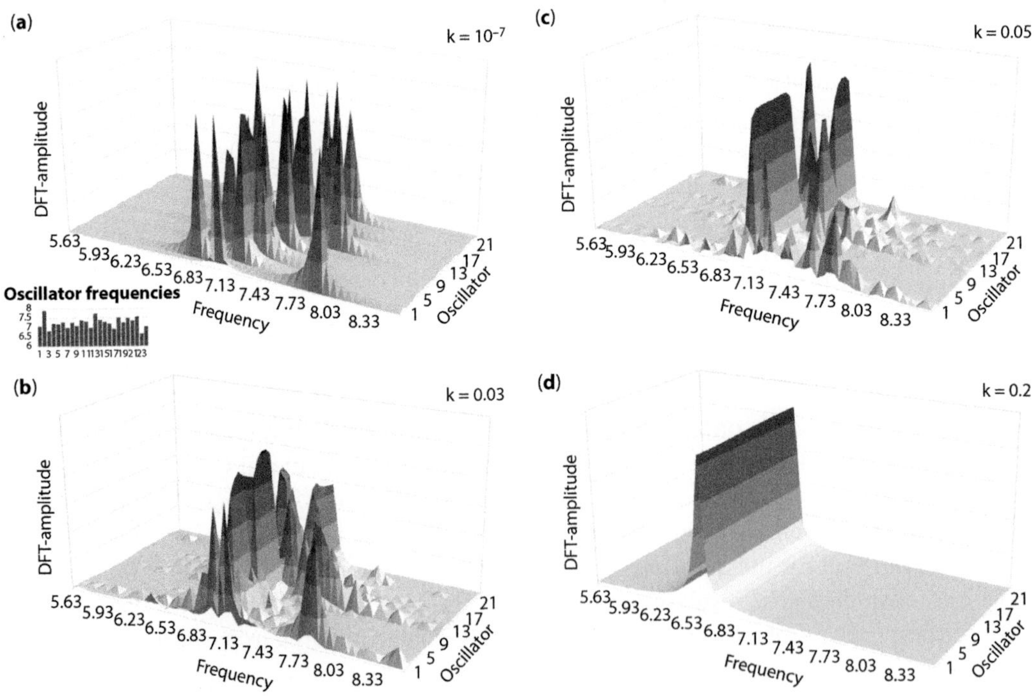

Figure 8.13 Results of a discrete Fourier transform, where $N = 2048$, $\tau = 0.04\,T'(T' = 2\pi/\bar{\omega})$ and empirical standard deviation of the sample $0.044\,\bar{\omega}$. (a) $k = 10^{-7}\,\bar{\omega}$. (b) $k = 0.03\,\bar{\omega}$. (c) $k = 0.05\,\bar{\omega}$. (d) $k = 0.2\,\bar{\omega}$ $(k > k_c)$.

frequencies were randomly selected according to a normal distribution. A bar chart with the frequencies of the oscillators is inserted in the lower left corner of Figure 8.13a. With negligible coupling, as shown in Figure 8.13a, there are many individual peaks corresponding to the frequencies of the uncoupled oscillators. The broadened peaks are a characteristic of the DFT of periodic functions. As the coupling increases, first small, then larger groups of oscillators combine (Figure 8.13b and Figure 8.13c) and move at the same frequency.

Above the critical coupling, the chain moves with uniform oscillation frequency (Figure 8.13d). Following the phases of such groups of oscillators with the same frequency in a diagram of circulating phases, it can be seen that they have a fixed phase difference except for minor fluctuations, which means that phase locking occurs. While the individual oscillators without coupling and the whole chain move periodically above the critical coupling strength, the coupled regions in between are subject to disturbances by the non-periodic oscillators. The Fourier transforms of the non-periodic oscillators provide a spectrum that depends on the selected time window and the initial conditions. Nevertheless, Figure 8.13 gives an impression of the transition to the synchronous state.

8.5.6 Coupling to a Periodic Light Source

As already mentioned, the movement activity of *B. paxillifera* is dependent on the intensity of light. This was observed on the time scale of the day-night rhythm [8.24]. The subject of the following considerations is the assumption that an influence on the activity on the time scale of the period of oscillations is also possible. For observation, a light source with a light intensity of the form $I(t) = I_0 sin(\omega_e + \vartheta_e) + I_1$ with $I_1 \geq I_0$ was positioned near the observed culture. The frequency of the light source ω_e was manually brought into coincidence with the frequency ω of a selected colony (approximately 60 s). Small colonies move from one elongated position to the next in half this time, so that, for example, the outward movement occurs in the brighter half-wave of illumination and the return movement in the darker half-wave. For the synchronization experiments, light intensities of typically 200 lx to 800 lx were used. An intensity $I_1 > I_0$ was selected to enable the camera to detect the motion sequences even in the range of the minimum of the periodic illumination. The minimal light intensity was about 30 lx at the minimum and was mostly realized by adding the microscope illumination.

When the intensity of the light source is sufficient, the oscillators synchronize with the periodic light source. As this occurs for most colonies, this synchronous state is easily recognized by the fact that the movements of the colonies in the area of observation are now synchronous. Therefore, a sufficiently large observation area is recommended for this purpose (see video at https://diatoms.de/en/periodic-illumination).

Synchronization with complete back and forth movement of the colonies is not always achieved. After a longer period of synchronicity, the motion stroke is often reduced, so that smaller asymmetric amplitudes occur. At high light intensity, the movement can come to a standstill after a few minutes. Obviously, frequency, light intensity and possibly also the light spectrum must fulfill certain conditions, whose determination need considerable further investigations. It is still unclear whether the motion follows the equations of motion (8.1) and (8.2) with the same parameters. A rest period obviously occurs.

Apparently, in this experiment the periodic external light source becomes the clock and the diatoms adapt to its frequency and phase. This can also be described mathematically by stating that the diatoms or oscillators attempt to minimize the phase difference to a reference. Matching to the neighboring oscillators and to the light field compete for this. In Eq. (8.37) this is taken into account by an additional term containing the phase difference to the periodic light

$$\frac{d\Theta_1(t)}{dt} = \omega_1 - k\sin(\Theta_1(t) - \Theta_2(t-\tau)) - L\sin(\Theta_1(t) - \Theta_e(t))$$
$$\frac{d\Theta_n(t)}{dt} = \omega_n - k\sin\left(\Theta_n(t) - \Theta_{n-1}(t-\tau)\right) - L\sin(\Theta_n(t) - \Theta_e(t))$$
$$\frac{d\Theta_j(t)}{dt} = \omega_j - k[\sin(\Theta_j(t) - \Theta_{j-1}(t-\tau)) + \sin(\Theta_j(t) - \Theta_{j+1}(t-\tau))] - L\sin(\Theta_j(t) - \Theta_e(t))$$
$$\text{for } 2 < j < n-1. \tag{8.37}$$

A coupling to a field in this form was introduced into the Kuramoto model by Sakaguchi [8.33] in 1988. The coupling constant L depends on the intensity of the periodic light field at the location of the diatoms. At exceptionally low intensity, L is so small that the motion is not noticeably affected. There is a strong influence when L and k are of the same order of magnitude. At a high light intensity there is a fast transition to a motion where all phases of the oscillators oscillate with the frequency ω_e. Let the phase of the periodic light source be given by

$$\Theta_e(t) = \omega_e t + \vartheta_e. \tag{8.38}$$

If one neglects the coupling of neighboring oscillators in the limiting case $L \gg k$, one obtains n decoupled differential equations ($j = 1..$ n) which no longer depend on connectivity

$$\frac{d\Theta_j(t)}{dt} = \omega_j - L\sin(\Theta_j(t) - \Theta_e(t)). \tag{8.39}$$

The assumption of a movement synchronous with the periodic light source

$$\Theta_j(t) = \omega_e t + \vartheta_j. \tag{8.40}$$

yields

$$\omega_e - \omega_j = L\sin(\vartheta_e - \vartheta_j) =: \mathrm{L}\sin(\Delta\vartheta_j). \tag{8.41}$$

The existence of a solution requires

$$\left|\frac{\omega_e - \omega_j}{L}\right| \leq 1\, f\ or\ j = 1..n. \tag{8.42}$$

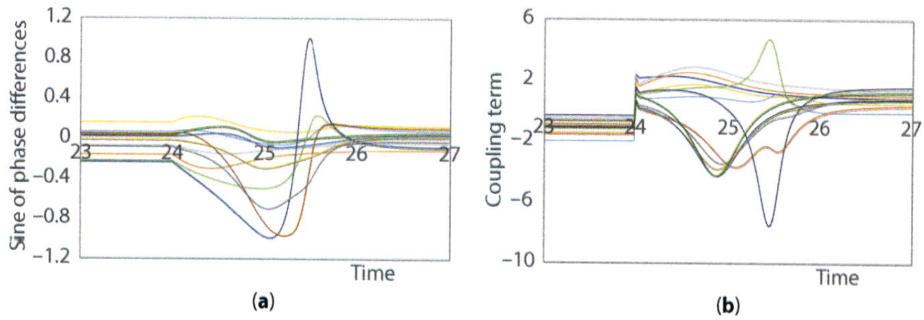

Figure 8.14 Short period of time from a simulation with $\tau = 0.04\,T'\,(T' = 2\pi/\bar{\omega})$, $k = 0.48\bar{\omega}$, $L = 0.64\bar{\omega}$, standard deviation of the angular frequencies is $0.07\bar{\omega}$ and the period of illumination $0.9\,T'(T' = 2\pi/\bar{\omega})$. The phase differences are constant as a periodic light source is switched on. After about two periods, the chain has settled and follows the periodic illumination. (a) Settling of the phase differences (b) Settling of the frequency correction term.

The entire colony thus moves synchronously with the light source if

$$L \geq \max_j |\omega_e - \omega_j|. \tag{8.43}$$

In the limiting case of exceptionally large L, one can experimentally determine the natural frequencies ω_j by measuring the phase differences $\Delta\vartheta_j$ using (8.41). Such observations and evaluations are subject to further investigation. It should be noted that synchronization could also be achieved with a squarewave light signal in the form of a bright period and a dark period of equal length.

If a colony is in a synchronous state of motion when the periodic illumination is switched on, it takes a few periods for the colony to follow the external clock. An exemplary simulation of Eq. (8.37) with 12 oscillators is illustrated in Figure 8.14 using the phase differences and frequency correction.

Figure 8.14a shows that, for the selected parameters, a changed sequence of phases is stably established approximately within two periods. The sudden switching on of the periodic light has an immediate effect on the frequency correction terms. The diagrams change considerably when the time of the onset of the illumination is varied.

In the simple model outlined, there is no slow switching on of the light. The oscillators "recognize" immediately the phase of the illumination, although they have no information about its amplitude. Eq. (8.37) is to be assessed as a very simple model, which seems primarily suitable to describe the synchronized state.

A similar experimentally successful synchronization of biological coupled oscillators was achieved by Meier [8.27] with blinking fireflies. A simulation of this can be found in [8.26].

8.6 Discussion

Systems theoretical partitioning and the hypothesis of the applicability of the Kuramoto model reproduce many aspects of the movement patterns of *Bacillaria* colonies. In particular, the existence of phase shift between the relative motions of oscillators, the sequence of motions, the partial — or more specifically "local" — synchronization in large colonies, the limited communication in diatoms at advanced stages of division, and the possibility of synchronization with periodic illumination support this approach. Demonstrating quantitative agreement of phase differences and transient response with theory is a major challenge for further studies.

A key assumption was the assignment of each oscillator to two adjacent diatoms. It should be mentioned that there is an alternative to the proposed model shown in Figure 8.6 and Figure 8.7 in which a central oscillator is assumed in each diatom, which sets the reference value for both raphe systems of the cell. This approach has methodological weaknesses and is therefore not discussed in more detail here.

The presented investigations provide only limited information about the physiology, in particular about the sensory function. Besides a mechanical recognition of the position, there is also the possibility of an optical perception. Each diatom shadows the light onto the neighboring diatom. This is also conceivable for colonies under the illumination of a microscope, as they are illuminated by a condenser of finite numerical aperture (NA = 0.4 for the inverted microscope used). In addition, there is usually ambient light. However, our observations at light intensities below 1 lx show that synchronous movement of colonies is maintained even at low intensities. On the other hand, the synchronization with a periodic illumination strengthens the assumption that the position recognition of diatoms is based on an optical process. In any other case, synchronization would have to use two physically different stimuli and different sensory systems. It can also not be excluded that the sensor detects the mechanical tensions that occur in the contact area.

In applying the Kuramoto model, assumptions specific to the model have been included, which are part of the hypothesis, such as the restriction to a nearest-neighbor interaction. Further degrees of freedom like a non-uniform delay, possibly even depending on the direction of the information flow, or a non-uniform coupling constant, are natural extensions to the model. One can argue that such non-uniform quantities should certainly exist, whereby the degree of their dispersion determines the relevance of their consideration. In a simulation, such variability does not represent a significant increase in complexity, but it certainly does in the mathematical treatment. Should such an increase in the variety of variables be essential for a quantitative comparison, it would be extremely difficult to quantitatively check the transient response or even the phase differences of the oscillators, because it cannot be assumed that all these quantities can be measured independently.

No statistical data were collected in the scope of this investigation. The theoretical considerations presented are intended as a basis for further observations. One assumption was the statistical independence of the natural frequencies of the oscillators. This can be well justified for oscillators associated with different individuals, such as frequencies of blinking fireflies. For *Bacillaria* colonies, however, this is likely to be only an approximation since neighboring oscillators contain functionalities of the same diatom. Statistical independence of the oscillators is only given if the two incomplete oscillators of a diatom have statistically independent frequencies. An alternative approach would be to assign to each diatom a frequency which occurs at both incomplete oscillators and to assume that the arithmetic mean value of the frequencies of both oscillators involved appears in the oscillators. Then the frequencies of the oscillators would be correlated.

A significant field for deeper understanding is the determination of the functional dependence of the parameters, especially amplitude, length of resting periods in stretched state and mechanical activity on environmental parameters such as light intensity, light spectrum and temperature. Only based on such results can we hope to be able to simulate and quantitatively examine the movements even under changing conditions, for example, when illumination sets in.

Often, a swarm is understood to be an assembly of animals that move collectively, where an assembly is a spatial grouping of individuals of the same species, sometimes of different species. Parunak [8.31] lists broader examples of swarming behavior, ranging from pattern generation in bacteria and slime molds, via synchronization in fireflies, to surrounding prey by wolves. Essentially, the behavior is based on self-organization through local interaction. In addition, coordinated behavior should result in group performance beyond what the member of the swarm is capable of achieving. Against the background of the coupled oscillators hypothesis, there are arguments in favor of interpreting a *B. paxillifer* colony as a swarm:

- *B. paxillifer* forms an aggregation of organisms that are also individually viable.
- There is a perception of the position of neighboring diatoms.
- Information about the position of the adjacent diatoms leads to a controlled motor response.
- A collective behavior is created that serves the colony. Unlike a single diatom, the colony, with synchronous movement, is able to protrude into the water body and achieve high velocities by summing the individual velocities. This presumably provides the individuals with benefits such as improved light exposure and possibly improved access to nutrients.
- There is a behavioral repertoire that besides oscillations includes a collective response to tactile stimuli (possibly an escape response) and formation of a resting configuration during long periods of darkness.

The group performance of a *B. paxillifer* colony initially increases linearly with the number of diatoms, both in terms of achievable range and maximum velocity. With increasing colony size, loss of synchronicity and mechanical stability, the group performance does not increase further and finally decreases. As with many swarms, there is an optimum at a certain number of individuals. The exceptionally large colonies often observed in cultures are not found in water samples according to our own observations. Presumably, environmental influences, such as grazing and water flow lead to fragmentation of large colonies, and thus potentially to an approach to optimal size.

Acknowledgment

Professor Richard Gordon deserves my special thanks. He drew my attention to the multifaceted behavior of *B. paxillifera* and stimulated my interest in the subject by asking inspiring questions. I gratefully received a sample of marine *B.* paxillifer from Matt Ashworth and was able to gain my first experience with its observation and cultivation. I thank

Norbert L. Brodtmann for the assembly and constant improvement of the function generator, with which the proof of a synchronization with a periodic illumination succeeded.

References

[8.1] Alicea, B., Gordon, R., Harbich, T., Singh, U., Singh, A., Varma, V. and Mehan, P. (2020) Towards a digital diatom: Image processing and deep learning analysis of *Bacillaria paradoxa* dynamic morphology. In: Diatom Gliding Motility [DIGM, Volume in the series: Diatoms: Biology & Applications, series editors: Richard Gordon & Joseph Seckbach, In preparation]. S.A. Cohn, K.M. Manoylov and R. Gordon, (eds.) Wiley-Scrivener, Beverly, MA, USA.

[8.2] Brown, D. (2012) Tracker - Free Video Analysis and Modeling Tool for Physics Education. Version 5.1.3. https://www.softpedia.com/get/Science-CAD/Douglas-Tracker.shtml.

[8.3] Buck, J., Buck, E., Hanson, F.E., Case J.F., Mets L. and Atta G.J. (1981) Control of flashing in fireflies. IV. Free Run Pacemaking in a Synchronic Pteroptyx. *Journal of Comparative Physiology* A 144, 287-98.

[8.4] Cai, J., Chen, M., Wang, Y., Pan, J., Li, A. and Zhang, D. (2013) Culture and motion analysis of diatom *Bacillaria* paradoxa on a microfluidic platform. *Current Microbiology* 67, 652–658, https://dx.doi.org/10.1007/s00284-013-0413-7.

[8.5] Choi, M., Kim, H., Kim, D. and Hong, H. (2000) Synchronization in a system of globally coupled oscillators with time delay. *Physical Review E* 61 (1) 371–381.

[8.6] Dekker, A.H. and Taylor, R. (2013) Synchronization properties of trees in the Kuramoto model. *SIAM Journal on Applied Dynamical Systems* 12(2), 596–617.

[8.7] Dorfler, F. and Bullo, F. (2014) Synchronization in complex networks of phase oscillators: A survey. *Automatica* 50 (6), 1539–1564.

[8.8] Drum, R.W. and Pankratz, H.S. (1965) Locomotion and raphe structure of the diatom *Bacillaria. Nova Hedwigia* 10(3/4), 315-317.

[8.9] Drum, R.W., Gordon, R., Bender, R. and Goel N.S. (1971). On weakly coupled diatomic oscillators *Bacillaria*'s paradox resolved. *Journal of Phycology* 7 (Suppl.), 13-14.

[8.10] Ermentrout, G. B. (1991) An adaptive model for synchrony in the firefly *Pteroptyx malaccae. Journal of Mathematical Biology* 29(6), 571–585.

[8.11] Funk, G. (1914) Beobachtungen über Bewegungen von Bacillariaceenkolonien und deren Abhängigkeit von äußeren Reizen. *Mitteilungen aus der zoologischen Station Neapel* 22, 45-58.

[8.12] Funk, G. (1919) Notizen über Meeresdiatomeen. *Berichte der Deutschen Botanischen Gesellschaft* 37, 187-192.

[8.13] Gall, W.G., Zhou, Y. and Salisbury, J.P. (2010) Synchronization of a Network With Piecewise-Linear Dynamics. In: *Proceedings of the ASME 2010 Dynamic Systems and Control Conference DSCC2010, September 12-15, 2010, Cambridge, Massachusetts, USA*, 1-7.

[8.14] Gebeshuber, I.C., Stachelberger, H. and Drack, M. (2005) Diatom Bionanotribology - Biological Surfaces in Relative Motion: their Design, Friction, Adhesion, Lubrication and Wear. *Journal of Nanoscience and Nanotechnology* 5 (1), 79-87.

[8.15] Gebeshuber, I.C., Stachelberger, H. and Drack, M. (2005) Diatom Biotribology. In: *Life Cycle Tribology* (48), D. Davidson *et. al.* (eds.) Elsevier B.V. Amsterdam 365-370.

[8.16] Gordon, R. (2016) Partial synchronization of the colonial diatom *Bacillaria "paradoxa". Research Ideas and Outcomes (RIO) 2*, e7869.

[8.17] Gordon, R. (2019) The whimsical history of proposed motors for diatom motility. In: *Diatom Gliding Motility [DIGM, Volume in the series: Diatoms: Biology & Applications*, series *editors: Richard Gordon & Joseph Seckbach*, [In preparation]. S.A. Cohn, K.M. Manoylov and R. Gordon, (eds.) Wiley-Scrivener, Beverly, MA, USA: In preparation.

[8.18] Guillard, R.R.L. (1975) Culture of phytoplankton for feeding marine invertebrates. In: *Culture of Marine Invertebrate Animals*. W.L. Smith and M.H. Chanley (eds.) Plenum Press, New York, 26–60.

[8.19] Jarosch, R. (1962). Gliding. In: *Physiology and Biochemistry of Algae*. R. A. Lewin (ed.) Academic Press, New York 573-581.

[8.20] Kaempfer, E. (1727) *The History of Japan (With a Description of the Kingdom of Siam)*. Sloane, London, Posthumous translation; or reprint by McLehose, Glasgow, 1906.

[8.21] Kapinga, M.R.M. (1987) The Motility of *Bacillaria paradoxa. IX North American Diatom Symposium [NADS]* Treehaven Field Station, Tomahawk, WI, October 7-10, 1987.

[8.22] Kapinga, M.R.M. (1989) Observations on the Growth and Motile Behavior of the Colonial Diatom *Bacillaria paradoxa* in Culture [M.Sc.Thesis, Supervisor: R. Gordon]. Winnipeg: University of Manitoba. https://mspace.lib.umanitoba.ca/bitstream/handle/1993/17075/Kapinga_Observations_on.pdf?sequence=1

[8.23] Kapinga, M.R.M. and Gordon, R. (1992) Cell attachment in the motile colonial diatom *Bacillaria paxillifer. Diatom Research* 7 (2), 221–225.

[8.24] Kapinga, M.R.M. and Gordon, R. (1992) Cell motility rhythms in *Bacillaria paxillifer. Diatom Research* 7(2), 221-225.

[8.25] Kuramoto, Y. (1975) Self-entrainment of a population of coupled nonlinear oscillators. In: *International Symposium on Mathematical Problems in Theoretical Physics, Lecture Notes in Physics* 39, H. Araki (ed.) Springer-Verlag, Berlin, 420–422.

[8.26] Meier, R. (2013) Firefly Sync Simulation. https://robinmeier.net/?p=960

[8.27] Meier, R. (2016) Synchronicity (Thailand) https://robinmeier.net/?p=2279

[8.28] Müller, O.F. (1782) Von einem sonderbaren Wesen im Meerwasser, welches aus kleinen Stäbgen, durch deren mancherley Stellung es verschiedene Gestalten bildet, zu bestehen scheint. In: *Otto Friedrich Müllers kleine Schriften aus der Naturhistorie*. J.A.E. Goeze (ed.), Dessau, 1(1), 1–14.

[8.29] Müller, O.F. (1783) Om et besoenderligt Vaesen i Strandvandet [On a peculiar being in the beach-water]. *Nye Samling af det Kongelige Danske Videnskabers Selskabs Skrifter (Copenhagen)/New Collection of the Proceedings of the Royal Danish Academy of Sciences and Letters*, Copenhagen (Pt. 2), 277-286 + plate.

[8.30] Ochab, J. and Gora, P. (2009) Synchronization of Coupled Oscillators in a Local One-Dimensional Kuramoto Model. *Acta Physica Polonica B, Proceedings Supplement* 3(2), 453–462.

[8.31] Parunak, H.V.D. (2003) Making Swarming Happen. In: *Proceedings of the Conference on Swarming and Network Enabled Command Control Communications Computers Intelligence Surveillance and Reconnaissance*, McLean, VA, USA.

[8.32] Pikovsky, A., Rosenblum, M. and Kurths, J. (2001) *Synchronization: A Universal Concept In Nonlinear Sciences*. Cambridge University Press, Cambridge, England.

[8.33] Sakaguchi, H. (1988) Cooperative phenomena in coupled oscillator systems under external fields. *Progress of Theoretical Physics*, 79(1), 39-46.

[8.34] Schmid, A. (2007) The "paradox" diatom *Bacillaria paxillifer* (Bacillariophyta) revisited. *Journal of Phycology* 43 (1), 139-155, doi: 10.1111/j.1529-8817.2006.00299.x.

[8.35] Stearns, S.D. and Hush D.R. (1990) *Digital Signal Analysis. Second edition*. Prentice Hall, EnglewoodCliffs, NJ, USA.

[8.36] Strogatz, S., and Stewart, I. (1993) Coupled Oscillators and Biological Synchronization. Scientific *American* 269(6), 102-109.

[8.37] Trojanowski, K. and Longa, L. (2013) Synchronization of Kuramoto oscillators with distance-dependent delay. *Acta Physica Polonica. B*, 44(5), 991–995.

[8.38] Ussing, A.P., Gordon, R., Ector, L., Buczko, K., Desnitskiy, A.G. and Vanlandingham, S.L. (2005) The colonial diatom "*Bacillaria paradoxa*": chaotic gliding motility, Lindenmeyer Model of colonial morphogenesis, and bibliography, with translation of O.F. Müller (1783), 'About a peculiar being in the beach-water'. *Diatom Monographs* Vol. 5. Koeltz Scientific Books, Koenigstein.

[8.39] Walne, P.R. (1970) Studies on the food value of nineteen genera of algae to juvenile bivalves of the genera Ostrea, Crassostrea, Mercenaria, and Mytilis. *Fishery Invest.* 26, 162.

[8.40] Yamaoka, N., Suetomo, Y., Yoshihisa, T. and Sonobe, S. (2016) Motion analysis and ultrastructural study of a colonial diatom, *Bacillaria paxillifer. Microscopy* 65(3), 211-221.

[8.41] Yang, H., Shao, L., Zheng, F., Wang, L. and Song, Z. (2011) Recent advances and trends in visual tracking: A review. *Neurocomputing* 74(18), 3823–3831.

[8.42] Yeung, M.K.S. and Strogatz, S.H. (1999) Time delay in the Kuramoto model of coupled oscillators. *Physical Review Letters* 82, 648–651.

9

The Psychophysical World of the Motile Diatom *Bacillaria paradoxa*

Bradly Alicea[1,2*], Richard Gordon[3,4] and Jesse Parent[2]

[1]*OpenWorm Foundation, Boston, Massachusetts, USA*
[2]*Orthogonal Research and Education Laboratory, Champaign-Urbana, Illinois, USA*
[3]*Gulf Specimen Marine Laboratory & Aquarium, Panacea, Florida, USA*
[4]*Department of Obstetrics and Gynecology, Wayne State University, Detroit, Michigan, USA*

Abstract

There are a large number of candidate hypotheses for the movement of diatoms, a taxonomic group of Algae. In particular, the genus *Bacillaria* exhibits highly interesting movement dynamics resembling a number of physical systems. None of these hypotheses account for the information processing that might result from the connection between environmental stimuli (e.g., photic, chemical, turbulence, and temperature gradients, and mechanical constraints) and motility. Even when there is no significant variation in terms of sensory input, the colonial structure is coordinated and mediated through information processing. Here, we propose a series of perspectives on how the concept of aneural cognition might explain the movement dynamics of individual colonies. More specifically, we can use analogies of neuronal mechanisms such as collective pattern generators, Hebbian learning, and predictive processing to understand and explore the potential for behavior generation. Quantitatively, these mechanisms can be summarized using psychophysical metrics, particularly the Weber-Fechner law. This analogy could be extended to out-of-equilibrium behaviors such as halting of oscillatory movement patterns, and further interpretations in terms of pseudo-intelligent behavior, which is simulated but not necessarily autonomous. Taken together, we can say that the psychophysical world can open up new avenues of exploration for understanding the regulation of diatom movement.

Keywords: Aneural cognition, diatoms, collective behavior, theoretical modeling, single-cell psychophysics, collective pattern generations

**Corresponding author*: balicea@openworm.org
Bradly Alicea: Google Scholar: https://scholar.google.com/citations?user=UaAdJi0AAAAJ&hl=en&oi=ao
ResearchGate: https://www.researchgate.net/profile/Bradly-Alicea
https://orcid.org/0000-0003-3869-3175
Richard Gordon: Google Scholar: https://scholar.google.com/citations?user=LnaynB0AAAAJ&hl=en&oi=sra
ResearchGate: https://www.researchgate.net/profile/Richard-Gordon-9
http://orcid.org/0000-0003-4970-9953
Jesse Parent: Google Scholar: https://scholar.google.com/citations?user=tSvD5eAAAAAJ&hl=en&oi=sra
ResearchGate: https://www.researchgate.net/profile/Jesse-Parent-2

Janice L. Pappas (ed.) The Mathematical Biology of Diatoms, (229–264) © 2023 Scrivener Publishing LLC

Abbreviations

SDT	Signal Detection Theory
FC	Fold-Change
R-density	Recognition Density
G-density	Generative Density
CoPG	Collective Pattern Generator
AIT	Algorithmic Information Theory
ROC	Receiver Operator Characteristic
dBV	Developmental Braitenberg Vehicle

9.1 Introduction

In the study of organismal behavior from across the tree of life, there are traditionally two types of explanatory mechanism. The first is behavior that is generated by neuronal processes (such as a network of neurons), and is restricted to organisms with a nervous system. The second is behavior that is characterized by loosely-relevant physical, mechanical, and social metaphors and generated by biophysical processes, including at the single cell level [9.34] [9.84] [9.191]. The biophysics of individual cell motility applied in parallel is assumed to be the primary explanatory paradigm. Yet there is an alternative explanatory framework that allows us to view the colonial diatom *Bacillaria* movement as a holistic, colony-level phenomenon: aneural cognition [9.14] [9.28] [9.63] [9.196] [9.209] and the possibility of light piping between cells [9.62], which could act as a synchronizing mechanism. While *Bacillaria paradoxa* has been divided into a few species [9.109] [9.193], we continue to use *B. paradoxa* as including all of them. This framework ascribes information processing and decision-making capabilities to systems that lack a formal nervous system structure. Rather, the information processing function occurs in the biophysical substrate itself. One definition of cognition is as an ill-defined cluster of capacities and mental (or psychophysical) phenomena [9.153] [9.175]. While aneural cognition can benefit from a representational structure in the form of neural codes [9.154] and other such higher-dimensional pattern generators [9.126], this does not need to be a part of a cognitive system.

While the diatom genus *Bacillaria* does not have a nervous system, it nevertheless exhibits organized behaviors and collective movement. *Bacillaria* consists of a line of long, narrow cells arrayed lengthwise, each cell perpendicular to the axis of the colony [9.206] in parallel, and the behaviors exhibited across these cells are temporally linked in a way that results in coordinated movement [9.89]. *Bacillaria* cells can exhibit coordinated behavior by forming a colony that oscillates from one end of the colony to another. The contraction and expansion movements resemble the contractile motion of an accordion (see Figure 9.1). One interpretation involves using a multitude of existing mechanical and physical metaphors to interpret *Bacillaria* movement [9.90]. As an alternative view, we propose that a model of minimal cognition [9.207] defined by the relationship between stimulus intensity and adaptive movement can be used instead. This model of cognition results in the

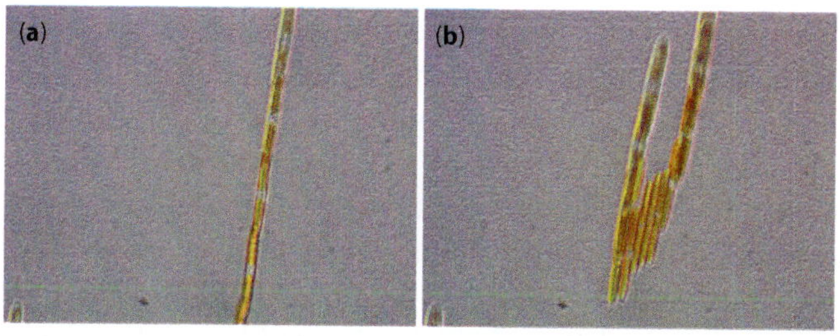

Figure 9.1 An image of the *Bacillaria* colony, captured in its elongated (a) and contractile (b) state.[1] (Thanks to Dr. Thomas Harbich.)

production of behaviors related to taxis [9.50], active materials [9.141], and even psychophysics [9.177], which can be distinguished from random behavior by separating an informational signal from noise.

Aneural cognition provides a context for diatom movement behaviors, particularly in response to environmental stimuli. Taking a broader view of diatom morphology, most diatoms are single-celled organisms that live in either a solitary or colonial context. Diatoms are eukaryotic algae that have chlorophyll and silicate cell walls [9.73]. In either the solitary or colonial setting, the cell cytoplasm and extracellular components are all important for facilitating a diatom's movement and other interactions with the environment [9.90]. In general, there are three types of diatom morphology: rod, flake, and three-dimensional shape. The raphid phenotype (which includes *Bacillaria*) is 300-500 μm long and typically 10-μm wide [9.73]. *Bacillaria* colonies in particular are held together by secreted polymers that enable extracellular adhesion and selective interactions with the external environment [9.47]. A similar kind of externalization is also observed in slime mold colonies [9.196]. The morphology of a *Bacillaria* cell chain and moving colony are shown in Figures 9.1a and 9.1b, respectively.

Ecologically, *Bacillaria* is native to estuary, littoral, and riverine environments. These aquatic environments provide three main perceptual challenges to the organism. The first involves changes in water chemistry as the colony moves through the water column. Attraction or repulsion from different chemical gradients (e.g., salinity or acidity) comes in the form of chemotaxis. Movement of cells through the water column (see examples in Figure 9.1) also introduces them to light and chemical gradients. These gradients can affect the mode of movement, the size of individual cells, and overall colonial size [9.199]. Encountering light gradients specifically leads to phototaxis, which can be positive (attractive) or negative (repulsive). In a like manner, chemical gradients lead to chemotaxis, and a combination of the two can lead to more complex movement behaviors. Cell movement over broader spatial scales introduces *Bacillaria* cells to temperature gradient exploration, which requires thermotaxis. If buried in benthic sediments, their movements may

[1]See also [9.113] [9.114] [9.115] [9.116] [9.117]. There are many online movies of *Bacillaria* movement at many magnifications [9.6] [9.15] [9.23] [9.40] [9.41] [9.48] [9.61] [9.92] [9.107] [9.111] [9.122] [9.129] [9.159] [9.160] [9.161] [9.164] [9.166] [9.201] [9.205] [9.219] [9.220] [9.10] [9.42] [9.60] [9.66] [9.70] [9.95] [9.121] [9.123] [9.143] [9.144] [9.146] [9.165] [9.167] [9.172] [9.173] [9.180] [9.210] [9.211] [9.218]. An earlier list is available in [9.206].

allow extrication towards sunlight. These three types of simple responses allow us to move towards a model of aneural cognition that characterizes trends in these behaviors.

The diatom cell and life cycles are also an important part of any adaptive behaviors and aneural cognition. Diatom cell division occurs in a manner different from other eukaryotes, and plays a role in shaping the relationship between cell cycle checkpoint and changes in physiological processes [9.108]. In general, diatom molecular processes play a key role in the perception of environmental stimuli and regulation of associated physiological processes [9.71]. The life cycle of a diatom [9.7] [9.67] [9.68] operates at a longer time scale, but is also affected by environmental factors such as changes in the local concentrations of phosphate, iron, and dissolved silicon [9.108]. Overall, variation of the life cycle is nonlinear and high-dimensional [9.185]. Returning to the effects of environmental stimuli, light not only results in reactive taxis, but can also trigger specific acclimation-specific physiological responses. Light variation is also a feature of day-night cycles [9.99] and relative position in the water column. More generally, response to environmental conditions (sensing and signaling) are tightly linked to cell cycle checkpoint activation [9.108]. This not only means that environmental factors affect simple output behaviors, but often affect organismal-level features such as transformations in shape, growth, and surface texture [9.185] [9.187].

We will now consider the aneural cognition and adaptive behavioral generation of *Bacillaria*. This will proceed by reviewing evidence for aneural cognition in a diverse range of species. Then, we introduce a formulation of psychophysics that is relevant to aneural cognition, particularly in the *Bacillaria* example. This will include considering the ways in which adaptive behavior can be generated in a cognitive system, including theories of neural function such as Hebbian learning and predictive processing. We will then consider whether or not aneural cognition is an epiphenomenon, followed by a review of potential measurement techniques. These include both singular equations that produce characteristic functions and networks that advance a connectionist approach to characterizing behavior [9.98]. To conclude, we discuss modeling the dynamical systems complexity of *Bacillaria* psychophysics, and further describe this model through the application of collective pattern generators (CoPGs).

9.1.1 Aneural Architecture of *Bacillaria*

In this section, we will introduce the aneural architecture of *Bacillaria*. This architecture generates behaviors from colony physiology. This includes both intracellular microstructural elements proposed to be involved in motility and extracellular polymers [9.90] [9.169]. The *Bacillaria* behavior-generation system features three architectural properties that define the coherent collective behavior of a colony. *Bacillaria* cells can respond to changes in light, water salinity, and temperature, but must regulate their response in accordance with the states of neighboring cells. This regulation and interdependence can occur in two ways. The first is that individual cells secrete polymers that both encase the cell and act as sites of adhesion [9.14]. Secondly, an extracellular space that acts as both a site of stigmergy and a memory medium [9.47]. Stigmergy is an important concept that is usually applied to the collective behavior of insects [9.105] [9.157]. It is a type of collective behavior that uses various features of the environment to guide autonomous behavior. When we talk about aneural cognition, stigmergy works in concert with physiological mechanisms to produce an output that supersedes what a single cell can achieve on its own [9.127].

From this behavior-generating architecture, there are several behavioral outcomes that provide a means for adaptive behavior. As we will see, these can take the form of potentially goal-directed or even intelligent behaviors. There are three such behaviors that can be observed among *Bacillaria*. 1) While *Bacillaria* colonies have no formal representation of space, they do exhibit limited goal-directed behavior. Behaviors are explicitly spatial in that multicellular chains move in certain directions (not simply via Brownian motion). 2) According to [9.93], movement is a source of intelligence. Simple movements, even of *Bacillaria* single cells attached to a killed neighboring cell, produce a set of rounded square waves [9.63]. A single cell being able to identify and track the simple movements of neighboring cells and other organisms is consistent with the processing of direct perceptual information [9.212]. 3) More complicated movement can result in other types of waveforms and augment the processing of direct perceptual information, and results in a basic oscillatory rhythm. For example, a colony (chain) of cells produces an oscillator lagged n degrees out-of-phase.

Viewing *Bacillaria* movement in this way provides an alternative to purely biophysical models of movement. While biophysics is the dominant mechanism for generating movement, the same can also be said for cognition in the brain. Neural cognition is generated by biophysical processes, and produces information that then translates into what is considered to be cognition. Our aneural perspective allows us to bridge the gap between purely biophysical models and analogies of cognition. Biophysical models also do not provide a means to characterize the diverse behaviors often observed across different life-history stages and environmental contexts. Furthermore, there are many interpretations of these behaviors that involve information processing and specific functional capacities. An aneural cognition model provides a potential means for understanding all of this diversity.

9.1.2 Aneural Cognition in a Broader Context

Observations of so-called psychic life in single cell organisms were initially proposed by Binet [9.22], while Verworn [9.208] first proposed the psychophysiological study of protists around the same time. At the time, the amoeba was thought to be an evolutionary linchpin that connected the behaviors of single-cell organisms with that of humans [9.178], cognitive capacity being no exception. A persistent series of studies on psychophysiology in unicellular organisms continued throughout the late 19th and early 20th centuries [9.192]. One example of such work involved studying stimulus-driven taxic responses in *Stentor* and *Paramecium* [9.110]. While this work establishes the existence of simple mechanisms for intelligent behavior, the highly metaphoric and analogical nature of this work obscures its potential to provide insight.

To establish the intentionality and intelligent nature of different behavioral and morphological functions, we can turn to examples ranging from marine invertebrates to synthetic systems. For example, the cnidarian species *Nematostella vectensis* uses a muscular hydrostatic mechanism to link morphogenesis and behavior during the larval stage [9.198]. This requires a tight integration between two biological processes. *Physarum* (slime molds) also demonstrate a form of learning called habituation [9.28]. *Physarum* colonies exhibit response decline (weaker effect over time) and spontaneous recovery (responding after a significant period of stimulus deprivation), both of which are hallmarks of habituation. We can further observe the ability to produce intelligent behavior in minimal and non-biological systems. Simple chemical systems, such as oil droplet protocells, can break their symmetry when exposed to water [9.96] [9.97]. This facilitates movement during the aqueous phase, which may provide clues as to the origins of learned movement. In a synthetic biology context, the interactions between protocells can be characterized using the language of control, cognition, and communication [9.20].

To gain a broader perspective of what aneural cognition looks like in single-cell organisms, both without brains and with simple brains, we can turn to two other model organisms: one being a single-celled organism from another taxonomic order, and the other a larval crustacean with a highly rudimentary nervous system. One such taxon is the genus *Paramecium*. *Paramecium* is able to regulate its swimming movements in an intracellular fashion [9.34]. Another type of adaptive behavior is associative learning [9.84], which is controversial yet demonstrable in a single-cell context. Many coordinated collective behaviors can result from the tuning of physical parameters, which has been demonstrated in agent-based models that are analogous to organisms such as the water flea (*Daphnia*: [9.135]). Although *Paramecium* does not have a nervous system and *Daphnia* has a simple brain, organisms in both groups nevertheless tend to minimize their free energy, resulting in relatively complex behaviors [9.78] [9.79]. Free energy is also more directly tied to cognition by enforcing sites of information encapsulation composed of specific networks and populations of cells and the source of cognitive function [9.16].

As the examples given here are phylogenetically disparate, the question arises as to the evolutionary origins of aneural cognitive mechanisms. The answer comes in two forms: the common ancestry of cell biochemistry and the polyphyly of specific mechanisms in different organisms. For example, both photoreceptors and mechanosensors share a deep molecular homology [9.80]. As a result, cells that express sensory receptors appear in locations far from the organism's periphery where such cells would normally be found [9.8]. Furthermore, shared developmental cascades [9.80] produce a wide variety of mechanisms and systems for information processing. Yet many information processing systems involve more than conserved homologs and paralogs. Particularly in single-cell organisms, these systems are often built upon simple anatomical structures reconfigured in different ways. Thus, it is proposed that aneural cognition is of polyphyletic origin, but is built upon structures and system components with a deep shared evolutionary history.

From the perspective of adaptive value, the likelihood of observing aneural cognition in other algae depends upon their environmental context. Sensory accommodation and even rudimentary forms of decision-making are important aspects of dealing with nutrient fluctuations and environmental stresses more generally [9.30] [9.151]. For diatoms in general, physiological responses involve a large number of parameters that are dependent on various forms of sensation [9.187]. For example, the diversity of chemical stimuli encountered in the aquatic environment requires multiple fitness imperatives to be satisfied simultaneously. Changes in *Seminavis robusta* behavior provide an optimal response to both nutrient gradients and reproductive needs [9.30], which implies an internal mechanism for temporal information integration [9.131]. Among intertidal populations of benthic diatoms, the ability to adapt behaviorally to changes in light intensity is distinct from physiological adaptation to the same [9.43]. In *Thalassiosira pseudonana*, the divergent response among temperature-dependent responses evolved from 450 generations of experimental evolution points to a role in evolutionary constraints and tradeoffs in shaping environment-specific information processing [9.155].

We will now revisit the basis for associative learning, habituation, and energy minimization by looking at their natural connections to more formal mathematical representations such as psychophysics, Hebbian learning, and predictive processing. These explanatory frameworks provide a link to the more conventional interpretation of biophysical mechanisms. We will also make the connection between these processes and the concept of minimal cognition [9.207]. Minimal cognition has an effect on behavioral dynamics of the individual and perhaps even natural selection [9.87]. This will be placed in the context of diatoms, which may share a number of information processing properties with systems as diverse as *Physarum* [9.28], *Paramecium* [9.34], and even protocells [9.96].

9.1.3 Psychophysics as Diatom Information Processing

Psychophysics is the investigation of the relationship between mental phenomena and physical stimuli, particularly the minimal differences in the variation of a stimulus detectable by a nervous system [9.94]. Various psychophysical behaviors are observed across a wide range of species, including mice, goldfish, and humans [9.186] [9.197]. While it is hard to demonstrate the mental phenomenology of *Bacillaria* (or of diatoms more generally), the response to physical stimuli has been previously demonstrated in the form of sensitivity to light intensity [9.51] [9.52] [9.54]. In general, diatom cells are photophobic in conditions of high light intensity [9.51] [9.53]. Cross-species comparative work [9.54] [9.55] reveal multiple details about this phenomenon. In general, multiple parameters can be used to understand the psychophysical properties of diatom information processing.

Several sensory mechanisms are commonly found across diatom species. For example, all species respond to high levels of light negatively by moving away from the light. These movements are similar to phototaxis that are generated by both artificial and simple nervous systems [9.91] [9.176]. There are also characteristic response times for changes in direction due to light exposure despite differences in species-specific optimal light frequencies that can be linked to ecological adaptations [9.171]. There is also evidence of habituation in the duration and direction of how a cell is irradiated. Using a light spot assay, Cohn *et al.* [9.52] exposes the end of diatom cells to a light source, which leads to accumulations of cells in particular spatial locations at a characteristic rate. This rate is species-specific, but in cases where members of different species are in close proximity, this rate can be attenuated.

This research up until now has not been interpreted using a theoretical framework. In this work, a suitable psychophysical model typically describes a systematic response in the form of a mathematical model. In the case of *Bacillaria* light response, this description is both curvilinear and multidimensional. This suggests that psychophysical models are not a sufficient means to describe *Bacillaria* aneural cognition. Indeed, we also introduce a number of complementary models that view *Bacillaria* aneural cognition as an input-output system constituting an internal model. It is these input-output models, rather than the mental phenomenon per se, that characterizes the stimulus-response relationship. Viewing aneural cognition in terms of inputs and outputs also allows us to view aneural cognition in terms of broader phenomena such as biological homeostasis [9.24] or active inference and perception [9.18] [9.204] [9.222] at the behavioral level. In a generic sense, these inputs-outputs are often referred to as internal models [9.142] [9.158]. Therefore, our goal will be to review suitable mathematical models that provide a reasonable description of behaviors relative to a potential stimulus. With this definition in mind, there are a number of psychophysical phenomena that are demonstrated by *Bacillaria*, and perhaps even diatoms more generally. These include, but are not limited to, flow detection, resonance detection, obstacle detection [9.90] and phototaxis thresholding. More generally, there are four basic regularities of psychophysics as discussed in Lubashevsky [9.134]. The first regularity involves range effects leading to the overestimation of weak stimuli and the underestimation of strong stimuli. Secondly, a predictable response to environmental stimuli can be found in the form of the Weber-Fechner law. Related to this, the effects of sequential presentation can serve as the structure for a memory-encoding mechanism. Finally, dynamic range allows for behaviors that adapt to a range of states between the lower- and upper-bound extremes of stimulus intensities or sensory magnitudes. Psychophysical models allow for dynamic range to be modeled with respect to a stimulus of variable intensities presented over time.

These regularities will apply broadly to systems both with and without a nervous system. This leads us to two assumptions when applying the cognition metaphor to *Bacillaria*. The first assumption is that movement of a *Bacillaria* colony is driven by information processing. This information processing is neuronal in scope, which means that *Bacillaria* behavior is both reactive and adaptive. A second assumption involves a distinction between the behavior of *Bacillaria* and behavior generated by a network of neurons that produce electrical potentials. However, as our multitude of models will demonstrate, this may be a distinction without a functional difference.

9.1.4 Information Processing and Aneural Cognition

Traditionally, information processing in nervous systems and organisms more generally has been assessed using an information theoretic approach. For our purposes, this means the production, transmission, and storage of various output processes are considered from the perspective of probabilistic channels [9.214]. Information theory is particularly useful for examining multiple realizations of random processes [9.190]. In models of the nervous system, the type of modulation determines the information-carrying capacity of specific links [9.137]. When extended to aneural systems, the properties of connective components within and immediate to the cells themselves determine the rate of transmission. Yet information is about more than simply transmission. Stimulus structure is of great importance, and goes well beyond activating the sensory apparatus [9.82]. Part of this structure involves redundancy, and in a biological context serves to enrich interactions of the organism beyond simple reactivity [9.82].

Sensory and internal states also determine the rate of information processing. This has been characterized in terms of active inference, where organisms must minimize their free energy to retain their integrity as individuals [9.124]. However, this also minimizes information content characterized in the form of surprise. The minimization of surprise, or unexpected information from external states [9.78], protects an organism's internal states from drastic changes. The relationship between surprise and organismal information processing relies upon a Bayesian formulation of information as opposed to a probabilistic one [9.12]. This type of Bayesian approach, commonly utilized in predictive processing models, also provides a dynamic perspective on causality [9.1] [9.106]. Using this formulation, continually updating the internal state with information has a cost. This leads to fluctuations in information processing over time, and results in some behaviors having a higher utility than others [9.223].

9.1.5 Hebbian Intelligence and Predictive Processing

The assumptions posed in this paper also allow us to borrow approaches from the neuroscience literature to more broadly understand how behaviors are generated among *Bacillaria* colonies. This section will review two candidates for interpreting the mechanisms responsible for a colony's information processing capacity. The first candidate approach is Hebbian learning, or more broadly, Hebbian interactivity. Hebbian interactivity allows for higher-order patterns such as dynamic inhibition between processing units [9.156], which in the *Bacillaria* context might be single cells or intracellular units. Hebbian associations occur when two or more units behave in a synchronized manner, thus strengthening future associations. A somewhat related approach is predictive processing, or the ability of an internal model to anticipate future environmental events. This also involves association between processing units that explicitly represents predictions of sensory inputs based on comparing prior and current observations [9.36].

We can apply a version of Hebbian learning where the site of action is cell coupling sites rather than synapses. Hebbian learning can be summarized as the causal association of activated neurons [9.103], often summarized as "neurons that wire together, fire together" [9.194]. In our case, cells that are adjoint also provide each other with a learning signal. This learning signal provides the basis for a coordinated, sliding motion. Once a pair of cells learn movements of their neighbors, their independent phase oscillations become entrained to each other [9.64]. While this seems to happen as a result of mechanical constraints with the onset of movement generation, we ask a more fundamental question: How does this behavior become organized in the development of a *Bacillaria* colony? The answer is through a form of differential Hebbian learning [9.49]. Differential Hebbian learning is concerned with the temporal difference between activation and response rather than the causality of the activation itself. The synchronization of *Bacillaria* cells may indeed be due to active associative processes.

While Hebbian learning is a convenient metaphor and heuristic mechanism, this type of mechanism is not observed in diatoms. While the components can synchronize, diatoms use something more similar to aneural connectionist learning. Connectionist learning [9.98] involves the temporal association of two connected components. Even in neural systems, but particularly as a metaphor for aneural systems, Hebbian learning presents many drawbacks to producing a capacity for cognition. Hebbian models are typically oversimplified models of neural processing and not precise with respect to specific connected components. While this may not be important for aneural purposes, in their most general context, Hebbian mechanisms have been shown not to implement learning on their own [9.183].

In Hebbian models, feedback is captured in the form of updated weights between connected components. This can be superior to other models of simple learning, such as classic conditioning. Classic conditioning has been shown to exist experimentally in rabbits [9.221], sea slugs [9.86] [9.132], and other organisms from across animal phylogeny. In this form of learning, a stimulus results in a predictable response in only a few exposures to the stimulus. Yet even in this simple form of learning, there are multiple neural processing steps [9.86] [9.101]. What is needed for many forms of aneural cognition is a feedback loop that provides a predictive path between interacting components, environmental stimuli, and future behavior. Therefore, a stronger mechanism that establishes more intentional behavior in a behavioral substrate is active inference and predictive processing.

Approaching aneural intelligence from a broader perspective, we can turn to predictive processing for clues as to how the proposed network is able to maintain its coordination across the colony. The mechanism behind predictive processing is active inference [9.77] [9.150], which in turn is based on a free-energy minimization approach to constraining behavioral outputs [9.75]. The predictive processing and active inference approaches are particularly useful in cases where the colony shifts orientation and behavioral mode. This combined approach offers us a mathematical framework for action and perception [9.37] that is compatible with psychophysical laws describing a more general perception-action relationship. More specifically, active inference offers a formal feedback mechanism for the relationship between perception, action, and even learning [9.27]. This can in turn be more fully understood using more formal tools such as a PID controller [9.13] or a Bayesian model [9.145]. Overall, active inference allows us to build upon *Bacillaria* aneural cognition in two ways. Active inference makes an explicit connection between our behavioral analogue and biophysical processes, particularly energy minimization. Secondly, the link between active inference and predictive processing also provides a regulatory rationale for understanding processes such as prediction and anticipation.

Returning to our *Bacillaria* network: How is information processed in a way that not only maintains coordination over seconds to hours, but enables the network to operate under a wide range of conditions? While we address this issue later on, one way to hypothesize about this is to turn to the adaptive aspects of the network's nodes and edges. Processing occurs through a perceptual output, which then drives cell behaviors. These cell behaviors must be coordinated in a way that does not require a leader [9.81]. This requires a decentralized structure that must integrate information on multiple timescales.

9.2 Measurement Techniques

In this section, we will discuss various measurement techniques that allow us to uncover details about how *Bacillaria* processes information and produces adaptive behavior without a centralized nervous system. We will begin by introducing the Weber-Fechner law [9.136], then move to more complex psychophysical laws. Then we will talk about connectionist models of behavior generation, from artificial neural systems to decentralized networks that coordinate coherent and adaptive behaviors.

9.2.1 Weber-Fechner Law

We can apply the Weber-Fechner law to demonstrate how *Bacillaria* may utilize psychophysics to process environmental information. To accomplish this, we introduce a generalized version of a psychophysical law for measuring the magnitude of stimulus intensity. The Weber-Fechner law is a mathematical function that describes the ability to distinguish between light sources of different intensities. It can be stated as shown in Eq. (9.1):

$$\mathrm{p} = \mathrm{k}\left(\ln\frac{s}{s_0}\right) \tag{9.1}$$

where the sensation (input of S) is proportional to the logarithm of the stimulus intensity. S_0 is the reference stimulus and S is the change from the reference to the target stimulus. The term $\ln\frac{s}{s_0}$ is the log difference between a model of the stimulus and the observed stimulus. This can be thought of as a predictive process, or even as a differential process that allows for more sophisticated computation.

Given that a model of the stimulus is involved, it may also be possible to modify the discrimination capacity of the colony by introducing it to a wide range of stimulus intensities. Particularly as information from the sensory gradient becomes highly unexpected or qualitatively unusual, the response itself may be able to be sharpened. This can be demonstrated through biased random walk behaviors exhibited by motile bacteria with flagella in response to chemical gradients [9.200]. In diatoms, a similar effect may result as a consequence of greater experience with a wider range of phenomena. Alternately, feedback resulting from changes in shape memory is stored in the extracellular structure or raphe secretions, and acts as a contributor to cognition.

The Weber-Fechner expression can be extended to a number of other models, including input-output relationship [9.128] and active perception in cells [9.222]. In these cases, our measurement is not simply a difference, but indicative of more sophisticated indicators such as signal detection theory (SDT) and fold-change detection [9.3] [9.112] [9.118] [9.120]. SDT is an important mathematical model for understanding the Weber-Fechner regularity as a signal-to-noise ratio. In particular, the ability to conduct magnitude estimation [9.170] provides us with a route to laws and principles [9.45] [9.46] [9.195].

In terms of an input-output relationship, let us consider the input of a light or chemical gradient, and an output of a movement response. The movement is activated when a difference in the stimulus intensity is detected from one sampling point to the next. Movement will also increase or decrease when the intensity increases by a n-fold difference over a fixed unit of time. Fold-change of the stimulus (FC) can be described mathematically as

$$\mathrm{FC} = \log_2\left(\frac{t_1}{t_0}\right) \tag{9.2}$$

where t_0 is the initial condition and t_1 is the sensory input at the time point to be compared. In the case of cell colonies where each unit moves in a correlated fashion, it is useful to have a measure of feedforward and feedback with regard to sensory information taken in from the environment at a single cell and propagated across the colony. While the Weber-Fechner law allows us to make a connection between stimulus and response, this does not account for noise that might degrade or otherwise mask the environmental stimulus. We can thus turn to SDT to characterize the signal-to-noise ratio. In aneural systems, SDT is conceived of in a slightly different way than that of neural systems. The mathematical formulation of SDT involves two (generally Gaussian) distributions: one representing pure signal, and the other representing pure noise. When completely separated, we have a signal in the absence of noise. As the two distributions overlap, the degree of noise interacting with the signal increases. The primary measurement of SDT is d', which can be defined in the following form

$$d' = z(FP) - z(TP) \tag{9.3}$$

where d' is the difference between the z-scores of all false positives and all true positives in a series of discrete sensory observations. A true positive is the coupling between a sensory signal and an output behavior, whereas a false positive is the lack of coupling between a sensory signal and behavior. This lack of coupling is characterized by the generation of a behavior without a corresponding sensory signal. In a neural system, the causal relationship between sensory input and behavioral output is provided by an internal information processing mechanism defined by a nervous system. In aneural systems, this internal information processing mechanism can be defined much more broadly, often defined using a physicochemical dynamical system.

In Buckley *et al.* [9.37], two different types of density models can be used to approximate both the production of internal states and the generation of output behaviors. In the case of the former, so-called recognition (or R-) density is the mechanism by which the brain encodes environmental states. R-density provides a means for which environmental information is encoded as a collection of internal states. This process can be captured through a Bayesian model that consists of a prior distribution, environmental samples (or empirical observations), and a posterior distribution that updates the prior distribution with the current observation. The Bayesian model is essentially a statistical feedback model. By contrast, generative (or G-) density uses a statistical distribution to produce behavior in a feedforward manner. In many cases, a Gaussian model is sufficient in producing a behavioral output representing oscillatory movement.

9.2.2 Connectionist Network

The network formed through the cytoskeleton and extracellular components of a *Bacillaria* colony can be modeled as a special class of connectionist network. Using a network as a means to connect cells and their components of movement allows us to build an internal model with inputs, outputs, processing units, and interactions. In this way, certain behaviors should yield a linear or curvilinear output function that corresponds in a systematic way to sensory input.

Figure 9.2 demonstrates the presence of light and the effects of this light source on the proposed aneural internal model. This is characterized by a hypothetical connectionist network of points across the surface of each cell in the colony. Rather than representing connections between synapses, however, our network is a set of interacting cytoplasms of individual cells. Given what we know about the generalized diatom response to light intensity, this model of interactions provides us with a way to find universal patterns of movement modalities over a variety of raphid diatoms that form a multicellular colony.

9.2.3 Algorithmic Information

To further characterize halting behavior in colonies, and to differentiate between continual movement and instances of halting, we can utilize a measure of algorithmic information. Algorithmic information can be quantified as the length of the shortest description of a sequence of events, and in various applications allows us to capture information content in

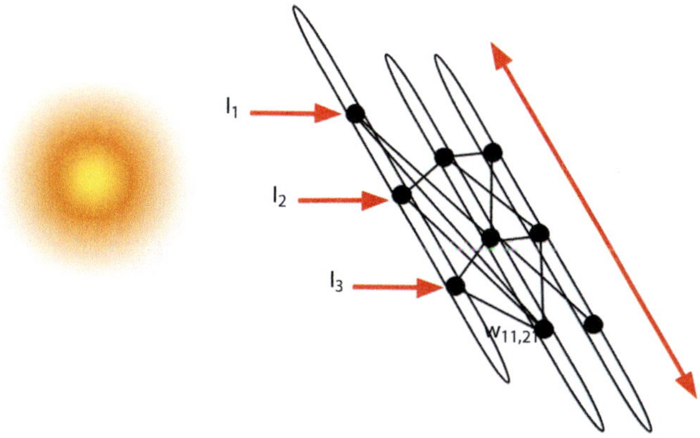

Figure 9.2 Sensory thresholds (just noticeable differences). Differences in light intensity measured by output movement weights ($W_{mm,nn}$). Network shown as edges and nodes that connect all cells in the colony.

both the probabilistic (Shannon) and dynamic (Bayesian) sense [9.58]. Mathematically, we can use the definition of Kolmogorov [9.125], stated as

$$k(x) = \min(l\,(p)) \tag{9.4}$$

where p is the length of a description in bits, and $\min(l(p))$ is the shortest lossless description, or minimal description retaining all the information of larger descriptions. To yield a description, we want to encode the string as a conventional program, a time-series measurement, or a short equation. The use of an equation allows us to plug in our formulation of the Weber-Fechner law, but descriptions involving sensory inputs and/or behavioral outputs can also be plugged into any number of psychophysical models.

In a computational context, the description measure must run on a machine that transforms the description code into something observable. In the context of diatom morphology, we can bridge the length of a binary string and a complex phenotype by treating cell colony movement as a model that exhibits the Principle of Computational Equivalence [9.152] [9.217]. Let us consider the string in question as a momentum vector measured along the trajectory of a single cell moving in concert with other cells in the colony. When considered as a series of discrete states, this sequence can be evaluated both in terms of its absolute minimum length (represented in terms of a pure sinusoid with no zero or negative momentum) and halting behavior.

9.2.4 Collective Pattern Generator

Figure 9.3 takes our connectionist network a bit further by proposing a specific circuit that operates as a collective pattern generator (CoPG). In general, signals between cells prompt pattern formation and other forms of intercellular coordination. This is demonstrated by Portegys [9.168] using a 2D cellular automata. A CoPG can be defined as a distributed network of oscillators coordinated by its periodic behavior. CoPGs are analogous to central pattern generators (CPGs;, [9.9] [9.139]), which allow for periodic signals to emerge from

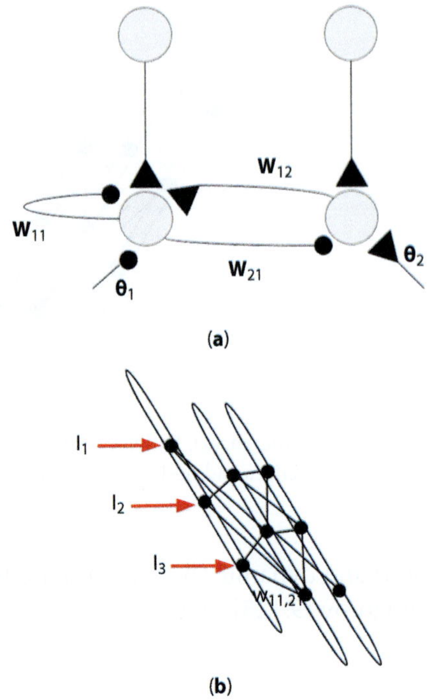

Figure 9.3 Deviations from the collective pattern generator (CoPG). Difference between a central pattern generator (CPG), characterizing a neural circuit, and collective pattern generator (CoPG), characterizing collective behavior.

distributed networks of excitable cells (Figure 9.3). Examples of CPGs can be found in the movement of insect locomotion or cardiac pacemaking [9.59] [9.138].

9.2.5 Dynamical States of the CoPG

One alternative to thinking about CoPGs as a connectionist model is to contrast them with CPGs. While CPGs are generally consistent across the lifespan, being disrupted only in times of injury or neurodegeneration, there are not a lot of good models for understanding their dissolution. In *Bacillaria* and their associated CoPGs, however, we consistently see degeneration of oscillatory behavior. We propose that this halting behavior is driven by a dynamical state transition occurring at points where the colony is fully elongated (Figure 9.1a) and/or fully contracted (Figure 9.1b). At either point in the oscillation, we observe either a gradual contraction at a characteristic rate, or halting behavior that freezes the colony in this position for a random interval of time (Figure 9.4).

9.3 CPGs vs. CoPGs

We now return to the notion of CoPGs as an analogy with generators of cyclical electrical activity (cf. [9.174]). While it has been established that electrophysiology plays a role in diatom cellular biology, it is unknown to what extent significant electrophysiological activity

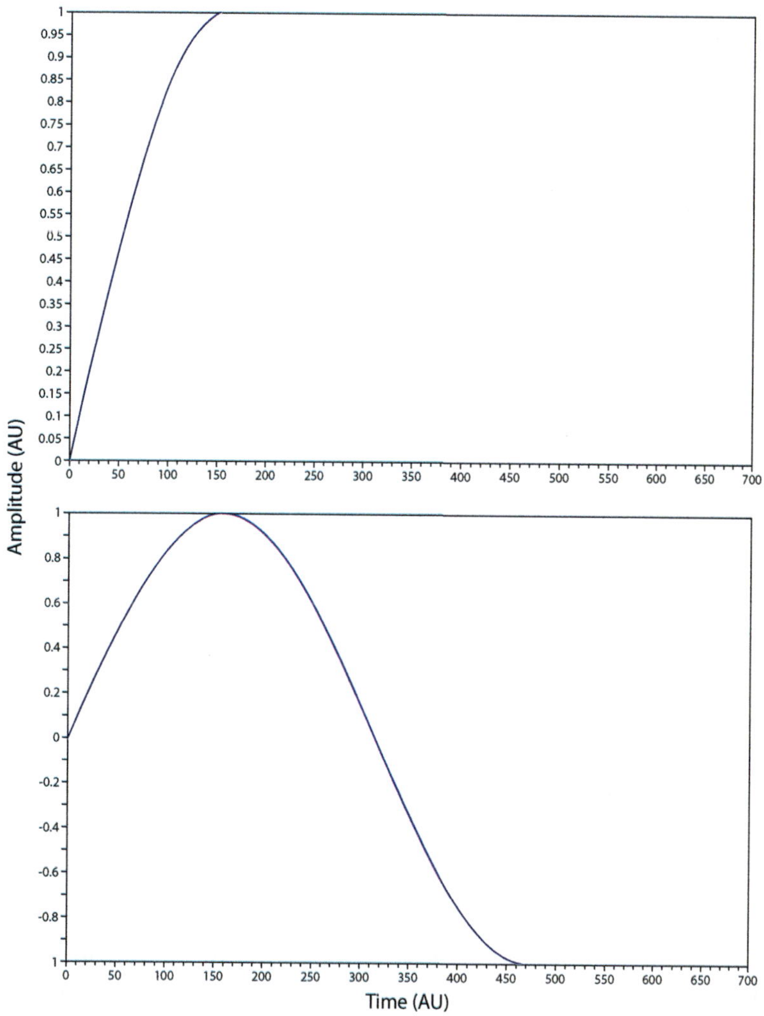

Figure 9.4 Movement trace of a single cell that halts at different points in a single oscillation. (top) Sine wave with tangent at quarter-phase; (bottom) Sine wave with tangent at three-quarter phase.

plays a role in *Bacillaria* rhythmic movement. In the diatom species *Coscinodiscus wailesii*, electrophysiological activity is enhanced after nitrogen starvation [9.32]. Unlike the case with coordinated CPGs, *Bacillaria* does not utilize action potentials. An example of how CoPGs might work comes from the pennate diatom *Pseudo-nitzschia fraudulenta* [9.181]. In this species, single cells use paracrine signaling to coordinate Ca^{2+} dynamics between cells. More generally, ion channel flux is driven by exposure to light, chemicals, motility, and temperature [9.29] [9.202]. This is similar to what we observe in a broader group of diatoms in terms of behavioral response to stimulus gradients.

Now that we have considered the potential role of electrophysiology in *Bacillaria* movement, as well as its connection to collective behavior and psychophysics, we can find the right level of analysis of this oscillatory behavior. While ion channel activity does not lead to action potentials, there are nonetheless relationships between electrophysiology and forms of aneural cognition in a variety of organisms and life-history contexts [9.31] [9.184]. Rather than including physiological detail in our model, we will proceed through the use of sine waves, sinusoidals, and hybrid sinusoidal-tangent functions. This maps to various features of the cell phenotype including image processing bounding boxes that define the edges of each cell and cell centroids that move in space relative to one another [9.5]. This provides a framework we call the digital *Bacillaria* that provides a quantitative framework for a dynamic model.

Each pair of digital *Bacillaria* cells acts as an oscillatory unit in a CoPG, in which a single cell oscillates. Oscillatory units overlap, so that a colony of three cells consists of two oscillatory units as shown in Figure 9.3b. While these oscillatory units (or pairs of cells) generally produce a sine wave, they can also stretch to a maximal value and stay there for long periods of time. Initially, we introduce chains of x interconnected oscillatory units. At subsequent time steps, we model the modes of movement behavior using a sinusoidal function as the baseline condition. We can also introduce an offset of phase in movement between neighboring oscillatory units, so that any two units move at different rates. Noise in the form of higher-order changes in velocity and acceleration can also be introduced, particularly at the extent of the oscillation.

How does this compare with models of CPGs and collective biological locomotion? We can look to stick insects [9.56] to understand synergistic CPGs (cf. [9.11] [9.38] [9.57] [9.69]). In this case of the genus *Phasmatodea*, the movement of multiple legs are integrated into a neuronal network. The activity of this network is generated by nerves in each leg, and the bilateral symmetry of the insect body results in an oscillatory electrical signal. The measurement of nerve activity is electrical activity exhibiting significant complexity. Not only is there information at multiple time scales, but noise can be both introduced through the regulatory capacity of peripheral sensory information [9.21] and dampened by ion channel activity [9.88]. Yet it is essentially oscillatory, and allows for movement to be coordinated for each pair of legs. In this sense, we can return to *Bacillaria* and our connectionist model of collective movement. In the CoPG case, we also see pairwise coordination between units of analysis. However, the oscillations can be much simpler, and the regulatory mechanisms less complex. Nevertheless, we can use an aneural model and psychophysical estimation to understand how this integration between oscillatory elements is coordinated over time, particularly in the face of environmental stimuli.

To demonstrate how we can model *Bacillaria* psychophysics, we introduce a series of characterizations of cell movement in a pairwise, nearest-neighbor cell comparison. The movement of two cells oscillating against one another is characterized by two overlapping sine waves. Yet before we discuss the overlapping case, let us look at the case of a single oscillator that halts after ¼ and ¾ phase of a single cycle (Figure 9.4). This halting behavior results in a tangent representing the cessation of anterior-posterior movement. Figure 9.5 features overlapping sine waves without halting between two neighboring cells for the ¼ and ¾ phase cases. We can add random noise to the sine wave to approximate noisy movement (Figure 9.6, top). The addition of noise (in this case Gaussian) represents changes in acceleration, orientation, and colonial integration over time. The non-halting pairwise case can be mapped to an attractor map. Figure 9.6 (bottom) shows the signal for an idealized movement trace (sine wave), which does not fully represent biological movement. Nevertheless, we can use such a comparison to understand the pairwise relationships that define a CoPG.

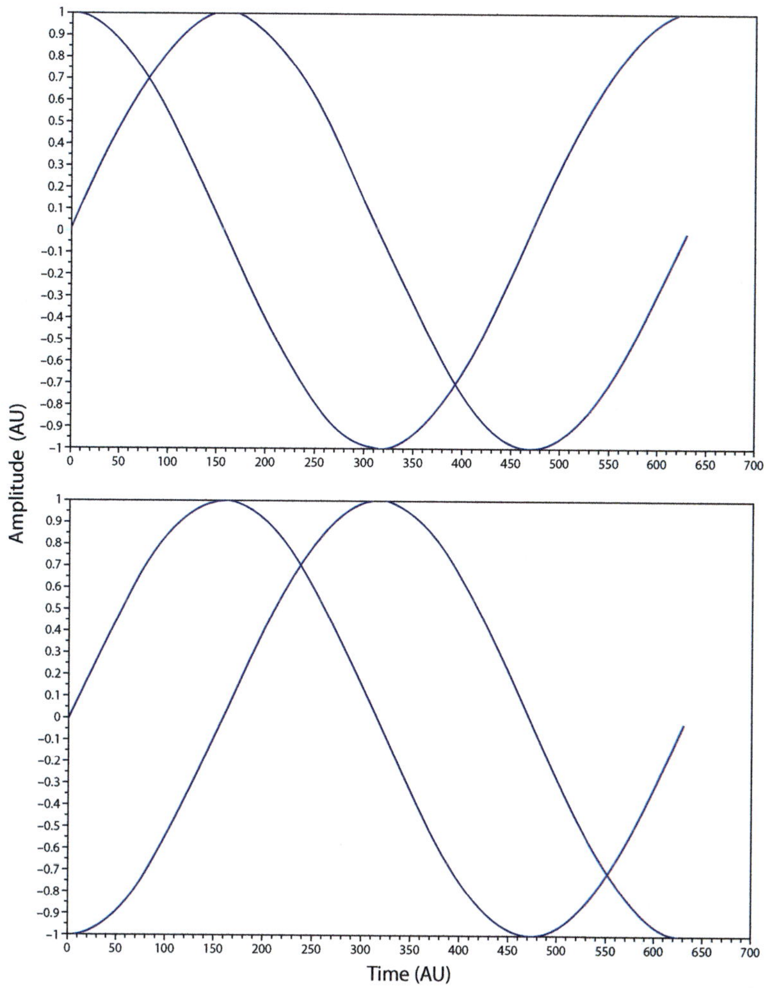

Figure 9.5 Movement trace of a single cell that halts at different points in a single oscillation. (top) Two adjacent and overlapping pairs of oscillatory cell movements (sine waves, three-quarter phase); (bottom) Two adjacent and overlapping pairs of oscillatory cell movements (sine waves, quarter-phase).

The non-halting case provides us with a full cycle that characterizes the information processing potential of a CoPG. While the full cycle is modeled using a pure sine wave, prior analysis of these cell tracking signatures shows instabilities and higher-order noise at both minimal and maximal amplitudes [9.5]. This may represent a regulatory role for noise in the stretch dynamics of each cell. From an information processing perspective, the interaction between cells results in an offset between the trace of neighboring cells. Changes in the strength of an environmental stimulus (cf. [9.19] [9.25] [9.102] [9.179] [9.224]) should change aspects of these traces. In some cases, the offset might be reduced either over time or sporadically. Alternatively, instability at the amplitude extremes with an additive effect of environmental signal might lead to halting in subsequent movement cycles.

Halting behavior and its variants can be captured through the application of Algorithmic Information Theory. Likewise. AIT can be used to better understand the relationship

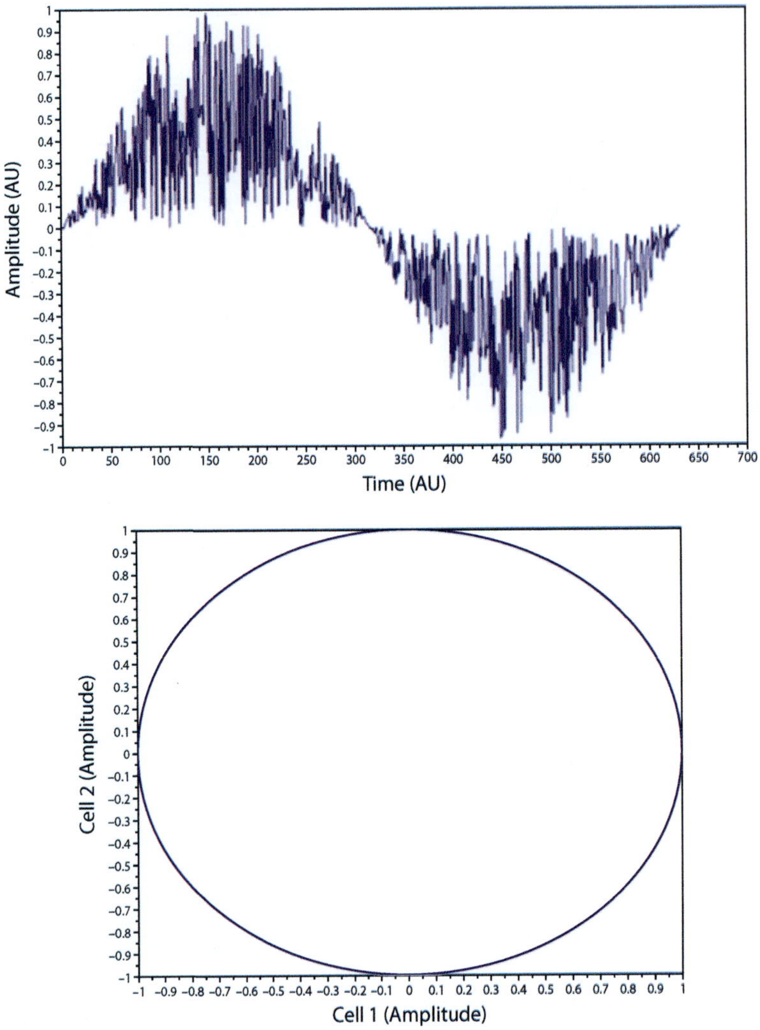

Figure 9.6 Single oscillations with noise and compared with another cell. (top) Sine wave with random (white) noise; (bottom) Attractor map (based on sine waves, quarter-phase).

between out-of-phase behavior and halting, particularly in cases where pairs of cells begin to oscillate at different phases and desynchronize [9.149]. *Bacillaria paxillifer* colonies demonstrate a very specific halting behavior related to a photokinetic response: during light periods, the colony is motile, while during dark periods, cells in the colony are arrhythmic and align in a stacked formation [9.116]. In some diatom species, it has been found that a chemical signal from predators can lead to a reduction in colony size (number of cells in colony) measured as chain length [9.179]. In solving this conundrum, AIT is helpful in two ways. First, AIT can describe the length of the colony in terms of optimal information given specific environmental conditions. Second, suppose that each cell in a colony has a set of instructions that guide behavior. When these instructions match (either in length or in type), cells are synchronized. When these instruction sets become mismatched, then desynchronization may occur. Reductions in colony size might become either a runaway or dampened process under the right environmental conditions, and one that is best explained by information processing dynamics.

In a case where pairs of cells begin to lag out-of-phase, this can have one of two outcomes. Either the instability continues and the colony becomes incoherent, or the instability is overcome, and the oscillations return to a stable state. A question for future research is whether colonial movement must be tightly synchronized or whether it can exist in a stable and persistent lagged state. If the former is true, excessive lag might provide a so-called tipping point for colony dissolution, perhaps based on immediate or temporally-integrated environmental cues. AIT might also help us understand this by understanding the set of behavioral instructions shared between cells that leads to stability.

9.3.1 Potential of Predictive Processing

Predictive processing provides us with more than a quantitative framework to analyze the relationship between brain and cognition. Theoretical frameworks such as the Free Energy Principle (FEP, [9.76]) may also provide a means to understand aneural systems. The connections between Markov blankets, energy minimization, and the processing of information by a morphological system that minimizes its energetic expenditures with respect to movement, are particularly interesting. This connects psychophysics and biological physics in a unique and informative manner.

For example, a Markov blanket can be defined as all variables which exclusively and conditionally predict a random variable of interest. This subset of variables is independent of all other variables in the system [9.76]. In the case of *Bacillaria* motility, the group of variables that predict oscillatory motion might include photic inputs, water chemistry, mechanical forces, predation, orientation to light [9.140] and resonance from other cells. Moreover, these variables act as conditional probabilities to predict the likelihood of movement in response to a stimulus at any given time. Formulating *Bacillaria* motility in this way also allows us to see how some variables are not at all predictive of coherent movement. This is connected to energy minimization in the sense that movement patterns will be optimized according to the strongest informational signal available. In other words, if the driving strength of resonance is stronger than that of the photic signal, then this should affect the phase, smoothness, and continuity of the movement output.

9.3.2 Phase Transitions in *Bacillaria* Movement

In the course of considering aneural cognition in diatoms (*Bacillaria*), we have considered not only the role of psychophysics, but the nature of colonial movement and intentionality of movements as well. We have proposed that *Bacillaria* movement can be characterized as a CoPG, which is the aneural analogue to a CPG. Like the CPG, the CoPG can be represented using a connectionist network. Unlike the CPG, the CoPG can undergo phase transitions [9.119] that enable halting behaviors without associated dissolution or separation phenotypes [9.17]. In the underlying representation, this is enabled by the structure of our connectionist network: the weights between connective components represent both intracellular microstructure and the extracellular milieu rather than synaptic connectivity driven by changes in the coupled ion channel activity. As such, CoPGs and their associated connectionist networks can represent non-equilibrium phenomena such as the decoupling of neighboring cells, damping of forward acceleration, and even halting of the entire colony at certain phases of movement.

The analogy between CPG and CoPG suggests that simple oscillators can provide a means for intelligent behavior, or at the very least adaptive information processing. In the case of CoPGs, the oscillatory behavior provides a deterministic signal that entrains behavior and thus is not directly associated with information content. On the other hand, phase transitions such as the switch from oscillation at a characteristic speed and halting at random intervals [9.174] may provide information content to individual cell and colony-wide behaviors. Moreover, different modes of dynamic and static behavior have different types of information content. While stretching and halting is part of a colony-wide phase transition, this behavior also enables symmetry breaking at the level of coupled cells, where the bonds between cells dissolve and irrevocably shape the future dynamics of a colony.

One alternative means to explore the rhythmicity of collective behavior is to understand cell movement at the colony level as an array of Kuramoto oscillators [9.2] [9.33]. While we will not explore this hypothesis in detail, a system of coupled oscillators [9.215] is an alternative to the connectionist models as introduced here, and allows us to model synchronized activity in a way that is consistent with the neurocognitive metaphor. This type of model stands in parallel to our CoPG approach, and describes synchronization rather than the rhythmicity of a cell colony. In our more generic connectionist model, nodes (neuronal units) act as aneural oscillators, but such oscillatory behavior is not the only type of coherent behavioral output.

Our aneural model of movement behavior has only one requirement: as the intensity of a sensory input varies by intensity, a corresponding intensity of behavioral response must also occur. We can also return to the concept of stigmergy to understand the relationship between extracellular components [9.180] and internal physiological mechanisms. The connectionist model in particular also reveals how extracellular components adjacent to different parts of the colony can serve a heterogeneous role in aneural cognition. This is an important feature that allows for a coherent psychophysical response that is regulated depending on the environmental context. Taken together, this provides a means to extrapolate even further, from aneural cognition to the regulation of aneural cognition.

9.4 Aneural Regulation

Now that we have considered the potential for aneural information processing in *Bacillaria* and demonstrated how we might understand and measure this phenomenon, we are ready to propose a potential path towards the regulation of aneural cognition. The first step involves Hebbian learning involving environmental (sensory) inputs. In aneural Hebbian learning, units that are active simultaneously couple their behavior. Over time, this synchronization is reinforced. In *Bacillaria*, each cell in the colony exhibits sliding motion. This sliding motion is coupled to neighboring oscillatory units and establishes Hebbian learning. This can be demonstrated computationally using our connectionist model. This coupling results in a rhythmic oscillation that may occur at multiple timescales and can be approximated by our CoPG model. The last step in our path is how feedback is regulated. In diatom sensory systems, feedback comes in a number of forms: from shape memories of gradient position to physical resonance between oscillating cells. Yet regulation often comes in more universal forms that are well-represented in models such as CoPGs, forms of Hebbian interactivity such as learning, and the free energy principle. The role of direct sensory information and feedback can be summarized by the concept of direct perception [9.85]. While direct perception has not typically been applied to aneural systems, the concept allows us to think of the relationship between cells, organisms, and their environment. The notion that the environment provides dynamic structure for goal-directed behaviors and information processing is attractive for describing behavior of organisms that do not have a nervous system.

Our approach describes only one mode of *Bacillaria* behavior, and only briefly draws from potential physiological mechanisms. In the initial step of the above model, neuronal behavior is generated by an analogy to a decentralized nervous system with electrical activity. This involves both intra- and extracellular mechanisms that leverage internal information processing and is augmented by stigmergy. We can explain a variety of behavioral modes using this model, including colony halting. This halting mode of behavior can be approximated using Algorithmic Information Theory. Eventually, this might lead to the development of a biological Turing test [9.74] [9.188] [9.189] for distinguishing neuronal-generated from aneural-generated behavior. Yet such a definitive standard for distinguishing neural behaviors from simulated ones requires us to address what exactly demarcates real behavior from simulated (but real-life looking behavior). We can generate realistic simulated behaviors using physics engines [9.130] [9.163] and a class of agents called animats [9.26] [9.213]. But the connection between the underlying mechanisms of such behaviors has not been properly established. Therefore, to better understand what kinds of behaviors are possible with aneural architectures and their relative complexity, we will discuss the phenomenon of pseudo-intelligence.

9.5 Broader Picture of Intelligence and Emergence

9.5.1 Pseudo-Intelligence

To place the contemporary concepts in neuroscience presented here in perspective, we return to the fact that some seemingly neuronal behaviors are generated by cellular motors along with other components of the colonial cellular matrix [9.90] [9.174]. The question might be asked: Is the synchronization of movements across cells in the colony simply the result of mechanical interactions, or is there a greater degree of autonomous behavior at work? If the former is the predominant explanation, then aneural cognition is merely a form of pseudo-intelligence. Yet if the latter holds true, then we face a dilemma of whether or not our output is cognitively realistic. Drawing this distinction relies on a definition of intelligence as a nonlinear and interactive process that is difficult to control using standard methodologies. *Bacillaria* colonies fall into the category of highly interactive, nonlinear systems [9.162]. But how do we understand the distinction between intelligent behavior and mimicry? In another context, Li [9.133] defines strong artificial life as simulated life that exhibits characteristics of real life. This description provides a standard for interpreting the generation of realistic neural behaviors using aneural systems. In investigating realistic vs. fake-looking artificial life [9.216], contestants used methods to distinguish between movement trajectories from living organisms (sharks, ants, spiders, and jaguars) and non-living

Table 9.1 A demonstration of how autonomy can be partitioned into generated and observed components.

		Generated by	
		Neural	**Aneural**
Instance of Behavior	**Neural**	Brain	*Bacillaria* chains
	Aneural	Jellyfish	Steam Engine

agents (simulated bird flocks and insect swarms, artificial chemical compounds). Among the examples listed above, living organisms generate their movement behaviors through biologically neuronal systems. This can be contrasted with the latter examples, which utilize either non-biological neural or biologically aneural systems. The general relation between neuronal and aneural systems can be partitioned as in Table 9.1.

Table 9.1 demonstrates that any single instance of behavior (feeding, taxis movements, quiescence) can be classified as a neural or aneural behavior, which can be generated by neural or aneural processes. This provides us with a classification problem that can be fit into an SDT model. For example, if a neural behavior is generated by a neural process, then it is a true positive. The true positive is produced entirely through the intentional component. *Bacillaria* occupy a unique position in this typology: a neuronal instance of behavior generated by aneural processes. This is considered a false positive, as the behavior is generated through both an intentional signal and an unintentional component (which roughly maps to the signal and noise components of the SDT model). Both neural and aneural behaviors can be simulated using deep learning and/or reinforcement learning techniques. This will also allow us to propose different neuronal-like cooperative behavioral states generated in our aneural context. Such behavioral states might include feedforward movement generation, density-dependent feedback, light-responsive feedback, and resonant feedback.

To further understand the role of halting and persistence of movement in *Bacillaria* colonies, it is worth expanding beyond the cognition metaphor and consider the connection between Turing machines and more formal mathematical formulations of algorithmic information. Specifically, algorithmic information theory [9.44] may allow us to approximate whether these patterns are random or something that approximates perceptual information [9.83]. If the entire sequence of behaviors observed is random, then the behaviors captured by the CoPG model might be due to exogenous forces such as colonial aging or hydrodynamics. If it is not random, then there is some endogenous control that might be due to connectionist network dynamics or a consequence of sensory inputs captured using a Weber-Fechner law analogue.

9.6 Discussion

We suggest that the motion of *Bacillaria* colonies is neuronal behavior generated by aneural processes. Some have suggested that physical principles can be used to understand how complex spatial information is integrated among slime molds and ciliates [9.191].

Here, we not only suggest that formal information processing mechanisms are important for understanding the generative outputs of internal processing, but also provide a series of formal approaches for modeling and analysis. While we begin with a reliance on simple psychophysical models such as the Weber-Fechner law and models that make a distinction between signal and noise (e.g., SDT), generative mechanisms such as Hebbian learning and predictive processing provide a more formal means to describe the production of a large number of degrees of freedom that enable adaptive behaviors [9.100]. Connectionist models also allow us to bridge between cells and the effects of sensory stimuli on the interactions between cells as they each respond internally to a stimulus from their own point of view. As a connectionist model allows for Hebbian interactivity and extracellular information processing, the adaptive behaviors that result represent alternate paths to the same problem. Ultimately, externalized network behavior (e.g., CoPG) unfolds over time in terms of the adjacent possible [9.35] [9.147] [9.203] and the output of such models as emergent phenomena [9.72] [9.148] [9.182]. We also consider oscillatory halting behaviors through the lens of algorithmic information, and consider how aneural information processing is a form of pseudo-intelligence.

Bacillaria colonies form by cells dividing as a clone in which cells adhere to one another, forming an accordion-like structure. As they age, *Bacillaria* colonies break. These colonies also undergo periods of extended elongation and/or quiescence in other positions. Despite periods of colonial breakage, we might be able to nonetheless utilize an aneural cognitive approach to understand if coherent behavior breaks down during this process. In particular, the uncoupling of predictive processing and the corresponding lack of tight feedback between cells in a colony may predict aperiodic oscillations by the proposed connectionist CoPG model. This leads us to an understanding of halting and dissolution behaviors using other techniques such as AIT. But in fact, colonies may simply break due to turbulence (as with a blender (Harbich, 2020, personal communication). Additional experimental validation and modeling may provide a definitive answer.

Now that we have posited a form of aneural cognition for diatoms and walked through a series of models demonstrating a collection of information-processing mechanisms, we can return to our original motivation. What does the psychophysical milieu of diatoms look like? *Bacillaria* psychophysics results from a light-dependent input/output curve for a single cell. This quantitative regularity determines how much input results in a corresponding degree of output. An alternate way to approach this is to apply one of our candidate psychophysical models to a behaving diatom colony, which yields predictable movement and even shape behaviors. One example of this is to build a receiver operator characteristics (ROC) curve from our SDT model, which yields the colony's ability to detect signal from noise. This might allow us to assess whether or not an aneural behavior is similar to neural processing. Yet a third way to answer this question is to consider the psychophysical milieu for different modes of colony (or collective cell) movement. This can be measured using an approach such as AIT, which enables us to measure halting behaviors and understand discrete temporal transitions in the internal state.

An alternate way to approach this is to utilize computational agents of simple neural systems as an analogue for what might be possible in aneural systems. We can base this on work involving developmental Braitenberg vehicles (dBVs [9.65]). These dBVs are computational agents where the brain is gradually constructed over the course of development. During this progression, behaviors come online as neurons and their connections emerge.

This in turn provides an indicator as to what constitutes the minimal components of a small nervous system [9.4] [9.104] as well what might be controllable with an aneural analogue. Using the dBV as a comparison also allows us to leverage the power of Braitenberg vehicles as they emerge in development in order to determine the complexity threshold between goal-directed neurally-produced behaviors [9.39] and incoherent behaviors.

In this study, we have taken the position that organisms in the genus *Bacillaria* exhibit information processing capabilities that resemble cognition, despite the lack of a formal nervous system. The evidence for this rests on two aspects of *Bacillaria* lifestyle: the ability to adaptively respond to sensory stimuli, and the collective coordination of behavioral output across cells constituting a colony. We bring several paradigms for adaptive information processing to bear on this problem. These include the application of psychophysical laws, Hebbian interactivity, connectionist models, and predictive processing. Taken together, these models, mathematical tools, and novel conceptual models provide us with new ways to understand the world of diatoms.

Acknowledgments

We would like to thank Dr. Thomas Harbich for his work on microscopy observations of *Bacillaria*, and the DevoWorm group for their feedback and suggestions. We would also like to thank the Neuromatch Conference for the chance to present early versions of this work. Thanks also go to Akshara Gopi, Ujjwal Singh, and Asmit Singh for their contribution to previous versions of this work.

References

[9.1] Aalen, O.O., Røysland, K., Gran, J.M. and Ledergerber, B. (2012) Causality, mediation and time: a dynamic viewpoint. *Journal of the Royal Statistical Society. Series A (Statistics in Society)* 175(4), 831-861.

[9.2] Acebrón, J.A., Bonilla, L.L., Vicente, C.J.P., Ritort, F. and Spigler, R. (2005) The Kuramoto model: A simple paradigm for synchronization phenomena. *Reviews of Modern Physics* 77, 137-185.

[9.3] Adler, M. and Alon, U. (2017) Fold-change detection in biological systems. *Current Opinion in Systems Biology* 8, 81-89.

[9.4] Alicea, B. (2020) Raising the connectome: The emergence of neuronal activity and behavior in *Caenorhabditis elegans*. *Frontiers in Cellular Neuroscience* 14, #524791.

[9.5] Alicea, B., Gordon, R., Harbich, T., Singh, U., Singh, A. and Varma, V. (2021) Towards a digital diatom: Image processing and deep learninganalysis of *Bacillaria paradoxa* dynamic morphology. In: *Diatom Gliding Motility [DIGM, Volume in the series: Diatoms: Biology & Applications, series editors: Richard Gordon & Joseph Seckbach]*. S.A. Cohn, K.M. Manoylov and R. Gordon, (eds.) Wiley-Scrivener, Beverly, MA, USA: 223-248.

[9.6] ALMuseum. (2009) クサリケイソウの舞; Dancing *Bacillaria* [movie]. https://www.youtube.com/watch?v=PUCSTiwZZ18

[9.7] Amato, A., Orsini, L., D'Alelio, D. and Montresor, M. (2005) Life cycle, size reduction patterns, and ultrastructure of the pennate planktonic diatom *Pseudo-nitzschia delicatissima* (Bacillariophyceae). *Journal of Phycology* 41, 542-556.

[9.8] Arendt, D., Tessmar-Raible, K., Snyman, H., Dorresteijn Adriaan, W. and Wittbrodt, J. (2004) Ciliary photoreceptors with a vertebrate-type opsin in an invertebrate brain. *Science* 306(5697), 869-871.

[9.9] Arshavsky, Y.I. (2003) Cellular and network properties in the functioning of the nervous system: from central pattern generators to cognition. *Brain Research Reviews* 41(2), 229-267.

[9.10] Arvetis. (2020) *Bacillaria* colony dancing. https://www.youtube.com/watch?v=r80s7fogkgM

[9.11] Ayali, A., Couzin-Fuchs, E., David, I., Gal, O., Holmes, P. and Knebel, D. (2015) Sensory feedback in cockroach locomotion: current knowledge and open questions. *Journal of Comparative Physiology a-Neuroethology Sensory Neural and Behavioral Physiology* 201(9), 841-850.

[9.12] Baldi, P. and Itti, L. (2010) Of bits and wows: A Bayesian theory of surprise with applications to attention. *Neural Netw* 23(5), 649-666.

[9.13] Baltieri, M. and Buckley, C.L. (2019) PID control as a process of active inference with linear generative models. *Entropy* 21(3), #257.

[9.14] Baluška, F. and Levin, M. (2016) On having no head: Cognition throughout biological systems. *Frontiers in Psychology* 7, #902.

[9.15] Basili, D. (2014) *Bacillaria paxillifer* [movie]. https://www.youtube.com/watch?v=SPfhjazILYA

[9.16] Beni, M.D. (2022) A tale of two architectures: Free energy, its models, and modularity. *Consciousness and cognition* 98, #103257.

[9.17] Bentley, K., Clack, C. and Cox, E.J. (2012) Diatom colony formation: A computational study predicts a single mechanism can produce both linkage and separation valves due to an environmental switch. *Journal of Phycology* 48(3), 716-728.

[9.18] Bentley, K., Harrington, K. and Regan, E. (2014) Can active perception generate bistability? Heterogeneous collective dynamics and vascular patterning. In: *ALIFE 14: The Fourteenth International Conference on the Synthesis and Simulation of Living Systems*. 328-335.

[9.19] Bergkvist, J., Thor, P., Jakobsen, H.H., Wangberg, S.A. and Selander, E. (2012) Grazer-induced chain length plasticity reduces grazing risk in a marine diatom. *Limnology and Oceanography* 57(1), 318-324.

[9.20] Bich, L. (2020) Interactive models in synthetic biology: Exploring biological and cognitive inter-identities. *Frontiers in Psychology* 11, #682.

[9.21] Bidaye, S.S., Bockemühl, T. and Büschges, A. (2018) Six-legged walking in insects: how CPGs, peripheral feedback, and descending signals generate coordinated and adaptive motor rhythms. *Journal of Neurophysiology,* 119(2), 459-475.

[9.22] Binet, A. (1888) *The Psychic Life of Micro-Organisms. A Study in Experimental Psychology.* Longmans, Green, and Company, New York, NY, USA.

[9.23] BirdWhisperer46. (2012) *Bacillaria paxillifer* - Is Not a Caterpilififer [movie]. https://www.youtube.com/watch?v=klMGTnqMwfg

[9.24] Biswas, D. and Iglesias, P.A. (2021) Sensitivity minimization, biological homeostasis and information theory. *Biological Cybernetics* 115, 103-113.

[9.25] Bjaerke, O., Jonsson, P.R., Alam, A. and Selander, E. (2015) Is chain length in phytoplankton regulated to evade predation? *Journal of Plankton Research* 37(6), 1110-1119.

[9.26] Boden, M.A. (2006) *Mind as Machine: A History of Cognitive Science.* Clarendon Press.

[9.27] Bogacz, R. (2017) A tutorial on the free-energy framework for modelling perception and learning. *Journal of Mathematical Psychology* 76(Pt B), 198-211.

[9.28] Boisseau, R.P., Vogel, D. and Dussutour, A. (2016) Habituation in non-neural organisms: Evidence from slime moulds. *Proceedings of the Royal Society B: Biological Sciences* 283(1829), #20160446.

[9.29] Bondoc, K.G., Heuschele, J., Gillard, J., Vyverman, W. and Pohnert, G. (2016) Selective silicate-directed motility in diatoms. *Nature Communications* 7, #10540.

[9.30] Bondoc, K.G.V., Lembke, C., Lang, S.N., Germerodt, S., Schuster, S., Vyverman, W. and Pohnert, G. (2019) Decision-making of the benthic diatom *Seminavis robusta* searching for inorganic nutrients and pheromones. *The ISME Journal* 13(2), 537-546.

[9.31] Bonzanni, M., Rouleau, N., Levin, M. and Kaplan, D.L. (2020) Optogenetically induced cellular habituation in non-neuronal cells. *PLoS One* 15(1), #e0227230.

[9.32] Boyd, C.N. and Gradmann, D. (1999) Electrophysiology of the marine diatom *Coscinodiscus wailesii* III. Uptake of nitrate and ammonium. *Journal of Experimental Botany* 50(333), 461-467.

[9.33] Breakspear, M., Heitmann, S. and Daffertshofer, A. (2010) Generative models of cortical oscillations: Neurobiological implications of the Kuramoto model. *Frontiers in Human Neuroscience* 4, doi: 10.3389/fnhum.2010.00190.

[9.34] Brette, R. (2021) Integrative neuroscience of *Paramecium*, a "swimming neuron". *Eneuro* 8(3), #ENEURO.0018-21.2021.

[9.35] Brockman, J. and Kauffman, S.A. (2003) The adjacent possible: A talk with Stuart A. Kauffman [11.9.03]. https://www.edge.org/conversation/stuart_a_kauffman-the-adjacent-possible

[9.36] Bubić, A., Cramon, D. and Schubotz, R. (2010) Prediction, cognition and the brain. *Frontiers in Human Neuroscience* 4, #25.

[9.37] Buckley, C.L., Kim, C.S., McGregor, S. and Seth, A.K. (2017) The free energy principle for action and perception: A mathematical review. *Journal of Mathematical Psychology* 81, 55-79.

[9.38] Buschges, A. (2012) Lessons for circuit function from large insects: towards understanding the neural basis of motor flexibility. *Current Opinion in Neurobiology* 22(4), 602-608.

[9.39] Busseniers, E., Veloz, T. and Heylighen, F. (2021) Goal directedness, chemical organizations, and cybernetic mechanisms. *Entropy* 23(8), #1039.

[9.40] Capilla, F.P. (2012) *Bacillaria paxillifer* diatomea - diatom [movie]. https://www.youtube.com/watch?v=lVg0lSKziys

[9.41] Capilla, F.P. (2013) *Bacillaria paxillifer* Diatomea [movie]. https://www.youtube.com/watch?v=xFfQ9piAzDo

[9.42] Capilla, F.P. (2021) *Bacillaria paxillifer* diatomea - diatom [movie]. https://www.youtube.com/watch?v=uK7wt4CdpUM

[9.43] Cartaxana, P., Ruivo, M., Hubas, C., Davidson, I., Serôdio, J. and Jesus, B. (2011) Physiological versus behavioral photoprotection in intertidal epipelic and epipsammic benthic diatom communities. *Journal of Experimental Marine Biology and Ecology* 405(1), 120-127.

[9.44] Chaitin, G. (2011) Life as evolving software. In: *A Computable Universe*. World Scientific: 277-302.

[9.45] Chater, N. and Brown, G.D. (2008) From universal laws of cognition to specific cognitive models. *Cogn Sci* 32(1), 36-67.

[9.46] Chater, N. and Brown, G.D.A. (1999) Scale-invariance as a unifying psychological principle. *Cognition* 69(3), B17-B24.

[9.47] Chen, L., Weng, D., Du, C., Wang, J.D. and Cao, S. (2019) Contribution of frustules and mucilage trails to the mobility of diatom *Navicula* sp. *Scientific Reports* 9, #7342.

[9.48] Cimellimay. (2011) *Bacillaria* [movie]. https://www.youtube.com/watch?v=M2f6y4aiU8s

[9.49] Cocchi, L., Gollo, L.L., Zalesky, A. and Breakspear, M. (2017) Criticality in the brain: A synthesis of neurobiology, models and cognition. *Progress in Neurobiology* 158, 132-152.

[9.50] Cohn, S.A. (2001) Photo-stimulated effects on diatom motility. In: *Photomovement*. D.P. Häder and M. Lebert, (eds.) Elsevier Science B.V.: 375-401.

[9.51] Cohn, S.A., Bahena, M., Davis, J.T., Ragland, R.L., Rauschenberg, C.D. and Smith, B.J. (2004) Characterisation of the diatom photophobic response to high irradiance. *Diatom Research* 19(2), 167-179.

[9.52] Cohn, S.A., Dunbar, S., Ragland, R., Schulze, J., Suchar, A., Weiss, J. and Wolske, A. (2016) Analysis of light quality and assemblage composition on diatom motility and accumulation rate. *Diatom Research* 31(3), 173-184.

[9.53] Cohn, S.A., Green, T., Payne, K.J. and Saini, V.S. (2004) Comparative analysis of irradiation-induced direction change in diatoms. *Molecular Biology of the Cell* 15(Suppl. S), 163a.

[9.54] Cohn, S.A., Halpin, D., Hawley, N., Ismail, A., Kaplan, Z., Kordes, T., Kuhn, J., Macke, W., Marhaver, K., Ness, B., Olszewski, S., Pike, A., Rice, E., Sbarboro, J., Wolske, A. and Zapata, Y. (2015) Comparative analysis of light-stimulated motility responses in three diatom species. *Diatom Research* 30(3), 213-225.

[9.55] Cohn, S.A., Manoylov, K.M. and Gordon, R. (eds.) (2021) *Diatom Gliding Motility [DIGM, Volume in the series: Diatoms: Biology & Applications, series editors: Richard Gordon & Joseph Seckbach]*. Wiley-Scrivener, Beverly, MA, USA.

[9.56] Daun, S., Mantziaris, C., Tóth, T., Büschges, A. and Rosjat, N. (2019) Unravelling intra- and intersegmental neuronal connectivity between central pattern generating networks in a multi-legged locomotor system. *PLoS One* 14(8), #e0220767.

[9.57] Delcomyn, F. (2004) Insect walking and robotics. *Annual Review of Entomology* 49, 51-70.

[9.58] Devine, S. (2009) The insights of algorithmic entropy. *Entropy* 11(1), 85-110.

[9.59] Dickinson, P. (2007) Neuromodulation of central pattern generators in invertebrates and vertebrates. *Current Opinion in Neurobiology* 16, 604-614.

[9.60] Diet Tom's Diatoms. (2021) Colonial Diatom *Bacillaria* Stretching and Glinting at 100x. https://www.youtube.com/watch?v=kAw6hxH5muM

[9.61] Do, R. (2012) *Bacillaria paradoxa* [movie]. https://www.youtube.com/watch?v=7ddWMC1KjPc

[9.62] Drum, R.W. and Gordon, R. (2003) Star Trek replicators and diatom nanotechnology. *Trends in Biotechnology* 21(8), 325-328.

[9.63] Drum, R.W., Gordon, R., Bender, R. and Goel, N.S. (1971) On weakly coupled diatomic oscillators: *Bacillaria*'s paradox resolved. *Journal of Phycology* 7(Suppl.), 13-14.

[9.64] du Bois, N. and Elliott, M.A. (2017) The temporal dynamics involved in object representation updating to predict change. In: *Progress in Brain Research*. C.J. Howard, (ed.) Elsevier. 236: 269-285.

[9.65] Dvoretskii, S., Gong, Z., Gupta, A., Parent, J. and Alicea, B. (2020) Braitenberg vehicles as developmental neurosimulation. *arXiv: Neurons and Cognition*.

[9.66] E.G Microscope. (2021) *Bacillaria Paxillifer*, Rotifers and a Water Flea:. https://www.youtube.com/watch?v=9fGKhIVTrw4

[9.67] Edlund, M.B. and Stoermer, E.F. (1993) Resting spores of the freshwater diatoms Acanthoceras andUrosolenia. *Journal of Paleolimnology* 9(1), 55-61.

[9.68] Edlund, M.B. and Stoermer, E.F. (1997) Ecological, evolutionary, and systematic significance of diatom life histories. *Journal of Phycology* 33(6), 897-918.

[9.69] Emanuel, S., Kaiser, M., Pflueger, H.J. and Libersat, F. (2020) On the role of the head ganglia in posture and walking in insects. *Frontiers in Physiology* 11, 11, #135.

[9.70] Enna, S.J. (2015) *Bacillaria paxillifer* (pennate diatom) sliding. https://www.youtube.com/watch?v=fH7LxWaRbmY

[9.71] Falciatore, A. and Bowler, C. (2002) Revealing the molecular secrets of marine diatoms. *Annual Review of Plant Biology* 53, 109-130.

[9.72] Fernandez-Leon, J.A. (2012) Behavioral robustness: An emergent phenomenon by means of distributed mechanisms and neurodynamic determinacy. *BioSystems* 107(1), 34-51.

[9.73] Frazer, J. (2012) Diatoms, or The Trouble with Life in Glass Houses. https://blogs.scientificamerican.com/artful-amoeba/diatoms-or-the-trouble-with-life-in-glass-houses/

[9.74] French, R.M. (2000) The Turing test: the first 50 years. *Trends in Cognitive Sciences* 4(3), 115-122.

[9.75] Friston, K. (2009) The free-energy principle: a rough guide to the brain? *Trends in Cognitive Science* 13(7), 293-301.

[9.76] Friston, K. and Kiebel, S. (2009) Predictive coding under the free-energy principle. *Philosophical transactions of the Royal Society of London. Series B, Biological sciences* 364(1521), 1211-1221.

[9.77] Friston, K.J., Daunizeau, J., Kilner, J. and Kiebel, S.J. (2010) Action and behavior: a free-energy formulation. *Biological Cybernetics* 102(3), 227-260.

[9.78] Friston, K.J., Heins, C., Ueltzhöffer, K., Da Costa, L. and Parr, T. (2021) Stochastic chaos and Markov blankets. *Entropy* 23, doi: 10.3390/e23091220.

[9.79] Friston, K.J. and Stephan, K.E. (2007) Free-energy and the brain. *Synthese* 159(3), 417-458.

[9.80] Fritzsch, B. and Piatigorsky, J. (2005) Ancestry of Photic and Mechanic Sensation? *Science* 308(5725), 1113-1114.

[9.81] Gao, R., Huang, J. and Wang, L. (2019) Leaderless consensus control of uncertain multi-agents systems with sensor and actuator attacks. *Information Sciences* 505, 144-156.

[9.82] Garner, W.R. (1974) *The Processing of Information and Structure*. Laurence Erlbaum Associates.

[9.83] Gauvrit, N., Singmann, H., Soler-Toscano, F. and Zenil, H. (2016) Algorithmic complexity for psychology: A user-friendly implementation of the coding theorem method. *Behavioral Research Methods* 48(1), 314-329.

[9.84] Gershman, S.J., Balbi, P.E.M., Gallistel, C.R. and Gunawardena, J. (2021) Reconsidering the evidence for learning in single cells. *Elife* 10, 15, #e61907.

[9.85] Gibson, J.J. (1979) *The Ecological Approach to Visual Perception*. Houghton Mifflin.

[9.86] Glanzman, D.L. (1995) The cellular basis of classical conditioning in *Aplysia californica* — it's less simple than you think. *Trends in Neurosciences* 18(1), 30-36.

[9.87] Godfrey-Smith, P. (2016) Individuality, subjectivity, and minimal cognition. *Biology and Philosophy* 31(6), 775-796.

[9.88] Goldammer, J., Mantziaris, C., Büschges, A. and Schmidt, J. (2018) Calcium imaging of CPG-evoked activity in efferent neurons of the stick insect. *PLoS One* 13, #e0202822.

[9.89] Gordon, R. (2016) Partial synchronization of the colonial diatom *Bacillaria "paradoxa"*. *Research Ideas and Outcomes (RIO) 2*, #e7869.

[9.90] Gordon, R. (2021) The whimsical history of proposed motors for diatom motility. In: *Diatom Gliding Motility [DIGM, Volume in the series: Diatoms: Biology & Applications, series editors: Richard Gordon & Joseph Seckbach]*. S.A. Cohn, K.M. Manoylov and R. Gordon, (eds.) Wiley-Scrivener, Beverly, MA, USA: 335-420.

[9.91] Gorostiza, E.A., Colomb, J. and Brembs, B. (2016) A decision underlies phototaxis in an insect. *Open Biol* 6(12).

[9.92] Gougi, S. (2007) クサリケイソウ（バキラリア）属の一種 - *Bacillaria* sp. [movie]. https://www.youtube.com/watch?v=sQU1nWZk_cE

[9.93] Graziano, M. (2009) *The Intelligent Movement Machine: An Ethological Perspective on the Primate Motor System*. Oxford University Press, USA.

[9.94] Grondin, S. (2016) *Psychology of Perception*. Springer International Publishing.

[9.95] Haggblom, K. (2021) *Bacillaria* dancing in the Indian River Lagoon. https://www.youtube.com/watch?v=SzOQiVqGbpw

[9.96] Hanczyc, M.M. (2014) Droplets: Unconventional protocell model with life-like dynamics and room to grow. *Life (Basel)* 4(4, Sp. Iss. SI), 1038-1049.

[9.97] Hanczyc, M.M. and Ikegami, T. (2010) Chemical basis for minimal cognition. *Artificial Life* 16(3), 233-243.

[9.98] Hanson, S.J. and Burr, D.J. (1990) What connectionist models learn: Learning and representation in connectionist networks. *Behavioral and Brain Sciences* 13(3), 471-489.

[9.99] Harbich, T. (2021) Some observations of movements of pennate diatoms in cultures and their possible interpretation. In: *Diatom Gliding Motility [DIGM, Volume in the series: Diatoms: Biology & Applications, series editors: Richard Gordon & Joseph Seckbach]*. S.A. Cohn, K.M. Manoylov and R. Gordon, (eds.) Wiley-Scrivener, Beverly, MA, USA: 1-32.

[9.100] Hart, Y., Goldberg, H., Striem-Amit, E., Mayo, A., Noy, L. and Alon, U. (2018) Creative exploration as a scale-invariant search on a meaning landscape. *Nature Communications* 9, #5411.

[9.101] Hawkins, R.D. (1984) A cellular mechanism of classical conditioning in *Aplysia*. *Journal of Experimental Biology* 112(1), 113-128.

[9.102] Hayakawa, T., Kudoh, S., Suzuki, Y. and Takahashi, M. (1994) Temperature dependent changes in colony size of the fresh water pennate diatom *Asterionella formosa* (Bacillariophyceae) and their possible ecological implications. *Journal of Phycology* 30(6), 955-964.

[9.103] Hebb, D.O. (2005) *The Organization of Behavior: A Neuropsychological Theory*. Taylor & Francis.

[9.104] Herrera-Rincon, C. and Levin, M. (2018) Booting up the organism during development: Pre-behavioral functions of the vertebrate brain in guiding body morphogenesis. *Communicative and Integrative Biology* 11(1), #e1433440.

[9.105] Heylighen, F. (2016) Stigmergy as a universal coordination mechanism I: Definition and components. *Cognitive Systems Research* 38, 4-13.

[9.106] Hipólito, I. (2022) Cognition without neural representation: Dynamics of a complex system. *Frontiers in Psychology* 12, doi: 10.3389/fpsyg.2021.643276.

[9.107] Hsiao, A. (2015) 奇異棍形藻 *Bacillaria paxillifera* （偏光、穿透光）[movie]. https://www.youtube.com/watch?v=PD7FKihgAjY

[9.108] Huysman, M.J.J., Vyverman, W. and De Veylder, L. (2014) Molecular regulation of the diatom cell cycle. *Journal of Experimental Botany* 65(10), 2573-2584.

[9.109] Jahn, R. and Schmid, A.-M.M. (2007) Revision of the brackish-freshwater diatom genus *Bacillaria* Gmelin (Bacillariophyta) with the description of a new variety and two new species. *European Journal of Phycology* 42(3), 295-312.

[9.110] Jennings, H.S. (1976) *Behavior of the Lower Organisms*. Indiana University Press, Bloomington, IN, USA.

[9.111] Journey to the Microcosmos. (2020) *Bacillaria*: Distractingly Beautiful Crystal Colonies. https://www.youtube.com/watch?v=Nysdq_plUTg

[9.112] Kamino, K., Kondo, Y., Nakajima, A., Honda-Kitahara, M., Kaneko, K. and Sawai, S. (2017) Fold-change detection and scale invariance of cell-cell signaling in social amoeba. *Proceedings of the National Academy of Sciences of the United States of America* 114(21), E4149-E4157.

[9.113] Kapinga, M. (1987) The motility of *Bacillaria paradoxa*. In: *IX North American Diatom Symposium, Treehaven Field Station, Tomahawk, WI 54487, October 7-10, 1987*. T. Roeder, (ed.) University of Wisconsin, Stevens Point: abstract #18.

[9.114] Kapinga, M. and Gordon, R. (1987) Cell to cell communication in the gliding diatom *Bacillaria*. In: *Proceedings of the Microscopical Society of Canada, Fourteenth Annual Meeting, 17-19 June 1987, University of Manitoba, Winnipeg, Manitoba*. Microscopical Society of Canada, Toronto: 12.

[9.115] Kapinga, M.R.M. (1989) *Observations on the Growth and Motile Behavior of the Colonial Diatom Bacillaria paradoxa in Culture [M.Sc. Thesis, Supervisor: R. Gordon]*. University of Manitoba, Winnipeg.
[9.116] Kapinga, M.R.M. and Gordon, R. (1992) Cell attachment in the motile colonial diatom *Bacillaria paxillifer. Diatom Research* 7(2), 215-220.
[9.117] Kapinga, M.R.M. and Gordon, R. (1992) Cell motility rhythms in *Bacillaria paxillifer. Diatom Research* 7(2), 221-225.
[9.118] Kello, C.T., Brown, G.D.A., Ferrer-i-Cancho, R., Holden, J.G., Linkenkaer-Hansen, K., Rhodes, T. and Van Orden, G.C. (2010) Scaling laws in cognitive sciences. *Trends in Cognitive Sciences* 14(5), 223-232.
[9.119] Kelso, J.A.S. (2010) Instabilities and phase transitions in human brain and behavior. *Frontiers in human neuroscience* 4, 23-23, #23.
[9.120] Kerkhoff, A.J. and Enquist, B.J. (2009) Multiplicative by nature: Why logarithmic transformation is necessary in allometry. *Journal of Theoretical Biology* 257(3), 519-521.
[9.121] Kersulis, D. (2012) *Bacillaria paxillifer.* https://www.youtube.com/watch?v=2np5EhefZKo
[9.122] Kersulis, D. (2012) *Bacillaria paxillifer* [movie]. https://www.youtube.com/watch?v=2np5EhefZKo
[9.123] Kimmich, S. (2016) *Bacillaria.* https://www.youtube.com/watch?v=rQ1Oea92J3U
[9.124] Kirchhoff, M.D., Parr, T., Palacios, E.R., Friston, K.J. and Kiverstein, J. (2018) The Markov blankets of life: autonomy, active inference and the free energy principle. *Journal of the Royal Society Interface* 15, #20170792.
[9.125] Kolmogorov, A.N. (1968) Three approaches to the quantitative definition of information. *International Journal of Computer Mathematics* 2(1-4), 157-168.
[9.126] Kriegeskorte, N. and Kievit, R.A. (2013) Representational geometry: integrating cognition, computation, and the brain. *Trends in Cognitive Science* 17(8), 401-412.
[9.127] Kuyucu, T., Tanev, I. and Shimohara, K. (2015) Superadditive effect of multi-robot coordination in the exploration of unknown environments via stigmergy. *Neurocomputing* 148, 83-90.
[9.128] Lan, G.H. and Tu, Y.H. (2016) Information processing in bacteria: memory, computation, and statistical physics: a key issues review. *Reports on Progress in Physics* 79(5), 17, #052601.
[9.129] Lanier, W. and Hidden Communities. (2006) *Bacillaria paxillifera* Just Slides Along [movie]. https://www.youtube.com/watch?v=Ol6hbX5mhPc
[9.130] Larson, S.D., Gleeson, P. and Brown, A.E.X. (2018) Connectome to behaviour: modelling *Caenorhabditis elegans* at cellular resolution. *Philosophical Transactions of the Royal Society B-Biological Sciences* 373(1758), 3, #20170366.
[9.131] Lavoie, I., Campeau, S., Darchambeau, F., Cabana, G. and Dillon, P.J. (2008) Are diatoms good integrators of temporal variability in stream water quality? *Freshwater Biology* 53(4), 827-841.
[9.132] Lechner, H.A., Baxter, D.A. and Byrne, J.H. (2000) Classical conditioning of feeding in *Aplysia*: I. Behaviorala analysis. *The Journal of Neuroscience* 20(9), 3369-3376.
[9.133] Li, J. (2018) On the possibility of strong artificial life. *Open Journal of Philosophy* 08, 495-505.
[9.134] Lubashevsky, I. (2019) Psychophysical laws as reflection of mental space properties. *Physics of Life Review* 31, 276-303.
[9.135] Mach, R. and Schweitzer, F. (2007) Modeling vortex swarming In *Daphnia. Bulletin of Mathematical Biology* 69(2), 539-562.
[9.136] MacKay, D.M. (1963) Psychophysics of perceived intensity: A theoretical basis for Fechner's and Stevens' Laws. *Science* 139(3560), 1213-1216.
[9.137] MacKay, D.M. and McCulloch, W.S. (1952) The limiting information capacity of a neuronal link. *The Bulletin of Mathematical Biophysics* 14(2), 127-135.

[9.138] Mantziaris, C., Bockemühl, T. and Büschges, A. (2020) Central pattern generating networks in insect locomotion. *Developmental Neurobiology* 80(1-2), 16-30.

[9.139] Marder, E. and Bucher, D. (2001) Central pattern generators and the control of rhythmic movements. *Current Biology* 11(23), R986-996.

[9.140] McFarland, M., Nayak, A.R., Stockley, N., Twardowski, M. and Sullivan, J. (2020) Enhanced light absorption by horizontally oriented diatom colonies. *Frontiers in Marine Science* 7, 10, #494.

[9.141] McGivern, P. (2019) Active materials: Minimal models of cognition? *Adaptive Behavior* 28(6), 441-451.

[9.142] McNamee, D. and Wolpert, D.M. (2019) Internal models in biological control. *Annual Review of Control, Robotics, and Autonomous Systems* 2(1), 339-364.

[9.143] microBizkaia. (2021) *Bacillaria paxillifera.* https://www.youtube.com/watch?v=sdnPCt2l1Bw

[9.144] Microscope Safari - Journey to the Unseen World. (2018) *Bacillaria paradoxa* diatom. https://www.youtube.com/watch?v=y-UUddbK4Mk

[9.145] Millidge, B., Seth, A.K. and Buckley, C.L. (2021) Predictive coding: A theoretical and experimental review. *ArXiv*:2107.12979.

[9.146] Mitch, W. (2012) *Bacillaria paxillifer* - Is Not a Caterpilififer. https://www.youtube.com/watch?v=klMGTnqMwfg

[9.147] Monechi, B., Ruiz-Serrano, Ã., Tria, F. and Loreto, V. (2017) Waves of novelties in the expansion into the adjacent possible. *PLoS One* 12(6), #e0179303.

[9.148] Morowitz, H.J. (2002) *The Emergence of Everything, How the World Became Complex.* Oxford University Press, Oxford, UK.

[9.149] Nanjundiah, V. (1986) How rapidly do uncoupled oscillators desynchronize? *Journal of Theoretical Biology* 121, 375-379.

[9.150] Nave, K., Deane, G., Miller, M. and Clark, A. (2020) Wilding the predictive brain. *Wiley Interdiscip Rev Cogn Sci* 11(6), #e1542.

[9.151] Nayaka, S., Toppo, K. and Verma, S. (2017) Adaptation in algae to environmental stress and ecological conditions. In: *Plant Adaptation Strategies in Changing Environment.* V. Shukla, S. Kumar and N. Kumar, (eds.) Springer Singapore, Singapore: 103-115.

[9.152] Nguyen, J. (2017) Scientific representation and theoretical equivalence. *Philosophy of Science* 84(5), 982-995.

[9.153] Nosofsky, R.M. (1992) Similarity scaling and cognitive process models. *Annual Review of Psychology* 43(1), 25-53.

[9.154] Nover, H., Anderson, C.H. and DeAngelis, G.C. (2005) A logarithmic, scale-invariant representation of speed in macaque middle temporal area accounts for speed discrimination performance. *Journal of Neuroscience* 25(43), 10049-10060.

[9.155] O'Donnell, D.R., Hamman, C.R., Johnson, E.C., Kremer, C.T., Klausmeier, C.A. and Litchman, E. (2018) Rapid thermal adaptation in a marine diatom reveals constraints and trade-offs. *Global Change Biology* 24(10), 4554-4565.

[9.156] O'Reilly, R. (2001) Generalization in Interactive Networks: The Benefits of Inhibitory Competition and Hebbian Learning. *Neural Computation* 13, 1199-1241.

[9.157] Oberst, S., Lai, J.C.S., Martin, R., Halkon, B.J., Saadatfar, M. and Evans, T.A. (2020) Revisiting stigmergy in light of multi-functional, biogenic, termite structures as communication channel. *Computational and Structural Biotechnology Journal* 18, 2522-2534.

[9.158] Orban de Xivry, J.J. and Ethier, V. (2008). Neural Correlates of Internal Models. *The Journal of Neuroscience,* 28(32), 7931-7932.

[9.159] Ott, D. (2012) *Bacillaria paxillifer* - a diatom with a spectacular gliding movement [movie]. https://www.youtube.com/watch?v=NvHF-YjDZBo

[9.160] Otyatatemushi. (2010) イカダケイソウの運動／*Bacillaria* sp. [movie]. https://www.youtube.com/watch?v=GUdDI_QhaoI

[9.161] Otyatatemushi. (2013) イカダケイソウ*Bacillaria paradoxa*の運動 [movie]. https://www.youtube.com/watch?v=SyYjuMhKU3Y

[9.162] Pagliarini, L. (2008) Polymorphic intelligence. *Artificial Life and Robotics* 12(1), 24-28.

[9.163] Palyanov, A., Khayrulin, S. and Larson, S.D. (2018) Three-dimensional simulation of the *Caenorhabditis elegans* body and muscle cells in liquid and gel environments for behavioural analysis. *Philosophical Transactions of the Royal Society B: Biological Sciences* 373(1758), #20170376.

[9.164] PeacefulProtist. (2013) The Amazing Sliding Diatom (*Bacillaria paxillifer*) [movie]. https://www.youtube.com/watch?v=s5LQbh4dVoM

[9.165] Phykos Algologia. (2022) *Bacillaria paxillifer.* https://www.youtube.com/shorts/Gcd4A63wSx4

[9.166] Pierre microscope. (2014) *Bacillaria paradoxa* [movie]. https://www.youtube.com/watch?v=3Mk1xi6FMXE

[9.167] plaenerdd. (2017) Bacillaria paxillifer. https://www.youtube.com/watch?v=07MTrdMq_Lg

[9.168] Portegys, T.E. (2002) An abstraction of intercellular communication. In: *Proceedings of the Eighth International Conference on Artificial Life.* MIT Press: 75–78.

[9.169] Poulsen, N.C., Spector, I., Spurck, T.P., Schultz, T.F. and Wetherbee, R. (1999) Diatom gliding is the result of an actin-myosin motility system. *Cell Motility and the Cytoskeleton* 44(1), 23-33.

[9.170] Poulton, E.C. (1968) The new psychophysics: Six models for magnitude estimation. *Psychological Bulletin* 69, 1-19.

[9.171] Prins, A., Deleris, P., Hubas, C. and Jesus, B. (2020) Effect of light intensity and light quality on diatom behavioral and physiological photoprotection. *Frontiers in Marine Science* 7, doi: 10.3389/fmars.2020.00203.

[9.172] Protist Lab Films. (2022) *Bacillaria* J.F. Gmelin, 1791. https://www.youtube.com/watch?v=hP513UiGMnk

[9.173] Purple Amoeba. (2017) *Bacillaria paxillifer* - The Carpenter's Rule Diatom. https://www.youtube.com/watch?v=OsGlXd5mspY

[9.174] Raj Vansh Singh, S., Krishna, K. and Gordon, R. (2022) RAPHE: Simulation of the dynamics of diatom motility at the molecular level: The domino effect model with concerted diffusion. In: *The Mathematical Biology of Diatoms[DMTH, Volume in the series: Diatoms: Biology& Applications, series editors: Richard Gordon& Joseph Seckbach].* J.L. Pappas and R. Gordon, (eds.) Wiley-Scrivener,, Beverly, Massachusetts, USA: THIS VOLUME.

[9.175] Ramsey, W. (2017) Must cognition be representational? *Synthese* 194(11), 4197-4214.

[9.176] Rano, I. (2009) A steering taxis model and the qualitative analysis of its trajectories. *Adaptive Behavior* 17, 197-211.

[9.177] Reina, A., Bose, T., Trianni, V. and Marshall, J.A.R. (2018) Psychophysical laws and the superorganism. *Scientific Reports* 8(1), #4387.

[9.178] Reynolds, A. (2008) Amoebae as exemplary cells: The protean nature of an elementary organism. *Journal of the History of Biology* 41(2), 307-337.

[9.179] Rigby, K. and Selander, E. (2021) Predatory cues drive colony size reduction in marine diatoms. *Ecology and Evolution* 11(16), 11020-11027.

[9.180] Rines, J. (2001) Plankton Theater, *Bacillaria paxillifera*, a motile, colonial pennate diatom, Narragansett Bay, Rhode Island, Fall [movie]. http://lin.irk.ru/nanochem/diatom_movies/films.html

[9.181] Rocha, P., D. Silva, A., Godinho, L., Dane, W., Estrela, P., Vandamme, L., Pereira-Leal, J., Leeuw, D. and Leite, R. (2018) Collective electrical oscillations of a diatom population induced by dark stress. *Scientific Reports* 8, #5484.
[9.182] Rosas, F.E., Mediano, P.A.M., Jensen, H.J., Seth, A.K., Barrett, A.B., Carhart-Harris, R.L. and Bor, D. (2020) Reconciling emergences: An information-theoretic approach to identify causal emergence in multivariate data. *PLoS Computational Biology* 16(12), #e1008289.
[9.183] Rosenblatt, F. (1958) The perceptron: A probabilistic model for information storage and organization in the brain. *Psychology Review* 65(6), 386-408.
[9.184] Rutenberg, J., Cheng, S.M. and Levin, M. (2002) Early embryonic expression of ion channels and pumps in chick and *Xenopus* development. *Developmental Dynamics* 225(4), 469-484.
[9.185] Sánchez, C., Cristóbal, G. and Bueno, G. (2019) Diatom identification including life cycle stages through morphological and texture descriptors. *PeerJ* 7, #e6770.
[9.186] Sarris, V. (2007) *Relational Psychophysics in Humans and Animals: A Comparative-Developmental Approach*. Taylor & Francis.
[9.187] Sarthou, G., Timmermans, K.R., Blain, S. and Treguer, P. (2005) Growth physiology and fate of diatoms in the ocean: a review. *Journal of Sea Research* 53(1-2), 25-42.
[9.188] Saygin, A.P., Cicekli, I. and Akman, V. (2000) Turing Test: 50 years later. *Minds and Machines* 10, 463–518.
[9.189] Saygin, A.P., Cicekli, I. and Akman, V. (2003) Turing test: 50 years later. In: *The Turing Test*. Springer: 23-78.
[9.190] Sayood, K. (2018) Information theory and cognition: A review. *Entropy (Basel)* 20(9).
[9.191] Schenz, D., Nishigami, Y., Sato, K. and Nakagaki, T. (2019) Uni-cellular integration of complex spatial information in slime moulds and ciliates. *Current Opinion in Genetics & Development* 57, 78-83.
[9.192] Schloegel, J.J. and Schmidgen, H. (2002) General physiology, experimental psychology, and evolutionism - Unicellular organisms as objects of psychophysiological research, 1877-1918. *Isis* 93(4), 614-645.
[9.193] Schmid, A.-M.M. (2007) The "paradox" diatom *Bacillaria paxillifer* (Bacillariophyta) revisited. *Journal of Phycology* 43, 139-155.
[9.194] Shatz, C.J. (1992) The developing brain. *Scientific American* 267(3), 61-67.
[9.195] Shepard, R.N. (1987) Toward a universal law of generalization for psychological science. *Science* 237(4820), 1317-1323.
[9.196] Sims, M. and Kiverstein, J. (2022) Externalized memory in slime mould and the extended (non-neuronal) mind. *Cognitive Systems Research* 73, 26-35.
[9.197] Stebbins, W.C. (2013) *Animal Psychophysics: the design and conduct of sensory experiments*. Springer US.
[9.198] Stokkermans, A., Chakrabarti, A., Wang, L., Moghe, P., Subramanian, K., Steenbergen, P., Mönke, G., Hiiragi, T., Prevedel, R., Mahadevan, L. and Ikmi, A. (2021) Ethology of morphogenesis reveals the design principles of cnidarian size and shape development. *bioRxiv*, 2021.2008.2019.456976.
[9.199] Svensson, F., Norberg, J. and Snoeijs, P. (2014) Diatom cell size, coloniality and motility: Trade-offs between temperature, salinity and nutrient supply with climate change. *PLoS One* 9(10), #e109993.
[9.200] Thar, R. and Kuhl, M. (2003) Bacteria are not too small for spatial sensing of chemical gradients: An experimental evidence. *Proceedings of the National Academy of Sciences of the United States of America* 100(10), 5748-5753.
[9.201] Tonysharks. (2015) イカダケイソウ（*Bacillaria paxilifera*） [movie]. https://www.youtube.com/watch?v=N6XV849ZAmQ

[9.202] Trewavas, A. and Malhó, R. (1997) Signal perception and transduction: The origin of the phenotype. *The Plant Cell* 9, 1181-1195.
[9.203] Tria, F., Loreto, V., Servedio, V.D.P. and Strogatz, S.H. (2014) The dynamics of correlated novelties. *Scientific Reports* 4, #5890.
[9.204] Ueltzhöffer, K. (2018) Deep active inference. *Biological Cybernetics* 112(6), 547-573.
[9.205] uni2 さんのチャンネル. (2012) *Bacillaria paradoxa*.AVI [movie]. https://www.youtube.com/watch?v=a864NCFBsDs
[9.206] Ussing, A.P., Gordon, R., Ector, L., Buczkó, K., Desnitskiy, A.G. and VanLandingham, S.L. (2005) *The Colonial Diatom "Bacillaria paradoxa": Chaotic Gliding Motility, Lindenmeyer Model of Colonial Morphogenesis, and Bibliography, with Translation of O.F. Müller (1783), "About a peculiar being in the beach-water"*. A.R.G. Ganter Verlag Kommanditgesellschaft, Ruggell. 5.
[9.207] van Duijn, M., Keijzer, F. and Franken, D. (2006) Principles of minimal cognition: Casting cognition as sensorimotor coordination. *Adaptive Behavior* 14(2), 157-170.
[9.208] Verworn, M. (1889) *Psycho-physiologische Protisten-Studien*. G. Fischer.
[9.209] Vogel, D. and Dussutour, A. (2016) Direct transfer of learned behaviour via cell fusion in non-neural organisms. *Proceedings of the Royal Society B: Biological Sciences* 283(1845), #20162382.
[9.210] Wagner, R. (2017) *Bacillaria paradoxa*. https://www.youtube.com/watch?v=-rYE8_Cowvo
[9.211] Wagner, R. (2021) *Bacillaria paxillifer* (*paradoxa*). https://www.youtube.com/watch?v=Or85v9t1pP8
[9.212] Warren, W. (2005) Direct perception. *Philosophical Topics* 33, 335-361.
[9.213] Watts, J. (1998) Animats: Computer-simulated animals in behavioral research. *Journal of Animal Science* 76, 2596-2604.
[9.214] Wibral, M., Lizier, J.T. and Priesemann, V. (2015) Bits from brains for biologically inspired computing. *Frontiers Robotics AI* 2, doi: 10.3389/frobt.2015.00005.
[9.215] Winfree, A.T. (1967) Biological rhythms and the behavior of populations of coupled oscillators. *Journal of Theoretical Biology* 16(1), 15-42.
[9.216] Witkowski, O. (2020) Christin Puthur & Tom Froese are our Fake Life Recognition Contest Winners. CrossLabs Blog, April 17. https://www.crosslabs.org/blog/christin-puthur-tom-froese-are-our-fake-life-recognition-contest-winners [cobweb site]
[9.217] Wolfram, S. (2001) *A New Kind of Science*. Wolfram Media, Champaign, Illinois.
[9.218] Yalcin, C. (2020) Colony of Diatom *Bacillaria*. https://www.youtube.com/watch?v=D5Zv-VwfJnQ
[9.219] Yanase, R. (2015) The gliding motion of diatom, *Bacillaria paradoxa* Part 2 [movie]. https://www.youtube.com/watch?v=4cJb9jCN2oI
[9.220] Yanase, R. (2015) The ground swell produced by gliding motion of diatom, *Bacillaria paradoxa* (うごめくイカダケイソウ) [movie]. https://www.youtube.com/watch?v=HyGoK7WfoS8
[9.221] Yeo, C.H., Hardiman, M.J. and Glickstein, M. (1985) Classical conditioning of the nictitating membrane response of the rabbit. *Experimental Brain Research* 60(1), 99-113.
[9.222] Zakirov, B., Charalambous, G., Thuret, R., Aspalter, I.M., Van-Vuuren, K., Mead, T., Harrington, K., Regan, E.R., Herbert, S.P. and Bentley, K. (2021) Active perception during angiogenesis: filopodia speed up Notch selection of tip cells *in silico* and *in vivo*. *Philosophical transactions of the Royal Society of London. Series B, Biological sciences* 376(1821), #20190753.

[9.223] Zénon, A., Solopchuk, O. and Pezzulo, G. (2019) An information-theoretic perspective on the costs of cognition. *Neuropsychologia* 123, 5-18.
[9.224] Zohary, T., Fishbein, T., Shlichter, M. and Naselli-Flores, L. (2017) Larger cell or colony size in winter, smaller in summer - a pattern shared by many species of Lake Kinneret phytoplankton. *Inland Waters* 7(2), 200-209.

10

Pattern Formation in *Diatoma vulgaris* Colonies: Observations and Description by a Lindenmayer-System

Thomas Harbich

Independent Researcher, Am Brüdenrain, Weissach im Tal, Germany

Abstract

The araphid diatom *Diatoma vulgaris* forms zig-zag-shaped colonies whose structures change spatially and temporally with cell divisions. To describe pattern formation, a notation of possible divisions is introduced, based on four symbols, which enables analysis of observations. The observed processes provide evidence for an asymmetry with respect to the mirroring on the transapical plane. An extension of the symbols by elements that consider this asymmetry allows description of colony morphogenesis by a non-deterministic Lindenmayer system with only two basic elements and one parameter. The space of possible patterns is strongly reduced by the observed production rules tenuis and their temporal distribution. Sections on the chain in which divisions occur synchronously show characteristic chain formations and transitional states between them. Considering only deterministic processes, patterns of the formed structures can be derived at the level of connections, which are frequently encountered and represent a confirmation of the theory. In addition, they have implications for the curvature of filamentous colonies. This is discussed in the context of evolutionary advantage.

***Keywords*:** Colony structure, *Diatoma vulgaris*, Lindenmayer system, pattern formation

10.1 Introduction

Diatoms form a class of unicellular algae of great ecological importance. They are found in almost all aquatic habitats as well as in the soil. Diatoms live in an exoskeleton (frustule) of hydrated silicate coated with organic substance. The frustule consists of two halves of different sizes, the larger epitheca and the smaller hypotheca. These are each formed by a valve and the cingulum, a number of associated girdle bands (see Round *et al.* [10.30]). Diatoms reproduce primarily by asexual division. In this process, the thecae of the parent cells are retained and each of the daughter cells forms the smaller hypotheca. These relationships

Email: thomas.harbich@diatoms.de

Thomas Harbich: https://orcid.org/0000-0002-9151-4052, https://www.researchgate.net/profile/Thomas_Harbich, https://diatoms.de/en/

Janice L. Pappas (ed.) The Mathematical Biology of Diatoms, (265–290) © 2023 Scrivener Publishing LLC

were discovered by MacDonald [10.19] and Pfitzer [10.25] and have been proven for many species (e.g., Locker [10.18]).

In many cases, diatoms separate completely after a vegetative division. However, if they remain connected, for example by extracellular polymeric substances (EPS), for a time that is significantly longer than the generation time, clonal colonies are formed [10.30]. An overview on the forms of such colonies can be found in Rimet and Bouchez [10.29]. Star-shaped colonies and zig-zag colonies develop in pennate diatoms, such as *Diatoma* spp., when they split apart after cell division but remain connected at their opposed poles.

Diatoma vulgaris Bory de Saint-Vincent, 1824 [10.2] is a widespread araphid benthic species living in fresh or brackish water [10.27]. It forms zig-zag-shaped chains that often adhere to the substrate at one end [10.4]. In long colonies, as they are found in nature as well as in cultures, several such attachments are common. Due to the longitudinal expansion as a result of the cell divisions, structures are formed that protrude into the water body. Figure 10.1 shows illustrations by Bory de Saint-Vincent [10.2], in which various colonies are shown in their natural environment.

In most cases, one sees the diatoms connected to each other at diagonally opposite apices. However, in Figure 10.1b, one finds two cases of diatoms whose connection positions are at the same end of the diatom. No conclusion can be drawn about the diatoms at the ends of the two chains. Furthermore, in some cases, non-separated groups of four diatoms are shown. The accompanying description by Bory de Saint-Vincent states: « Ces filamens, quand ils sont disposés en zig-zag, atteignent une ligne de longueur, mais l'oeil désarmén'y saurait distinguier aucune organisation. » (Translation: These filaments, when they are arranged in zig-zag, become one long line, but the unarmed eye cannot discern any organization in them.) Apparently, this observation refers to the fact that the concatenation of diatoms takes place at changing apices and no simple pattern such as a periodic repetition of a basic pattern can be detected in it.

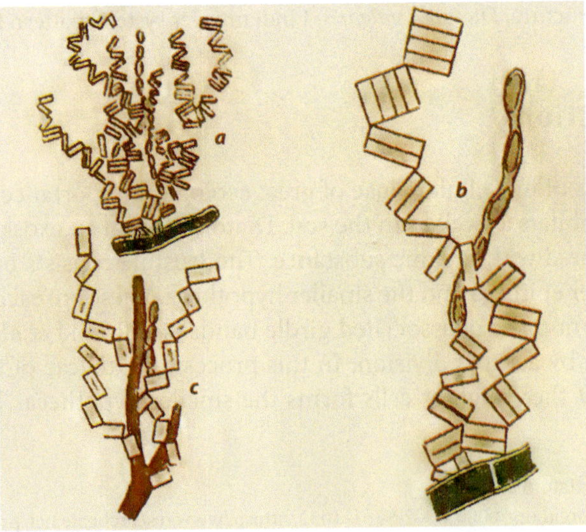

Figure 10.1 *Diatoma vulgaris* after a drawing by Bory de Saint-Vincent [10.2]. Colonies in epiphytic growth are shown at three different scales. In addition to the girdle view, the valve view is also sketched in each partial image.

As pattern formation is based on division processes of diatoms, it is of central importance whether these occur synchronously. If this is the case within a colony or a culture, the description becomes considerably easier, because then the temporal development of the patterns can be parameterized by an index for numbering the generations. Synchronization can have intrinsic origins or be caused by periodic influence of environmental parameters. For example, periodic changes in brightness and darkness can lead to a certain extent to a synchronization of the divisions [10.15]. More successful is the targeted control of the photoperiod (see Pirson and Lorenzen [10.26] and Paasche [10.21] [10.22]). Synchronization can also be achieved by dosed periodic silicate addition in cultures that are kept under silicate deprivation (see Lewin *et al.* [10.15], Busby and Lewin [10.3], Darley and Volcani [10.5]; Davis *et al.* [10.6]). Species studied include colony-forming diatoms such as *Ditylum brightwelli*, *Nitzschia turgidula* [10.22] and *Skeletonema costatum* [10.6].

A periodic variation of environmental parameters affects all diatoms exposed to it. Besides, it is to be expected that the divisions of cells starting from a mother cell occur synchronously over a certain time. All observable daughter cells belong to the same generation until loss of synchronicity.

Doubling time is defined as the period between successive cell separations. The doubling time is subject to statistical fluctuations and can be described by a density function [10.9] which is characterized by its expected value μ and its standard deviation σ. The average doubling time is called generation time. For the following considerations, it is assumed that successive doubling times are statistically independent. This is not the case in general [10.13]. Furthermore, it will be presumed that there are no differences in the statistical properties of daughter cell doubling times, an assumption that does not always hold [10.14]. If we approximate the density function of the doubling time by a normal distribution, then the density function for the duration that k completed divisions take is also a normal distribution, where the expected value is $k * \mu$ and the standard deviation is $\sigma\sqrt{k}$. The density functions become continuously broader as the number of generations becomes larger, so that they increasingly overlap. The probability that a cell under consideration has already completed k divisions at time $t_k = (k + 1/2)\,\mu$ and has not yet completed the $k + 1$th division reduces as k increases. This can be determined numerically [10.12]. For example, a cell is in that division state at time t_k with a probability of about 0.95 if $\sqrt{k}\sigma/\mu = 0.25$ holds. If the ratio σ/μ is known, k can be calculated. For example, if $\mu/\sigma = 8$, then $k = 4$ follows. Because of the assumed statistical independence of divisions, two diatoms originating from the same parent cell and developed independently by divisions have this division state at time t_k with a probability of 0.95^2 and are synchronous at least with this probability. The generation time depends on environmental parameters, such as light intensity, temperature, and nutrient concentrations, but it is assumed that these can be considered identical for a studied colony.

In case of a diatom colony, the daughter cells of a parent cell are adjacent to each other, so that a synchronous section of consecutive cells in a colony results in case of temporary, local synchronicity. To illustrate the effect on the synchronous sections of the colony, a binary tree is shown in Figure 10.2, beginning with a newly formed pair of diatoms from a colony generated by cell division and showing the generational sequence over four divisions.

Between the nodes of the fourth generation, which represent the diatoms immediately after division, the number of generations back to the first common pair is noted. If, for example, the synchronicity is maintained over three generations, then a subtree is synchronous

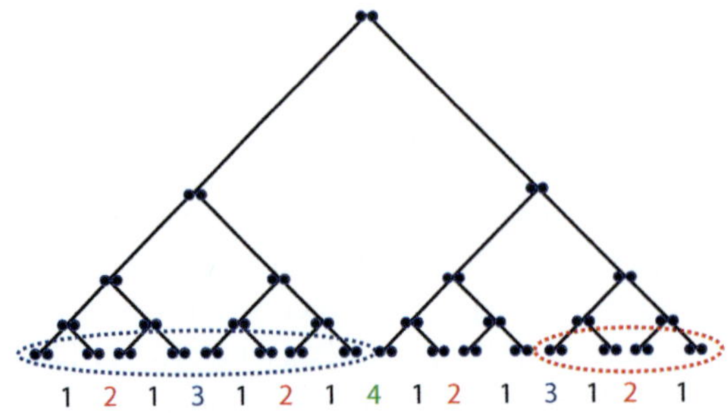

Figure 10.2 Tree structure of binary divisions starting from a pair of diatoms in a chain. Each diatom forms another pair after an individual doubling time. Diatom pairs are represented as adjacent dots. The diatoms of each generation are on a horizontal straight line. Shown is the logical structure, not the doubling times. Thus, the cells of one of a selected generation can be formed at different times.

whose common root goes back three divisions. Because of the doubling in each generation, this section has a width of $2^3 = 8$ diatom pairs (dotted line on the left in the fourth generation in Figure 10.2). In case of synchronicity over two generations, the synchronous section comprises only 4 diatom pairs (e.g., dotted line on the right in the fourth generation in Figure 10.2). The two neighboring diatom pairs in the middle have a mother cell four generations back, so loss of synchrony is most likely to occur here. In the case of synchronous sections of 32 diatom pairs, one can assume synchronous divisions over 5 generations. However, no certainty about this can be obtained from the appearance. The observation of synchronous sections only provides a necessary, but not sufficient criterion.

Passy has studied the morphological complexity of colony-forming diatoms, including *D. vulgaris*, as a function of flow conditions [10.24]. For this purpose, the fractal dimension was determined from images of such chains. The characteristics of the patterns and the pattern formation itself were not the subject of the publication. This work, in contrast, will focus on the question of which patterns are formed and whether they are generated by stochastic processes or whether deterministic processes also play a role. The implications of the patterns on higher-level structures of the chains, such as arcs or entangled balls, shall be considered.

10.2 Materials and Methods

10.2.1 Cultivation and Recording of Images

Freshwater diatoms of the species *D. vulgaris* were sampled two times from the Neckar River in Germany (49°04'41.8 N 9°09'17.9 E). During each sampling, several clonal strains were cultured in polystyrene Petri dishes, kept in an f/2 medium according to Guillard [10.10], and exposed to light intensities ranging from 200 lx to 800 lx. An artificial day-night cycle with a light phase of 11 hours and a dark phase of 13 hours was used. The room temperature was 19 °C to 21 °C. Every two weeks, diatoms were harvested and used to inoculate a Petri dish with fresh nutrient solution (batch cultures). The typical apical length was 40 μm.

Overview images of longer colonies were taken directly in culture using an inverted microscope (Zeiss Axiovert A1), while details were obtained using an upright microscope (Zeiss Axioplan).

Long-term recordings over several division phases were made using the inverted microscope, sometimes with reduced microscope illumination and a fluorescent lamp (color temperature 6000 K) mounted next to the microscope stage to enable photosynthesis. A time interval between shots of 10 minutes proved to be sufficient to follow the development of the colony. Time lapse was not used when observing the individual separation processes.

10.2.2 Formal Notation of Types of Concatenation and Splitting Processes

To study the sequence of connections, one needs a notation that allows the description of the connections that occur and the splitting processes. It is a prerequisite for the evaluation of the observations made. Each diatom, forming zig-zag-shaped colonies, has exactly one extracellular polymeric substance pad on the opposite valve faces. If a chain lies in front of the viewer in a horizontal orientation in a girdle band view, a simple method is to classify the connection points, e.g., from left to right, according to whether they lie at the upper or lower edge. This resulting binary sequence of "above" and "below" is free of redundancy and allows for analysis but proves to be impractical. Instead, we want to consider the diatoms with their connections. With the aforementioned orientation of the chain, four possibilities exist for the diatoms, as shown in Figure 10.3a.

The proportions are not intended to provide information about division states and are therefore not relevant. Only the EPS pads shown as small semicircles with their assignment to an apex are of importance. Each type has been assigned an identifier (*a*, *b*, *c*, *d*), which can be considered as characters of an alphabet. A colony or a section of it is represented at a given time by a finite sequence of characters of the alphabet, i.e., a "word" or "string" of these 4 letters. A colony with the representation *a c b d* is shown in Figure 10.3b.

Only those types can be adjacent that have the common EPS pad at the opposite end. Possible are the sequences with two elements *a b*, *a c*, *b a*, *b d*, *c c*, *c b*, *d d* and *d a*. The mentioned binary sequence of the positions of the connections would work without such restrictions. If one considers with a horizontally unfolded colony the apices lying at the upper and lower edge and the existing EPS pads, then the connection point changes only by means of the diagonal elements between upper and lower edge. Therefore, a longer colony containing several diagonal elements can be regarded as an alternating sequence of

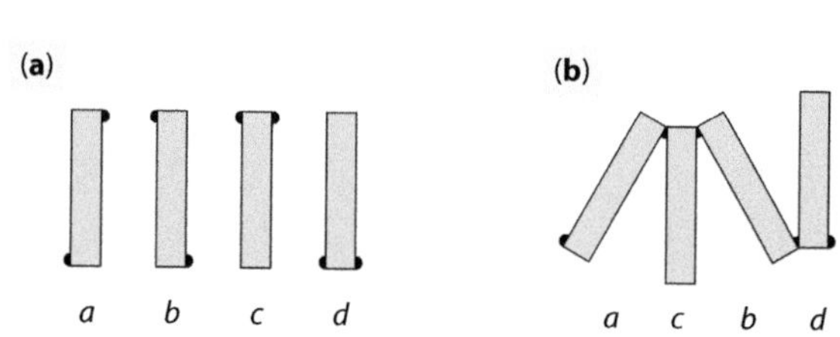

Figure 10.3 (a) The four types shown with their identifiers form the basis for a formal notation of the transitions after cell division. In the "diagonal" types *a* and *b*, the contact points are located at the most distant apices. Types *c* and *d* are referred to as "non-diagonal." (b) Example of a small colony in which each type appears once.

elements *a* and *b* (.. *a b a b* ...), into which elements of types *c* and *d* are inserted according to the possibilities mentioned.

When a pair of daughter diatoms splits apart, there are two possibilities for each type because the position of the outer pads must be preserved (Figures 10.4 and 10.5). The processes are numbered by Roman numerals.

A case to be considered separately occurs when the two diatoms do not separate after a completed cell division, but a further cell division occurs. In this case, four diatoms lie parallel in a compact structure, as can be seen in Figure 10.1. The unfolding then does not necessarily have to start in the middle, so that two pairs of diatoms are formed, but can also take place between an outer and the neighboring diatom. The order of the generations is then not kept. Nevertheless, for notation purposes, the relevant separation is the one that lies between the two pairs, because this would have occurred if the diatoms had not divided a second time before separating. The same applies to even larger aggregates of unseparated diatoms. These were rarely found in our cultured strains. However, their frequency increased considerably with the age of the strains.

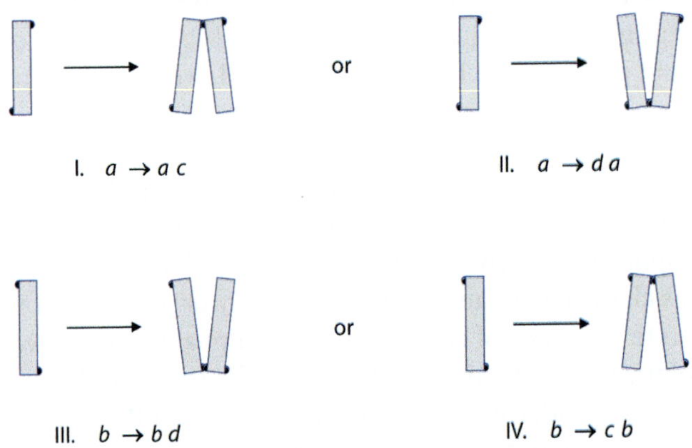

Figure 10.4 The diagonal types split into the same type and a non-diagonal type.

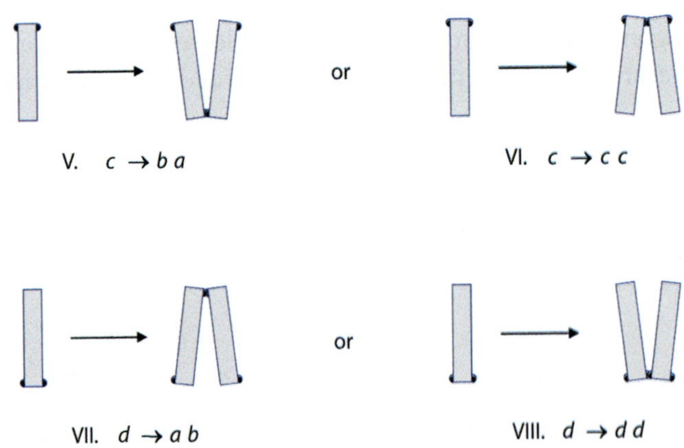

Figure 10.5 The non-diagonal types can split into two different diagonal types or two identical non-diagonal types.

It should be emphasized that this notation does not suggest or imply any statements about a possible mechanism. It only gives the possibility of describing the divisions for a chosen horizontal position of the chain.

The assignment of types in an observed colony is done with a fixed viewing direction. To understand the processes defined in Figures 10.4 and 10.5, it proves useful to introduce geometric transformations. This allows one to determine how the describing string $g_1g_2g_3 \dots g_j$ of a colony changes and how processes I, II, … VIII transform when the colony is turned ("upside down") or mirrored ("viewed from below"). If the plane of observation is rotated by 180 degrees perpendicular to the microscope axis, this transformation can be described by an operator T for which:

$$T\,(g_1\, g_2\, g_3 \dots g_j) = T\,(g_j)\; T\,(g_{j-1}) \dots T\,(g_2)\; T\,(g_1), \tag{10.1}$$

where the characters of the alphabet transform, according to:

$$\mathrm{T}(a) = a;\ \mathrm{T}(b) = b;\ \mathrm{T}(c) = d;\ \mathrm{T}(d) = c \tag{10.2}$$

The operator M mirrors a string with respect to a horizontal line corresponding to the orientation of the colony. The order of the elements is thereby preserved.

$$\mathrm{M}(g_1\, g_2\, g_3 \dots g_j) = \mathrm{M}(g_1)\; \mathrm{M}(g_2)\; \mathrm{M}(g_3) \dots \mathrm{M}(g_j) \tag{10.3}$$

with

$$\mathrm{M}(a) = b;\ \mathrm{M}(b) = a;\ \mathrm{M}(c) = d;\ \mathrm{M}(d) = c \tag{10.4}$$

The result of the sequential execution of T and M, written as S = T*M (with commutation T*M = M*T) produces a reflection with respect to the vertical.

If we apply the rotation T on process I according to Figure 10.4, i.e., T(a) → T(ac), we obtain using Eqs. (10.1) and (10.2) process II, that is $a \rightarrow da$. Correspondingly, the process II changes into the process I by rotation. Similarly, when the chain is rotated, processes III

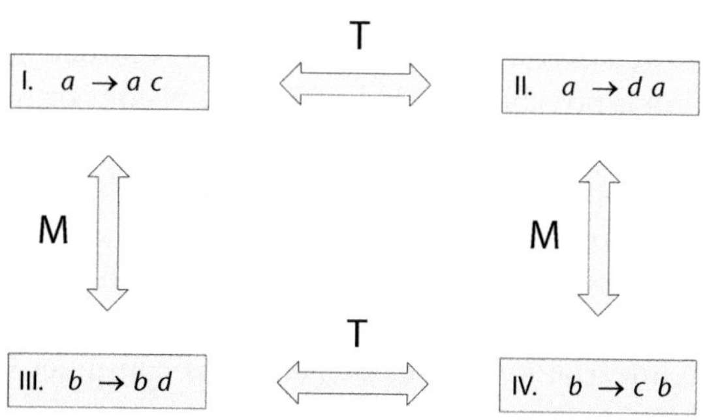

Figure 10.6 Effect of rotation and mirroring on processes I to IV. The execution of S = T*M interchanges processes I and IV, as well as II and III.

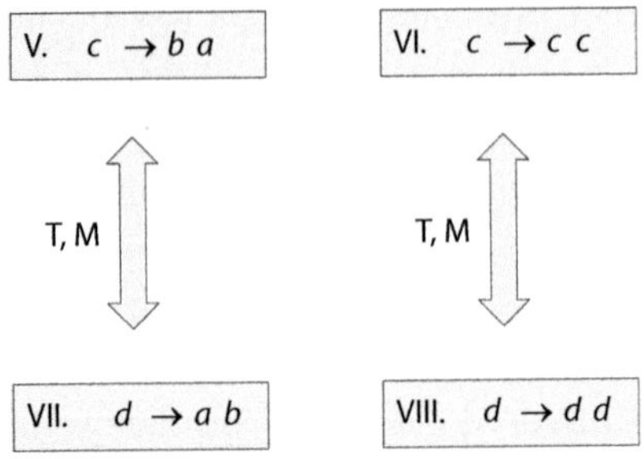

Figure 10.7 Effect of rotation and mirroring on the processes V to VIII. There is no geometric transformation that interchanges V and VI or VII and VIII.

and IV exchange their roles. Note that the rotation T changes the assignment of the resulting diatoms according to Eq. (10.1).

By applying Eqs. (10.3) and (10.4) for the reflection M on the horizontal, we find that processes I and III, as well as II and IV, transform into each other (see Figure 10.6).

By mirroring and by rotation, V and VII, as well as VI and VIII change into each other (Figure 10.7).

10.2.3 Methods to Observe the Processes

10.2.3.1 Basic Options

Short-term observations allow conclusions to be drawn regarding the process during separation. Two approaches are used to study the transitions that occur in colonies, on the one hand long-term observations of the formation of colonies and on the other hand the analysis of single recordings of chains, occasionally in combination with a short-term observation of separations of neighboring cells.

10.2.3.2 Long-Term Observations

In this approach, short fragments with a similar physiological state of development of their diatoms are searched. A simple measure of the state of development is the distance between the valves in girdle band view. This thickness in the direction of the pervalvar axis is minimal after the separation of the diatoms and maximal just before the separation. These fragments are followed over a few cell divisions and recorded in a series of images so that subsequent evaluation is possible. Not only can the divisions be indicated in the formal notation, but the onset of synchrony loss becomes apparent. Such a loss of synchrony is particularly evident when developing pairs of diatoms divide before all diatoms of the previous generation have done so. In such a case, the chain typically splits into smaller synchronized sections. In all cases, it is not possible to identify to which type the outer diatoms at the ends of the initial colonies belong, because EPS pads at the outer valves of a fragment can only rarely be identified. The uncertainty in the outer diatoms continues through the observed generations.

The strength of the method lies in the certainty about the synchronicity of the divisions. Assignment to processes succeeds even in case of loss of synchronicity. The greatest difficulty of such long-term observations results from the constantly changing spatial positions of the colony because of the division processes. If the colony takes a position that is not approximately perpendicular to the optical axis, sections of the chain may be so far from the focal plane that evaluation becomes impossible. It also happens to twist so that it can be seen in parts in valve view. Sometimes the shape changes in such a way that the colony overlaps in the image itself, allowing only limited observability. Long-term observations were typically systematically analyzable over three to four subsequent generations (5 to 7 observation days).

10.2.3.3 Analysis of Single Images

Under certain conditions, one can try to infer the previous division from a single image of a colony. The prerequisite is that a section is available in which all diatoms, not just the neighboring ones, have close division states. This is again evaluated on the basis of the distance between the valves. Then the section is divided into pairs for which the type of the parent cell is determined according to the defined processes. The difficulty lies in the choice of the pairing. To do this, one can look for indications. If there is a pair that has not yet been split within the chain or at the edge of the chain, the grouping can be carried out on the basis of this pair. It is convenient to observe the chain until the separation of this pair, i.e., to combine static and dynamic observation. On the basis of a division into pairs, the occurring processes can be counted according to Figures 10.4 and 10.5.

Because of the unknown history, an evaluation of a synchronicity is uncertain even if closely spaced division stages are present in one section. Results are therefore to be evaluated critically. If there is a section appropriate for analysis, it can be assumed that the last division occurred synchronously in it, but it cannot be concluded that this was also the case in the preceding divisions. A further backtracking according to this procedure does not make much sense, particularly as the problem of a correct pairing arises again, but one does not have an image of the preceding generation.

10.3 Results

10.3.1 Observation of Elementary Splitting Processes

If one observes a single *D. vulgaris* chain or a whole network over a longer period, there are short-time twitches in these formations at random intervals, when at one place a pair of diatoms after completion of cell division splits apart except for the remaining connection point (see video at https://diatoms.de/en/twitches-of-the-colony). At a frame rate of 30 fps, the splitting always occurred between two frames, so that a shorter duration for the splitting than 1/30 s can be assumed. This separation happens spontaneously without the need for external forces such as those exerted on the chains by water flow. Even with very short, freely moving chains, the diatoms suddenly unfold into a V-shape. Therefore, it cannot be assumed that the apex with the remaining connection is determined by external forces. It is more likely that the EPS secreted at one apex through the apical pore field (APF) build up mechanical tension over time, which then leads to sudden detachment of the diatoms. The angle taken by the diatoms then increases only slowly to a final value. The resulting EPS pad

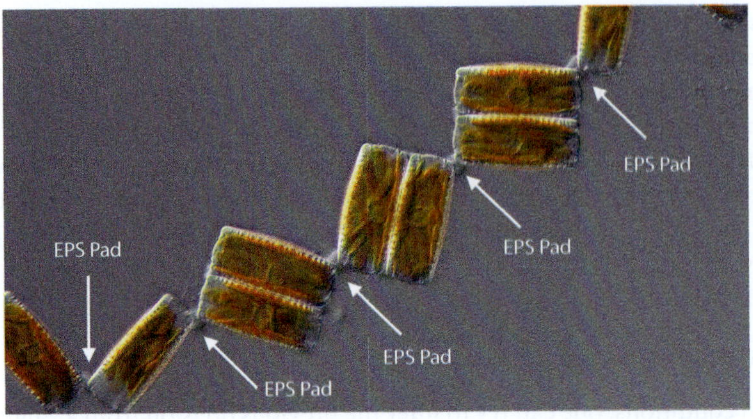

Figure 10.8 A short section from a colony in the DIC. EPS pads are only visible at the connecting points of adjacent diatoms.

represents a spring element in an elastic hinge. Forms of colonies and mechanical properties, especially in flows, were studied by Celler *et al.* [10.4].

Another observation supports this hypothesis. With the help of phase-sensitive microscopic contrast techniques such as phase contrast or differential interference contrast (DIC), by suitable staining and by fluorescence microscopy [10.1], the mucilage pads can be recognized at the connection points of the diatoms. They are also found at the breakpoints of fragments, but not at the unconnected apices, as shown by the DIC image in Figure 10.8. Staining with Alcian blue also gives no evidence of EPS pads in the area of the uncovered apical pore fields.

These observations motivate the search for rules for the selection of the apex at which the connection is established and thus for the pattern formation for a clonal chain.

10.3.2 Observation of Synchronism

No synchronization was observed in the *D. vulgaris* cultures at the chosen day-night rhythm. Attempts to induce synchronization were not made, since the purpose was to describe and explain the natural pattern formation of the colonies.

In all observations of chains with more than about 30 or 35 diatoms, clearly different division stages could be seen. It is not uncommon that even neighboring diatoms have a maximally different developmental state, which is half of a generation time (see Figure 10.2). When assessing division states, it should be noted that diatom pairs just before splitting and diatoms just after splitting have a close division state, so the appearance of a colony can be very different even in a nearly synchronous state.

On the other hand, it is striking that there are usually sections on a colony in which all diatoms show a development close to each other, i.e., presumably developed by synchronous division at least over the previous generation. This allows segmentation into sections of similar cell cycle stage, although not always with clearly identifiable boundaries. Specific examples with indications of the size of the sections are presented in the context of the appearances of colonies (see Figure 10.9).

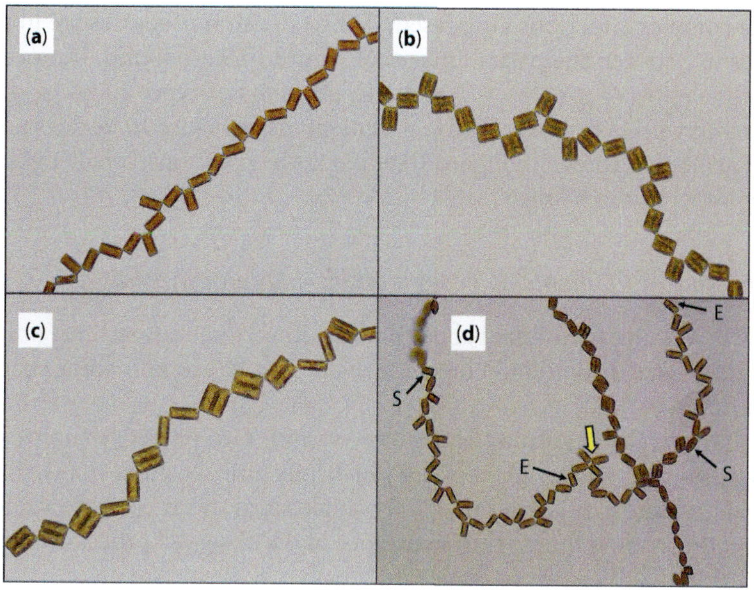

Figure 10.9 Images (a), (b) and (c) show phases of colony development in synchronous sections in chronological order. In image (d), a long chain is displayed with different stages of division, a section with overlap, an area in valve view, and an area outside the focal plane. In many cases, a focus stack helps.

10.3.3 Observation of the Processes and Appearance of Colonies

10.3.3.1 Splitting of Elements of Types c and d

A central statement, which is already evident from many short-term observations, concerns the processes of separation of types *c* and *d*. Processes V or VII were always observed, never VI or VIII. Thus, a diatom in which both EPS pads are on the side of the same apex always forms the new EPS pad at the distant apex after cell division. The number of observed events supporting this conclusion is in the range of several hundred. Although one can also see sequences of the form *cc* or *dd* in colonies, these never emerge from a type *c* or *d* in a single division step. Furthermore, it is striking that processes V or VII, measured from the time of cell division, occur very early, usually before the processes I to IV. If a section of a colony is in very similar division states, so that a parallel development of the next generation can be anticipated, the types *c* and *d* split first. Often, their separation in the section is completed before the first division processes of elements *a* and *b* take place. This is reflected in the appearance of the colony. Figures 10.9a to 10.9c shows three typical phases of synchronous development as examples.

Figure 10.9a illustrates a section that has developed synchronously and in which all separation processes have been completed. About one generation later, the chain is as shown in Figure 10.9b. No separation has yet taken place. After processes V and VII are completed and before the other processes begin, there are only diagonal elements *a* and *b*, because only diagonal element types are formed in processes V and VII. An image is shown in Figure 10.9c. In this form, "zig-zag" is an accurate term in a strict sense. This pattern does not last long, because shortly after its formation, processes I to IV begin, leading again to an appearance as in Figure 10.9a.

Often more complex situations are present. When division processes are in progress, all transitional forms between the partial images of Figure 10.9 are found. If processes I to IV start before completion of processes V and VII, elements of types *c* and/or *d* are formed and an appearance with exclusively diagonal elements as in Figure 10.9c does not occur. An image of a long colony is shown in Figure 10.9d. In it, the mentioned challenges in observation and interpretation can be seen.

10.3.3.2 Splitting of Elements of Types a and b – Dynamic Analysis

Because of the difficulties mentioned above and the long observation time, only a few successful long-term observations could be made. Figure 10.10 presents some snapshots from such an observation.

Figure 10.10 shows the beginning of the observation. Two pairs of diatoms are about to separate, which has occurred in (2). Loss of synchrony already starts in (4). The last completely analyzable image is (5). After that, only single separations could be captured due to overlaps and sections not in focus. The occurrence of the observed processes is summarized in Table 10.1.

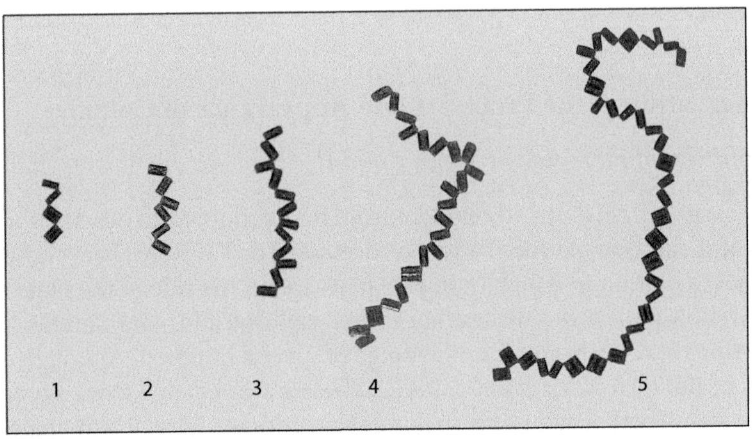

Figure 10.10 A few images from the development of a colony. Image 5 was taken 130 hours after the first image. The colony was exposed to a light-dark cycle with 13 hours of light phase. For analysis, the chains are viewed from the right.

Table 10.1 Observed processes and their frequencies.

Processes	**Number of events**
Process I $a \to a\ c$	7
Process II $a \to d\ a$	1
Process III $b \to b\ d$	6
Process IV $b \to c\ b$	2

Here, as well as in comparable observations, no indications of a strict rule could be found that would allow a prediction, according to which processes a certain diatom will split up. Neither neighborhood correlations, in which processes accumulated, nor a chronological dependence were noticeable. Therefore, the hypothesis of independent random processes is assumed (Bernoulli trial).

Process I occurs more often in the observation than the alternative process II, process III more often than the alternative process IV, although these processes merge into each other by rotation T. This is to be analyzed with the help of a hypothesis test, which is formulated below for the processes I and II.

- *Null hypothesis:* An obvious assumption is that processes I and II occur with equal frequency.
- *Alternate hypothesis:* There is an asymmetry, i.e., the probability of processes I and II to occur is different.
- *Test statistic:* It is assumed that each splitting of two diatoms can be understood as a Bernoulli trial, i.e., a binomial distribution is present. The number of trials is $n = 8$. The Bernoulli trials yielded process I seven times and process II one time. The probability for each individual process is ½ according to the null hypothesis.
- *Determining the p-value:* Assuming the null hypothesis, the probability that process II occurs at most once or not at all can be calculated (p-value). The cumulative Bernoulli distribution (with $n = 8$ and $p = ½$) provides the probability $Pr(x \leq 1) \sim 0.0352$.
- *Significance:* If 0.05 is chosen as the significance level, the null hypothesis can be rejected.

Performing this hypothesis test for processes III and IV, we obtain a p-value of approximately 0.144 according to the table above. While this cannot be considered sufficient to reject the null hypothesis, it should be noted that all observations yielded comparable results.

10.3.3.3 Separation of Elements of Types a and b – Static Analysis

In order to obtain meaningful results, only those recordings that meet two criteria were taken into account. There must be a chain with many (preferably more than 20) diatoms that have a hardly distinguishable division state. Furthermore, there should be a clear indication pointing to a specific pairing. A section of a colony suitable for static analysis is shown in Figure 10.11. The analysis is performed for the entire marked section from the start (S) to the end (E). For the determination of the pairing, the assumption is made that the still connected pair consists of two diatoms of the same generation as the neighboring diatoms. This establishes the pairing and yields the sequence (pairs grouped by parentheses):

$$(ba)\ (cb)\ (da)\ (cb)\ (ab)\ (da)\ (ba)\ (xy)\ (da)\ (cb)\ (da)\ (cb)\ (da)\ (ba)\ (cb)\ (da)\ (cb)\ (ab)$$

Thereby, the unseparated pair of the form b was described by the unknown pair (xy) after separation, where (xy) is given by either (cb) or (bd).

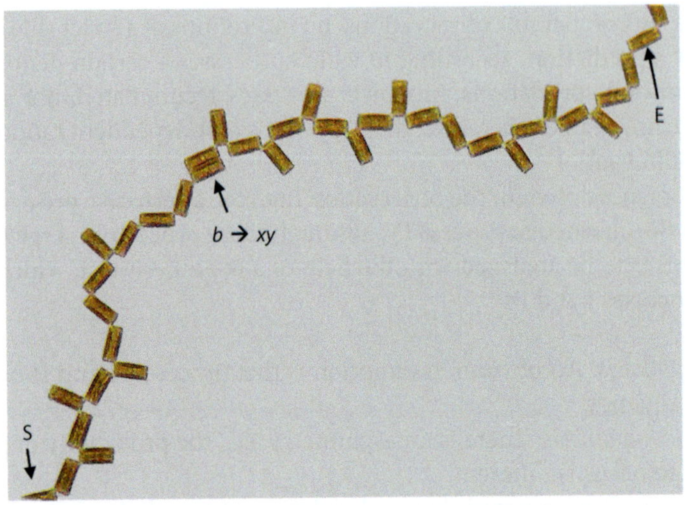

Figure 10.11 Section of a colony where an indication of pairing is recognizable. As the types of diatoms are determinable at the image edge, these are included, although they are cut.

The previous generation is obtained by combining the pairs according to the rules: *c b a b d a c* ***b*** *a b a b a c b a b d*, where the not-separated diatom pair was copied (written in bold). There are 6 substitutions $a \to da$ and 7 substitutions $a \to cb$. No other replacements of *a* and *b* occur. Carrying out the hypothesis test, p-values are 0.016 and 0.0078 for the respective process pairs.

10.3.3.4 Dependence on Environmental Parameters

Both the analysis of the images of the development of a colony and the analysis of single images consistently showed the asymmetry of the frequencies, but the occurring frequencies exhibited large fluctuations. These are not only a consequence of the short colonies considered, but also seem to depend on environmental parameters. Estimated values for the probabilities of the occurrence of the processes are therefore not given. Observations under intensive LED microscope illumination (approx. 3000 to 6000 lx, approx. 6000 K color temperature) seem to lead to higher p-values according to current knowledge, i.e., the difference in the frequencies of the processes is less pronounced. On the other hand, when evaluating single images, sections of up to 30 diatoms were found that showed deterministic behavior. These dependencies are likely complex and beyond the scope of this work.

10.3.4 Theory Formation

10.3.4.1 Description of the Asymmetry

The analysis showed that processes I and II, as well as processes III and IV, occurred with different probabilities. In all observations, a high proportion of processes I compared to alternative II was combined with a high proportion of processes III compared to alternative IV. No significant differences were found between the frequencies of occurrence of process I and III, and II and IV, in the same observation (same environmental conditions), so they are assumed to occur with the same probability. Apparently, an asymmetry is found in colonies. There is a view in which the processes I and III are frequent and a rotated view (operation T) in which these processes are rare. If one places the chain horizontally oriented so that processes I and III are the more likely processes, then this establishes an orientation in the direction of the apical plane. At the same time, an orientation of the pervalvar axis (axis connecting top valve and bottom valve) is also defined, because the processes of splitting elements *a* and *b* are not symmetrical to the vertical. It is reasonable to assume that the asymmetry is not an emergent property that occurs only in the aggregate of diatoms, i.e., a collective phenomenon of the diatoms involved, but exists at the level of individual diatoms. Otherwise, one would have to assume a functional coupling of the diatoms in a colony. Therefore, an asymmetry at the level of diatoms is assumed. This asymmetry is not included in the symbols (Figure 10.3) and therefore also not in the processes (Figures 10.4 and 10.5). Since only a formal description for an arbitrarily chosen assignment to diatoms in a selected view was determined, this is not a contradiction. One can take the asymmetry into account in extended symbols. For this purpose, the representation of the element *a* is completed by an arrow pointing to the location where the EPS pad is more likely to be formed at separation. Because of the orientation of the colony in the pervalvar direction, the symbol *c* is also extended by an arrow expressing this asymmetry (see Figure 10.12).

These two types are sufficient to describe all processes. To notate a colony, we label the directed elements in the spatial position shown as $\bar{a}$ and as $\bar{c}$. Such position-dependent notation is required to describe a colony by a string. Carrying out the transformation M, we obtain the elements $\bar{b}$ and $\bar{d}$ (see Figure 10.13).

These are not new types, but only other views. This allows the processes to be represented with asymmetry taken into account. In the following illustration (Figure 10.14), the more probable (probability $p > ½$) and less probable (probability $1\text{-}p$) process for the splitting of $\bar{a}$ are shown.

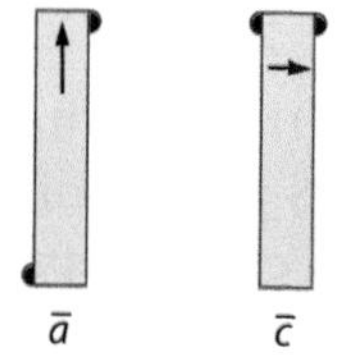

Figure 10.12 Real types that contain information about asymmetry.

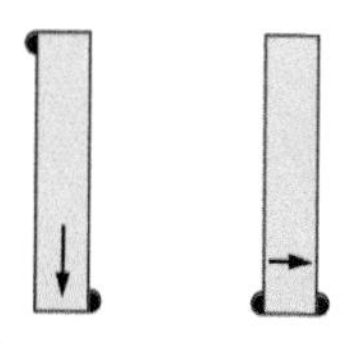

Figure 10.13 Elements created by mirroring on the horizontal line.

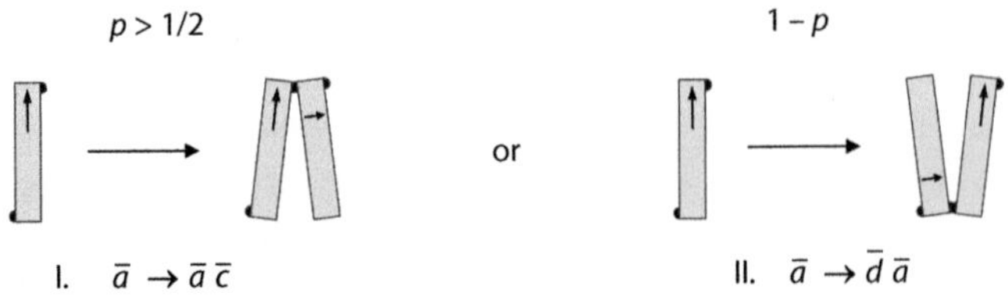

Figure 10.14 Illustration of the processes of splitting $\bar{a}$. The process identifiers have been kept.

Figure 10.15 Processes derived by the operation of mirroring M.

Since, according to the observation of dynamic behavior, the frequency of the processes occurs in each generation, the directional property must be inherited, as shown in the products of separation in Figure 10.14. It is hypothesized that the description by the extended symbols fully describes the possible arrangements, so that the graphical process representations in Figure 10.14 can be placed in any necessary position to obtain the remaining processes. Thus, by applying the mirror operation M, processes III and IV are derived, as shown in Figure 10.15.

In this model, processes I and III (or II and IV) are just different views of the same underlying process. Therefore, the probability of their occurrence p is also identical. Process V is assigned to the arrows as shown in Figure 10.16 on the left. Process VII is obtained by mirroring.

Here, the requirement of the indicator for alignment in the pervalvar direction becomes apparent, because the products of separation are not mirror symmetric with respect to the vertical. So, this cannot hold for $\bar{c}$ and $\bar{d}$ either.

An essential aspect of this concept is that the direction indicator points to the location of the probable position of the next EPS pad. Its direction is maintained in the daughter cells of this concept even when the rarer case of separation occurs at the opposite side. An alternative concept would be based on the assumption that the arrow always points to the location where the pad is formed. To model the less likely processes, one must then assume

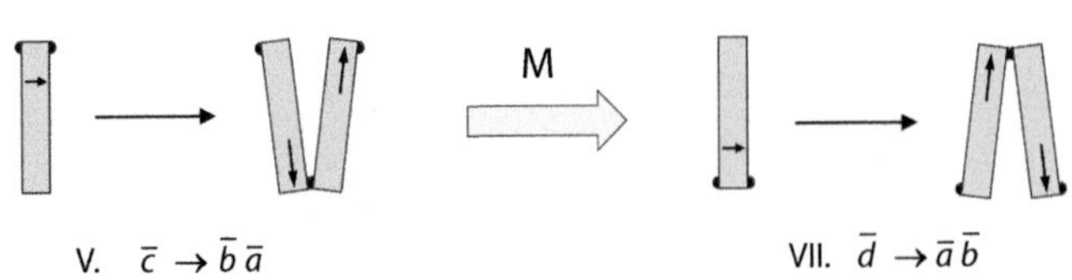

Figure 10.16 Processes starting from non-diagonal elements. The processes merge into each other by mirroring M.

that when a separation occurs, the direction of the arrow is not necessarily preserved, but flips with probability 1 − *p*, so that the alternative less likely process occurs in the subsequent generation. The concepts differ in their characteristics. If the arrow direction can change during a division, it is highly probable that this change will also affect further subsequent generations. Even within a synchronous section, contiguous zones would be formed, within which the frequencies of the types of division would be reversed. This could not be observed. In the presented analysis of the colony development, the subsequent processes of the rarer processes II and IV were again the frequent processes I and III. Nevertheless, the existence of such a flipping of the orientation cannot be excluded in principle with the data situation, especially as both possibilities could appear independently of each other.

10.3.4.2 Lindenmayer System

10.3.4.2.1 Formulation as L-System and Its Properties

In the following, synchronicity is assumed on a considered section of the colony. The presented theory then is a Lindenmayer system (L-system). Such an L-system is composed of a triple ([10.16] [10.17]) of

- Alphabet
- Replacement rules (production rules)
- Initial sequence ω (Axiom)

In the case of *D. vulgaris* colonies, the alphabet consists of the set $\{\overline{a},\overline{b},\overline{c},\overline{d}\}$. The replacement rules are given by the processes shown in Figures 10.14 to 10.16. The initial sequence ω represents a word (string, sequence of letters of the alphabet) with which the calculation of possible subsequent generations is started. It represents the first generation under consideration. The initial sequence can vary depending on the application, i.e., it can also be a single letter of the alphabet and thus represent a single cell. To determine a subsequent generation, the replacement rules are applied to all letters of the previous generation. The processes are chosen randomly and independently of each other according to the probability *p*. Thus, it is a one-dimensional, non-deterministic (i.e., stochastic) L-system. It is context-free, as the evolution depends only on one element, but not on its neighbors. Since the elements have no parameters, it is called “non-parametric.” There is one global parameter, the probability *p*.

A one-dimensional L-system was successfully used already in 1987 by Koster and Lindenmayer for describing the size sequence in the cyanobacterium *Anabaena catenula* [10.7]. Versatile applications in the field of plant structures can be found in Prusinkiewicz and Lindenmayer [10.28]. In the context of diatoms, it is used to generate the size sequence of diatoms in clonal colonies [10.31] [10.12].

10.3.4.2.2 Stochastic Model

The calculation of possible generation sequences can be carried out numerically with the help of a Monte Carlo simulation based on a start string. An exact comparison with observations cannot be made because of the randomness of the selection of the processes.

Based on the processes, rate equations can be formulated which allow the relative proportions of the different types $\overline{a},\overline{b},\overline{c}$ and $\overline{d}$ to be calculated. For longer development, the proportions of types $\overline{a}$ and $\overline{b}$ each converge toward 1/3, and types $\overline{c}$ and $\overline{d}$ each converge toward 1/6. These ratios are independent of probability p. Therefore, the expected values of these quantities do not allow the determination of p.

10.3.4.2.3 Idealized Deterministic Model

It turns out to be useful to consider the idealized L-system obtained by the choice $p = 1$. The system is completely given by the processes $\overline{a} \to \overline{a}\,\overline{c}$ and $c \to \overline{b}\,\overline{a}$, and their mirrored processes $\overline{b} \to \overline{b}\,\overline{d}$ and $\overline{d} \to \overline{a}\,\overline{b}$. The processes II and IV do not occur then. With these one-limit rules, the Lindenmayer system becomes deterministic and context-free and is referred to as a D0L-system (see, for example, [10.20]).

The generations are calculated iteratively according to the four processes listed above. One can also specify a simple calculation scheme which only requires concatenating strings as an operation (see Appendix 10A). It is more effective in computing a large number of generations than running the rules for each diatom. More important is a statement that follows directly from the algorithm. If a given pattern appears for the first time in a certain generation, it will also be found in all following generations. If a fragment or a section of a colony is found, for which the agreement with the theoretical sequence is to be checked, then one must look therefore only up to the first appearance of the associated sequence and knows that it appears in higher generations further. Thereby the number of matches increases. This is a consequence of the self-similarity of the sequences. This can be visualized by assigning colors to the letters (see Chapter 11 of this book, "RAPHE: Simulation of the Dynamics of Diatom Motility at the Molecular Level: The Domino Effect Hydration Model with Concerted Diffusion."

The restriction to a deterministic system allows conclusions about easily recognizable structures in the sequence of types. With the theorem presented below, one has a tool to quickly evaluate colonies according to whether they are candidates for fully deterministic development. It was already stated at the beginning that a longer colony can be described as a sequence of alternating elements a and b, into which elements c and d are inserted at certain positions, which applies analogously to the oriented elements. Between two equal elements of type $\overline{c}$ or $\overline{d}$ there is an even number of diagonal elements, between elements of type $\overline{c}$ and $\overline{d}$ it is an odd number. For the considered D0L-system, the theorem holds:

> "Let any letter of the alphabet be taken as the axiom. From a generation on, which has several elements of the types $\overline{c}$ or $\overline{d}$, it holds that $\overline{c}$ or $\overline{d}$ occur alternatingly and that between these elements there are always one or three elements of the diagonal types $(\overline{a},\overline{b})$.

Every real colony and every fragment of it goes back to the initial cell, the type of which is unknown. Therefore, the axiom was not specified more precisely. The proof is made by complete induction. A complete or strong induction is a proof technique that is described in textbooks, such as [10.11].

The base clause consists of the existence of a string of the mentioned preconditions, which arises from the letters of the alphabet. To arrive at a string containing at least two elements of the types $\overline{c}$ or $\overline{d}$, few iterations are necessary:

$$\bar{a} \rightarrow \bar{a}\,\bar{\mathbf{c}} \rightarrow \bar{a}\,\bar{\mathbf{c}}\,\bar{b}\,\bar{a} \rightarrow \bar{a}\,\bar{\mathbf{c}}\,\bar{b}\,\bar{a}\,\bar{b}\,\bar{\mathbf{d}}\,\bar{a}\,\bar{\mathbf{c}} \tag{10.5}$$

$$\bar{b} \rightarrow \bar{b}\,\bar{\mathbf{d}} \rightarrow \bar{b}\,\bar{\mathbf{d}}\,\bar{a}\,\bar{b} \rightarrow \bar{b}\,\bar{\mathbf{d}}\,\bar{a}\,\bar{b}\,\bar{a}\,\bar{\mathbf{c}}\,\bar{b}\,\bar{\mathbf{d}} \tag{10.6}$$

$$\bar{\mathbf{c}} \rightarrow \bar{b}\,\bar{a} \rightarrow \bar{b}\,\bar{\mathbf{d}}\,\bar{a}\,\bar{\mathbf{c}} \tag{10.7}$$

$$\bar{\mathbf{d}} \rightarrow \bar{a}\,\bar{b} \rightarrow \bar{a}\,\bar{\mathbf{c}}\,\bar{b}\,\bar{\mathbf{d}} \tag{10.8}$$

As (10.5) to (10.8) show, sequences arise which satisfy the alternation and spacing conditions. For easier recognition of the structure, the types $\bar{c}$ or $\bar{d}$ are written bold.

Inductive step: We now assume that the induction hypothesis is true for the generation n. So let there be a string of the form .... $\bar{a}\,\bar{b}\,\bar{a}\,\bar{b}\,\bar{a}$ ... in which elements of the types $\bar{c}$ and $\bar{d}$ are inserted alternatingly at intervals of one or three diagonal elements.

One can carry out the formation of the next generation in two steps. The first one applies the productions to the chain without the elements of types $\bar{c}$ and $\bar{d}$ and in the second step one inserts the products of these elements. The first step gives the chain:

$$\ldots\bar{a}\,\bar{\mathbf{c}}\,\bar{b}\,\bar{\mathbf{d}}\,\bar{a}\,\bar{\mathbf{c}}\,\bar{b}\,\bar{\mathbf{d}}\,\bar{a}\,\bar{\mathbf{c}}\,\bar{b}\,\bar{\mathbf{d}}\ldots \tag{10.9}$$

In it, $\bar{c}$ and $\bar{d}$ alternate with exactly one "diagonal" element in between. (The sequence $a\ c\ b\ d$ is shown in Figure 10.3b). Since the elements of the types $\bar{c}$ and $\bar{d}$, which have not yet been considered, produce only diagonal elements, the generation n+1 has the property of the alternating sequence of non-diagonal elements. If an element $\bar{c}$ is considered in the second step, it must be embedded between diagonal elements according to the induction hypothesis. In the generation $n+1, \bar{a}\,\bar{\mathbf{c}}\,\bar{b}$ becomes the sequence $\bar{a}\,\bar{\mathbf{c}}\,\bar{b}\,\bar{a}\,\bar{b}\,\bar{\mathbf{d}}$. Consideration of $\bar{c} \rightarrow \bar{a}\,\bar{b}$ increases the distance between $\bar{c}$ and $\bar{d}$ to three diagonal elements. The same applies to the consideration of $\bar{d}$. At places where a $\bar{c}$ or $\bar{d}$ exists in generation n, this increased distance is formed in the next generation. In all other positions the distance remains unchanged 1 as in the sequence (10.9).

Apparently, the proof also succeeds if one uses the processes II and IV instead of the processes I and III. When checking whether in a real colony there is a structure arising from a single cell according to the D0L-assumptions, it is not relevant from which side one notes the sequence of the elements. If there is a coincidence with the theoretical sequence, there is also a coincidence of the sequence transformed by T. Therefore, marking the orientation is not necessary.

10.3.4.2.4 Observation of the Idealized L-System

If $p \ll 1/2$, sections can often be found on a longer colony that correspond to the sequence of the deterministic L-system and possibly developed from a common cell without the processes II or IV. This justifies the consideration of the deterministic limit case. To find such sections, one regards areas with similar division states. The existence of a sequence which has the above properties of alternating elements $\bar{c}$ and $\bar{d}$ with distances of one or three diagonal elements is necessary for the section we are looking for, but not sufficient. Deviations can be caused by the occurrence of processes II or IV, as well as by loss of synchronization. Having determined the sequence of types from an image, one can compare it with the theoretical sequence. It is not important which character of the alphabet was chosen as axiom, because from each of these characters sequences arise in a few iterations which contain all characters of the alphabet, so that in the generation sequence the complete sequences to each character are also found (see also Appendix 10A).

The examples of chains in Figure 10.9 were deliberately chosen to allow successful analysis. To demonstrate the methodology and results, this analysis is now performed for Figure 10.9a. The visible section of the colony meets the required criteria because, in addition to similar developmental states, the orientation and distances of the non-diagonal elements fit the statements of the theorem. The diatoms cut in the image are considered because their type is recognizable. The sequence of types to the 24 visible diatoms when viewed from the right reads: *a b d a b a c b d a c b a b d a b a c b a b d a*. In the 7th generation, which arises from axiom a, this sequence is found for the first time as a subsequence (bold):

a c b a b d a c b d a b a c b a b d a b a c b d a c b a b d a c b d a b a c b d a c b a b d a b
a c b a b d a c b d a b a c b ***a b d a b a c b d a c b a b d a b a c b a b d a*** c b d a b ...

The analyses of the other colonies shown in Figure 10.9 can be found in Appendix 10B.

10.3.5 Outer Shape of the Colonies

Depending on population density, the appearance of *D. vulgaris* colonies can vary from short segments to an entangled ball. Common to these forms are elongated filaments that rarely show local strong changes in direction. To understand this qualitatively, the angles between successive diatoms are considered. These are dependent on external forces because of the elastic properties. Therefore, in real situations, it is usually not easy to determine the angles of the rest position. Non-adherent colonies are exceptions. In addition, the angles appear to increase slowly for some time after the rapid opening, so that they vary over a wider range. In practice, angle measurement in cultures is complicated by deviations from the exact valve view.

It is necessary to distinguish between three possibilities of connection between the defined types and thus three angles to be considered. Let $\alpha = \sphericalangle (a, b)$ be the smaller angle between the elements *a* and *b*. Accordingly, let $\beta = \sphericalangle (c, a) = \sphericalangle (b, c) = \sphericalangle (d, b) = \sphericalangle (a, d)$ and $\gamma = \sphericalangle (c, c) = \sphericalangle (d, d)$ be defined. The angles were determined from Figure 10.9 as an example. Mean values and standard deviations can be found in Table 10.2.

Table 10.2 Measured angles and standard deviations.

Figure	$\alpha = \sphericalangle (a, b)$	$\beta = \sphericalangle (c, a) = \sphericalangle (b, c) = \ldots$
10.9a	MV = 119°, SD = 14.7°	MV = 73°, SD = 23.3°
10.9b	MV = 98°, SD = 10.0°	MV = 88°, SD = 9.6°
10.9c	MV = 105°, SD = 23.9°	-

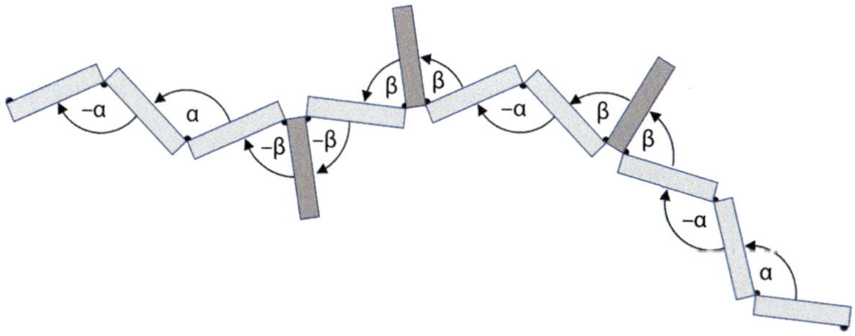

Figure 10.17 Exemplary colony projected into the plane.

The angle ∢ (*c*, *c*) can only be found in Figure 10.9d (see yellow arrow). Since the measurement of this angle is not reasonable because of the location of the colony and the resolution of the image, it was averaged from a number of other images. Typically, it is 95° with a wide range of variation (standard deviation ~ 25°).

For the following consideration, the statistical fluctuations of the mentioned angles are neglected. The orientation of a diatom in a colony with respect to the first considered diatom is obtained by partial summation of the difference angles between the diatoms. The signs of the angles have to be taken into account, as can be seen in Figure 10.17. If the rotation is mathematically positive (counterclockwise), the angles are positive, otherwise negative. This consideration assumes a sufficiently planar structure, as is usually observed.

Chains of diagonal elements, as shown in Figure 10.9c, have only the angle α with alternating sign and thus an alternating change of direction. In the absence of external forces, their curvature is given on a larger scale by the fluctuations of the angles. Such chains are only weakly curved. In the idealized deterministic case within synchronous sections, this is also true. The reason lies in the rules of pattern formation. In the band of diagonal elements, successive elements *c* and *d* are combined. Such a pair does not lead to a change of direction, because the whole contribution to the direction disappears. In the left part of Figure 10.17 this case is shown. In real, non-deterministic systems there are deviations, as shown in the right part of Figure 10.17. It has already been mentioned that on average the elements *c* and *d* are equally frequent. Thus, the chains are less smooth than in the D0L-system, but still usually have the shape of stretched filaments. It is essential for a stretched form of the filaments that the rules of the L-system do not lead to the accumulation of angle changes in the same direction. The abrupt dynamic change of the shape at individual divisions, i.e., the transitions between the patterns, has already been described.

10.4 Discussion

The structure of a *D. vulgaris* colony shows characteristics at different scales, from large, entangled structures of long filaments to the patterns of separations of diatoms. On a longer filament there are almost always sections of similar stages of division, sometimes with clear boundaries. Depending on the stage of development, basic types of patterns and less ordered transitional stages can be seen in these sections. Looking at the sequences of the described types, one is able to find the typical patterns generated by the L-system.

The knowledge about the deterministic limit case allows an evaluation of the order structure with the naked eye. In view of the variability of the appearances, the statement of Bory de Saint-Vincent is nevertheless understandable, that no organization is recognizable in the chains of *D. vulgaris*.

The methods presented can also be applied to other diatoms forming zig-zag colonies, i.e., other species of the genus *Diatoma* or species of the genus *Tabellaria*. Particularly, simple rules are found in *Diatoma tenuis*, as all detachments observed by the author are deterministic (to be published elsewhere).

Theodosius Dobzhansky's article, "Nothing in Biology Makes Sense Except in the Light of Evolution," is probably the most common citation in biology [10.8]. Pattern formation in *D. vulgaris* colonies based on selection of possible processes also requires interpretation in terms of evolutionary advantage. This is even more true as *Diatoma* spp. generate different patterns associated with different ways of life. For example, *D. tenuis* forms small colonies in the form of stars and connected stars, which are advantageous in their planktonic way of life [10.23] by causing a low sinking speed. The formation of filamentous colonies, which adhere to the bottom at the ends, allows *D. vulgaris* good access to nutrients and light in flowing water. This is even more relevant as *Diatoma* spp. is not motile, i.e., can only find sites with more favorable living conditions by drifting. The EPS pads at the connection points stabilize the chains, give them stiffness, and allow moderate flows without immediate loss of structure [10.4]. An extended structure protruding into the water body requires the chains to be sufficiently elongated. This is achieved by the morphogenesis of the chains based on the described L-system. A purely random selection of contact positions would not accomplish this. Figure 10.18 shows an example of this.

Questions remain open that leave room for further study. A central question concerns the completeness of the description. Are the processes described as nondeterministic actually stochastic, or is there an unknown deterministic process behind it? Against the background of ecology, the question remains whether deviations from the idealized D0L-system also provide an evolutionary advantage.

The model presented has only one parameter, the probability p. Although it was always clearly smaller than 1/2, it seems to be dependent on environmental parameters. Suspected, but not quantitatively proven, is a correlation with light intensity. Higher values were observed when division processes were observed over several generations under microscope illumination than in static image evaluations. Other relevant parameters could be the

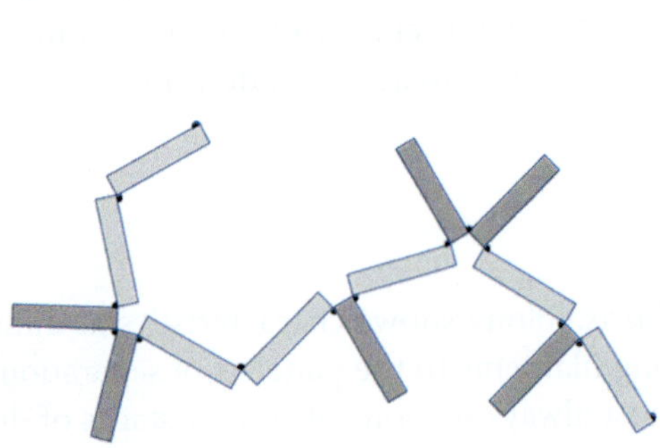

Figure 10.18 Random structure not formed according to the production rules of the D0L system. In a planar structure and with the angles chosen, a sequence of three consecutive non-diagonal elements would lead to a collision of parts of the chain with itself.

temperature and the age of the strain. The age of the strain could be parameterized with the mean apical cell length.

Diatoma vulgaris is isopolar, i.e., bilaterally symmetrical. Nevertheless, exactly one rimoportula exists positioned close to one of the apical pore fields [10.27]. The asymmetry described here with respect to the transapical plane in the pattern formation of the colonies is apparently not the only such asymmetry. This raises the question whether there is a connection of the morphogenesis of the colony with this morphological asymmetry.

The details of the formation of the EPS pads were not addressed, although it could be related to pattern formation. Does it start from both diatoms at both adjacent APFs at the same time or just from one pore field? What is the relationship with the rules of the L-system if it is formed by only one side? The asymmetry would allow a fixed assignment. With light microscopic methods the formation of the pads and an assignment to the excreting diatom could not be observed.

The mechanism of separation of neighboring diatoms raises further questions. The initially fluid extracellular polymeric substances apparently develop a torque in the EPS pad, which leads to detachment. In addition to explaining the chemical processes, this also gives rise to the question of dependence on environmental parameters. A striking feature is an immediate rise in the frequency of the separations when the light intensity is increased. As has been described, diatoms with neighboring EPS pads usually separate earlier with respect to the division state. This also requires a physiological explanation.

Acknowledgment

I am grateful to Claudia M. K. Leon for her help in translating Bory de Saint-Vincent's text.

Appendix 10A: Calculation Scheme

To derive a calculation scheme for the deterministic limit case, it is advantageous to describe the production rules by an operator P. It is applied to all letters of a string $G = g_1\, g_2\, g_3 \cdots g_j$:

$$\mathrm{P}\,(G) = \mathrm{P}\,(g_1\, g_2\, g_3 \cdots g_j) = \mathrm{P}\,(g_1)\,\mathrm{P}\,(g_2)\,\mathrm{P}\,(g_3) \ldots \mathrm{P}\,(g_j) \tag{A10.1}$$

The production rules of the D0L-system read:

$$\mathrm{P}(\overline{a}) = \overline{a}\,\overline{c} \tag{A10.2}$$

$$\mathrm{P}(\overline{b}) = \overline{b}\,\overline{d} \tag{A10.3}$$

$$\mathrm{P}(\overline{c}) = \overline{b}\,\overline{a} \tag{A10.4}$$

$$\mathrm{P}(\bar{d}) = \bar{a}\,\bar{b} \tag{A10.5}$$

Assuming that we know the sequences formed by the axioms $\bar{a}, \bar{b}, \bar{c}$ and $\bar{d}$ after n generations, i.e., $\mathrm{P}^n(\bar{a}), \mathrm{P}^n(\bar{b}), \mathrm{P}^n(\bar{c})$ and $\mathrm{P}^n(\bar{d})$, then (A10.1) and (A10.2) yield the sequences of generation n + 1 by:

$$\mathrm{P}^{n+1}(\bar{a}) = \mathrm{P}^n(\mathrm{P}(\bar{a})) = \mathrm{P}^n(\bar{a}\,\bar{c}) = \mathrm{P}^n(\bar{a})\mathrm{P}^n(\bar{c}) \tag{A10.6}$$

and accordingly, one obtains:

$$\mathrm{P}^{n+1}(\bar{b}) = \mathrm{P}^n(\bar{b})\mathrm{P}^n(\bar{d}) \tag{A10.7}$$

$$\mathrm{P}^{n+1}(\bar{c}) = \mathrm{P}^n(\bar{b})\mathrm{P}^n(\bar{a}) \tag{A10.8}$$

$$\mathrm{P}^{n+1}(\bar{d}) = \mathrm{P}^n(\bar{a})\mathrm{P}^n(\bar{b}) \tag{A10.9}$$

The calculation scheme consists in starting with the first generation of the products of the letters as seen in Eqs. (A10.2) to (A10.5). From each generation, one proceeds to the next by concatenating them according to the right-hand sides in Eqs. (A10.6) to (A10.9). If one would instead apply the production rules to each element, one would have to perform 2^n operations in the nth generation. In case one chooses as axiom a string with several characters, one composes the nth generation from the iteration of its characters.

Appendix 10B: Accordance with the D0L-System

Figure 10.9b: Including the cut diatoms at the edge of the image (23 diatoms) gives the sequence *a c b d a b a c b d a c b a b d a b a c b a b*, which satisfies the D0L approximation.
Figure 10.9c: Reading from bottom left to top right (the diatom cut in the image at the right edge is included) yields the sequence *a b a (b a) b (a b) a b a (b a) b a*, where the recently separated diatoms are grouped together with parentheses. Using the production rules $d \rightarrow a\ b$ and $cd \rightarrow b\ a$, the sequence *a b a c b d a b a c b a* is obtained by combining the parentheses. Testing for agreement with the D0L approximation yields no agreement. However, shortening the chain by the first two diatoms results in a sequence of 10 elements that conforms to the D0L model. At this end of the section considered, there was either a deviation from D0L-system or a loss of synchronicity.
Figure 10.9d: There are two analyzable sections marked in the image (start at "S", end at "E"). The longer section on the left (26 diatoms) is bounded by a sequence *b d d a* that apparently does not conform to the D0L model. With the boundaries given, the sequence is *a b d a c b d a b a c b d a c b a b d a c b d a b a* conforms to the D0L-system. This is also true for the shorter sequence on the right, which describes the connections of 14 diatoms: *a c b d a b a c b d a c b a*.

References

[10.1] Bahulikar, R.A., and Kroth, P.G. (2007) Localization of EPS components secreted by freshwater diatoms using differential staining with fluorophore-conjugated lectins and other fluorochromes. *European Journal of Phycology*, *42*(2), 199-208.

[10.2] Bory de Saint-Vincent, J.B.G.M. (1824). Diatome. *Diatoma*. In: *Dictionnaire Classique d'Histoire Naturelle* Tome 5ème CRA-D, I. Audouin et al. (eds), Rey et Gravier; Baudouin Freres Paris, 461

[10.3] Busby, W.F. and Lewin, J. (1967) Silicate uptake and silica shell formation by synchronously dividing cells of the diatom *Navicula pelticutosa* (Breb.) Hilse. *Journal of Phycology* 3, 127-131.

[10.4] Celler, K., Hodl, I., Simone, A., Battin and T.J., Picioreanu, C. (2014) A mass-spring model unveils the morphogenesis of phototrophic *Diatoma* biofilms. *Scientific Reports* 4, 3649

[10.5] Darley, W. M. and Volcani, B. E. (1969) Role of silicon in diatom metabolism. A silicon requirement for desoxyribonucleic acid synthesis in the diatom *Cylindrotheca fusiformis* Reimann and Lewin. *Experimental Cell Research* 58, 334-343.

[10.6] Davis, C.O., Harrison, P.J. and Dugdale, R.C. (1973) Continuous culture of marine diatoms under silicate limitation I. Synchronized life cycle of *Skeletonema costatum*. *Journal of Phycology* 9, 175-80.

[10.7] De Koster, C. G. and Lindenmayer, A. (1987) Discrete and continuous models for heterocyst differentiation in growing filaments of blue-green bacteria. *Acta Biotheoretica* 36, 249-273.

[10.8] Dobzhansky, T. (1973) Nothing in biology makes sense except in the light of evolution. *The American Biology Teacher* 35, 125-129.

[10.9] Engelberg, J. and Hirsch, H.R. (1966) On the Theory of Synchronous Cultures. In: *Cell Synchrony*. I.L. Cameron and G.M. Podilla (eds.) Academic Press, New York, 14-37

[10.10] Guillard, R.R.L. (1975) Culture of phytoplankton for feeding marine invertebrates. In: *Culture of Marine Invertebrate Animals*. W.L. Smith and M.H. Chanley (eds.) Plenum Press, New York, 26-60.

[10.11] Gunderson, D. S., and Rosen, K. H. (2010). *Handbook of mathematical induction*. Chapman and Hall/CRC, 2010.

[10.12] Harbich, T. (2021) On the Size Sequence of Diatoms in Clonal Chains In: *Diatom Morphogenesis [DIMO, Volume in the series: Diatoms: Biology & Applications, series editors: Richard Gordon & Joseph Seckbach]* V. Annenkov, J. Seckbach and R. Gordon, (eds.) Wiley-Scrivener, Beverly, MA, USA.

[10.13] Hejblum, G., Costagliola, D., Valleron, A.J. and Mary, J.Y. (1988) Cell-Cycle Models and Mother Daughter Correlation. *Journal of Theoretical Biology* 131(2), 255-262.

[10.14] Laney, S.R., Olson, R.J. and Sosik, H.M. (2012) Diatoms favor their younger daughters. *Limnology and Oceanography* 57(5), 1572-1578.

[10.15] Lewin, J.C., Reimann, B.E., Busby, W.F. and Volcani, B.E. (1966) Silica shell formation in synchronously dividing diatoms. In: *Cell Synchrony*. I.L. Cameron and G.M. Podilla (eds.) Academic Press, New York, 169-188

[10.16] Lindenmayer, A. (1968) Mathematical models for cellular interactions in development: I. Filaments with one-sided inputs. *Journal of Theoretical Biology* 18, 290-299.

[10.17] Lindenmayer, A. (1968) Mathematical models for cellular interactions in development: II. Simple and branching filaments with two-sided inputs. *Journal of Theoretical Biology* 18, 300-315.

[10.18] Locker, F. (1950) Beiträge zur Kenntnis des Formwechsels von Diatomeen an Hand von Kulturversuchen. Österreichische botanische Zeitschrift 97, *322-2*.

[10.19] Macdonald, J.D. (1869) On the structure of the diatomaceous frustule, and its genetic cycle. *Annals and Magazine of Natural History*. 4, 1–8.

[10.20] Mishra, J. and Mishra, S. (2007). *L-system Fractals*. Elsevier, 2007.

[10.21] Paasche, E. (1967) Marine Plankton Algae Grown with Light-Dark Cycles. I. *Coccolithus huxleyi*. *Physiologia Plantarum* 20, 946-956.

[10.22] Paasche, E. (1968) Marine plankton algae grown with light-dark cycles. 2. *Ditylum brightwellii* and *Nitzschia turgidula*. *Physiologia Plantarum* 21, 66–77.

[10.23] Padisák, J., Soróczki-Pintér, É., and Rezner, Z. (2003) Sinking properties of some phytoplankton shapes and the relation of form resistance to morphological diversity of plankton—an experimental study. *Hydrobiologia* 500(1), 243-257.

[10.24] Passy, S.I. (2002) Environmental randomness underlies morphological complexity of colonial diatoms. *Functional Ecology* 16, 690–695

[10.25] Pfitzer, E. (1869) Über den Bau und die Zellteilung der Diatomeen. Bot. Zeitung 27, 774–776.

[10.26] Pirson, A. and Lorenzen, H. (1966) Synchronized dividing algae. *Annual Rev. Plant Physiol.* 17, 439–58.

[10.27] Potapova, M. Diatoms of North America (2009) https://diatoms.org/species/diatoma_vulgaris

[10.28] Prusinkiewicz, P. and Lindenmayer, A. (1990) *The Algorithmic Beauty of Plants*. Springer, New York.

[10.29] Rimet, F. and Bouchez, A. (2012). Life-forms, cell-sizes and ecological guilds of diatoms in European rivers. *Knowledge and management of Aquatic Ecosystems*, (406), 01.

[10.30] Round, F.E., Crawford, R.M. and Mann, D.G. (1990) *The diatoms: biology and morphology of the genera*. Cambridge University Press, Cambridge.

[10.31] Ussing, A.P., Gordon, R., Ector, L., Buczko, K., Desnitskiy, A.G. and Vanlandingham, S.L. (2005) The colonial diatom "*Bacillaria paradoxa*": chaotic gliding motility, Lindenmeyer Model of colonial morphogenesis, and bibliography, with translation of O.F. Müller (1783), 'About a peculiar being in the beach-water'. *Diatom Monographs* Vol. 5. Koeltz Scientific Books, Koenigstein.

11

RAPHE: Simulation of the Dynamics of Diatom Motility at the Molecular Level – The Domino Effect Hydration Model with Concerted Diffusion

Shruti Raj Vansh Singh[1], Krishna Katyal[1] and Richard Gordon[2,3]*

[1]OpenWorm Foundation, Jammu, India
[2]Gulf Specimen Marine Laboratory & Aquarium, Panacea, Florida, USA
[3]C.S. Mott Center for Human Growth & Development, Department of Obstetrics & Gynecology, Wayne State University, Detroit, Michigan, USA

Abstract
Recent work has proved that the jerky gliding motility of diatoms known for centuries is present even at high spatial and temporal resolutions. Here we attempt a molecular level computer simulation using concerted diffusion, concerted rotational Brownian motion, and concerted hydration to account for the jerkiness, which includes occasional short excursions backwards from the mean direction of movement. The alignment of the mucopolysaccharide (which we call "raphan") that appears to be the major component of the material in the raphe and the diatom trail, perpendicular to its direction of movement, is in a mechanical instability on ejection from the cell to the raphe, which could tilt it either way. Here we confirm that a slight bias, perhaps attributable to shear from the general forward motion of the diatom, qualitatively accounts for the jerky motion. Electrophoresis of raphan in the raphe is an alternative possible source of bias. This is an example of Janus-faced causality: top-down causality is the diatom's movement influencing the direction of thrust by the raphan, bottom-up causality is the direction of thrust of the raphan influencing the direction the diatom moves. A domino effect is postulated for hydration of the raphan, possibly accounting for independence of a diatom's average speed from diatom size, for a given species. In our simulated model, the force for propulsion comes from radial expansion of the raphan as it enters the diatom trail. Direction of the diatom, but not speed, is imparted by direction of movement of the cell membrane, presumed to be carrying raphan synthase. While qualitative simulation of jerky diatom movement has been achieved, there is a large parameter space that needs to be specified experimentally to achieve quantitatively accurate simulations. Our approximations have been made explicit. Detailed comparison with computer simulations of other models is encouraged.

Keywords: Jerky gliding motility, diatoms, concerted diffusion, raphe, electrophoresis, domino effect, one-dimensional Ising model

**Corresponding author*: dickgordoncan@protonmail.com
Shruti Raj Vansh Singh: shrutirajvanshsingh@protonmail.com
Krishna Katyal: krishna.katyal@protonmail.com
Richard Gordon: https://www.researchgate.net/profile/Richard-Gordon-9, http://orcid.org/0000-0003-4970-9953

Janice L. Pappas (ed.) The Mathematical Biology of Diatoms, (291–342) © 2023 Scrivener Publishing LLC

11.1 Introduction

We have reviewed the long history of models for pennate diatom gliding motility and produced a new model [11.66]. We disregard here the popular model of cytoplasmic motor-driven polysaccharide molecules in the raphe, because it seems physically impossible, as that model predicts an exponential falloff in force from its postulated motor molecules through the raphe [11.66] [11.87] [11.88]. In this "adhesion complex" model [11.47] [11.141], the distance from membrane to substrate can be "upwards of 2 μm [11.91] [11.242] (0.3-3 μm [11.67]), reducing the transmission of force from motor proteins inside the cell, and thus reducing the efficiency of the diatom motor below its estimated 99% [11.63]. Furthermore, there is no microscopy evidence that the adhesion complex exists [11.11] [11.15].

Our purpose here is to demonstrate one instantiation of our new model, which contains many presently unknown parameters. Its principal features are:

1. The polysaccharide fibers in raphes (called "raphans" [11.66]) are assumed to be the source of the energy of motility; we assume that raphan is the major component of the material in the raphe [11.26] [11.66] and the diatom trail; the roles of minor components such as water, proteins and carbohydrates [11.24] [11.120] [11.141] [11.232], are not considered;
2. This comes about via swelling of the raphans on contact with water (hydrolysis is the likely means of raphan swelling, perhaps accompanied by uncoiling [11.63]);
3. The raphan units are inserted via the cell membrane mostly at the leading end of the raphe;
4. The raphans are carried by a postulated membrane-bound raphan synthase [11.66], analogous to cellulose synthase [11.49] [11.199] [11.236] [11.252];
5. A gradient of raphans available to enter the raphe is generated by membrane flow (perhaps amoeboid [11.66] [11.164] or cytoplasmic streaming [11.66]); that flow otherwise makes no direct contribution to motility);
6. The raphan synthase in the membrane is the source of raphan, with diffusion of the raphan into the raphe as its sink;
7. The pair of microfilament bundles adjacent to the raphe [11.63] [11.66] may be involved in directing the membrane flow to the raphe (Figure 14.50 of [11.55]) and may drive membrane motion;
8. The raphans in the raphe are modeled as being in single file;
9. In order to account for the average constant speed of a pennate diatom, we invoke concerted diffusion and concerted swelling of consecutive raphans, which act analogously to falling dominoes, which fall at near constant speeds;
10. An interaction is assumed between the average motion of the diatom and a slight tilt of the raphans, also assumed to be subject to propagation via a slight but concerted rotational Brownian motion.

Details of terminal raphe anatomy [11.114], motions of diatoms other than forward and back [11.13] [11.85], are not simulated, nor colonial behavior [11.2]. We make the implicit assumption that the S-shaped cross section of raphes [11.45] [11.93] [11.96] [11.185] [11.233] is not important for diatom motility. (Perhaps it strengthens the valve from crushing forces, as might the fibulae [11.176] of canal raphes [11.177] and the raphe sternum [11.204].) The raphe's axially bilateral asymmetry is also ignored here, though axially curved raphes do cause some diatom trajectories to be curved [11.84] [11.85]. The curvature at the ends of raphes (helictoglossa) in valve view is ignored, except that the ends are assumed to be barriers for raphan diffusion in the raphe. Thus, any raphe curvature is assumed here as irrelevant to the mechanism of motility. The generally slight shortening of the raphe relative to the diatom length, due to the positioning of the helictoglossa [11.176], is ignored. Other raphe asymmetries [11.74] and details of anatomy are ignored. Most motile pennate diatoms have a hylalin area in the middle of the raphe [11.74] [11.132]. How raphan gets across this silica barrier has not been addressed nor considered here. No attempt is made here to calculate the viscosity or adhesion [11.23], properties of the diatom trail [11.24] [11.34] [11.40] [11.67] [11.79] [11.88] [11.92] [11.97] [11.120] [11.129] [11.147] [11.186] [11.237] [11.240] [11.241]. We do not deal with the force of adhesion of a diatom to a substrate (methods applied to the pulling apart of plates adhered to one another by a viscous fluid [11.6] may be applicable). The raphe is thus simplified as a one-dimensional channel carrying raphan single file.

11.2 Parameters

Choices are made for the values of the many parameters, such as rate constants, that have to be specified in a computer simulation of this model. These can only be guesses, until experimentally measured. All rates are normalized relative to the rate of raphan hydrolysis (variable: raphanRateOfHydration), which may vary between species, and with temperature and other environmental parameters. Along the way we suggest options for more detailed simulations than that given here, and alternatives. Construction of an artificial, robotic, motile "diatom" might even be in the offing (cf. [11.159]). We hope that our modeling will either lead to experimental tests and/or rigorous simulations and testing of alternative or refined models.

Since this is the first attempt at a molecular level computer simulation of diatom gliding motility, we have made our assumptions explicit, as present-day computers do exactly what one tells them to do, not what one intends. In our computer program, for which we used Python 3.9.5 [11.170], variable names in the code are verbose, in an attempt at clarity (Table 11.1). The code is open source.

We assume that raphan occupies discrete positions approximately perpendicular (cf. [11.35] [11.45] [11.46] [11.224]) to their direction of flow [11.66] in the raphe (Figure 11.1). Our lattice approximation to continuum fluid mechanics can be justified [11.158]. It becomes a Markov simulation rather than one of continuum Brownian dynamics [11.101] (cf. [11.50] [11.193]). Note that, while a ratio of molecular width to lattice width of 10:1 is recommended for accurate lattice simulation of a continuum fluid [11.146], our ratio of 1:1 is suboptimal if the raphan positions in the raphe prove to be not discrete. At high raphan densities a phase transition to a nematic (in our case smectic) phase transition is anticipated [11.14] [11.153], which may be important in raphans retaining their orientation.

Table 11.1 The set of parameters that require values for a run of the RAPHE computer program. Except for the number of events simulated, we have 9 independent input variables that could vary between individual diatoms and species. A variable that changes with each event is labeled "none" under Default value.

Variable name	**Default value**
bias	0.6
distanceIncrementContributedBySwelling	3
directionForDiatom	none: value +1 or -1
directorDiffusionRatio	1
event	none: 0 to events
events	60
groupCharacteristic	10
(iChosen, jChosen)	none: (row, column) of event chosen in table
membraneCharacteristic	10
membraneOccupancy	0.25
raphanRateOfHydration	1
rapheLength	50
rateOfHopping	1
ratePerMembraneToRapheStep	1
table	none: real array (0: rapheLength-1,0: totalColumns)
totalColumns	8
totalRate	none: sum of all rates in table
Output Variables	**Type**
distanceDiatomMoves	none: real array (0:events)
rapheString	none: string of fixed width characters (1:rapheLength)
timeIncrement	none: real array (0:events)

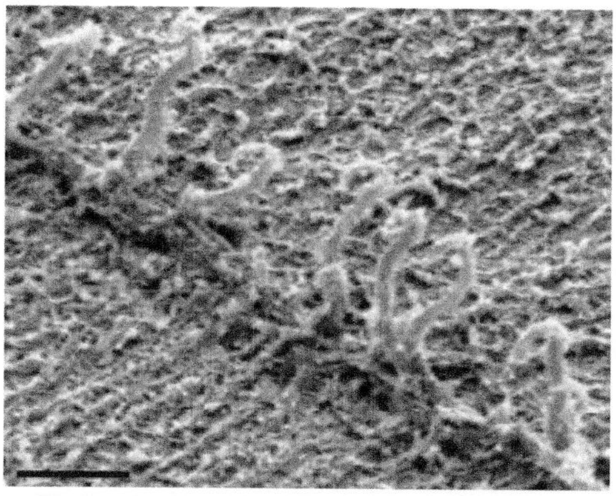

Figure 11.1 Scanning electron micrograph showing a single file row of secreted fibrils in *Pinnularia viridis*. Scale bar 500 nm. (Reproduced from [11.93] with permission of John Wiley & Sons/Blackwell Publishing).

11.3 Ising Lattice Modeling

We use a one-dimensional "lattice-gas" model, as has been developed for other systems of this and higher dimensions [11.58] [11.59] [11.60] [11.62] [11.216]. Each lattice site along the raphe is either occupied by raphan or not. Our model for the raphe becomes an example of a one-dimensional Ising system [11.1] [11.4] [11.52] [11.59] [11.62] [11.105] [11.255].

We assume a slight tilt from perpendicular orientation of each raphan. This is exaggerated by representing orientations as /////.\\\, where right (/) and left (\) tilted raphans are indicated, in this example separated by an empty (.) lattice site. If two groups "collide" by diffusion, we make the assumption that the concatenated group /////\\\ takes on the orientation of the larger of the two groups //////// by an unknown concerted mechanism, perhaps cooperative, rotational Brownian motion. We assume that concatenation is so fast that it takes little time relative to all other possible events. Thus, concatenation is not included as an event in Table 11.2. Concatenation, when the adjacent subgroups are of equal length, such as ///\\\, depends on a biased random effect, described below (variable: bias). One refinement would be to give concatenation a finite rate. The concatenation includes the possibility of reversal of the direction of the whole group according to the same bias. Our model becomes a 3-state one-dimensional Ising system with interactions extending to whole groups, instead of just nearest neighbor.

While the computer simulation focuses on what is happening inside the raphe, it is best thought of as involving three layers:

```
|.||.|..   membrane -> (carrying raphan synthase making raphan and
           moving left to right)
.\\\\.\\   raphe
--------   trail (adhering to the substrate)
```

considering the first 8 lattice sites. Note that a fixed width font has been used here, so that vertical alignments are accurately depicted.

Table 11.2 States and rates. All rates in columns 4 through 7 (variable: total_columns-1) are set to zero at the beginning of each computational loop for an event. For those events that can happen, the rates are then calculated and become positive real numbers, indicated here by a + in those columns for this example of a raphe configuration. These are recalculated for each event after concatenation. The actual numbers depend on the values for the parameters in Table 11.1. If groups contain oppositely directed raphans, they are first concatenated, as described in the text. Only already concatenated groups are shown here. States of lattice sites are shown both numerically and via symbols. -1 / means hydrates to the left, 0 means empty, +1 \ means hydrates to the right. The diatom is presumed to move in the opposite direction (variable: directionForDiatom) of the hydration when hydration occurs, by a fixed amount (variable: distanceIncrementContributedBySwelling). For readability of the code, the column names correspond to variable names, which are assigned the column number.

0	1	2	3	4	5	6	7
site no	**site state**	**run no**	**run length**	**fill**	**diffuse left**	**hydrate**	**diffuse right**
0	+1 /	1	7	0	0	+	+
1	+1 /	0	0	0	0	0	0
2	+1 /	0	0	0	0	0	0
3	+1 /	0	0	0	0	0	0
4	+1 /	0	0	0	0	0	0
5	+1 /	0	0	0	0	0	0
6	+1 /	0	0	0	0	0	0
7	0 .	2	3	+	0	0	0
8	0 .	0	0	+	0	0	0
9	0 .	0	0	+	0	0	0
10	+1 /	3	2	0	+	+	+
11	+1 /	0	0	0	0	0	0
12	0 .	4	2	+	0	0	0
13	0 .	0	0	+	0	0	0
14	-1 \	5	3	0	+	+	0
15	-1 \	0	0	0	0	0	0
16	-1 \	0	0	0	0	0	0

While we don't know where raphan is synthesized in the cell, we assume that it is transported to the cell membrane, which is represented above by a 1D lattice for its portion adjacent to the inner aspect of the raphe (variable: membraneOccupancy).

The membrane motion, one step to the right, may be visualized as:

```
| | • | | • | •    membrane
• \ \ \ \ • \ \    raphe
— — — — — — — —    trail
```

Here we have permitted one raphan synthase with attached raphan (|) to be added at the left of the membrane above the raphe.

To simplify the computer simulation, an average gradient in the membrane of raphan synthase in the membrane above the raphe is assumed, which declines exponentially along the raphe (variable: membraneCharacteristic, i.e., the "characteristic" value is $x = c$ at which $e^{-x/c} = e^{-1}$). In other words, the probability of a membrane lattice site being occupied is $e^{-x/c}$. This is a mean field approximation [11.221]. Nevertheless, in our lattice representation, a raphan can only jump from the membrane to the raphe if the corresponding lattice site in the raphe is unoccupied, as shown here for the first raphan, which then concatenates with the group of raphans in the raphe that it joins, including concerted tilting:

```
• | • | | • | •    membrane
\ \ \ \ \ • \ \    raphe
— — — — — — — —    trail
```

We presume that fluctuations in the raphan synthase gradient are of secondary importance. A more detailed simulation would generate the raphan synthase gradient as raphan departed from raphan synthase into the raphe and/or could model transport of the raphan to the membrane, perhaps via "crystalloid bodies" [11.35] [11.66] [11.98]. A canal raphe could be added to the simulation and interrupt this gradient, more or less periodically blocking access of raphan from the membrane to the raphe [11.177].

In other organisms, membrane speeds are 1.0-1.2 μm/min or 0. 017-0.02 μm/sec [11.150], but cytoplasmic streaming in diatoms has been observed at 5-10 μm/sec in diatoms and in the green alga *Chara* up to 100μm/sec [11.66], a wide range. However, we assume there is no direct coupling mechanism of membrane speed to diatom speed, so the membrane speed is not represented here by an explicit variable. On the other hand, membrane speed clearly affects the gradient of raphan synthase along the raphe, and so could be important for future simulations.

We assume that diffusion of raphan from the membrane to the raphe (variable: ratePerMembraneToRapheStep) is irreversible. This means that when a raphan moves from the membrane to the raphe to the diatom trail at each step, it never goes backwards. The above configuration cannot, for example, result in the following motion from the raphe to the membrane:

```
• | | | | • | •    membrane
\ \ • \ \ • \ \    raphe
— — — — — — — —    trail
```

11.4 Allowing Bias

We permit the possibility that the diffusion of raphan is biased, such as by electrophoresis of the raphan in the raphe (variable: bias). For example, capillary electrophoresis of polysaccharides [11.133] might prove to be analogous to what occurs in the raphe, especially if motile diatoms generate electric currents, of which they may be capable [11.16] [11.17] [11.75] [11.76] [11.77] [11.175] [11.217]. Extracellular glycoproteins of diatoms have been separated by electrophoresis [11.228], so this assumption warrants further investigation, perhaps by investigating the effects of external electric fields on diatom motility, which affect the motility of many other kinds of eukaryotic cells [11.33] [11.54] [11.89] [11.100] [11.148] [11.174] [11.197] [11.198] [11.203] [11.210] [11.251] [11.253]. At the molecular level, an electric field would cause the raphan to undergo a biased, bounded random walk [11.168] [11.220]. This assumption permits the backwards motion of a diatom, which is occasionally observed [11.179]. The same bias value is used for all events subject to bias.

What causes the hypothesized alternation of directional flow of the cell membrane is not known nor specified here (cf. [11.196]). One could speculate that it is driven by a reversible electric current (see below) and/or cytoplasmic streaming, which might involve motor molecules such as myosin [11.66] on the pair of microfilament bundles adjacent to the raphe [11.66] [11.71]. Whether or not the two microfilament bundles have opposite polarities has not yet been determined (cf. [11.124]).

For the reason explained below, we assume that diffusion of groups of raphan in the raphe occurs at a rate that depends on the size of the group (variable: groupCharacteristic), reduction or periodic variation of rate with size of the group has been observed in other systems [11.195], which is a plausible refinement.

We assume that irreversible, directional, explosive, concerted hydration of raphan groups into the diatom trail provides thrust and a nearly constant speed, independent of diatom length. This assumed phenomenon is analogous to falling dominoes as described below (cf. [11.162]). In detail, we programmed:

1. initial hydration of each group of raphans in the raphe at the leading end of the group;
2. global/local (Janus-faced [11.65] [11.73]) feedback between the mean speed of the diatom and the direction of thrust of a raphan group via a slight tilt of the raphan in the raphe, presumed reversible via rotational Brownian motion;
3. lengths of pennate diatoms can be 2 to 2000 μm [11.51] [11.154] (variable: raphanRateOfHydration), but we chose a length of only 50 lattice sites for display purposes.

We assume that the major component of raphe fluid, raphan [11.66], occupies discrete positions perpendicular (cf. [11.35] [11.45] [11.224]) to their direction of flow [11.66] in the raphe (Figure 11.1). This lattice approximation to continuum fluid mechanics can be justified [11.158]. Lattice models have been compared with continuum models in a similar situation [11.194]. However, periodic indentations in the diatom trail [11.232] might be interpreted as suggesting that there actually are discrete positions of the raphan, despite indications that the raphe lining is atomically smooth [11.66] [11.246]. (On the other hand, they might be Rayleigh instabilities of the viscous diatom trail [11.68] [11.69].) As lattice simulations are an order of magnitude faster on present computers than continuum molecular dynamics, and we don't yet know if there is an actual lattice in the raphe, we will nevertheless use the lattice approach here. Also, while a ratio of molecular width to lattice width of 10:1 is recommended for accurate lattice simulation of a continuum fluid [11.146], it should be noted that our ratio of 1:1 is suboptimal if the raphan positions in the raphe prove not discrete. In any case, at high raphan densities a phase transition to a nematic (in our case smectic [11.128] [11.205] [11.225] [11.243] [11.248] [11.249]) liquid crystal phase transition is anticipated [11.14] [11.153]. This assumption presents the problem that smaller diatoms of the same species probably have shorter raphe cross-sectional profiles, requiring a presumed adjustment of the raphan length to diatom size, or raphan shortening by folding or coiling. If raphan is multimolecular, this could be accomplished by sliding of individual molecules in each raphan bundle along the raphe depth. These speculations are testable by TEM of clonal colonies. The two ends of each raphan might be different, by analogy with cellulose, which has chitin at one end [11.66].

11.5 Computer Representation

We found it easiest to think and write the computer program using a table containing details of the current state of the raphe and all possible stochastic events it could undergo (variable: table) (Table 11.2). The table contains the state of each lattice site and the rates for each possible event bringing that state to another state. Table 11.2 shows the state after a typical event or the initial random setup of the raphe has occurred and been concatenated. As we are invoking group diffusion and hydration, we use a method called "run-length encoding" to define the groups [11.32] [11.61] [11.239]. The lattice sites are allowed three states:

1. / full, with a raphan that can hydrate to the left, pushing the diatom an increment to the right, and becoming empty in the process;
2. . empty, but capable of being filled by a raphan coming from the raphan synthase in the cell membrane;
3. \ full, with a raphan that can hydrate to the right, pushing the diatom an increment to the left, and becoming empty in the process.

Introduction of Boltzmann factors [11.163] and activation energies [11.72], presently unknown, might allow kinematics and thus temperature dependence, to be simulated. The "totally asymmetric simple exclusion process" (TASEP) [11.163] is a subset of our model, except that we do not allow end dissociation of raphan from groups, i.e., except for hydration, group formation and concatenation are assumed irreversible. Scarce data on temperature dependence of diatom speeds is shown in Figure 11.2 as an incentive for future work. As the motility mechanism appears to break down at higher temperatures [11.30] [11.55], the full biochemistry of diatom motility may have to be understood for modeling its temperature dependence. However, as noted in [11.30], "The uniformity of shape between the speed vs. temperature curves for the four species was striking, particularly when the data were standardized relative to the maximal cell speed," suggesting some universality of the motility mechanism amongst diatoms.

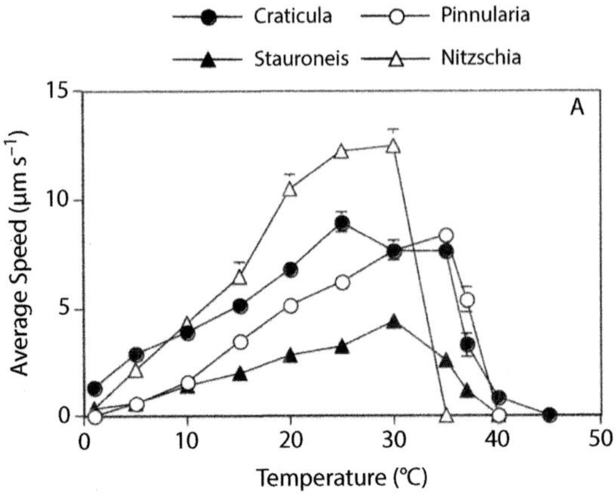

Figure 11.2 Temperature dependence of average speed for four genera. (Reproduced from [11.30] with permission of Taylor & Francis).

11.6 The Roles of the Cell Membrane, Canal Raphes, and the Diatotepum

Raphan synthase with attached raphan may have actually been seen in the membrane: "Ultrastructural studies of diatoms have revealed bundles of 7-nm microfilaments running longitudinally in the cortical cytoplasm parallel to the raphe systems [11.39] [11.46]" [11.91]. (The observations in [11.35], with fibrils pictured parallel to the raphe rather than perpendicular to it in their Figure 14, may have been of what are now known to be actin microfilament bundles [11.167], but similarly oriented mucopolysaccharides were found by [11.41]). This parallel orientation might be due to alignment by membrane flow over the raphe, but is not included in our simulation. It could imply that raphan rotates -90° when moving from the membrane to its vertical position in the raphe (possibly caused by the electric field of the cell membrane, by analogy with magnetic alignment of paramagnetic ellipsoids [11.215]), followed by the apparent +90° rotation during exit from the raphe to the trail, where 0° is along the raphe axis.

We make the implicit assumption that release of raphan from raphan synthase is not the rate-limiting step for its diffusion into the raphe. We make no assumptions about how raphan is released from raphan synthase into the raphe, except that a lattice site in the raphe must be unoccupied to receive a raphan.

Although we do not use it here, note that the diffusion coefficient of raphan synthase in membrane can be estimated at 0.125 $\mu m^2/s$, based on the average for membrane proteins in [11.131]. While simple membrane diffusion could be modeled, domains in the plasma membrane may produce much more complicated diffusion patterns [11.99] [11.119] [11.127]. Similarly, the typical membrane occupancy by proteins is 0.25 [11.213] (variable: membraneOccupancy). We assume that all membrane proteins are raphan synthase.

Partial plasmolysis does not hinder diatom motility [11.63] [11.165] [11.166], and could be simulated by intervals of contact between the cell membrane and the inner surface of the raphe, with an effect analogous to canal raphes. The same might be true of the diatotepum whose function appears to be unknown, but is usually missing over the raphe [11.129] [11.136] [11.139] [11.187] [11.219] [11.233].

11.7 Raphan and the Raphe

It is possible, as in cellulose [11.66], that cach raphan is actually made of multiple molecules. This would imply that we are making an assumption that raphan molecules are not exchanged between raphans if they do contain multiple molecules, nor get entangled in the raphe, nor together occupy more than one lattice site. Our raphe width = raphan width assumption has been widespread in models for raphe fluid [11.66], and single file movement appears correct from SEMs [11.66] [11.93] (Figure 11.1). A caveat is that the swirling patterns of particles attached to the dorsal (unused) raphe may suggest that raphans can get past one another, and thus may not be aligned single file in the raphe [11.47] [11.141] [11.149]. On the other hand, for some unknown reason this only occurs on the dorsal raphe, i.e., the raphe not attached to a substrate [11.39] [11.47] [11.91].

Note that the swollen raphan width is reported as 2 μm = 2000 nm in [11.66] [11.107], and 12.5 nm in [11.39], the latter used in [11.254]. Widely varying determinations of swollen raphan widths may also be due to "variation in specimen preparation" (Ryan Drum, personal communication, 2021), an objection equally applicable to anhydrous raphan width measurements, since most TEM preparation methods involve aqueous solutions of reagents. Determination of raphe widths by TEM may be artifactual, due to stresses placed on the raphe during specimen preparation. Focused ion beams may produce more reliable estimates [11.246].

11.8 The Jerky Motion of Diatoms

In [11.179] we confirmed the observation of large frame to frame one-dimensional accelerations of gliding diatoms between movie frames observed by [11.44] at 10 frames per second. To our surprise, similar observations of even larger one-dimensional accelerations occurred at 890 frames per second [11.179]. This result is to be expected, in retrospect, because we showed that diatom motion could be fit by Brownian motion with drift [11.179]. One way of looking at this is that the higher the rate of sampling of a Brownian motion over a given time interval, the higher is the probability of capturing a rare large excursion and not averaging it out with the more common small excursions before and after it (cf. the long tails in low mass distributions in [11.122]). See Appendix. (For observations at even slower frame rate than [11.44], see [11.106] [11.144] [11.145]).

As our previous study [11.179] was done at the limits of bright field light microscopy, with subpixel resolution made possible by the rigidity of the valve in the species studied, it became clear that other approaches are required to resolve the events that occur between frames. An attempt to listen to moving diatoms at best set an upper bound to the expected staccato noise [11.254]. Atomic force microscopy (AFM) with an AFM tip attached to a motile diatom, perhaps via materials incorporated into live diatom frustules [11.25] [11.172] [11.173] [11.227] [11.229], might resolve the motion.

Two mechanisms contributing to the observed jerky motion were proposed [11.179]:

1. explosive hydration of anhydrous raphan [11.66];
2. snapping of raphan stuck to the valve as the diatom leaves it behind in the diatom trail.

They are not mutually exclusive. Here we only simulate the first proposal, nor do we consider the suggestion of a tug-of-war between dorsal and ventral raphes [11.141].

11.9 Diffusion and Concerted Diffusion of Raphan

Our assumption that the raphan width = raphe width results in single file motion, a major constraint we place on raphan diffusion in the raphe, and we make the assumption that it holds. It also enhances our assumption of concerted diffusion of raphan, a phenomenon that has been studied and occurs for atoms and small molecules in both 1- and 2-dimensions [11.6] [11.113] [11.137] [11.146] [11.195], and which we assume is extendable to large molecules such as raphan. In concerted diffusion, local groups of atoms or molecules move together. If they are in single file, they move along a one-dimensional line, simplifying computation.

The self-diffusion constant of raphan in the raphe is difficult to estimate. We can start with values of $D = O(10^7 \text{nm}^2/\text{s})$ for polysaccharides of molecular weight about 4×10^5 daltons in aqueous solution [11.20], but these are for hydrated, coiled polysaccharides with axial ratios of at most 20. For diffusion of pure polymers, we turn to polymer melts, with $D = O(10^3 \text{ to } 10^6 \text{nm}^2/\text{s}))$ [11.140] with molecular weight (MW) $O(10^4)$ for a linear polyethylene. For polymer melts confined to thin films we get $D = O(10^2 \text{nm}^2/\text{s}))$ [11.112]. For unentangled polymers the diffusion constant decreases with increasing chain length [11.36]. If we assume that raphan is a linear polymer of sucrose (MW 342), taking the square root of the cross-sectional area of sucrose (as we do not know its 3D orientation in raphan), as an example, we get 1.3 nm per monomer in a polysaccharide [11.169]. Thus, a raphan of 160 nm to 3 μm in length, if it is a single molecule, would consist of 120 to 2300 monomers (equivalent to axial ratios), and have MWs of 41,000 to nearly 800,000. In one case, the raphe cross-sectional (curved) height is 1 μm and the raphan appears to be 0.3 to 0.5 μm in length (Figures 10–13 in [11.45]), falling into this range.

Note that the general diffusion equation yields:

$$<(x^2)> = 2dDt \tag{11.1}$$

where d is the dimension (1, 2, or 3) to which a molecule is confined, x is the raphan width, and the distance traversed in time t is $<x^2>$ [11.38], where $< >$ means "expectation of." The rate is $1/t$ in raphans per time, or in one-dimension a "speed" of x/t, where x is the width of a raphan. In our case, $d = 1$. As the estimates for the diffusion constant D above have a huge range, we only use relative rates per site and do not attempt to estimate absolute times or distances.

11.10 Shear and Janus-Faced Causation: Motility and Raphan Tilting

We hypothesize that the raphans are tilted in the direction of motion of the diatom by shear due to the motion of the diatom in a given direction. Such a tilt could help provide a component of force parallel to the direction of the diatom as the raphan discharges. Such a tilt could be looked for by TEM of rapidly frozen [11.41], moving diatoms. There is presently no evidence for tilt during motion, but no one has looked. Tilt would change the raphan in a raphe from a single layer of a smectic A phase to a single layer of a smectic C phase [11.18]. In [11.24] a distinction is made between "mucilage strands" and "mucilage trails," the latter containing polymers that are "highly-coiled" and entangled. The dynamics of injection of hydrated raphan into the diatom trail could thus be a quite complicated process. The effective component of force in the direction of the diatom's motion depends on many unknowns, perhaps species-specific factors, including raphan rigidity, structure, hydration, reptation [11.181], interaction with other raphe fluid components, viscosity of the trail, adhesion to the substrate, etc.

Shear is an example of Janus-faced feedback between a global and a local phenomenon [11.65] [11.73], in this case biasing the direction of "fall" of the raphan from its vertical orientation in the raphe to its horizontal position in the diatom trail, a form of symmetry breaking from an unstable configuration, as it could "fall" either way. This is analogous to the instability present when one tries to balance a pencil on its tip [11.130] [11.135], which is in an unstable equilibrium [11.7] (Figure 11.3). Even when the tilt angle due to shear is not 0, rotational Brownian motion [11.111] [11.231] permits "falling" either way. The shear is presumed to come from the average motion of the diatom. It is possible that the raphe walls play a role.

Figure 11.3 A pencil balanced on its tip could fall in any direction. A perfect balance of raphan can be designated by the symbol |, perpendicular to the axis of the raphe, whereas tilting left or right can be indicated by \ or /. On hydration, such raphans would push the diatom right or left, respectively (with the pivot point corresponding to the "eraser" end). We assume tilting occurs instantaneously, upon injection from the membrane into the raphe. Groups are allowed to flip back and forth (variable: bias). (Reproduced from [11.171] with open access permission from SlideServe).

11.11 The Domino Effect Causes Size Independence of Diatom Speed

Setting up a row of dominoes and knocking the first one over provides an analogy to one-dimensional groups of raphan in the raphe: the stored energy (hydration swelling) is analogous to the potential energy of gravity [11.160]. Consider a row of standing dominoes, which tend to fall in the same direction [11.64] (Figure 11.4). If raphans in a group usually swell left (the leading end in our simulations) to right, then swelling could push some raphans off into the trail at the trailing (right) end. Each swelling takes a certain amount of time (perhaps similar to the 0.1 μsec for discobolocyst discharge [11.63]). Note that the rate of propagation of falling of dominoes is nearly independent of the total number of dominoes, which thus provides a possible explanation for the independence of average speed from the length of the diatom. The limited data on this independence of diatoms is summarized in Table 11.3 and Figure 11.5. Independence of speed from cell volume has been observed for 100 species [11.12]. Near independence of speed versus cell size was found by [11.29]. The speed of a mother cell and its daughters after cell division was reported as "similar" [11.106]. On the other hand, Thomas Harbich (personal communication, 2021) reports: "In general, it seems to be true that shorter individuals of the same species move more slowly. This was also visible with the naked eye in other species." Nevertheless, assuming size independence, the main difference between diatom raphan and falling dominoes is that the diatom, so to speak, repeatedly sets up "dominoes" (raphans) to replace those that have "fallen" (swollen, exiting to the diatom trail). An analogy would be a machine that sets up new bowling pins as some fall. The spacing of raphans is presumably set by their noncovalent interactions, whereas the average speed of falling dominoes varies with their spacing, which is determined by hand when they are set up [11.56]. However, despite the elimination of transients (perhaps occurring in raphan groups too), empirical measurements [11.121] show average speeds for dominoes that are nearly independent of their spacing. Later work taking into account "cooperative groups" of neighboring dominoes found that

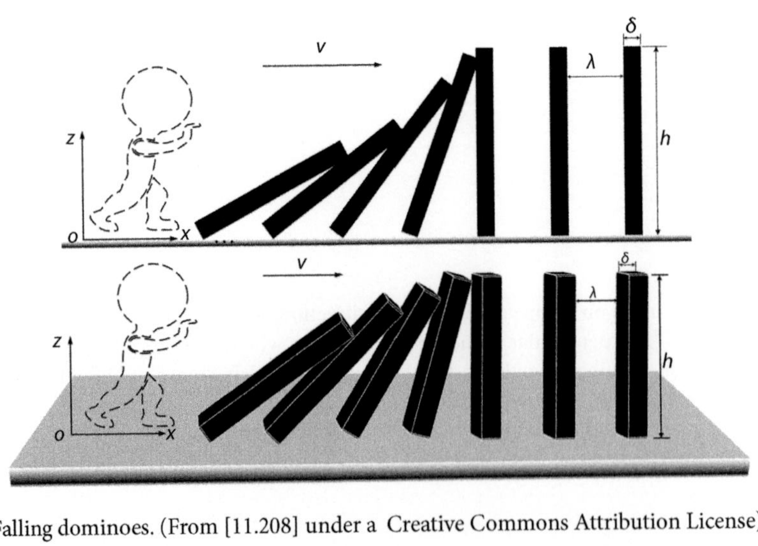

Figure 11.4 Falling dominoes. (From [11.208] under a Creative Commons Attribution License).

Table 11.3 Published data on pennate diatom speeds and their lengths. Speeds in three-dimensional substrates are not compiled here. As in our simulation, data for the typical smooth, clean microscope slide is assembled here and all other properties related to the substrate are ignored, such as adhesion, roughness, gaps, or obstacles.

Genus	Species	Average speed (μm/s)	Maximum speed (μm/s)	Length (μm)	References
Craticula		10.0	20	109.9 ± 0.8	[11.31]
Gyrosigma	*spencerii*	4.7±0.2	7.1		[11.90]
Navicula		0.5-1.7	2.7-5	15	[11.144] [11.145]
Nitzschia		10.4	20	129.9 ± 3.3	[11.31]
Nitzschia	*communis*	7	9	30	[11.80]
Nuvicula	*oblongu*	11		86-176	[11.86]
Phaeodactylum	*tricornutum*	0.017±0.001	0.05	3.5-8.5	[11.106] [11.218]
Pinnularia		5.3	13	166.4 ± 1.4	[11.31]
Pleurosigma	*angulatum*		9	65-75	[11.48] [11.83]
Stauroneis		4.6	12	139.6 ± 1.3	[11.31]

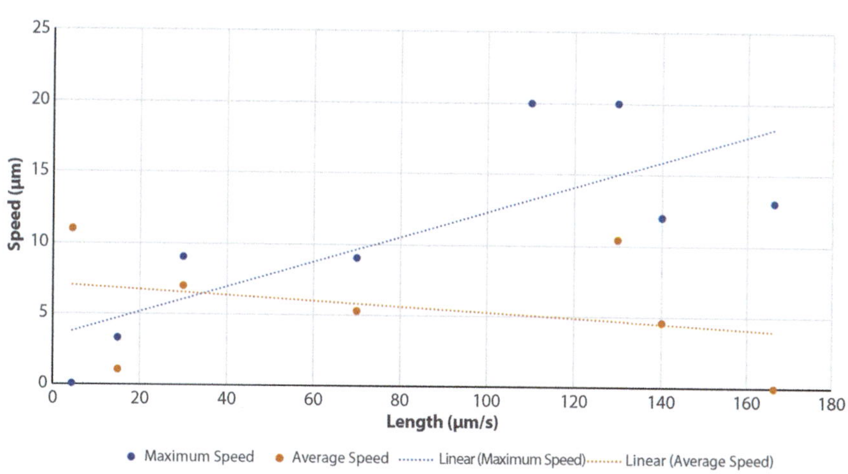

Figure 11.5 Speed of diatoms (μm/s) in Table 11.3 versus length (μm) for various diatoms of many species. The linear fits have R2 values of 60.7% for the maximum speed (blue) and 4.1% for average speed (green). While these low values, despite being below or near the standard cutoff for significance of 50%, might contradict our and other's assumptions of independence of speed from length (see text), we hope this will inspire more definitive experiments. Given the decreasing size vs. generation in the vegetative phase of diatoms, a more definitive test would be to compare speeds of diatoms of different sizes from the same clone.

for strings of 4 to 6 [11.190] [11.192] [11.209] or more dominoes the "domino wave" speed or "asymptotic" speed [11.191] [11.206] [11.207] [11.223] is nearly constant (Figure 3 in [11.191] [11.192]). Propagation is as a "solitary wave" [11.192].

We have one more implicit assumption, namely that "front" and "rear" are defined by the average direction of motion (not counting brief reversals of motion [11.179]) and the membrane gradient of raphan synthase. Whether or not any trail is deposited at the "front" of a diatom during its brief backwards excursions has not been investigated. Short reversals in direction could serve to lay down the trail for subsequent forward motion. Total internal reflectance microscopy [11.244] might help here.

The domino effect is quite widespread in nature and artifices. Single-walled carbon nanotubes collapse via a domino effect [11.21] [11.22], photoinduced phase transitions can occur via the domino effect [11.152], water transport in Spanish moss uses the domino effect [11.82], as do popsicle stick structures [11.184] and other macroscopic phenomena [11.81] [11.110]. Sparklers [11.3] [11.102] [11.104], nerve propagation [11.78], burning cigarettes, cigars and fuses for fireworks and dynamite fuses which propagate their reactions at nearly constant average speed, probably also exhibit the domino effect. Domino effects occur through interactions at the intra- and intermolecular level [11.103] [11.151] [11.155] [11.188] [11.202].

11.12 Quantitating the Swelling of Raphan in the Diatom Trail

Consider a 10 μm long diatom moving at 30 μm/sec. With an unhydrolyzed raphan width of 7 nm [11.66] [11.107], a raphe would be occupied by at most 1430 raphans. In one second the diatom would expel a trail 30 μm long, suggesting that each raphan contributes 21 nm in length to the trail after it swells. Thus, we use a maximum factor of 3 in estimating the thrust or movement per raphan (variable: distanceIncrementContributedBySwelling). Note that the time per discharge is $1/1430 = 7\times10^{-4}$ seconds, much slower than the 10^{-7}s for discharge of a discobolocyst [11.63]. In other words, when one raphan in the raphe hydrates, an adjacent raphan is triggered to hydrate, but this domino effect, relaying of triggering may take 7000× longer than hydration of a raphan itself. We assume that this is the time per domino effect propagation of hydration from one raphan to the next in a concerted group. Note that if the raphe is full, it would take a long time (a full second in this case) to discharge, during which it would start to fill up again. This is the reason we reduce the rate of hydration with group length (variable: groupCharacteristic), so that the smaller raphan groups contribute most of the motility. Also, we hydrate a group sequentially, allowing other events of filling or diffusion to occur in the time in between hydrations.

We can also roughly estimate the distance a diatom moves per raphan released into the trail in the following way. From [11.24], the cross section of the trail under one *Navicula* sp. diatom was 120 nm high and 400 nm wide or a cross-sectional area of 4800 nm^2, with a swollen raphan length of 160 nm (a short raphan). Let's take the swollen diameter as 12.5 nm [11.39] [11.254], so that the projected area of one raphan is 2000 nm^2. This gives about 40% of the expansion of a raphan contributing to the motion of the diatom and the rest goes "sideways," making the diatom trail wider than an individual raphan. We do not take into account the orientation of the raphan in estimating its thrust, just noting that, perhaps due to shear after expansion, raphan orientation changes from vertical in the raphe to horizontal in the diatom trail left behind [11.34] [11.40] [11.67] [11.79] [11.88] [11.92] [11.97] [11.129] [11.147] [11.186] [11.234] [11.241] [11.245]. Variations in raphan length, swelling, coiling and the ratio of projected area of a raphan to the size of the trail beneath a diatom might account for variations in diatom speeds between species of 0.17 to 30.5 μm/s [11.29] [11.134].

One set of TEMs suggests that the raphan retains its vertical alignment in the trail beneath the raphe [11.45] [11.141]. This may indicate that the stiffness of raphan varies between species. Note that here anhydrous raphan is presumed to uncoil, with an axial ratio of up to nearly 430, for a 3 μm thick raphe ("raphe depth" 1 μm [11.24], 0.3-3 μm [11.67]) and unhydrated raphan width of 7 nm. The value 0.4 we take as a lower bound, with an upper bound of 3 (variable: distanceIncrementContributedBySwelling) estimated above for the factor giving the motion per swollen raphan (output variable: array distanceDiatom-Moves (event)).

This presumably explosive process could involve both swelling [11.107] and elongation [11.63] of the raphan: "secreted fibrils experience hydration and begin to swell and elongate" [11.47]. (Swelling in length, although such is possible [11.109], does not explicitly enter in our model.) If raphan is similar to hyaluronan, hydrophilic swelling may correlate with "a highly negative fixed charge density" [11.230], which could also vary between species.

Of course, if a large raphan group discharges backwards, under our assumptions, this would be tantamount to a reversal of direction of the whole diatom, also reversing the tilt effect. This might, therefore, explain reversals of diatoms when there are no apparent obstacles [11.145].

11.13 A Schematic of Raphan Discharge

Each group of raphans therefore is hypothesized to have one of two configurations:

.\\\\\\\\\\\\\\\\\\\\\\\\\\\\\\\\\. for forward propulsion
or
.//////////////////////////////////. for backwards propulsion

The discharge of the raphan is in the opposite direction to subsequent motion of the diatom. Since we presume that water for hydration of the raphan comes in from below, the group termini with .\ or /. are more likely to be hydrated than the opposite ends (taken as probability = 1 in the simulation, though it could be less, suggesting the need for another parameter). Thus, the domino effect should proceed from left to right or right to left, respectively, yielding forward or backward propulsion of the diatom. Any bias is interpreted using the same parameter as the biasing due to electrophoresis (variable: bias). Each raphan is presumed to swell in response to its neighbor swelling, so the amount of motion, on average, should be proportional to the length of the group.

We can visualize the one-dimensional wave of swelling of a group as follows:

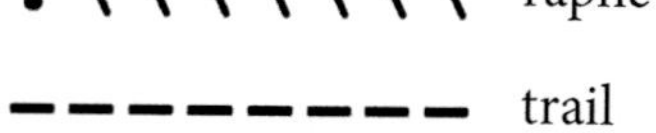

where the - signs below the raphe represent the diatom trail in a fixed width font.

On hydration of the leftmost raphan from the raphe we get (with our default variable: distanceIncrementContributedBySwelling = 3),

```
..\\\\\\        raphe
-----------     trail
```

where it has swollen, rotated, and been removed to the trail from the anhydrous raphe. This representation oversimplifies the phenomena we expect in the diatom trail. As hydration propagates:

```
...\\\\\         raphe
--------------   trail
```

After expulsion of the whole group to the trail we have:

```
........                           raphe
--------------------------------   trail
```

during which raphan from the membrane will, asynchronously, at least partially fill in the raphe and diffuse:

```
.\\/.\\\                           raphe
--------------------------------   trail
```

The extension of the trail via the swelling of the raphans on exit from the raphe into the trail indicates movement of the diatom, in this example, to the left, assuming that the trail adheres to the substrate.

11.14 Transitions of Raphan

We started by assuming that each individual raphan in the raphe can only have a discrete set of states, and that only the states of the nearest neighbor lattice sites can alter the state of a given raphan. This approach was inspired by [11.94] [11.95]. This is a situation analogous to the rules in Wolfram patterns [11.247]. However, the number of states is complicated by our assumption of concerted diffusion and concerted hydration of groups that can be bigger than the 3 consecutive sites used for Wolfram rules (Figure 11.6). Another significant difference from Wolfram patterns is that our rules are applied asynchronously [11.8] [11.178] [11.226]. Nevertheless, consecutive states of the raphe produce Wolfram-like patterns (Figure 11.7, output variable: rapheString). Note that any 1-dimensional Ising lattice system, followed over time, will generate a Wolfram-like chaotic pattern. We are investigating whether or not the Wolfram rules, when generalized, lead to new patterns, or that the rules are closed in the sense of being sufficient to generate all such patterns.

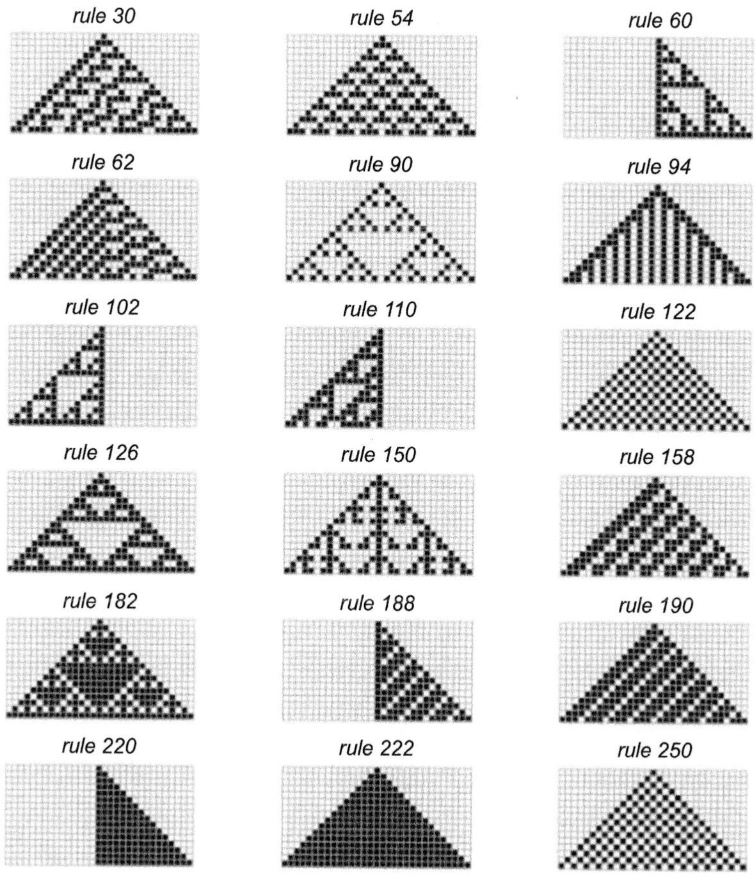

Figure 11.6 A sampling of patterns generated by the Wolfram rules. Rule 30 is chaotic. (Reproduced from [11.238] with permission of Wolfram Research).

BIAS - 0.8

```
//./\.\...\/../\///./\\./\\\.\.\//\/.///....//\\. 0
//.\.\./...//../////.\\\.////./..////.\\\....\\\\. 1
\\.\.\.\...//..\\\\\.///.\\\\.\..\\\..///....\\\\. 2
\\.\.\.\...\\.\\\\\\.///.\\\\.\..\\\..\\\....\\\\. 3
\\.\...\...\\.\\\\\\.\\\.\\\\.\..///..///....\\\\. 4
\\.\...\...//.\\\\\\.\\\\\\\..\..\\\..\\\....\\\\. 5
\\./...\..\\..\\\\\\.\\\\\\\..\..///..\\\....\\\\. 6
\\.\..\...\\..\\\\\\.\\\\\\\../..\\\..\\\....\\\\. 7
\\.\../...\\..\\\\\\.///////../..\\...\\\....////. 8
\\./..\...//..\\\\\\.\\\\\\...\..\\...\\\....\\\\. 9
\\.\../...\\..\\\\\..\\\\\\.../..\\...\\\....\\\\. 10
\\.\..\...\\../////..\\\\\\...\..\....///....////. 11
\\./......\\../////..\\\\\\...\..\....\\\....////. 12
\\.\......//..\\\\\..\\\\\\...\..\....\\\...\\\\.. 13
\\.\......//..\\\\\..\\\\\\......\....\\\...\\\\.. 14
```

```
\\.\......\\..\\\\\.\\\\\\.......\....\\\...\\\\.. 15
\\.\......\\..\\\\\\\\\\\......../..../// ...\\\\.. 16
//./......\\..\\\\\\\\\\\........\....\\\...\\\... 17
\..\......\\..///////////........\..../// ...\\\... 18
\..\.....\\...\\\\\\\\\\\......../..../// ...\\\... 19
\..\.....\\...///////////./......\....\\\...///... 20
\../.....\\...\\\\\\\\\\\......../....\\\...\\\... 21
\..\.....\\...\\\\\\\\\\\.........\....\\\...\\\... 22
/..\.....\\...\\\\\\\\\\\......../...../// ...\\\... 23
\../...\.\\...\\\\\\\\\\\........\.....\\\...\\\... 24
\..\...\.\\...//////////........\......//...\\\... 25
\..\...\.\\...//////////........\......\....\\\... 26
\...\..\.\\...\\\\\\\\\\........\....../....\\\... 27
\......\.\\...\\\\\\\\\\........\....../....\\\... 28
\./....\.//...//////////........\......\....///... 29
\\.....\.\\...//////////........\....../....\\\... 30
//.....\.//...\\\\\\\\\\........\...../.....\\\... 31
\\.....\.\\...//////////........\.....\.....\\.... 32
\\.......//...\\\\\\\\\\........\.....\.....\\.... 33
//.......\\...\\\\\\\\\\........./....\.....\\.... 34
\........\\...\\\\\\\\\\.........\....\.....\\.... 35
\........\\...\\\\\\\\\\............../.....\\.... 36
\.........\\..\\\\\\\\\\.............\.....\\.... 37
\.........\\...\\\\\\\\\\............./.....\\.... 38
\\........\\...\\\\\\\\\\.............\.....\\.... 39
//........\....\\\\\\\\\\.............\.....\\.... 40
\\...\..../....\\\\\\\\\\............./.....\\.... 41
//...\..../.....//////////.............\.....\\.... 42
\\...\..../.....\\\\\\\\\\.............\.....\..... 43
\..../..../.....\\\\\\\\\\.............\.....\..... 44
\....\.../......//////////.............\.....\..... 45
\....\...\......//////////............\....../..... 46
\....\...\......\\\\\\\\\\............\............ 47
\....\../.......\\\\\\\\\\............\............ 48
...../..\.......\\\\\\\\\\............\............ 49
.....\../.......//////////......................... 50
```

11.15 The Roles of the Diatom Trail

The trail is not simulated. As the time increments are calculated (1/(chosen rate), from the variable table (Table 11.1), these are recorded (output variable: array timeIncrement (event)). But the trail increases in length with each raphan hydrated to the right (variable: distanceIncrementContributedBySwelling).

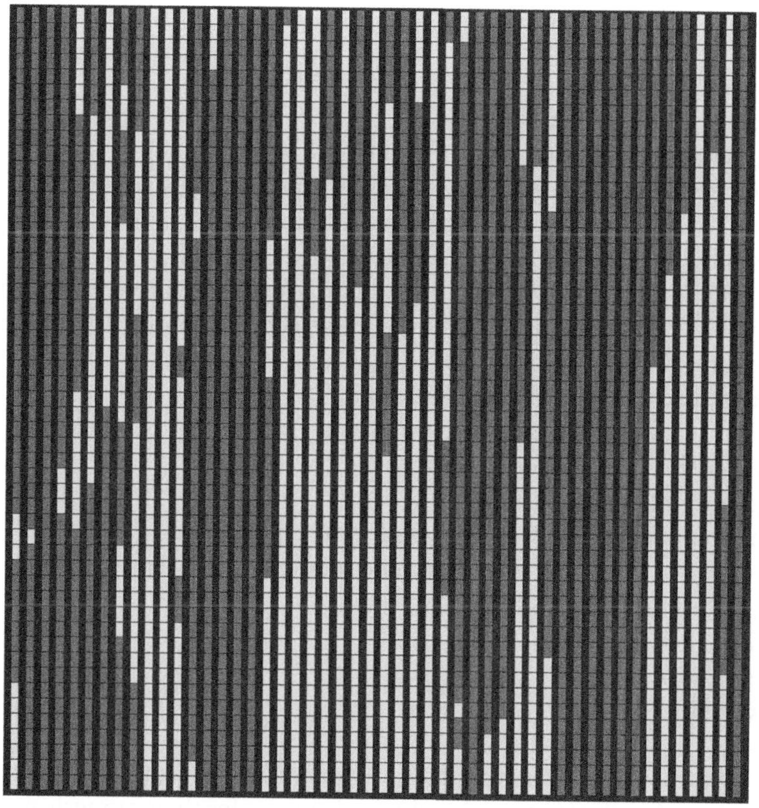

Figure 11.7 Top: Details of the first 50 events for bias = 0.8. Bottom: Wolfram-like representation of the consecutive states, ignoring tilt of the raphans. This emphasizes the diffusion events, which do not directly contribute to motility, as the hydration events do this.

Note that for a smectic liquid, diffusion out of one layer (the "director" direction in which nematic molecules are aligned) is often much faster than diffusion within the layer [11.5] [11.142] [11.235], though others conclude the opposite [11.157], depending on temperature [11.180]. Typical ratio is 0.3 to 3 [11.180] (cf. 1.2 to 4 in [11.118]) (variable: directorDiffusionRatio).

Each hydration/discharge moves the diatom the same distance: we ignore the likely statistical variation in this parameter, which could also contribute to jerkiness of diatom movement.

11.16 Outline of the Simulation

We used the continuous time Monte Carlo method. It requires simultaneous calculation of all possible events at a given moment, random selection of one event to occur (probability = its rate in the table divided by totalRate, Table 11.1), and calculation of the time step [11.62]. The disadvantage of the classical Monte Carlo approach [11.138] is that sometimes no events happen per its fixed time step, requiring more computer time.

The general flow of the RAPHE program is:

1. Setup input parameters and initialize derived variables
2. Initialize output variables
3. Loop for a designated number of events
   a. concatenate collided groups, with the larger changing the orientation of the smaller, a process assumed to be instantaneous compared to other events
   b. list all possible subsequent molecular events and their rates
   c. possible events are in 5 categories: filling from the membrane, diffusion of groups left or right, and hydration of the leading raphan of a group to the left or right
   d. normalize rates to the sum of all rates
   e. select at random which event is to occur
   f. change the state of affected lattice sites
   g. update all output variables
   h. save output data, such as symbols for the raphe state
4. Go to step #3 unless the pre-set number of events has been reached
5. Output distance diatom moved vs. time

11.17 Results

All inputs to the computer program are physical quantities, presently unknown, so the best we can do here is find analogies or make educated guesses for their numerical values, as we

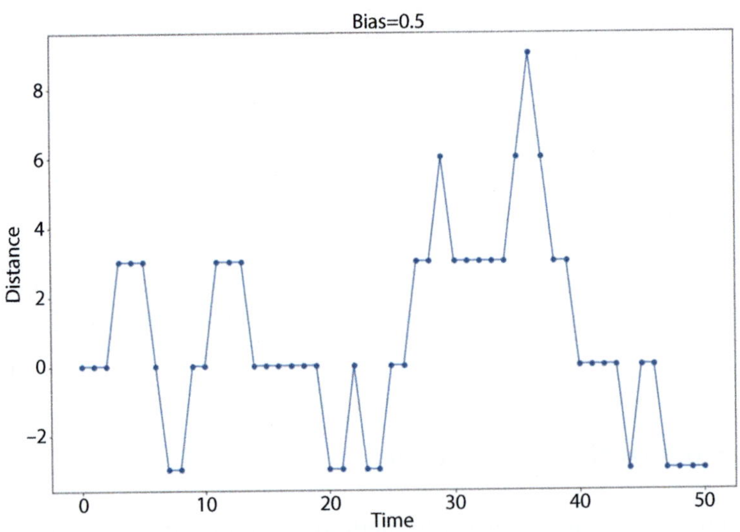

Figure 11.8 Sample runs with the bias parameter incremented from 0.5 (no bias) to 1 (maximum bias). Note: 1) even with no bias, the source of the raphan being highest at one end is sufficient, so that the net motion is forwards; 2) in all cases, except bias = 1, some backwards motion occurs. More or less steady forward progress of the diatom may require bias > 0.7. All other variables were set to the defaults (Table 11.1). For intuitive clarity, in this figure the mean motion of the simulated diatom is to the right. Note that in all cases with bias > 0.5 there are long periods of nearly constant speed. *(Continued)*

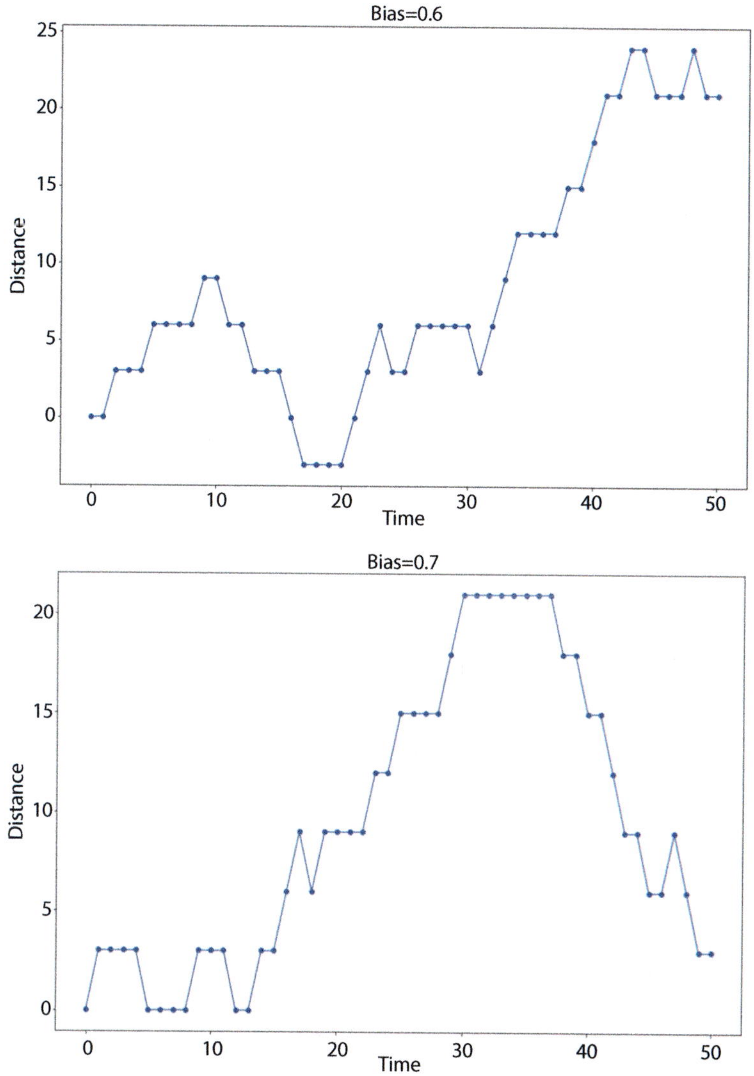

Figure 11.8 (Continued) Sample runs with the bias parameter incremented from 0.5 (no bias) to 1 (maximum bias). Note: 1) even with no bias, the source of the raphan being highest at one end is sufficient, so that the net motion is forwards; 2) in all cases, except bias = 1, some backwards motion occurs. More or less steady forward progress of the diatom may require bias > 0.7. All other variables were set to the defaults (Table 11.1). For intuitive clarity, in this figure the mean motion of the simulated diatom is to the right. Note that in all cases with bias > 0.5 there are long periods of nearly constant speed. *(Continued)*

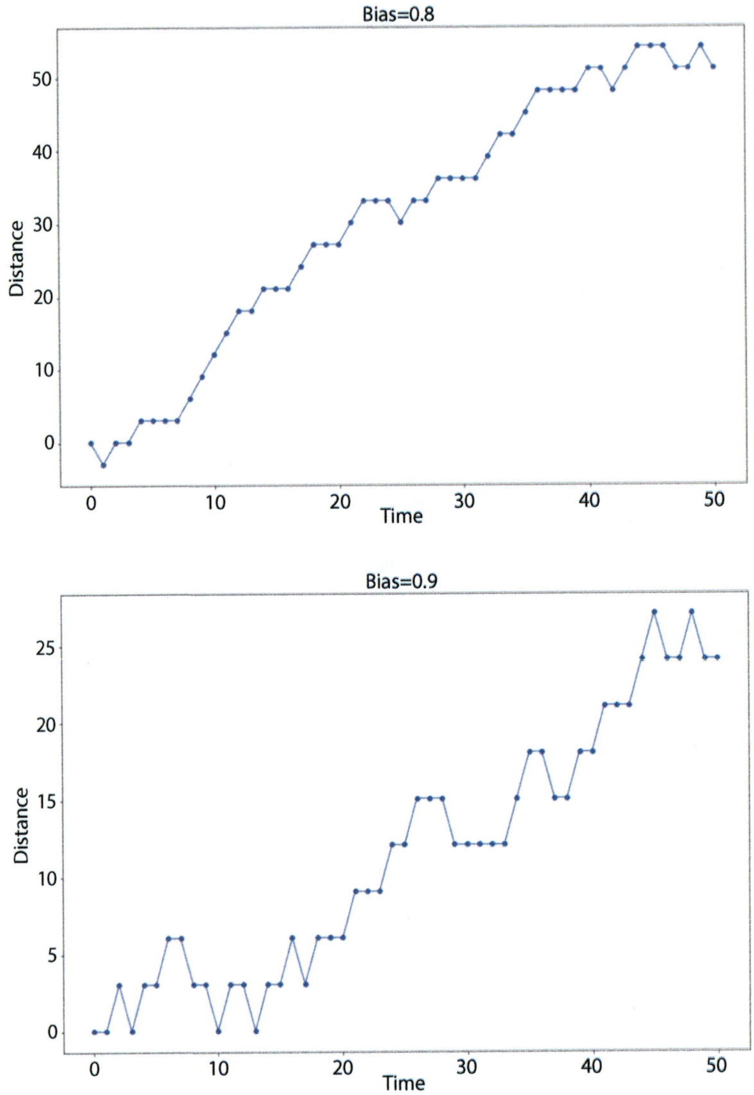

Figure 11.8 (Continued) Sample runs with the bias parameter incremented from 0.5 (no bias) to 1 (maximum bias). Note: 1) even with no bias, the source of the raphan being highest at one end is sufficient, so that the net motion is forwards; 2) in all cases, except bias = 1, some backwards motion occurs. More or less steady forward progress of the diatom may require bias > 0.7. All other variables were set to the defaults (Table 11.1). For intuitive clarity, in this figure the mean motion of the simulated diatom is to the right. Note that in all cases with bias > 0.5 there are long periods of nearly constant speed. *(Continued)*

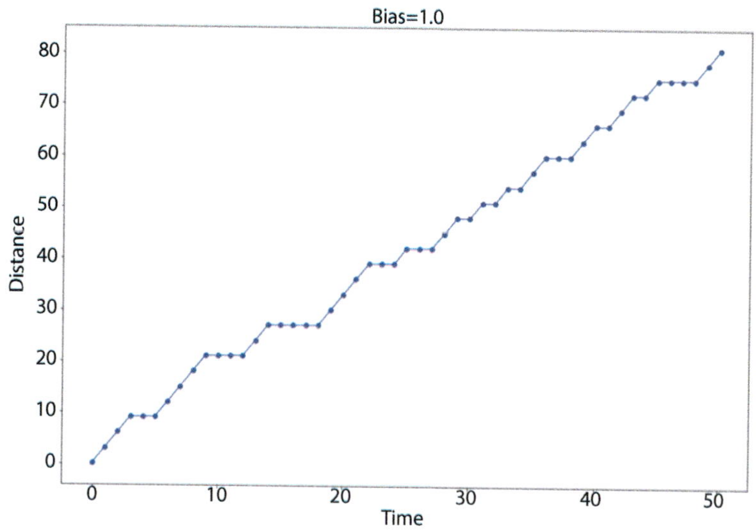

Figure 11.8 (Continued) Sample runs with the bias parameter incremented from 0.5 (no bias) to 1 (maximum bias). Note: 1) even with no bias, the source of the raphan being highest at one end is sufficient, so that the net motion is forwards; 2) in all cases, except bias = 1, some backwards motion occurs. More or less steady forward progress of the diatom may require bias > 0.7. All other variables were set to the defaults (Table 11.1). For intuitive clarity, in this figure the mean motion of the simulated diatom is to the right. Note that in all cases with bias > 0.5 there are long periods of nearly constant speed.

have tried in our defaults (Table 11.1). Output is distance versus time. One could also tabulate internal gradients of the hypothesized raphan and the distributions of group sizes and their positions, which might be observable by flash freezing [11.41], but these are presently unobserved.

Because of the large number of variables (Table 11.1), we show just a few sample runs (Figure 11.8), varying only the bias, from 0.5 (no bias) to 1 (maximum bias), and do not attempt to fit the model to real data [11.179].

11.18 Discussion

The substrate could be represented as a periodic or irregular surface against which raphan release can occur, in which case one would like to account for its degree of hydrophobicity [11.10] [11.19] [11.108] [11.214]. The lower speed of diatoms moving through sediments [11.90] requires explanation (cf. [11.134]), not attempted here, as we do not calculate the force generated by a moving diatom [11.67] [11.88]. The trail is not simulated per se. Other motions of the diatom [11.13] are not simulated. Feedback from "touch" by the substrate or nearby diatoms is not simulated, but as the cell membrane raphan synthase profile over the raphe and thus its occupancy reflects the movement, the current model could be expanded to allow some sort of "communication" between diatom motion and its cell despite the generally rigid valve, by altering the gradient of raphan synthase in the membrane because of the "backup," a sign to the cell that movement has been obstructed (cf. [11.189]). It has been observed that *Pinnularia* stop and reverse if they hit an object perpendicular to its surface, but not at an angle to it (Thomas Harbich, personal communication, 2021), suggesting that in this situation flow in the raphe may halt, and thus presumably transport of raphan from the membrane to the raphe halts. Perhaps the cell can detect this situation. Reversal can be triggered by a "wall of light" [11.28], which followed by flash freezing electron microscopy might indicate if the groups of raphans in the raphe or membrane of a stopped diatom are different from those of cruising in one direction. The compact, nonmoving configuration of *Bacillaria* in the dark [11.222] might also be useful in this regard.

Alternatives to concerted diffusion are the phenomenon of jerky movement of drops [11.211] [11.212], growing crystals [11.116] [11.143], percolation through porous media [11.156], and "chaotic jerk functions" [11.201]. The difference in forces involved in the advancing versus the retracting edge of a drop [11.219] is analogous to the snapping of raphan at the trailing edge, and might also be involved in the jerkiness of diatom motility. Backwards motion is also sometimes observed [11.125] [11.126] [11.200] [11.211] [11.212], and we would then attribute the jerkiness of diatom motility to such a wetting phenomenon of the diatom trail against its substrate, in addition to or rather than phenomena inside the raphe. The apparent lack of jerkiness of *Bacillaria* colonies (Thomas Harbich, personal communication, 2021, [11.250]) might be due to lack of jerkiness of wetting against the neighbouring valve or insufficient time resolution. For now, we will not consider this alternative, until the motion of *Bacillaria* is quantified at high temporal resolution. Observations of single motile diatoms using total internal reflectance might help resolve whether the above phenomena contribute to the jerky movement. The other components of the raphe fluid, other than raphan, could be involved as wetting surfactants [11.117]. The motion of diatoms crossing recent trails of other diatoms might also help understand the motility mechanism, if wetting is involved.

There is also a possible role of diatom pores in entangling raphan [11.123], though this may only apply to adhesive polymer on the valve [11.24] [11.141] [11.182].

One remarkable phenomenon that may be involved is that the raphan may be a "reentrant nematic," with up to an order of magnitude increase in diffusion rate [11.43]. If there is an electric field parallel to the raphe, it could bias diffusion [11.9]. A "slip-stick" mechanism is also a possibility [11.37] [11.53].

11.19 Conclusion

While this model represents a first molecular level attempt at simulating diatom motility, some of its many assumptions may prove incorrect, or get modified by further knowledge and attempts to incorporate other diatom movements. The basic assumption behind jerkiness is that consecutive groups of raphe mucopolysaccharides (raphans) can hydrate in either direction. Three mechanisms are simulated that might explain sustained motility of a diatom, mostly in one direction:

1. Membrane motion along the raphe can provide a gradient of raphan to the raphe along its length (variables: rapheLength, membraneCharacteristic);
2. Electrophoresis along the raphe could bias diffusion of raphan within the raphe (variables: rateOfHopping, bias);

3. A global/local Janus-faced feedback of the average motion of a diatom, causing a slight tilt of the raphan, could bias the direction (but not so much as to prevent short backwards excursions, possibly initiated by rotational Brownian motion).

In the simulation, these three mechanisms could be tested separately or together, as they are not mutually exclusive. The second and third cannot yet be distinguished in their effects, and are summarized here via use of biased random walks [11.27] [11.57] [11.115] [11.161] [11.183].

Three mechanisms are proposed that involve concerted actions of raphan:

1. In the raphe, concerted diffusion of groups of raphan;
2. Concerted slight tilting of groups (with concatenation of collided groups);
3. Concerted hydration of each group in falling domino order.

The second involves concerted Brownian rotation, which is here assumed to be instantaneous. If it actually occurs at a rate comparable to other events, another rate column would have to be added to Table 11.1 and another rate variable would be needed. The raphan at the leading edge of a group is assumed to hydrate first, triggering hydration (variable: raphanRateOfHydration) of the other raphans in the group consecutively, by analogy with falling dominoes. The nearly uniform speed at which this propagation occurs for small groups is assumed to set the maximum speed of a diatom, fixing it for a given species, explaining the possible independence of speed on the size of an individual diatom (Figure 11.4, Table 11.3). While the membrane is assumed to be in motion to bring raphan synthase to the raphe, only its direction, not its speed, is assumed to set the average direction of motion of the diatom. The two bundles of microfilaments adjacent to a raphe may be involved in producing this membrane motion, perhaps via motor proteins, and occasionally reversing the membrane flow.

Because of the number of unknowns, we make no attempt to "fit" the parameters to any particular diatom, except to visually check if the behavior of the simulation looks at all like the observed behavior (Figure 11.6). We hope that our attempt will be applied to alternative models [11.66], requiring them in the process to be examined in detail for their assumptions and the possibility of experimentally testing them. In the course of our modeling, alternative phenomena became apparent that might aid in explaining diatom motility.

Some of our assumptions may be testable by high spatial and temporal resolution methods in fluorescent microscopy and transmission electron microscopy. The easiest and most important parameter that needs immediate experimental study is speed vs. length for many species of raphid pennate diatoms.

We have suggested that the motility mechanism in diatoms may contain two examples of top-down and bottom-up Janus-faced logic [11.65] [11.73]:

1. a slight tilt of the raphan caused by the average direction of motility, and vice versa;
2. an interaction between the membrane concentration gradient of raphan synthesis and whether or not the raphan in the raphe is moving, allowing the cell to obtain feedback on obstacles hindering its motion, i.e., a sense of "touch."

In retrospect, we could have used a simpler 2-state model, in which each lattice site in the raphe is either empty or full. In this case, we would have had a model equivalent to the 1-dimensional Ising lattice [11.1] [11.4] [11.52] [11.59] [11.62] [11.255] with a distributed source (membrane raphan synthase) and sink (diatom trail), and an applied force, represented by the bias. While in a way more elegant, such a model would not have led to the intuition about possible raphan tilt and its effect on motility.

Our open source computer program is stored at:

https://ln5.sync.com/dl/da35e3480/88sawk4m-27t8kmg3-khqpnsv8-4bjtg9qv: RAPHE code v28.docx

This chapter elaborates on a preliminary video presentation [11.70] also archived at:

https://ln5.sync.com/dl/da35e3480/88sawk4m-27t8kmg3-khqpnsv8-4bjtg9qv: Gordon2021 Video-20210808_170504-Meeting Recording.mp4

Dedication

In dedication to two loving fathers, Jack Gordon and Anil Katyal, who loved unconditionally and by their good examples have taught us to work hard for what we aspire to achieve. We are proud to be their children, yet heartbroken that they are no more.

Appendix 11.1

Suppose we put a square grid over a simulation of two-dimensional Brownian motion, such as [11.42]. Let the squares correspond to spatial resolution, i.e., move each consecutive point in the trajectory of a particle to the center of the square in which it occurs. Now calculate the speed and two-dimensional acceleration for consecutive positions of the particle using those centers. Our hypothesis is that as the squares get smaller, those values will increase, and are perhaps unbounded. Certainly, the latter occurs if consecutive positions are within the same resolution square, as the distance apart is estimated as 0 due to pixelation.

Appendix 11.2

Our open source computer program can be downloaded from here in the electronic version of this book or obtained on request from the authors.

```
# Computer program v27 Y2021m10d27 in Python: RAPHE: Simulation of the
# dynamics of diatom motility at the molecular level: The domino effect model with concerted diffusion
# Shruti Raj Vansh Singh
# Openworm Foundation
# House Number 44
```

```
# New Plot, Jammu
# Jammu and Kashmir, India
# shrutirajvanshsingh@protonmail.com
# Krishna Katyal
# Openworm Foundation
# House Number 30, Sector 5
# Lower Roop Nagar, Jammu
# Jammu and Kashmir, India
# krishna.katyal@protonmail.com
# Richard Gordon
# DickGordonCan@protonmail.com

"""
FLOW OF THE PROGRAM
1. Set up and initialize variables
2. Set up functions
3. Set up an initial array of raphans
4. Begin of Loop
5. Call concatenation of groups of raphans of opposite
   orientations
6. Initialize the table showing the state of the array
7. Initialize the rates of all possible events
8. Choose which event is to happen
9. Execute that event
10. End of Loop
11. Display results
"""

# Importing Python Libraries
# Numpy for array processing [https://numpy.org]
import numpy as np
# Random for random integer generation [https://docs.python.
org/3/library/random.html]
import random

# random.seed(42) #FOR DEBUGGING: REMOVE FOR RUNNING
# Math for mathematical operations [https://docs.python.
org/3/library/math.html]
import math
# Itertools for efficient loops [https://docs.python.org/3/
library/itertools.html]
from itertools import groupby
# Sys allows exiting from the program in case an error is
encountered
import sys
```

```
#The function RunLengthEncode emulates:
#Weisstein, Eric W. (2021) Run-Length Encoding.
#From MathWorld--A Wolfram Web Resource. https://mathworld.
wolfram.com/Run-LengthEncoding.html
def RunLengthEncode(s_list):
   return [[len(list(group)), key] for key, group in
groupby(s_list)]

# naming columns
total_columns = 8
site_no = 0
site_state = 1
run_no = 2
run_length = 3
fill = 4
diffuse_left = 5
hydrate = 6
diffuse_right = 7

# Usage of default or user input variables
# using user input for the variable values
bias = float(input("Bias value (default: 0.6): ") or '0.6')
directionForDiatom = 0
directionFractionContributedBySwelling = float(input("Enter
Direction Fraction Contributed by Swelling (default: 3): ")
or '3')
directorDiffusionRatio = float(input("Direction Diffusion
Ratio (default: 1): ") or '1')
events = int(input("Number of Events (default: 50): ") or
'50')
# groupCharacteristic = int(input("Group Characteristic
(default: 10): ") or '10')
membraneCharacteristic = int(input("Membrane Characteristic
(default: 10): ") or '10')
membraneOccupancy = float(input("Membrane Occupancy (default:
0.25): ") or 0.25)
raphanRateOfHydration = float(input("Raphan Rate of Hydration
(default: 1): ") or '1')
rapheLength = int(input("RapheLength (default: 50): ") or
'50')
rateOfHopping = float(input("Rate of Hopping (default: 1):
") or '1')
ratePerMembraneToRapheStep = int(input("Rate PerMembrane
To Raphe Step (default: 1): ") or '1')
```

```
# def means "define a function" delimited by : and a blank
line

# function to update "run no"
def           calculate_run_count(table,            site_no,
rapheLength,site_state,run_no):
  table[site_no][run_no] = 1
  run=1
  for site in range (1, rapheLength):
    if table[site-1][site_state]!=table[site][site_state]:
      run+=1
      table[site][run_no] = run

def concatenate(table, runList, site_state, site_no, run_
no, run_length):
  pos=0
  state = table[pos][site_state]
  while True:
    if pos<rapheLength and table[pos][site_state]==0:
      pos+=1
    else:
      break

  while pos<rapheLength:
    if table[pos][site_state] != 0:
    start = pos
    while pos<rapheLength and table[pos][site_state]!=0:
      pos += 1
    end = pos
    randomNumber = random.random()
    if randomNumber < bias:
      for i in range(start,end):
        table[i][1] = 1
    else:
      for i in range(start,end):
        table[i][1] = -1
 else:
   pos+=1
 def  calculate_length(table,  rapheLength,  site_state,
 run_no):
   for site in range(rapheLength):
     count=1
     pos=site
     while site!=rapheLength-1 and table[site][site_state]
     == table[site+1][site_state]:
       count+=1
```

```
        site+=1
      table[pos][run_length]=count
      site+=1
    for site in range(rapheLength):
      if table[site][run_no] == 0:
        table[site][run_length] = 0

    def flip_config():
      rightEndOfGroup=int(iChosen+table[iChosen][run_length])
      leftEndOfGroup = iChosen
      if leftEndOfGroup>=0 and rightEndOfGroup<rapheLength:
        if table[iChosen][site_state] == -1:
          for i in range(leftEndOfGroup, rightEndOfGroup+1):
            if table[i][site_state] != 0:
              table[i][site_state] = 1
        elif table[iChosen][site_state] == 1:
          for i in range(leftEndOfGroup, rightEndOfGroup+1):
            if table[i][site_state] != 0:
              table[i][site_state] = -1

# initialize output variables:
# Python starts array indices with 0:
# Creating a numpy array(a list ) filled with zeros
table = np.zeros((rapheLength, total_columns),dtype=np.
float)

#initialization for output
distanceDiatomMoves = np.zeros(events+1)
configuration = []
timeIncrement = np.zeros(events+1)
distance=0
cumulativeDistance = np.zeros(events+1)
cumulativeTime = np.zeros(events+1)

timeIncrement[0] = 0
# Column 0: "Site" in Tables 1,2 in the manuscript
# The first column (indexed 0)contains the lattice site
number:
for site in range (0,rapheLength):
  table[site][site_no] = site

# Column 1: "State" in Tables 1,2 in the manuscript
# lattice states are (-1, 0, +1) internally to this program,
corresponding to (<, :, >) in a fixed width font on output.
# Their meanings are (hydrate left, empty, hydrate right),
where the diatom moves in the opposite direction
```

```
# The initial states are chosen at random:
for site in range(0,rapheLength):
  # The random.randint function generates different numbers
  each time it is run
  table[site][site_state] = random.randint(-1,1)
  # table[site][site_state] = 1
present = ''
#Add code to plot symbols < : > for - 1, 0, +1 from
table2[[site, 2]] using a fixed width font .
for site in range(len(table)):
  if table[site][site_state]==0:
    present += ":"
  elif table[site][site_state]==1:
    present += "<"
  elif table[site][site_state]==-1:
    present += ">"
  configuration.append(present)

# Column 2: "Run #" in Tables 1,2 in the manuscript
# populating run, i.e., column 2, containing the run num-
ber, left to right, if the site is the beginning of a run,
0 otherwise:
# The first site starts from run=1, no matter what the state
is
calculate_run_count(table, site_no, rapheLength, site_
state, run_no)

# Column 3: "Length" of run in Tables 1,2 in the manuscript
# loop starts
calculate_length(table, rapheLength, site_state, run_no)

#main loop
for event in range(1,events+1):
  print("event",event)
  #Extract the states from table, states will be the second
  column
  states = table[:,site_state]

# some print statement for debugging are of this form
if event==1:
  print(">",states,"<")

runList = RunLengthEncode(states)

runs = len(runList)
  # Concatenate all groups of raphans that touch:
```

```
# starts with run=2 since run=1 has nothing to its left.
# Consider consecutive pairs of raphan groups:
# due to initial random configuration or changes due to the
previous
# event, RunLengthEncode must be run again, as some groups
may have collided:
# Note that words "run" means the sequential sites are of
the same state and "group" refers to runs of +1 or -1
concatenate(table, runList, site_state, site_no, run_no,
run_length)

# reinitialized the run Length data
table[:,run_no:fill]=0

# Column 2: "Run #" in Tables 1,2 in the manuscript
# populating run, i.e., column 2, containing the run num-
ber, left to right, if the site is the beginning of a run,
0 otherwise:
  calculate_run_count(table, site_no, rapheLength, site_
  state, run_no)

# Column 3: "Length" of run in Tables 1,2 in the manuscript
calculate_length(table, rapheLength, site_state, run_no)

# Reset all rates to 0 :
table[:,fill:total_columns]=0
# Assign rates to the 5 possible events at each site :
# For each group of empty sites, each site in the group can
receive a randomly oriented raphan from the membrane at a
rate that depends on
# the raphan concentration in the membrane at the distance
of the site from the left .

for site in range(rapheLength):
  if table[site][site_state] == 0:
    table[site][fill] = ratePerMembraneToRapheStep*membra-
    neOccupancy*math.exp(-site/membraneCharacteristic)
    continue

  # Group events:
  # Each group, except for those at the very ends, can dif-
  fuse both left or both right or hydrate left or right

  # The rates are diminished with increasing length of the
  group:
```

```
    # lengthDiminishment = math.exp(-table[site][run_length]/
    groupCharacteristic)

    ran = random.random()
    diffusionLeft = 1
    diffusionRight=1
    # Diffusion left
    if site == 0:
      diffusionLeft=0
      # can't diffuse past end of raphe
    else: # elif means else if
      table[site][diffuse_left] = bias*rateOfHopping*diffusionLeft

    diffusionRight=1
    # Diffusion right
    if site == rapheLength-1:
      diffusionRight=0
      # can't diffuse past end of raphe
    else:
      table[site][diffuse_right] = (1-bias)*rateOfHopping*dif-
      fusionRight

    ran = random.random()
    table[site][hydrate] = raphanRateOfHydration

for site in range(rapheLength):
  if table[site][run_no]==0:
    table[site, diffuse_left:total_columns] = 0
# Sum all of the rates:
totalRate = 0
for site in range(rapheLength):
  for j in range(fill, total_columns):
    totalRate += table[site][j]
np.set_printoptions(precision=3, suppress=True)

print("Total Rate: ", totalRate)
# Select which event occurs :
eventSelected = random.random()*totalRate
print("Event Selected: ", eventSelected)

# Find the event that has been selected:
cumulativeEventRates = 0

# the flag variable is used to break out of the nested loop
flag=False
for iChosen in range(rapheLength):
```

```
    # python does not include the last value of the range, the
    loop will go from 4 to 8
    for jChosen in range(fill, total_columns):
      cumulativeEventRates = cumulativeEventRates + table[iCho-
      sen][jChosen]
      if cumulativeEventRates > eventSelected:
        flag = True
        break # selects the element of the table (iChose,jCho-
        sen) to be carried out
    if flag:
      break
  print("Total event rate:", totalRate)
  print("Cumulative Event rate:", cumulativeEventRates)
  print(iChosen, jChosen)
  action=""
  if iChosen<(rapheLength):
    # Change column 2, updating the raphe to the new configu-
    ration, depending on the chose event .

    # Fill
    if jChosen == fill and table[iChosen][jChosen]!=0:
      # filling a site does not cause any motion of the diatom:
      directionForDiatom = 0
      action = 'f'
      # The raphan arriving from the membrane can be oriented
      either way:
      if random.random()<bias:
        table[iChosen][site_state] = 1
      else:
        table[iChosen][site_state] = -1

    # Diffusion left
    if jChosen == diffuse_left and table[iChosen][jChosen]!=0:
      # diffusion does not cause any motion of the diatom:
      directionForDiatom = 0

      # Move the group left by one site
      table[iChosen-1][site_state] = table[iChosen][site_state]
      y = int(iChosen+table[iChosen][run_length])
      if y < rapheLength:
          table[y-1][site_state] = 0
          action="dl"

    # Diffusion right
    if jChosen == diffuse_right and table[iChosen][jChosen]!=0:
        # diffusion does not cause any motion of the diatom:
```

```
      directionForDiatom = 0
      # Move the group right by one site:
      rightEndOfGroup=int(iChosen+table[iChosen][run_length])
      if rightEndOfGroup < rapheLength:
        table[rightEndOfGroup][site_state] = table[iCho-
        sen][site_state]
        table[iChosen][site_state] = 0
        action="dr"

    # hydrate
    if jChosen == hydrate and table[iChosen][jChosen]!=0:
      random_no = random.random()
      if random_no < (1-bias):
        flip_config()
      # only the lead raphan in the group hydrates, leav-
      ing the raphe:
      if table[iChosen][site_state] == 1:
        rightEndOfGroup = int(iChosen+table[iChosen][run_length])
        if rightEndOfGroup < rapheLength:
          table[rightEndOfGroup-1][site_state] = 0
          directionForDiatom = -1
          distanceDiatomMoves[event] = directionFraction-
          ContributedBySwelling * directionForDiatom
          action='h'
      elif table[iChosen][site_state] == -1:
        table[iChosen][site_state] = 0
        directionForDiatom = 1 # Diatom moves right
        distanceDiatomMoves[event] = directionFraction-
        ContributedBySwelling * directionForDiatom
        action = 'h'
      cumulativeDistance[event]=cumulativeDistance[event-1]
      + distanceDiatomMoves[event]
      table[:,[site_no,run_length]] = table[:,[site_
      no,run_length]].astype(int)
      present = ''
  #Add code to plot symbols < :> for - 1, 0, +1 from
  table2[[site, 2]] using a fixed width font .
  for site in range(len(table)):
    if table[site][site_state] == 0:
      present += ":"
    elif table[site][site_state] == 1:
      present += "<"
    elif table[site][site_state] == -1:
  present += ">"
  present += " "
  present += action
```

```
      # present += " "
      # present += str(jChosen)
      # present += " "
      # present += str(table[iChosen][jChosen])
      configuration.append(present)
    #End of event loop

    for event in range(events+1):
      print(configuration[event], event)

  # 11. Plot diatom position vs time

  import matplotlib.pyplot as plt

  plt.figure(figsize=(15,10))
  # For visualization the diatom moves in the positive direction
  plt.scatter(range(events+1), -cumulativeDistance)
  plt.plot(range(events+1), -cumulativeDistance)
  plt.title("Bias="+str(bias))
  plt.xlabel("Time")
  plt.ylabel("Distance", labelpad=5)
  plt.show()

  # 12 Wolfram pattern
  from termcolor import colored

  for event in range(events+1):
    config = configuration[event]
    for site in range(len(config)):
      if config[site]=='<' or config[site]=='>':
        print(colored(" ", "grey", "on_grey"), end=" ")
      elif config[site]==':':
        print(colored(" ", "white", "on_white"), end=" ")
    print()
```

References

[11.1] Achiam, Y. (1980) Diffusion in the one-dimensional Ising-model. *Journal of Physics A-Mathematical and General* 13(5), 1825-1833.

[11.2] Alicea, B., Gordon, R., Harbich, T., Singh, U., Singh, A. and Varma, V. (2021) Towards a digital diatom: Image processing and deep learninganalysis of *Bacillaria paradoxa* dynamic morphology. In: *Diatom Gliding Motility [DIGM, Volume in the series: Diatoms: Biology & Applications, series editors: Richard Gordon & Joseph Seckbach]*. S.A. Cohn, K.M. Manoylov and R. Gordon, (eds.) Wiley-Scrivener, Beverly, MA, USA: 223-248.

[11.3] Allain, R. (2014) The Awesome Physics in a Simple Sparkler. https://www.wired.com/2014/07/the-awesome-physics-in-a-simple-sparkler/

[11.4] Amar, J.G. and Family, F. (1990) Diffusion annihilation in one dimension and kinetics of the Ising-model at zero temperature. *Physical Review A* 41(6), 3258-3262.

[11.5] Aoki, K.M., Fujiwara, S., Soga, K., Ohnishi, S. and Yamamoto, T. (2013) One-, two-, and three-dimensional hopping dynamics. *Crystals* 3(2), 315-332.

[11.6] Apajalahti, T., Niemela, P., Govindan, P.N., Miettinen, M.S., Salonen, E., Marrink, S.J. and Vattulainen, I. (2010) Concerted diffusion of lipids in raft-like membranes. *Faraday Discussions* 144, 411-430.

[11.7] autodidude. (2012) Why does a pen fall over when placed on its tip? , https://www.physicsforums.com/threads/why-does-a-pen-fall-over-when-placed-on-its-tip.618856/

[11.8] Bandman, O. (2016) Asynchronous cellular automata simulating complex phenomena. In: *Designing Beauty: The Art of Cellular Automata*. A. Adamatzky and G.J. Martínez, (eds.) Springer International Publishing: 111-115.

[11.9] Barlett, V.R., Bigeón, J.J., Hoyuelos, M. and Mártin, H.O. (2009) Differences between fixed time step and kinetic Monte Carlo methods for biased diffusion. *Journal of Computational Physics* 228(16), 5740-5748.

[11.10] Bauer, S., Arpa-Sancet, M.P., Finlay, J.A., Callow, M.E., Callow, J.A. and Rosenhahn, A. (2013) Adhesion of marine fouling organisms on hydrophilic and amphiphilic polysaccharides. *Langmuir* 29(12), 4039-4047.

[11.11] Bedoshvili, Y.D. and Likhoshway, Y.V. (2021) Chapter 14. The effects of cytoskeletal inhibitors in diatom valve morphogenesis. In: *Diatom Morphogenesis [DIMO, Volume in the series: Diatoms: Biology & Applications, series editors: Richard Gordon & Joseph Seckbach]*. V. Annenkov, J. Seckbach and R. Gordon, (eds.) Wiley-Scrivener, Beverly, MA, USA: 349-364.

[11.12] Bertrand, J. (1990) La vitesse de déplacement des diatomées [Speed of diatom motion]. *Diatom Research* 5(2), 223-239.

[11.13] Bertrand, J. (2021) Movements of diatoms VIII: Synthesis and hypothesis. Translation of: Bertrand, J. (2008). Mouvements des diatomées VIII: synthèse et hypothèse. Diatom Research 23(1), 19-29, by: Richard Gordon, Martin Laviale & Karen K. Serieyssol in consultation with Jean Bertrand. In: *Diatom Gliding Motility [DIGM, Volume in the series: Diatoms: Biology & Applications, series editors: Richard Gordon & Joseph Seckbach]* S.A. Cohn, K.M. Manoylov and R. Gordon, (eds.) Wiley-Scrivener, Beverly, MA, USA: 283-294.

[11.14] Bolhuis, P. and Frenkel, D. (1997) Tracing the phase boundaries of hard spherocylinders. *Journal of Chemical Physics* 106(2), 666-687.

[11.15] Bondoc-Naumovitz, K.G. and Cohn, S.A. (2021) Motility of biofilm-forming benthic diatoms. In: *Diatom Gliding Motility [DIGM, Volume in the series: Diatoms: Biology & Applications, series editors: Richard Gordon & Joseph Seckbach]. Wiley-Scrivener, Beverly, MA, USA*. S.A. Cohn, K.M. Manoylov and R. Gordon, (eds.) Wiley-Scrivener, Beverly, Massachusetts, USA: 77-109.

[11.16] Boyd, C.M. and Gradmann, D. (1999) Electrophysiology of the marine diatom *Coscinodiscus wailesii* - I. Endogenous changes of membrane voltage and resistance. *Journal of Experimental Botany* 50(333), 445-452.

[11.17] Boyd, C.N. and Gradmann, D. (1999) Electrophysiology of the marine diatom *Coscinodiscus wailesii* - III. Uptake of nitrate and ammonium. *Journal of Experimental Botany* 50(333), 461-467.

[11.18] Britannica. (2021) Smectic-A phase. https://www.britannica.com/science/smectic-A-phase

[11.19] Calabrese, D.R., Wenning, B.M., Buss, H., Finlay, J.A., Fischer, D., Clare, A.S., Segalman, R.A. and Ober, C.K. (2017) Oligopeptide-modified hydrophobic and hydrophilic polymers as antifouling coatings. *Green Materials* 5(1), 31-43.

[11.20] Callaghan, P.T. and Lelievre, J. (1986) The influence of polymer size and shape on self-diffusion of polysaccharides and solvents. *Analytica Chimica Acta* 189(1), 145-166.
[11.21] Chang, T. (2008) Dominoes in carbon nanotubes. *Physical Review Letters* 101(17), #175501.
[11.22] Chang, T. (2010) Temperature-induced reversible dominoes in carbon nanotubes. *Nano Letters* 10(9), 3490-3493.
[11.23] Chen, C.S., Shiu, R.E.F., Hsieh, Y.Y., Xu, C., Vazquez, C.I., Cui, Y.J., Hsu, A., Quigg, A., Santschi, P.H. and Chin, W.C. (2021) Stickiness of extracellular polymeric substances on different surfaces via magnetic tweezers. *Science of the Total Environment* 757, #143766.
[11.24] Chen, L., Weng, D., Du, C., Wang, J.D. and Cao, S. (2019) Contribution of frustules and mucilage trails to the mobility of diatom *Navicula* sp. *Scientific Reports* 9, #7342.
[11.25] Cicco, S.R., Vona, D., De Giglio, E., Cometa, S., Mattioli-Belmonte, M., Palumbo, F., Ragni, R. and Farinola, G.M. (2015) Chemically modified diatoms biosilica for bone cell growth with combined drug-delivery and antioxidant properties. *Chempluschem* 80(7), 1104-1112.
[11.26] Ciglenečki, I., Plavšic, M., Vojvodić, V., Ćosović, B., Pepi, M. and Baldi, F. (2003) Mucopolysaccharide transformation by sulfide in diatom cultures and natural mucilage. *Marine Ecology-Progress Series* 263, 17-27.
[11.27] Codling, E.A., Bearon, R.N. and Thorn, G.J. (2010) Diffusion about the mean drift location in a biased random walk. *Ecology* 91(10), 3106-3113.
[11.28] Cohn, S.A., Bahena, M., Davis, J.T., Ragland, R.L., Rauschenberg, C.D. and Smith, B.J. (2004) Characterisation of the diatom photophobic response to high irradiance. *Diatom Research* 19(2), 167-179.
[11.29] Cohn, S.A. and Disparti, N.C. (1994) Environmental factors influencing diatom cell motility. *Journal of Phycology* 36(5), 818-828.
[11.30] Cohn, S.A., Farrell, J.F., Munro, J.D., Ragland, R.L., Weitzell, R.E. and Wibisono, B.L. (2003) The effect of temperature and mixed species composition on diatom motility and adhesion. *Diatom Research* 18(2), 225-243.
[11.31] Cohn, S.A. and Weitzell, R.E. (1996) Ecological considerations of diatom cell motility .1. Characterization of motility and adhesion in four diatom species. *Journal of Phycology* 32(6), 928-939.
[11.32] Cookson, J.P. and Thoma, G.R. (1988) X-ray image compression using run length coding. *J Med Syst* 12(4), 201-209.
[11.33] Cooper, M.S. and Schliwa, M. (1986) Motility of cultured fish epidermal-cells in the presence and absence of direct-current electric-fields. *Journal of Cell Biology* 102(4), 1384-1399.
[11.34] Daniel, G.F., Chamberlain, A.H.L. and Jones, E.B.G. (1987) Cytochemical and electron microscopical observations on the adhesive materials of marine fouling diatoms. *British Phycological Journal* 22, 101-118.
[11.35] de Beer, S., Mensink, L.I.S. and Kieviet, B.D. (2016) Geometry-dependent insertion forces on particles in swollen polymer brushes. *Macromolecules* 49(3), 1070-1078.
[11.36] Deng, B.H., Huang, L.P. and Shi, Y.F. (2017) Solvent effect on the diffusion of unentangled linear polymer melts. *Langmuir* 33(42), 11845-11850.
[11.37] Denn, M.M. (2001) Extrusion instabilities and wall slip. *Annual Review of Fluid Mechanics* 33, 265-287.
[11.38] Dror, R. (2016) Diffusion. https://web.stanford.edu/class/archive/cs/cs279/cs279.1172/lectures/lecture10.pdf
[11.39] Drum, R.W. and Hopkins, J.T. (1966) Diatom locomotion: An explanation. *Protoplasma* 62(1), 1-33.
[11.40] Du, C., Wang, J., Cao, S. and Chen, D. (2013) Adhesion characterization of 'trails' mucilage of diatom *Navicula* sp. *Protoplasma*, submitted preprint.

[11.41] Dubochet, J. (2007) The physics of rapid cooling and its implications for cryoimmobilization of cells. In: *Cellular Electron Microscopy*. J.R. McIntosh, (ed.). 79: 7-21.

[11.42] Duffy, A. (2018) Brownian motion. https://physics.bu.edu/~duffy/HTML5/brownian_motion.html

[11.43] Dvinskikh, S.V. and Furo, I. (2012) Anisotropic self-diffusion in nematic, smectic-A, and reentrant nematic phases. *Physical Review E* 86(3), #031704.

[11.44] Edgar, L.A. (1979) Diatom locomotion: Computer assisted analysis of cine film. *British Phycological Journal* 14(1), 83-101.

[11.45] Edgar, L.A. (1983) Mucilage secretions of moving diatoms. *Protoplasma* 118, 44-48.

[11.46] Edgar, L.A. and Pickett-Heaps, J.D. (1983) The mechanism of diatom locomotion. I. An ultrastructural study of the motility apparatus. *Proceedings of the Royal Society B: Biological Sciences* 218, 331-343.

[11.47] Edgar, L.A. and Pickett-Heaps, J.D. (1984) Diatom locomotion. *Progress in Phycological Research* 3, 47-88.

[11.48] Edlund, M. and Burge, D. (2015) *Pleurosigma* W. Smith 1852. https://diatoms.org/genera/pleurosigma

[11.49] Emons, A.M., Akkerman, M., Ebskamp, M., Schel, J. and Mulder, B. (2007) *How cellulose synthase density in the plasma membrane may dictate cell wall texture*. Springer, Dordrecht, Netherlands: 183-197.

[11.50] Erban, R. (2014) From molecular dynamics to Brownian dynamics. *Proceedings of the Royal Society a-Mathematical Physical and Engineering Sciences* 470(2167), 20, #20140036.

[11.51] Fact File. (2016) 10 Facts About Diatoms. https://factfile.org/10-facts-about-diatoms

[11.52] Family, F. and Amar, J.G. (1991) Diffusion-annihilation and the kinetics of the Ising-model in one dimension. *Journal of Statistical Physics* 65(5-6), 1235-1246.

[11.53] Ferras, L.L., Afonso, A.M., Nobrega, J.M. and Pinho, F.T. (2017) A numerical and theoretical study on viscoelastic fluid slip flows. *Physics of Fluids* 29(5), #053102.

[11.54] Finkelstein, E., Chang, W., Chao, P.H.G., Gruber, D., Minden, A., Hung, C.T. and Bulinski, J.C. (2004) Roles of microtubules, cell polarity and adhesion in electric-field-mediated motility of 3T3 fibroblasts. *Journal of Cell Science* 117(8), 1533-1545.

[11.55] Fizz-Looney, F. and Linderstrøm-Lang, K. (1962) Thermodynamic activity of the male housefly. In: *Kaj Linderstrøm-Lang. Selected Papers*. H. Holter, H. Neurath and M. Ottesen, (eds.) Danish Science Press, Copenhagen, Denmark: 573-584.

[11.56] Fujii, F., Inoue, Y. and Nitta, T. (2012) Modeling the domino wave propagation in contact mechanics. *Transactions of the Japan Society of Mechanical Engineers Series C* 78(788), 1133-1142.

[11.57] Gauthier, M.G. and Slater, G.W. (2005) A new set of Monte Carlo moves for lattice random walk models of biased diffusion. *Physica A-Statistical Mechanics and Its Applications* 355(2-4), 283-296.

[11.58] Gordon, R. (1968) Adsorption isotherms of lattice gases by computer simulation. *J. Chem. Physics* 48(3), 1408-1409.

[11.59] Gordon, R. (1968) Steady-state properties of Ising lattice membranes. *Journal of Chemical Physics* 49, 570-580.

[11.60] Gordon, R. (1970) On Monte Carlo algebra. *Journal of Applied Probability* 7, 373-387.

[11.61] Gordon, R. (1977) High-speed reconstruction of the finest details available in x-ray projections. In: *Reconstruction Tomography in Diagnostic Radiology and Nuclear Medicine [Proceedings of a Workshop on Reconstruction Tomography in Diagnostic Radiology and Nuclear Medicine, San Juan, Puerto Rico. April 17-19, 1975]*. M.M. Ter-Pogossian, M.E. Phelps, G.L. Brownell, J.R. Cox Jr., D.O. Davis and R.G. Evens, (eds.) University Park Press, Baltimore: 77-83.

[11.62] Gordon, R. (1980) Monte Carlo methods for cooperative Ising models. In: *Cooperative Phenomena in Biology*. G. Karreman, (ed.) Pergamon Press, New York: 189-241 + errata.

[11.63] Gordon, R. (1987) A retaliatory role for algal projectiles, with implications for the mechanochemistry of diatom gliding motility. *Journal of Theoretical Biology* 126, 419-436.

[11.64] Gordon, R. (1999) *The Hierarchical Genome and Differentiation Waves: Novel Unification of Development, Genetics and Evolution*. World Scientific & Imperial College Press, Singapore & London. 2 vols.

[11.65] Gordon, R. (2021) Are we on the cusp of a new paradigm for biology? The illogic of molecular developmental biology versus Janus-faced control of embryogenesis via differentiation waves. *BioSystems*(Waves in Fertilization, Cell Division and Embryogenesis, Guest Editors: Jack Tuszynski, Luigia Santella & Richard Gordon), #104367.

[11.66] Gordon, R. (2021) The whimsical history of proposed motors for diatom motility. In: *Diatom Gliding Motility [DIGM, Volume in the series: Diatoms: Biology & Applications, series editors: Richard Gordon & Joseph Seckbach]*. S.A. Cohn, K.M. Manoylov and R. Gordon, (eds.) Wiley-Scrivener, Beverly, MA, USA: 335-420.

[11.67] Gordon, R. and Drum, R.W. (1970) A capillarity mechanism for diatom gliding locomotion. *Proceedings of the National Academy of Sciences of the United States of America* 67(1), 338-344.

[11.68] Gordon, R., Goel, N.S., Steinberg, M.S. and Wiseman, L.L. (1972) A rheological mechanism sufficient to explain the kinetics of cell sorting. *Journal of Theoretical Biology* 37(1), 43-73.

[11.69] Gordon, R., Goel, N.S., Steinberg, M.S. and Wiseman, L.L. (1975) A rheological mechanism sufficient to explain the kinetics of cell sorting. In: *Mathematical Models for Cell Rearrangement*. G.D. Mostow, (ed.) Yale University Press, New Haven: 196-230.

[11.70] Gordon, R., Krishna, K. and Raj Vansh Singh, S. (2021) An attempted simulation of diatom motility at the molecular level: The domino effect hydration model with concerted diffusion. https://isdr.org/wp-content/uploads/2021/08/Online-IDS-Oral-program.pdf

[11.71] Gordon, R., Losic, D., Tiffany, M.A., Nagy, S.S. and Sterrenburg, F.A.S. (2009) The Glass Menagerie: Diatoms for novel applications in nanotechnology. *Trends in Biotechnology* 27(2), 116-127.

[11.72] Gordon, R., Raj-Vansh Singh, S., Krishna, K. and Gordon, N.K. (2021) The fish ladder toy model for a thermodynamically at equilibrium origin of life in a Lipid World in an endoreic lake. In: *Conflicting Models for the Origin of Life [COLF, Volume in the series Astrobiology Perspectives on Life of the Universe, Eds. Richard Gordon & Joseph Seckbach*. S.K. Smoukov, J. Seckbach and R. Gordon, (eds.) Wiley-Scrivener, Beverly, Massachusetts, USA, 425-458.

[11.73] Gordon, R. and Stone, R. (2021) A short tutorial on the Janus-faced logic of differentiation waves and differentiation trees and their evolution. *BioSystems* 205(Waves in Fertilization, Cell Division and Embryogenesis, Guest Editors: Jack Tuszynski, Luigia Santella & Richard Gordon), #104414.

[11.74] Gordon, R. and Tiffany, M.A. (2011) Possible buckling phenomena in diatom morphogenesis. In: *The Diatom World*. J. Seckbach and J.P. Kociolek, (eds.) Springer, Dordrecht, The Netherlands: 245-272.

[11.75] Gradmann, D. and Boyd, C.M. (1995) Membrane voltage of marine phytoplankton, measured in the diatom *Coscinodiscu radiatus*. *Marine Biology* 123(4), 645-650.

[11.76] Gradmann, D. and Boyd, C.M. (1999) Electrophysiology of the marine diatom *Coscinodiscus wailesii* II. Potassium currents. *Journal of Experimental Botany* 50(333), 453-459.

[11.77] Gradmann, D. and Boyd, C.M. (1999) Electrophysiology of the marine diatom *Coscinodiscus wailesii* IV: types of non-linear current-voltage-time relationships recorded with single saw-tooth voltage-clamp experiments. *European Biophysics Journal with Biophysics Letters* 28(7), 591-599.

[11.78] Griff, E.R. (2006) How neurons work: An analogy & demonstration using a sparkler & a frying pan. *American Biology Teacher* 68(7), 412-417.

[11.79] Grove, D. (1984) *Diatom locomotion: Variations in methods of locomotion and colonial coupling in Bacillaria paradoxa*. Miller Library, Stanford University, Stanford.

[11.80] Gutierrez-Medina, B., Guerra, A.J., Maldonado, A.I.P., Rubio, Y.C. and Meza, J.V.G. (2014) Circular random motion in diatom gliding under isotropic conditions. *Physical Biology* 11(6), #066006.

[11.81] Guttenberg, N., Abbot, D.S., Amundson, J.M., Burton, J.C., Mac Cathles, L., MacAyeal, D.R. and Zhang, W.W. (2011) A computational investigation of iceberg capsize as a driver of explosive ice-shelf disintegration. *Annals of Glaciology* 52(59), 51-59.

[11.82] Ha, N., Park, J., Park, S.H., Seo, E., Lim, J.H. and Lee, S.J. (2021) Domino-like water transport on *Tillandsia* through flexible trichome wings. *New Phytologist* 231(5), 1906-1922.

[11.83] Happey-Wood, C.M. and Jones, P. (1988) Rhythms of vertical migration and motility in intertidal benthic diatoms with particular reference to *Pleurosigma angulatum*. *Diatom Research* 3(1), 83-93.

[11.84] Harbich, T. (2019) Change in the trajectories of Cymatopleura solea as a result of cell size reduction. https://www.diatoms.de/en/influence-of-reduced-cell-size

[11.85] Harbich, T. (2021) Observations of Diatoms: Curvature of trajectories I, II, III. https://diatoms.de/en/curvature-of-the-trajectories-i; https://diatoms.de/en/curvature-of-the-trajectories-ii; https://diatoms.de/en/curvature-of-trajectories-iii

[11.86] Harper, M.A. (1969) Movement and migration of diatoms on sand grains. *British Phycological Journal* 4(1), 97-103.

[11.87] Harper, M.A. (1977) Movements. In: *The Biology of Diatoms*. D. Werner, (ed.) Blackwell Scientific Publications, Oxford, UK: 224-249.

[11.88] Harper, M.A. and Harper, J.T. (1967) Measurements of diatom adhesion and their relationship with movement. *British Phycological Bulletin* 3(2), 195-207.

[11.89] Hart, F.X. and Palisano, J.R. (2017) Glycocalyx bending by an electric field increases cell motility. *Bioelectromagnetics* 38(6), 482-493.

[11.90] Hay, S.I., Maitland, T.C. and Paterson, D.M. (1993) The speed of diatom migration through natural and artificial substrata. *Diatom Research* 8(2), 371-384.

[11.91] Heintzelman, M.B. (2006) Cellular and molecular mechanics of gliding locomotion in eukaryotes. *International Review of Cytology - A Survey of Cell Biology* 251, 79-129.

[11.92] Higgins, M.J., Crawford, S.A., Mulvaney, P. and Wetherbee, R. (2000) The topography of soft, adhesive diatom 'trails' as observed by atomic force microscopy. *Biofouling* 16(2-4), 133-139.

[11.93] Higgins, M.J., Molino, P., Mulvaney, P. and Wetherbee, R. (2003) The structure and nanomechanical properties of the adhesive mucilage that mediates diatom-substratum adhesion and motility. *Journal of Phycology* 39(6), 1181-1193.

[11.94] Hill, T.L. (1976) Diffusion frequency factors in some simple examples of transition-state rate theory. *Proceedings of the National Academy of Sciences of the United States of America* 73(3), 679-683.

[11.95] Hill, T.L. (1985) *Cooperative Theory in Biochemistry, Steady-State and Equilibrium Systems*. Springer-Verlag, New York.

[11.96] Hoagland, K.D., Rosowski, J.R., Gretz, M.R. and Roemer, S.C. (1993) Diatom extracellular polymeric substances: function, fine structure, chemistry, and physiology. *Journal of Phycology* 29(5), 537-566.

[11.97] Hopkins, J.T. (1967) The diatom trail. *Microscopy* 30, 209-217.

[11.98] Hopkins, J.T. and Drum, R.W. (1966) Diatom motility: An explanation and a problem. *British Phycological Bulletin* 3(1), 63-67.

[11.99] Hosy, E., Martiniere, A., Choquet, D., Maurel, C. and Luu, D.T. (2015) Super-resolved and dynamic imaging of membrane proteins in plant cells reveal contrasting kinetic profiles and multiple confinement mechanisms. *Molecular Plant* 8(2), 339-342.

[11.100] Huang, Y.J., Samorajski, J., Kreimer, R. and Searson, P.C. (2013) The influence of electric field and confinement on cell motility. *PLoS One* 8(3), #e59447.

[11.101] Huber, G.A. and McCammon, J.A. (2019) Brownian dynamics simulations of biological molecules. *Trends in Chemistry* 1(8), 727-738.

[11.102] Ichimura, Y., Inoue, T., Sibutani, R., Horioka, M. and Watanabe, M. (2006) Spark of Japanese Sparkler "senkoh-hanabi" under Microgravity. *International Journal of Microgravity Science and Application* 23(1), 44-45.

[11.103] Inai, Y., Komori, H., Takasu, A. and Hirabayashi, T. (2003) Noncovalent chiral domino effect on one-handed helix of nonapeptide containing a midpoint L-residue. *Biomacromolecules* 4(1), 122-128.

[11.104] Inoue, C., Koshi, M., Himeno, T. and Watanabe, T. (2016) Investigation on liquid atomization mechanism in Japanese sparkler "Senko-hanabi". *Science and Technology of Energetic Materials* 77(3-4), 51-58.

[11.105] Ising, E. (1925) Beitrag zur Theorie des Ferromagnetismus [Report on the theory of ferromagnetism]. *Zeitschrift für Physik* 31, 253-258.

[11.106] Iwasa, K. and Shimizu, A. (1972) Motility of the diatom, *Phaeodactylum tricornutum*. *Experimental Cell Research* 74(2), 552-558.

[11.107] Jahn, T., Bovee, E. and Drum, R.W. (1967) Motility in algae. *Journal of Phycology* 3(s1), 7-8.

[11.108] Jeong, Y., Ko, S. and Kang, S.M. (2017) Superhydrophilic conversion of stainless steel surfaces by biomimetic silica coating and its effect on marine fouling. *Bulletin of the Korean Chemical Society* 38(8), 972-975.

[11.109] Joshi, G., Okeyoshi, K., Okajima, M.K. and Kaneko, T. (2016) Directional control of diffusion and swelling in megamolecular polysaccharide hydrogels. *Soft Matter* 12(25), 5515-5518.

[11.110] Jujuly, M.M., Rahman, A., Ahmed, S. and Khan, F. (2015) LNG pool fire simulation for domino effect analysis. *Reliability Engineering & System Safety* 143, 19-29.

[11.111] Kalmykov, Y.P., Titov, S.V. and Coffey, W.T. (2011) Inertial and bias effects in the rotational Brownian motion of rodlike molecules in a uniaxial potential. *Journal of Chemical Physics* 134(4), #044530.

[11.112] Katsumata, R., Dulaney, A.R., Kim, C.B. and Ellison, C.J. (2018) Glass transition and self-diffusion of unentangled polymer melts nanoconfined by different interfaces. *Macromolecules* 51(19), 7509-7517.

[11.113] Khantha, M., Cordero, N.A., Alonso, J.A., Cawkwell, M. and Girifalco, L.A. (2008) Interaction and concerted diffusion of lithium in a (5,5) carbon nanotube. *Physical Review B* 78(11), #115430.

[11.114] Kobayasi, H. (1997) Comparative studies among four linear-lanceolate *Achnanthidium* species (Bacillariophyceae) with curved terminal raphe endings. *Nova Hedwigia* 65(1-4), 147-163.

[11.115] Korosec, C.S., Sivak, D.A. and Forde, N.R. (2020) Apparent superballistic dynamics in one-dimensional random walks with biased detachment. *Physical Review Research* 2(3), #033520.

[11.116] Korzhenevskii, A.L., Bausch, R. and Schmitz, R. (2008) Surface instability of growing binary crystals. *Journal of Statistical Mechanics-Theory and Experiment*, #P04013.

[11.117] Kovalchuk, N.M. and Simmons, M.J.H. (2021) Surfactant-mediated wetting and spreading: Recent advances and applications. *Current Opinion in Colloid & Interface Science* 51, 101375.

[11.118] Krüger, G.J. (1982) Diffusion in thermotropic liquid crystals. *Physics Reports-Review Section of Physics Letters* 82(4), 229-269.

[11.119] Kure, J.L., Andersen, C.B., Mortensen, K.I., Wiseman, P.W. and Arnspang, E.C. (2020) Revealing plasma membrane nano-domains with diffusion analysis methods. *Membranes* 10(11), #314.

[11.120] Lachnit, M., Buhmann, M.T., Klemm, J., Kröger, N. and Poulsen, N. (2019) Identification of proteins in the adhesive trails of the diatom *Amphora coffeaeformis*. *Philosophical Transactions of the Royal Society B-Biological Sciences* 374(1784), #20190196.

[11.121] Larham, R. (2008) Validation of a Model of the Domino Effect?, https://arxiv.org/abs/0803.2898v1

[11.122] Le Doussal, P., Petkovic, A. and Wiese, K.J. (2012) Distribution of velocities and acceleration for a particle in Brownian correlated disorder: Inertial case. *Physical Review E* 85(6), #061116.

[11.123] Lee, J., Park, E., Fujisawa, A. and Lee, H. (2021) Diatom silica/polysaccharide elastomeric hydrogels: Adhesion and interlocking synergy. *ACS Applied Materials & Interfaces* 13(18), 21703-21713.

[11.124] Lenartowska, M. and Michalska, A. (2008) Actin filament organization and polarity in pollen tubes revealed by myosin II subfragment 1 decoration. *Planta* 228(5), 891-896.

[11.125] Li, J., Li, Y., Zheng, H., Liu, M., Gu, H., Lu, K., Zhou, X. and Wang, Z. (2021) Strengthening unidirectional liquid pumping using multi-biomimetic structures. *Extreme Mechanics Letters* 43, #101144.

[11.126] Li, J. and Wang, L. (2020) Coalescence-induced transition between unidirectional and bidirectional propagation of droplets. *Materials Horizons* 7(8), 2078-2084.

[11.127] Li, X.J., Luu, D.T., Maurel, C. and Lin, J.X. (2013) Probing plasma membrane dynamics at the single-molecule level. *Trends in Plant Science* 18(11), 617-624.

[11.128] Liang, H.L., Schymura, S., Rudquist, P. and Lagerwall, J. (2011) Nematic-smectic transition under confinement in liquid crystalline colloidal shells. *Physical Review Letters* 106(24), #247801.

[11.129] Lind, J.L., Heimann, K., Miller, E.A., vanVliet, C., Hoogenraad, N.J. and Wetherbee, R. (1997) Substratum adhesion and gliding in a diatom are mediated by extracellular proteoglycans. *Planta* 203(2), 213-221.

[11.130] Lynch, P. (2014) The Not-so-simple Pendulum: Balancing a Pencil on its Point. https://arxiv.org/abs/1406.1125

[11.131] Manley, S., Gillette, J.M., Patterson, G.H., Shroff, H., Hess, H.F., Betzig, E. and Lippincott-Schwartz, J. (2008) High-density mapping of single-molecule trajectories with photoactivated localization microscopy. *Nature Methods* 5(2), 155-157.

[11.132] Mann, D.G. (1982) The use of the central raphe endings as a taxonomic character. *Plant Systematics and Evolution* 141, 143-152.

[11.133] Mantovani, V., Galeotti, F., Maccari, F. and Volpi, N. (2018) Recent advances in capillary electrophoresis separation of monosaccharides, oligosaccharides, and polysaccharides. *Electrophoresis* 39(1), 179-189.

[11.134] Marques da Silva, J., Duarte, B. and Utkin, A.B. (2020) Travelling expenses: The energy cost of diel vertical migrations of epipelic microphytobenthos. *Frontiers in Marine Science* 7(433).

[11.135] MathematicalPhysicist. (2004) Which direction will the pencil fall https://www.physicsforums.com/threads/which-direction-will-the-pencil-fall.20811/

[11.136] McConville, M.J., Wetherbee, R. and Bacic, A. (1999) Subcellular location and composition of the wall and secreted extracellular sulphated polysaccharides/proteoglycans of the diatom *Stauroneis amphioxys* Gregory. *Protoplasma* 206(1-3), 188-200.

[11.137] Merikoski, J., Vattulainen, I., Heinonen, J. and AlaNissila, T. (1997) Effect of kinks and concerted diffusion mechanisms on mass transport and growth on stepped metal surfaces. *Surface Science* 387(1-3), 167-182.

[11.138] Metropolis, N., Rosenbluth, A.W., Rosenbluth, M.N., Teller, A.H. and Teller, E. (1953) Equation of state calculations by fast computing machines. *Journal of Chemical Physics* 21(6), 1087-1092.

[11.139] Moffat, M.C. (1994) An ultrastructural study of *Didymosphenia geminata* (Bacillariophyceae). *Transactions of the American Microscopical Society* 113(1), 59-71.

[11.140] Mohoric, A., Lahajnar, G. and Stepisnik, J. (2020) Diffusion spectrum of polymer melt measured by varying magnetic field gradient pulse width in PGSE NMR. *Molecules* 25(24), #5813.

[11.141] Molino, P.J., Chiovitti, A., Higgins, M.J., Dugdale, T.M. and Wetherbee, R. (2016) Diatom adhesives: Molecular and mechanical properties. In: *Biological Adhesives*. A.M. Smith, (ed.) Springer International Publishing, Cham: 57-86. 2nd.

[11.142] Morillo, N., Patti, A. and Cuetos, A. (2019) Brownian dynamics simulations of oblate and prolate colloidal particles in nematic liquid crystals. *Journal of Chemical Physics* 150(20), #204905.

[11.143] Müller, P. and Kern, R. (2000) Equilibrium nano-shape changes induced by epitaxial stress (generalised Wulf-Kaishew theorem). *Surface Science* 457(1-2), 229-253.

[11.144] Murase, A., Kubota, Y., Hirayama, S., Kumashiro, Y., Okano, T., Mayama, S. and Umemura, K. (2011) Two-dimensional trajectory analysis of the diatom *Navicula* sp. using a micro chamber. *Journal of Microbiological Methods* 87(3), 316-319.

[11.145] Murase, A., Kubota, Y., Hori, S., Hirayama, S., Mayama, S. and Umemura, K. (2012) Importance of observation interval in two-dimensional video analysis of individual diatom cells. *European Biophysics Journal* 41, 545-550.

[11.146] Nakamura, Y. and Ohno, T. (2012) Single-domain ferroelectric water and its concerted diffusion in nanotubes. In: *Advanced Materials and Nanotechnology*. B.J. Ruck and T. Kemmitt, (eds.) Trans Tech Publications Ltd, Durnten-Zurich. 700: 108-111.

[11.147] Novarino, G. (1993) Presence of minerals in the mucilage stalk of the diatom *Achnanthes longipes*. *Diatom Research* 8(1), 199-202.

[11.148] Nuccitelli, R. and Erickson, C.A. (1983) Embryonic cell motility can be guided by physiological electric fields. *Experimental Cell Research* 147(1), 195-201.

[11.149] Nultsch, W. (1962) Über das Bewegungsverhalten der Diatomeen. *Planta (Berlin)* 58(1), 22-33.

[11.150] O'Neill, P.R., Castillo-Badillo, J.A., Meshik, X., Kalyanaraman, V., Melgarejo, K. and Gautam, N. (2018) Membrane flow drives an adhesion-independent amoeboid cell migration mode. *Developmental Cell* 46(1), 9-22.

[11.151] Oelschlaeger, P., Schmid, R.D. and Pleiss, J. (2003) Modeling domino effects in enzymes: Molecular basis of the substrate specificity of the bacterial metallo-β-lactamases IMP-1 and IMP-6. *Biochemistry* 42(30), 8945-8956.

[11.152] Ogawa, T. (2001) Domino mechanisms in photoinduced phase transitions. *Phase Transitions* 74(1-2), 93-128.

[11.153] Olivier, Y., Muccioli, L. and Zannoni, C. (2014) Quinquephenyl: The simplest rigid-rod-like nematic liquid crystal, or is it? An atomistic simulation. *Chemphyschem* 15(7), 1345-1355.

[11.154] Olney, M. (2002) Diatoms. https://www.ucl.ac.uk/GeolSci/micropal/diatom.html

[11.155] Ousaka, N., Inai, Y. and Okabe, T. (2006) Chiral interaction in Gly-capped N-terminal motif of 3_{10}-helix and domino-type induction in helix sense. *Biopolymers* 83(4), 337-351.

[11.156] Pachepsky, Y., Timlin, D. and Rawls, W. (2003) Generalized Richards' equation to simulate water transport in unsaturated soils. *Journal of Hydrology* 272(1-4), 3-13, #Pii s0022-1694(02)00251-2.

[11.157] Palermo, M.F., Muccioli, L. and Zannoni, C. (2015) Molecular organization in freely suspended nano-thick 8CB smectic films. An atomistic simulation. *Physical Chemistry Chemical Physics* 17(39), 26149-26159.

[11.158] Panagiotopoulos, A.Z. (2000) On the equivalence of continuum and lattice models for fluids. *The Journal of Chemical Physics* 112(16), 7132-7137.

[11.159] Panda, A., Reddy, A.S., Venkateswarlu, S. and Yoon, M. (2018) Bio-inspired self-propelled diatom micromotor by catalytic decomposition of H_2O_2 under low fuel concentration. *Nanoscale* 10(34), 16268-16277.

[11.160] Pant, A. (2014) The Domino Effect Physics Can be Pretty Incredible. http://awesci.com/domino-effect-physics-can-pretty-incredible/

[11.161] Patlak, C.S. (1953) Random walk with persistence and external bias. *Bulletin of Mathematical Biophysics* 15((3)), 311-338.

[11.162] Patti, A., El Masri, D., van Roij, R. and Dijkstra, M. (2010) Collective diffusion of colloidal hard rods in smectic liquid crystals: Effect of particle anisotropy. *Journal of Chemical Physics* 132(22), #224907.

[11.163] Pierobon, P., Parmeggiani, A., von Oppen, F. and Frey, E. (2005) Dynamic correlation functions and Boltzmann-Langevin approach for driven one-dimensional lattice gas. *Physical Review E* 72(3), 10, #036123.

[11.164] Pollock, F., Pickett-Heaps, J. and Schmid, A.M. (1998) Diatom protoplasts are amoeboid during recovery from osmotic shock. *Journal of Phycology* 34(3, Suppl.), 48.

[11.165] Pollock, F.M. and Pickett-Heaps, J.D. (2005) Spatial determinants in morphogenesis: Recovery from plasmolysis in the diatom *Ditylum*. *Cell Motility and the Cytoskeleton* 60(2), 71-82.

[11.166] Pollock, F.M. and Pickett-Heaps, J.D. (2006) Valve formation without mitosis in the diatom *Ditylum* recovering from plasmolysis. *Nova Hedwigia*(Suppl. 130), 119-125.

[11.167] Poulsen, N.C., Spector, I., Spurck, T.P., Schultz, T.F. and Wetherbee, R. (1999) Diatom gliding is the result of an actin-myosin motility system. *Cell Motility and the Cytoskeleton* 44(1), 23-33.

[11.168] Prasad, M.A. and Unnikrishnan, K. (1984) Biased random walk on a multidimensional lattice with absorbing boundaries. *Physica A-Statistical Mechanics and Its Applications* 127(3), 659-666.

[11.169] PubChem. (2021) Sucrose (Compound). https://pubchem.ncbi.nlm.nih.gov/compound/Sucrose#section=Collision-Cross-Section

[11.170] Python.org. (2021) Python 3.9.5. https://www.python.org/downloads/release/python-395/

[11.171] Quijano, J. (2914) Balancing Pencils Lesson 2. https://www.slideserve.com/juan/balancing-pencils-lesson-2

[11.172] Ragni, R., Cicco, S.R., Vona, D. and Farinola, G.M. (2017) Multiple routes to smart nanostructured materials from diatom microalgae: A chemical perspective. *Advanced Materials*, #1704289.

[11.173] Ragni, R., Scotognella, F., Vona, D., Moretti, L., Altamura, E., Ceccone, G., Mehn, D., Cicco, S.R., Palumbo, F., Lanzani, G. and Farinola, G.M. (2018) Hybrid photonic nanostructures by in vivo incorporation of an organic fluorophore into diatom algae. *Advanced Functional Materials* 28(24), #1706214.

[11.174] Rio, M., Bola, S., Funk, R.H.W. and Gerlach, G. (2016) Microfluidic measurement of cell motility in response to applied non-homogeneous DC electric fields. *Journal of Sensors and Sensor Systems* 5(2), 237-243.

[11.175] Rocha, P.R.F., Silva, A.D., Godinho, L., Dane, W., Estrela, P., Vandamme, L.K.J., Pereira-Leal, J.B., de Leeuw, D.M. and Leite, R.B. (2018) Collective electrical oscillations of a diatom population induced by dark stress. *Scientific Reports* 8, #5484.

[11.176] Round, F.E., Crawford, R.M. and Mann, D.G. (1990) *The Diatoms, Biology & Morphology of the Genera*. Cambridge University Press, Cambridge.

[11.177] Ruck, E.C. and Theriot, E.C. (2011) Origin and evolution of the canal raphe system in diatoms. *Protist* 162(5), 723-737.

[11.178] Ruivo, E.L.P., Montalva-Medel, M., de Oliveira, P.P.B. and Perrot, K. (2018) Characterisation of the elementary cellular automata in terms of their maximum sensitivity to all possible asynchronous updates. *Chaos Solitons & Fractals* 113, 209-220.

[11.179] Sabuncu, A.C., Gordon, R., Richer, E., Manoylov, K.M. and Beskok, A. (2021) The kinematics of explosively jerky diatom motility: A natural example of active nanofluidics. In: *Diatom Gliding Motility [Volume in the series: Diatoms: Biology & Applications, series editors: Richard Gordon & Joseph Seckbach]*. S.A. Cohn, K.M. Manoylov and R. Gordon, (eds.) Wiley-Scrivener, Beverly, MA, USA: 33-63.

[11.180] Saielli, G., Voth, G.A. and Wang, Y.T. (2013) Diffusion mechanisms in smectic ionic liquid crystals: insights from coarse-grained MD simulations. *Soft Matter* 9(24), 5716-5725.

[11.181] Sakaguchi, H. and Ishihara, T. (2011) Locomotive and reptation motion induced by internal force and friction. *Physical Review E* 83(6), #061903.

[11.182] Sanka, I., Suyono, E.A. and Alam, P. (2017) The effects of diatom pore-size on the structures and extensibilities of single mucilage molecules. *Carbohydrate Research* 448, 35-42.

[11.183] Sarvaharman, S. and Giuggioli, L. (2020) Closed-form solutions to the dynamics of confined biased lattice random walks in arbitrary dimensions. *Physical Review E* 102(6), #062124.

[11.184] Sautel, J., Bourges, A., Caussarieu, A., Plihon, N. and Taberlet, N. (2017) The physics of a popsicle stick bomb. *American Journal of Physics* 85(10), 783-790.

[11.185] Schmid, A.-M.M. (2007) The "paradox" diatom *Bacillaria paxillifer* (Bacillariophyta) revisited. *Journal of Phycology* 43, 139-155.

[11.186] Schröfel, A., Kratošová, G., Bohunická, M., Dobročka, E. and Vávra, I. (2011) Biosynthesis of gold nanoparticles using diatoms - silica-gold and EPS-gold bionanocomposite formation. *Journal of Nanoparticle Research* 13(8), 3207-3216.

[11.187] Schulz, D., Drebes, G., Lehmann, H. and Jank-Ladwig, R. (1984) Ultrastructure of *Anaulus creticus* Drebes & Schulz with special reference to its reduced ocelli. *European Journal of Cell Biology* 33(1), 43-51.

[11.188] Sebban, P., Maroti, P., Schiffer, M. and Hanson, D.K. (1995) Electrostatic dominoes: Long-distance propagation of mutational effects in photosynthetic reaction centers of *Rhodobacter capsulatus*. *Biochemistry* 34(26), 8390-8397.

[11.189] Sharov, A., Gordon, R. and Mikhailovsky, G. (eds.) (2023) *Pathways to the Origin and Evolution of Meanings in the Universe [Volume in the series: Astrobiology Perspectives on Life of the Universe book series (Series editors: Richard Gordon & Joseph Seckbach, Wiley-Scrivener)]*. Wiley-Scrivener, Beverly, Massachusetts, USA.

[11.190] Shaw, D.E. (1978) Mechanics of a chain of dominoes. *American Journal of Physics* 46(6), 640-642.

[11.191] Shi, T.F., Liu, Y., Wang, N.N. and Liu, C.S. (2018) Toppling dynamics of regularly spaced dominoes in an array [erratum: 35(6),]. *Journal of Applied Mechanics-Transactions of the ASME* 85(4), #041008.

[11.192] Shi, T.F., Liu, Y., Wang, N.N. and Liu, C.S. (2019) Toppling dynamics of a mass-varying domino system. *Nonlinear Dynamics* 98(3), 2261-2275.

[11.193] Shin, H.K., Kim, C., Talkner, P. and Lee, E.K. (2010) Brownian motion from molecular dynamics. *Chemical Physics* 375(2-3), 316-326.

[11.194] Shinto, H. (2012) Computer simulation of wetting, capillary forces, and particle-stabilized emulsions: From molecular-scale to mesoscale modeling. *Advanced Powder Technology* 23(5), 538-547.

[11.195] Sholl, D.S. and Fichthorn, K.A. (1997) Concerted diffusion of molecular clusters in a molecular sieve. *Physical Review Letters* 79(19), 3569-3572.

[11.196] Sikora, J., Wasik, A. and Zajaczkowska, M. (1991) Cytoplasmic streaming direction reverses in dividing *Paramecium bursaria*. *European Journal of Protistology* 27(4), 352-356.

[11.197] Simpson, M.J., Lo, K.Y. and Sun, Y.S. (2017) Quantifying the roles of random motility and directed motility using advection-diffusion theory for a 3T3 fibroblast cell migration assay stimulated with an electric field. *BMCSystems Biology* 11, #39.

[11.198] Siwy, Z., Mycielska, M.E. and Djamgoz, M.B.A. (2003) Statistical and fractal analyses of rat prostate cancer cell motility in a direct current electric field: comparison of strongly and weakly metastatic cells. *European Biophysics Journal with Biophysics Letters* 32(1), 12-21.

[11.199] Slabaugh, E., Davis, J.K., Haigler, C.H., Yingling, Y.G. and Zimmer, J. (2014) Cellulose synthases: New insights from crystallography and modeling. *Trends in Plant Science* 19(2), 99-106.

[11.200] Soltani, M. and Golovin, K. (2020) Anisotropy-induced directional self-transportation of low surface tension liquids: a review. *RSC Advances* 10(66), 40569-40581.

[11.201] Sprott, J.C. (1997) Some simple chaotic jerk functions. *American Journal of Physics* 65(6), 537-543.

[11.202] Stahl, M. and Sieber, S.A. (2017) An amino acid domino effect orchestrates ClpP's conformational states. *Current Opinion in Chemical Biology* 40, 102-110.

[11.203] Stern, C.D. (1981) Behavior and motility of cultured chick mesoderm cells in steady electrical fields. *Experimental Cell Research* 136(2), 343-350.

[11.204] Sterrenburg, F.A.S., Tiffany, M.A. and Meave del Castillo, M.E. (2005) Valve morphogenesis in the diatom genus *Pleurosigma* W. Smith (Bacillariophyceae): Nature's alternative sandwich. *Journal of Nanoscience and Nanotechnology* 5(1), 140-145.

[11.205] Stewart, I.W. and Stewart, F. (2009) Shear flow in smectic A liquid crystals. *Journal of Physics-Condensed Matter* 21(46), #465101.

[11.206] Stronge, W.J. (1987) The domino effect: A wave of destabilizing collisions in a periodic array. *Proceedings of the Royal Society of London Series A-Mathematical Physical and Engineering Sciences* 409(1836), 199-208.

[11.207] Stronge, W.J. and Shu, D. (1988) The domino effect: Successive destabilization by cooperative neighbors. *Proceedings of the Royal Society of London Series A-Mathematical Physical and Engineering Sciences* 418(1854), 155-163.

[11.208] Sun, B.-H. (2020) An Explicit Scaling Law for the Speed of Falling Dominoe. https://www.preprints.org/manuscript/201912.0017/v1

[11.209] Sun, B.H. (2020) Scaling law for the propagation speed of domino toppling. *Aip Advances* 10(9), #095124.

[11.210] Sun, Y.H., Sun, Y.X., Zhu, K., Reid, B., Gao, X., Draper, B.W., Zhao, M. and Mogilner, A. (2018) Electric fields accelerate cell polarization and bypass myosin action in motility initiation. *Journal of Cellular Physiology* 233(3), 2378-2385.

[11.211] Tadmor, R., Baksi, A., Gulec, S., Jadhav, S., N'Guessan, H.E., Sen, K., Somasi, V., Tadmor, M., Wasnik, P. and Yadav, S. (2019) Drops that change their mind: Spontaneous reversal from spreading to retraction. *Langmuir* 35(48), 15734-15738.

[11.212] Tadmor, R., Baksi, A., Gulec, S., Jadhav, S., N'Guessan, H.E., Somasi, V., Tadmor, M., Tang, S., Wasnik, P. and Yadav, S. (2020) Defying gravity: Drops that climb up a vertical wall of their own accord. *Journal of Colloid and Interface Science* 562, 608-613.

[11.213] Takamori, S., Holt, M., Stenius, K., Lemke, E.A., Grønborg, M., Riedel, D., Urlaub, H., Schenck, S., Brügger, B., Ringler, P., Müller, S.A., Rammner, B., Gräter, F., Hub, J.S., De Groot, B.L., Mieskes, G., Moriyama, Y., Klingauf, J., Grubmüller, H., Heuser, J., Wieland, F. and Jahn, R. (2006) Molecular anatomy of a trafficking organelle. *Cell* 127(4), 831-846.

[11.214] Tan, J., Wang, X.L., Zhang, J., Wang, J., Wang, J., Fu, X., Li, J.B., Cao, Y. and Yu, X.L. (2015) Influence of positive charge and surface free energy on the adhesion of diatom. *Journal of Biomaterials and Tissue Engineering* 5(3), 219-225.

[11.215] Tan, X., Chen, Y., Wang, H., Zhang, Z. and Ling, X.S. (2021) 2D isotropic–nematic transition in colloidal suspensions of ellipsoids. *Soft Matter* 17(24), 6001-6005.

[11.216] Tarasenko, A. and Jastrabik, L. (2015) The studies of particle diffusion on a heterogeneous one-dimensional lattice. *Surface Science* 641, 266-268.

[11.217] Taylor, A.R. (2009) A fast $Na^+/Ca2^+$-based action potential in a marine diatom. *Plos One* 4(3), #e4966.

[11.218] Tesson, B., Gaillard, C. and Martin-Jézéquel, V. (2009) Insights into the polymorphism of the diatom *Phaeodactylum tricornutum* Bohlin. *Botanica Marina* 52(2), 104-116.

[11.219] Tesson, B. and Hildebrand, M. (2013) Characterization and localization of insoluble organic matrices associated with diatom cell walls: Insight into their roles during cell wall formation. *PLoS One* 8(4), #e61675.

[11.220] Torres, F.A., Gauthier, M.G. and Slater, G.W. (2008) Biased random walks on a lattice: Exact numerical method to study the effect of alternating fields in disordered and asymmetric systems of obstacles. *Physical Review E* 78(6), #065701.

[11.221] Tuszynski, J.A. and Gordon, R. (2012) A mean field Ising model for cortical rotation in amphibian one-cell stage embryos. *BioSystems* 109(3, Special Issue on Biological Morphogenesis), 381-389.

[11.222] Ussing, A.P., Gordon, R., Ector, L., Buczkó, K., Desnitskiy, A.G. and VanLandingham, S.L. (2005) *The Colonial Diatom "Bacillaria paradoxa": Chaotic Gliding Motility, Lindenmeyer Model of Colonial Morphogenesis, and Bibliography, with Translation of O.F. Müller (1783), "About a peculiar being in the beach-water"*. A.R.G. Ganter Verlag Kommanditgesellschaft, Ruggell. 5.

[11.223] van Leeuwen, J.M.J. (2010) The domino effect. *American Journal of Physics* 78(7), 721-727.

[11.224] van Roij, R., Dijkstra, M. and Evans, R. (2000) Orientational wetting and capillary nematization of hard-rod fluids. *Europhysics Letters* 49(3), 350-356.

[11.225] Varga, S., Gurin, P., Armas-Perez, J.C. and Quintana-H, J. (2009) Nematic and smectic ordering in a system of two-dimensional hard zigzag particles. *Journal of Chemical Physics* 131(18), #184901.

[11.226] Vielhaber, M. (2012) Computing by temporal order: Asynchronous cellular automata. *Electronic Proceedings in Theoretical Computer Science*(90), 166-176.

[11.227] Vinayak, V., Joshi, K.B., Gordon, R. and Schoefs, B. (2018) Nanoengineering the diatom surface for plausible applications. In: *Diatom Nanotechnology: Progress and Emerging Applications*. D. Losic, (ed.) Royal Society of Chemistry, London: 55-78.

[11.228] Volpi, N., Maccari, F. and Linhardt, R.J. (2008) Capillary electrophoresis of complex natural polysaccharides. *Electrophoresis* 29(15), 3095-3106.

[11.229] Vona, D., Cicco, S.R., Ragni, R., Leone, G., Lo Presti, M. and Farinola, G.M. (2018) Biosilica/polydopamine/silver nanoparticles composites: new hybrid multifunctional heterostructures obtained by chemical modification of T*halassiosira weissflogii* silica shells. *MRS Communications* 8(3), 911-917.

[11.230] Voutouri, C., Polydorou, C., Papageorgis, P., Gkretsi, V. and Stylianopoulos, T. (2016) Hyaluronan-derived swelling of solid tumors, the contribution of collagen and cancer cells, and implications for cancer therapy. *Neoplasia* 18(12), 732-741.

[11.231] Waldron, J.T., Kalmykov, Y.P. and Coffey, W.T. (1994) Rotational Brownian motion and dielectric relaxation of polar molecules subjected to a constant bias field: Exact solution. *Physical Review. E. Statistical Physics, Plasmas, Fluids, and Related Interdisciplinary Topics* 49(5), 3976-3989.

[11.232] Wang, J., Weng, D. and Cao, S. (2021) Locomotion of benthic pennate diatoms: Models and thoughts. In: *Diatom Gliding Motility [DIGM, Volume in the series: Diatoms: Biology & Applications, series editors: Richard Gordon & Joseph Seckbach] [In press]*. S.A. Cohn, K.M. Manoylov and R. Gordon, (eds.) Wiley-Scrivener, Beverly, MA, USA: 295-333.

[11.233] Wang, Y., Chen, Y., Lavin, C. and Gretz, M.R. (2000) Extracellular matrix assembly in diatoms (Bacillariophyceae). IV. Ultrastructure of *Achnanthes longipes* and *Cymbella cistula* as revealed by high-pressure freezing/freeze substituton and cryo-field emission scanning electron microscopy. *Journal of Phycology* 36(2), 367-378.

[11.234] Wang, Y., Lu, J.J., Mollet, J.C., Gretz, M.R. and Hoagland, K.D. (1997) Extracellular matrix assembly in diatoms (Bacillariophyceae).II. 2,6-dichlorobenzonitrile inhibition of motility and stalk production in the marine diatom *Achnanthes longipes*. *Plant Physiology* 113(4), 1071-1080.

[11.235] Watanabe, G., Saito, J., Fujita, Y. and Tabe, Y. (2013) Molecular dynamics simulation study of two-dimensional diffusion behavior in smectic liquid crystalline monolayers. *Journal of the Physical Society of Japan* 82(8), #084603.

[11.236] Watanabe, Y., Meents, M.J., McDonnell, L.M., Barkwill, S., Sampathkumar, A., Cartwright, H.N., Demura, T., Ehrhardt, D.W., Samuels, A.L. and Mansfield, S.D. (2015) Visualization of cellulose synthases in Arabidopsis secondary cell walls. *Science* 350(6257), 198-203.

[11.237] Webster, D.R. (1984) *Biochemical and Morphological Analysis of Cytoskeletal Proteins and Trail Secretion During Gliding Motility in the Marine Benthic Diatom, Amphora coffeaeformis [PhD thesis]*. Department of Cell Biology and Anatomy, University of Miami, Coral Gables, Florida, USA.

[11.238] Weisstein, E.W. (2021) Elementary Cellular Automaton. https://mathworld.wolfram.com/ElementaryCellularAutomaton.html

[11.239] Weisstein, E.W. (2022) Run-Length Encoding. https://mathworld.wolfram.com/Run-LengthEncoding.html

[11.240] Wenderoth, K., Marquardt, J. and Rhiel, E. (2004) The big trail: Many migrate at the expense of a few. *Diatom Research* 19(1), 115-122.

[11.241] Wetherbee, R. (2004) Diatom Trail Formation [movie]. http://lin.irk.ru/nanochem/diatom_movies/films.html

[11.242] Wetherbee, R., Lind, J.L., Burke, J. and Quatrano, R.S. (1998) The first kiss: Establishment and control of initial adhesion by raphid diatoms. *Journal of Phycology* 34(1), 9-15.

[11.243] Wikipedia. (2020) Liquid crystal: Smectic phases. https://en.wikipedia.org/wiki/Liquid_crystal#Smectic_phases

[11.244] Williams, S.K., Sarfati, M.R. and McDonagh, P.F. (1994) Direct observation of leukocyte migration across endothelial monolayers using total internal reflectance fluorescence microscopy. *FASEB Journal* 8(4), A321.

[11.245] Windler, M., Bova, D., Kryvenda, A., Straile, D., Gruber, A. and Kroth, P.G. (2014) Influence of bacteria on cell size development and morphology of cultivated diatoms. *Phycological Research* 62(4), 269-281.

[11.246] Witkowski, A., Płociński, T., Grzonka, J., Zgłobicka, I., Bąk, M., Dąbek, P., Gomes, A.I. and Kurzydłowski, K.J. (2019) Application of focused ion beam technique in taxonomy-oriented research on ultrastructure of diatoms [Chapter 6]. In: *Diatoms: Fundamentals & Applications [DIFA, Volume 1 in the series: Diatoms: Biology & Applications, series editors:*

Richard Gordon & Joseph Seckbach]. J. Seckbach and R. Gordon, (eds.) Wiley-Scrivener, Beverly, MA, USA: 115-127.

[11.247] Wolfram, S. (2001) *A New Kind of Science*. Wolfram Media, Champaign, Illinois.

[11.248] Xing, X.J. (2008) Topology of smectic order on compact substrates. *Physical Review Letters* 101(14), #147801.

[11.249] Xing, X.J. (2009) Topology and geometry of smectic order on compact curved substrates. *Journal of Statistical Physics* 134(3), 487-536.

[11.250] Yamaoka, N., Suetomo, Y., Yoshihisa, T. and Sonobe, S. (2016) Motion analysis and ultrastructural study of a colonial diatom, *Bacillaria paxillifer*. *Microscopy* 65(3), 211-221.

[11.251] Yang, G., Long, H.Y., Ren, X.M., Ma, K.L., Xiao, Z.H., Wang, Y. and Guo, Y.Q. (2017) Regulation of adipose-tissue-derived stromal cell orientation and motility in 2D-and 3D-cultures by direct-current electrical field. *Development Growth & Differentiation* 59(2), 70-82.

[11.252] Zhang, W.W., Cai, C. and Staiger, C.J. (2019) Myosins XI are involved in exocytosis of cellulose synthase complexes. *Plant Physiology* 179(4), 1537-1555.

[11.253] Zhao, M., Dick, A., Forrester, J.V. and McCaig, C.D. (1999) Electric field-directed cell motility involves up-regulated expression and asymmetric redistribution of the epidermal growth factor receptors and is enhanced by fibronectin and laminin. *Molecular Biology of the Cell* 10(4), 1259-1276.

[11.254] Zischka, F., Kratochvil, H., Noll, A., Gordon, R., Harbich, T. and Gebeshuber, I.C. (2021) Diatom triboacoustics. In: *Diatom Gliding Motility [DIGM, Volume in the series: Diatoms: Biology & Applications, series editors: Richard Gordon & Joseph Seckbach]. Wiley-Scrivener, Beverly, MA, USA*. S.A. Cohn, K.M. Manoylov and R. Gordon, (eds.) Wiley-Scrivener, Beverly, Massachusetts, USA: 249-282.

[11.255] Zwerger, W. (1981) Critical slowing down of diffusion in a one-dimensional kinetic Ising-modeL. *Physics Letters A* 84(5), 269-270.

Part IV

DIATOM ECOLOGICAL AND ENVIRONMENTAL ANALYSIS

12

Following the Photons Route: Mathematical Models Describing the Interaction of Diatoms with Light

Edoardo De Tommasi[1*], Alessandra Rogato[2,3], Diego Caratelli[4,5], Luciano Mescia[6] and Johan Gielis[4,7]

[1]*National Research Council, Institute of Applied Sciences and Intelligent Systems "Eduardo Caianiello", Unit of Naples, Via P. Castellino 111, Naples, Italy*
[2]*National Research Council, Institute of Biosciences and Bioresources, Unit of Naples, Via P. Castellino 111, Naples, Italy*
[3]*Stazione Zoologica Anton Dohrn, Department of Integrative Marine Ecology, Villa Comunale 1, Naples, Italy*
[4]*Department of Research and Development, The Antenna Company, High Tech Campus, AE Eindhoven, The Netherlands*
[5]*Department of Electrical Engineering, Eindhoven University of Technology, MB Eindhoven, The Netherlands*
[6]*Department of Electrical and Information Engineering, Polytechnic University of Bari, Via E. Orabona 4, Bari, Italy*
[7]*Department of Bioscience Engineering, University of Antwerp, Groenenborgerlaan, Antwerp, Belgium*

Abstract

The interaction of diatoms with sunlight is fundamental in order to deeply understand their role in terrestrial ecology and biogeochemistry, essentially due to their massive contribution to global primary production through photosynthesis and its effect on carbon, oxygen and silicon cycles. Following the journey of light through natural waters, its propagation through the intricate frustule micro- and nano-structure and, finally, its fate inside the photosynthetic machinery of the living cell requires several mathematical and computational models in order to accurately describe all the involved phenomena taking place at different space scales and physical regimes.

In this chapter, we review the main analytical models describing the underwater optical field, the essential numerical algorithms for the study of photonic properties of the diatom frustule seen as a natural metamaterial, as well as the principal models describing photon harvesting in diatom plastids and methods for complex EM propagation problems and wave propagation in dispersive materials with multiple relaxation times. These mathematical methods will be integrated in a unifying geometric perspective.

**Corresponding author*: edoardo.detommasi@na.isasi.cnr.it
Edoardo De Tommasi: https://orcid.org/0000-0003-2030-4557
Alessandra Rogato: https://orcid.org/0000-0002-0373-9076
Diego Caratelli: https://orcid.org/0000-0003-0969-884X
Luciano Mescia: https://orcid.org/0000-0002-2339-1214
Johan Gielis: https://orcid.org/0000-0002-4536-3839

Janice L. Pappas (ed.) The Mathematical Biology of Diatoms, (345–392) © 2023 Scrivener Publishing LLC

***Keywords*:** Diatoms, diatom chloroplasts and photoreceptors, diatom photosynthesis, fractional calculus, Gielis transformations, natural photonic crystals, optical properties of diatom frustules, underwater optical field

12.1 Introduction

Almost all aquatic habitats on Earth are colonized by diatoms [12.108]. Being the most common type of phytoplankton, they contribute massively to global primary production (up to 20-25%) [12.77], fixing large amounts of carbon dioxide in the oceans, in freshwater, and generally wherever water is at least occasionally present. Understanding the mechanisms by which diatoms interact with sunlight is the key to unveiling the origin of their extraordinary photosynthetic efficiency and evolutionary success. The exchanges between diatoms and external environment are mediated by the frustule, a porous, hydrated silica investment which encloses the cell and is characterized by the presence of regular patterns of pores. Frustules are at the basis of diatom taxonomy [12.104], and their huge variety of shapes and dimensions (ranging from some tens of microns to 1-2 millimeters) allows distinguishing up to more than 10^3 genera and about 10^5 species [12.111]. They basically consist in an *epitheca* overlapping a *hypotheca* in a "petri-dish-like" arrangement. Every *theca* is formed by a valve and one or more lateral bands (girdle bands) connected along the margins. Frustule functionalities comprise mechanical protection [12.61], gas exchange [12.37], sorting of nutrients from noxious agents such as viruses [12.60], and sinking rate control [12.120]. The impressive similarity of diatom frustules with artificial photonic crystals induced to also deepen their optical properties, including the ability to couple light in waveguided modes [12.45], to collect and focus photosynthetic active radiation (PAR, 400-700 nm) [12.23] [12.24], to support photonic pseudo-band gaps [12.56], and to protect the cell from detrimental ultraviolet radiation (UVR) [12.2] [12.28] [12.29]. All these features contribute to diatom high efficiency of sunlight harvesting also where it is not so easily accessible, and have been exploited in several technological fields where light manipulation and control are essential, such as super-resolution [12.25], enhancement of efficiency in dye-sensitized solar cells (DSSCs) [12.10], plasmonics and surface-enhanced Raman spectroscopy [12.83], optical sensing and biosensing [12.101], and random lasing [12.76], to name a few.

Here, our aim is to describe the journey of sunlight through the atmosphere, the water bodies, the intricate silica matrix of the frustule, and finally illustrate its interaction with the photosynthetic machinery of the cell. We will mainly focus on the mathematical depiction of all the involved phenomena, in particular analyzing the equations which rule underwater light propagation, describing frustule morphology and shape by means of innovative geometrical approaches based on Gielis superformula [12.52], introducing the main numerical methods used to describe light propagation through the frustule seen as a periodic dielectric medium, and deepening the behavior of plastids seen as efficient antennas.

12.2 The Underwater Light Field

Before analyzing the underwater optical field and its mathematical description, we will briefly summarize the main characteristics of the solar radiation incident on Earth's surface and transmitted by the air-water interface. We further qualitatively consider light absorption and scattering in a water body. Finally, the main equation governing underwater light propagation will be illustrated, together with a representative computational method aimed at its solution.

12.2.1 The Travel of Light from the Sun into Water Bodies

Light irradiated by the Sun and propagating through atmosphere is partially scattered by air molecules and dust particles and absorbed by water vapor, oxygen, ozone and carbon dioxide, among others [12.11]. Being much smaller than the wavelength of solar radiation, according to Rayleigh's law, air molecules mostly scatter radiation in the visible and ultraviolet (UV) range of the electromagnetic spectrum, while dust particles obey Mie scattering, which is not so strictly dependent on wavelength and is predominant in forward direction. A higher proportion of infrared radiation in comparison to photosynthetic active radiation (PAR) is absorbed by water vapors, thus PAR represents the most abundant solar radiation which reaches Earth's surface [12.9]. Also taking into account the contribution of clouds, it has been estimated that 34% of the incoming solar radiation incident on the northern hemisphere over a year is reflected to space by atmosphere (25% reflected by the clouds and 9% scattered out to space by other constituents of the atmosphere); another 19% is absorbed (10% within the clouds and 9% by other components) [12.47].

The amount of solar radiation reflected back to the atmosphere at the air-water interface depends on the zenith angle of the incident light in air. In first approximation, in particular considering a flat water surface and unpolarized incident light, it can be simply derived from the well-known Fresnel equation [12.15]:

$$R = \frac{1}{2}\left[\frac{\sin^2(\theta_a - \theta_w)}{\sin^2(\theta_a + \theta_w)} + \frac{\tan^2(\theta_a - \theta_w)}{\tan^2(\theta_a + \theta_w)}\right] \tag{12.1}$$

where R stands for reflectance and θ_a and θ_w stand for the zenith angle of incident light in air and the angle to the downward vertical of the transmitted light in water, respectively. When not considering absorption and scattering events, conservation of energy requires that the sum of the reflected and transmitted energy equals the incident energy, thus the quantity $T = 1 - R$ allows evaluating, in first approximation, the amount of light penetrating in the water body. Actually, reflectance is partially reduced by roughening of water surface induced by wind [12.6] and slightly increased by the formation of whitecaps [12.122]. The unreflected portion of an incident light beam is refracted by water according to Snell's law [12.15]:

$$\frac{\sin\theta_a}{\sin\theta_w} = \frac{n_w}{n_a} \tag{12.2}$$

with n_w and n_a refractive indices of water and air, respectively. Even though the refractive index of water is a function of temperature, salt content and wavelength of incoming radiation, for most practical analyses $\frac{n_w}{n_a}$ ratio can be considered equal to 1.33 for PAR radiation and at normal ambient temperatures. Taking into account this value, it can be determined that most of the light transmitted by water is refracted at an angle θ_w comprised between 0° and 49°. On the other side, considering the case of light propagating upward from water to air at an angle greater than 49°, total internal reflection takes place and all the light is reflected back by the water-air interface.

In general, diffusion of underwater light is enhanced by an increase in wind speed as expressed by the following equations empirically derived by Cox and Munk [12.22], starting from aerial data and the observation of Sun glittering patterns:

$$\sigma_u^2 = 0.003 \cdot U \pm 0.004 \tag{12.3}$$

$$\sigma_c^2 = 0.003 + 0.002 \cdot U \pm 0.002 \tag{12.4}$$

with σ_u^2 and σ_c^2 mean square slopes of the waves measured in parallel and cross directions of the wind, respectively, and U wind speed above the water surface.

When light passes through a medium, a certain fraction of it is attenuated whenever absorption or scattering events take place. Transmittance T through a homogeneous medium follows the well-known Lambert-Beer law:

$$T = \exp(-cl) \tag{12.5}$$

with c total attenuation coefficient taking into account the probability of a photon to be absorbed or scattered after a path length l. Pure water is basically transparent for wavelengths below 550 nm, while absorption starts to be significant in red and mainly in infrared regions of the spectrum, where the roto-vibrational bands, combination bands and overtone bands of water are located [12.62]. Salts present in seawater, such as nitrates and bromides, contribute to absorption only for wavelengths below 250 nm [12.93]. A natural source of light absorption, mainly in UV and blue regions, is given by dissolved organic matter originating from plant tissue decomposition (known as *humic substances*). It is estimated that at least 10% of the humic material present in ocean waters is terrestrially derived from river discharges [12.86]. Absorption in UV-blue spectral range is also partially ascribable to inanimate particulate matter (known as *tripton*). Phytoplankton contribute to total absorption of natural waters both in visible spectral range (mainly due to clorophylls, carotenoids and biliproteins) and in near UV (due to the presence of UV-absorbing pigments such as *mycosporine-like amino acids*, MAAs) [12.70]. Solar radiation penetrating in a water body is thus progressively depleted in those spectral regions where the medium mostly absorbs. In clear oceanic waters most of the visible light at depths below ~ 15 m comprises the blue-green spectral range (400-550 nm) and peaks in the blue region (440-490 nm), while UVB radiation (280-315 nm) has been detected in depths down to 60-70 m [12.28]. In turbid waters characterized by high levels of humic substances, blue light is rapidly removed within shallow depths and the available irradiance often results peaked around 580 nm (yellow) [12.70].

In general, the absorption coefficient a of a medium can be defined starting from the effective *absorption cross section* σ_a (which quantifies the probability of the absorption process) as:

$$a = \rho_a \sigma_a \tag{12.6}$$

with ρ_a volume density of the absorbing particles.

Vertical penetration of light into a water body is also obstructed by scattering, which increases the total path length of propagating radiation and raises the probability of photons to be absorbed by one of the components of the medium. Scattering events can originate from interaction of light with localized microscopic fluctuations of density in the medium. These inhomogeneities can be regarded as induced dipoles which emit radiation at the same frequency of the exciting one. In this case the angular distribution of scattered light is analogous to that given by Rayleigh theory for gases, as well as the wavelength dependence ($\propto \lambda^{-4}$). When the scattering centers are bigger than the wavelength of light (as in the case of mineral particles, phytoplankton, bacteria, dead cells and fragment of cells), Mie's theory of scattering has to be taken into account. Instead of considering the scattering centers as single oscillating dipoles, in Mie theory the scattering particle is modeled as the location of the additive contribution of a series of electrical and magnetic multipoles. In this case most of the scattered radiation takes place in forward direction and generally at small angles respect to the propagation axis. Analogous with absorption, the scattering coefficient b can be defined as:

$$b = \rho_s \sigma_s \tag{12.7}$$

with ρ_s volume density of the scattering particles in the medium and σ_s effective scattering cross section (quantifying the probability of a scattering event). Referring to Eq. (12.5), total attenuation coefficient c can thus be expressed by $c = a + b$.

12.2.2 Numerical Computation of the Underwater Optical Field

In order to compute the underwater optical field, we have to find solutions to the *radiative transfer equation*, which, in its time-independent, monochromatic and one-dimensional form can be written as [12.89] (see Table 12.1 for symbols definition):

$$\mu \frac{\partial L(\tau;\theta,\phi)}{\partial \tau} = -L(\tau;\theta,\phi) - \frac{b}{c}(\tau)$$
$$\iint_{(\theta',\phi')\in\Xi} L(\tau;\theta',\phi') \times \tilde{\beta}(\tau;\theta',\phi' \rightarrow \theta,\phi)\, \sin\theta' d\theta' d\phi' + S_{in}(\tau;\theta,\phi) \tag{12.8}$$

It is worth noticing the dependence of the total attenuation coefficient c by depth z and the consequent definition of the optical depth τ as its integral with respect to distance.

Table 12.1 Symbols, units and definitions of the quantities in Eq. 12.8.

Symbol	Units	Definition
θ	deg	Polar angle of photon travel
ϕ	deg	Azimuthal angle of photon travel
z	m	Depth (defined positive downward)
τ	-	Optical depth (positive downward), $\tau \equiv \int_0^z c(z)d(z)$
L	W m^{-2} sr^{-1} nm^{-1}	Unpolarized spectral radiance
Ξ	-	Set of all downward and upward directions
β	m^{-1} sr^{-1}	Volume scattering function
$\tilde{\beta}$	sr^{-1}	Scattering phase function, $\tilde{\beta} \equiv \beta / b$
S_{in}	W m^{-2} sr^{-1} nm^{-1}	Internal source of radiance

The radiance $L = L(\theta, \phi)$ leaving an elemental surface dS at polar angle θ and azimuthal angle ϕ is defined, at a fixed depth z and wavelength λ, as:

$$L(\theta,\phi) = \frac{d^2\Phi}{dS\cos\theta d\Omega} \tag{12.9}$$

where $\Phi = \frac{\partial Q_e}{\partial t}$ is the radiant flux leaving dS within the elemental solid angle $d\Omega$ and Q_e is the radiant energy. The involved geometry is schematized in Figure 12.1.

The scattering phase function $\tilde{\beta}(\theta,\phi)$ appearing in the double integral of Eq. 12.8 gives information on how light incident on an elemental volume is scattered in different directions [12.90]. It can be derived starting from the volume scattering function $\beta(\theta, \phi)$, which in turn is defined as [12.1]:

$$\beta(\theta, \phi) = dI / [L(\theta, \phi) \cdot dV] \tag{12.10}$$

i.e., as the radiant intensity $\left(I = \frac{d\Phi}{d\Omega} \right)$ emanating from an elemental volume dV in a given direction (θ, ϕ) per unit incident radiance L. The scattering phase function can be obtained simply dividing the volume scattering function by the scattering coefficient b, which can be defined, in alternative to Eq. 12.7 and for a given wavelength λ, as:

$$b = \int_{4\pi} \beta(\theta,\phi) d\Omega = 2\pi \int_0^{\pi} \beta \sin\theta d\theta \tag{12.11}$$

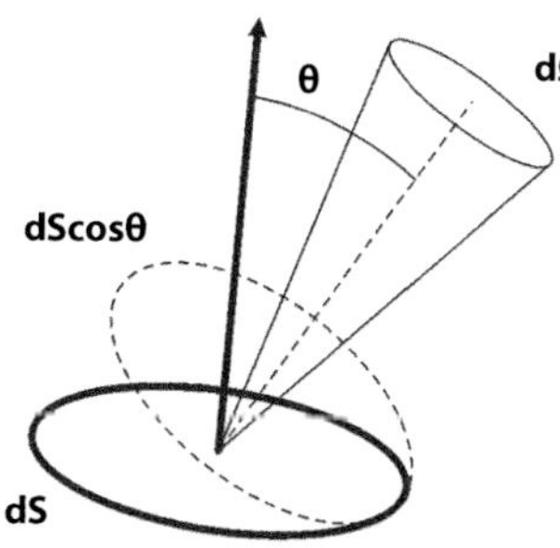

Figure 12.1 Visual representation of the quantities used in the definition of radiance $L = L(\theta, \phi)$: polar angle defining photon travel (θ), elemental surface (dS), its projection on the direction of propagation of light ($dS \cos \theta$), and elemental solid angle surrounding the same direction ($d\Omega$).

where the symmetry of the azimuthal angle ϕ has been taken into account.

In summary, Eq. 12.8 expresses the conservation of energy of an underwater beam of light: the first term on the right gives the Lambert-Beer loss, the second term is the gain from light scattered from the direction defined by angles (θ', ϕ') into the direction defined by (θ, ϕ), and the third term refers to internal sources of radiance (e.g., bioluminescence). Many numerical models aimed at the solution of Eq. 12.8 are based on a first partitioning of Ξ, the set of all downward and upward directions, into a grid of so-called *quads*, i.e., quadrilateral regions bounded by lines of constant μ and ϕ, plus two polar caps [12.89], with $\mu \equiv \cos \theta$. In this partitioned domain, the evaluated quantity is the *quad-averaged radiance*:

$$L(\tau;u,v) \equiv \frac{1}{\Omega_{uv}} \iint_{(\mu,\phi)\in Q_{uv}} L(\tau;\mu,\phi)d\mu d\phi \tag{12.12}$$

where Q_{uv} stands for the *uv*th quad subtending the solid angle Ω_{uv}, u labels μ bands and v labels ϕ bands. When applying the integral operator described in Eqs. 12.8 to 12.12, integration over all directions is replaced by summation over all quads and the phase function $\tilde{\beta}(\tau;\mu',\phi' \to \mu,\phi)$ is replaced by the quad-averaged quantity $\tilde{\beta}(\tau;r,s \to u,v)$ that specifies the radiance amount scattered from quad Q_{rs} into quad Q_{uv}. The equations for $L(\tau; u, v)$ can be transformed into a set of Riccati differential equations (first-order ordinary differential equations that are quadratic in the unknown function) describing the dependence of reflectance and transmittance functions by depth within the water body. This approach, described in detail in Mobley [12.88], allows the modelling of absorption and scattering events as sums of terms representing contributions by pure water, particles, and dissolved substances. Furthermore, reflectance and transmittance of the sea surface from above and below are evaluated making use of a Monte Carlo simulation of the surface itself, which is resolved into a grid of triangular wave facets characterized by random vertex elevations determined starting from a wave slope-speed spectrum (see, for example Eqs. 12.3 and 12.4) [12.96]. The atmospheric radiance incident onto the sea surface can be derived from an analytical model (e.g. a cardioidal distribution) or from a separate atmospheric radiative transfer model. On the other side, the bottom boundary can be schematized as a semi-infinite homogeneous layer of water starting at some depth τ_{max} or an opaque bottom at τ_{max}. In the first case, the bidirectional radiance reflectance properties of the infinite water layer, the asymptotic diffuse attenuation coefficient k_∞, and the asymptotic radiance distribution $L_\infty(\mu)$ can be obtained by an eigenmatrix analysis, as the one introduced by Preisendorfer [12.95]. In the second case, a Lambertian surface (i.e. an ideal matte surface characterized by an isotropic luminance) with a given irradiance reflectance is able to describe the reflectance properties of the bottom.

In conclusion, this approach (which represents only one of the possible ways to solve Eq. 12.8), has a strong analytical basis related to the solution of the Riccati differential equations for the retrieval of the radiance L. The only Monte Carlo fluctuations are restricted to the numerical simulation of the sea surface. The computation time is a linear function of depth, thus accurate radiance distributions are obtained for sufficiently high values of τ (typically for $\tau > 10$) [12.87].

12.3 Novel Geometrical Models for Diatoms

12.3.1 Gielis Transformations

Thus far, light has traveled through (relatively) isotropic media, with no preferred directions, but when it reaches the diatoms, matters become very different. Most methods in physics and mathematical physics assume isotropy in some sense, but frustules of diatoms have very characteristic shapes, and apart from centric diatoms, they are distinctly anisotropic. In addition, methods borrowed from mathematical physics and differential geometry are local methods, inherited from the pointwise method of determining curvature of curves developed by Newton.

Ultimately the shape of the frustules, the structure of valves and girdles, and the distribution of pores need to be studied via mathematics, with an emphasis on geometry, optimization and energy minimization, but the question is whether the methods borrowed from physics are adequate to study living organisms. In particular, there is a need for dedicated geometrical structures, describing living (and non-living) natural shapes in one coherent global framework. Gielis transformations are a recent development that allows one to study biological shapes and phenomena, generalizing Lamé's superellipses [12.51] [12.54] [12.53]. Its origin is in the study of plants and organisms, hence it is a scientific method *sui generis*, for what we can observe in our macroscopic world. In the past five years over 40 000 botanical specimen have been tested, and Gielis transformations and superellipses are found everywhere [12.53]. In contrast, circles and ellipses are very rare in biology (centric diatoms may be exceptions to this rule).

This geometrical treatment of diatoms from a global viewpoint, should be integral part of a systems biological approach, coupling insights on form and function to understand genomic and genetic diversity in diatoms, during evolution and development (see Figure 12.2). In diatoms this genetic toolbox is known to be very complex with a rich history, and it is thus important to understand form and function, since mathematical and physical laws are prevalent in our universe.

Gielis transformations are a generalization of the circle (as a constant function) and of the Pythagorean theorem: indeed, selecting $n_1 = n_2 = n_3 = 2$ and $f(\vartheta) = R = 1$ gives the unit circle and the Pythagorean theorem. Instead of only the classic Euclidean circle, Eq. (12.13) defines natural shapes at all levels, such as diatoms, starfish, flowers and molluscs. The exponents $n_{1,2,3}$ change the basic polygons defined by the symmetry parameter m. Parameters A and B are scaling parameters. If $A = B$, the basic shape is a circle, but when they differ, the basic shape is an ellipse. Since shape and size parameters are real numbers, a huge diversity and variability can be described in a very compact way:

$$\varrho(\vartheta, f(\vartheta), A, B, m, n_1, n_2, n_3) = \frac{1}{\sqrt[n_1]{|(1/A)\cos(m\vartheta/4)|^{n_2} + |(1/B)\sin(m\vartheta/4)|^{n_3}}} \cdot f(\vartheta) \quad (12.13)$$

In particular, diatoms and their substructures can be described in one coherent framework (Figure 12.2). With $m = 0$ (circle) radially symmetrical cylindrical frustules can be

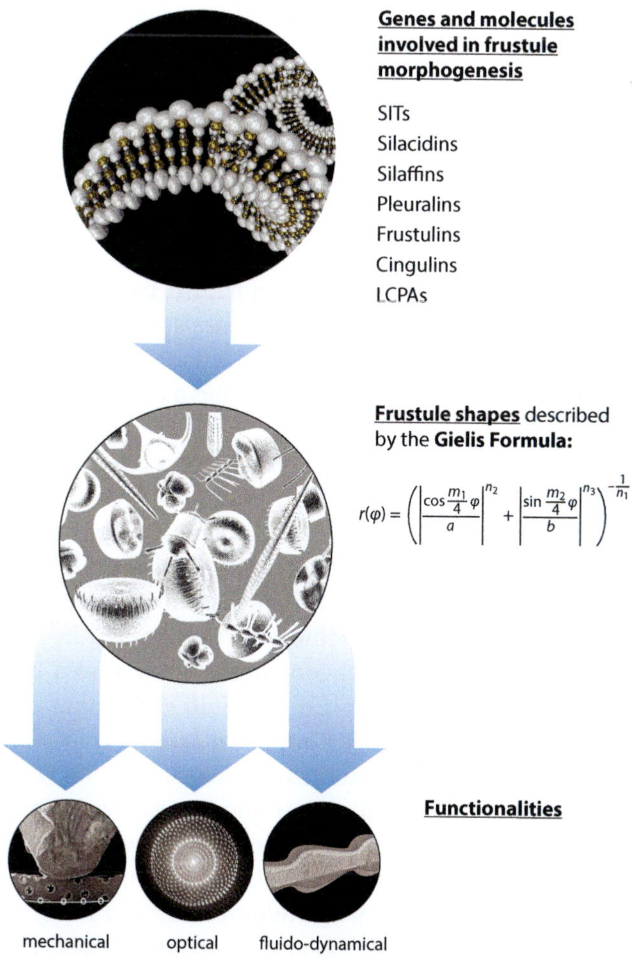

Figure 12.2 Schematic representation of the connections between frustule morphogenesis, geometry, and physical functions. (Reproduced with permission from De Tommasi *et al.* [12.27].)

described, and with increasing m various diatom shapes are modeled. Square shapes have $m = 4$ and benthic diatoms with pennate shapes have bilateral symmetry, defined by $m = 2$. *Stictodiscus* diatoms can be circular but also triangular ($m = 3$) or square. Higher symmetries can be observed in sectorized *Arachnoidiscus* or *Coscinodiscus*. Actually, the parameter m divides the plane into m sectors. Equation (12.14), a generalization of the sphere for $f(\vartheta, \varphi) = R$ (= the radius of the sphere), can describe the shape of the complete frustule:

$$\varrho(\vartheta,\varphi,f(\vartheta,\varphi),A,B,C,m_1,m_2,n_1,n_2,n_3,n_4)$$
$$= \frac{1}{\sqrt[n_1]{\left|\frac{1}{A}\sin\left(\frac{m_1}{4}\vartheta\right)\cdot\cos\left(\frac{m_2}{4}\varphi\right)\right|^{n_2}+\left|\frac{1}{B}\sin\left(\frac{m_1}{4}\vartheta\right)\cdot\sin\left(\frac{m_2}{4}\varphi\right)\right|^{n_3}+\left|\frac{1}{C}\cos\left(\frac{m_1}{4}\vartheta\right)\right|^{n_4}}}\cdot f(\vartheta,\varphi) \tag{12.14}$$

It describes the complete frustule and the 3D coordinate planes that correspond to the valvar, radial/transapical and the (per)apical planes in diatoms. The arrangement of the pores in fact provides a very specific coordinate system adapted to the shape, as a natural generalization of classical coordinate systems such as spherical and elliptical coordinates. Using the frustule and the pores as a natural coordinate system should then allow understanding the mathematical physics underlying species-specific and environmental developmental-specific distribution of pores. In this way we can also encompass different symmetries encountered in diatoms, and provide for a quantitative measure for qualitative terms such as elliptical or oval, (sub)circular and (sub-)spherical, crescent shapes, lanceolate, clavate, naviculoid, sigmoid and more.

A geometrization of biology, or more generally of nature, based on forms and formation of natural shapes (a geometrical theory of morphogenesis) requires the combination of a uniform description of shapes, coupled to differential equations within a coherent geometrical framework. This uniform description then generates a best adapted coordinate system for both the complete structure and its details.

The proposed uniform description in diatoms can be very effective in the study of global properties, rather than local. Beyond shape description, we have to understand the forces that generate the shapes. The nearly universal principle in the natural sciences is that the equilibrium configuration of a system can be found by minimizing its total energy among all admissible configurations. When considering the surface as interface between two (or more) immiscible materials, the surface geometry is determined by minimizing the surface tension subject to whatever additional constraints are imposed by the environment. There is a canonical equilibrium surface, called the Wulff shape, that can be characterized as the absolute minimizer of the free energy F among all surfaces enclosing the same three-dimensional volume. Diatoms can be considered as Wulff shapes, on which surface stresses act, and they are the "unit spheres" for an anisotropic energy. With Equations (12.13) and (12.14), we can study diatoms as Wulff shapes, minimizing certain anisotropic energy functionals [12.72]. From a geometrical point of view, diatoms are extremely interesting because the overall stable geometrical structure of frustules, costae and other structures, is combined with local optimization of the precise distribution of the pores. This distribution and the diversity make diatoms individually unique; they are the snowflakes of (the liquid state of) water and they are evolutionary very stable solutions.

12.3.2 Laplace and Fourier Revisited

The morphology of diatoms can then be studied in the same way as crystals, snowflakes, or soap films, based on the harmonic function theory. Solutions to boundary value problems can be found in an analytic way, not only computationally. So far analytic Fourier-like solutions have been restricted only to a few domains, but can not be extended to a wide range of domains in 2D and 3D, without the need for meshing and finite elements. Analytic Fourier-like solutions to Laplace, Helmholtz and wave equations have been found for various 2D and 3D domains, including shells. Fourier-like refers to analytic solutions combining Fourier series with special functions of Hankel and Bessel type. The domains or shells are normal with respect to a suitable spherical coordinate system so that the relevant boundary may be regarded as anisotropically stretched unit circles or spheres. In this way, accurate solutions can be obtained with very low orders of expansions of spherical harmonics. In Caratelli *et al.* [12.19], a number of boundary value problems have been solved in starlike domains, under different boundary conditions. Figure 12.3 shows one solution for the Robin problem for the Laplace equation on closed shells with fourfold symmetry in the *XY* plane, with a cross-sectional symmetry observed in *Amphitetras* genus. This method is generally applicable for all 3D shells and the results show that, depending on the boundary value problem and the boundary conditions chosen, the solutions reflect the way shapes deal with tensions on surfaces. In turn, these tensions are described by the mean curvature H, which is directly related to the Laplacian Δ, due to a theorem of Beltrami, namely $\Delta = 2H$ for surfaces.

These geometrical methods thus provide a more global view, which can be aligned in the future with existing CAD and Finite Element studies in diatom frustules; but such global geometric view is indispensable if we wish to understand form and function in diatoms.

It is clear that at present we are still far from using this in a complete way, and in studying the fate of photons entering the frustrule and into the protoplast – beyond the glass cage. In Section 12.4 several classical methods will be presented to simulate light propagation in the frustrule. In Section 12.5, novel mathematical methods are presented treating the whole diatom as one dielectric complex structure.

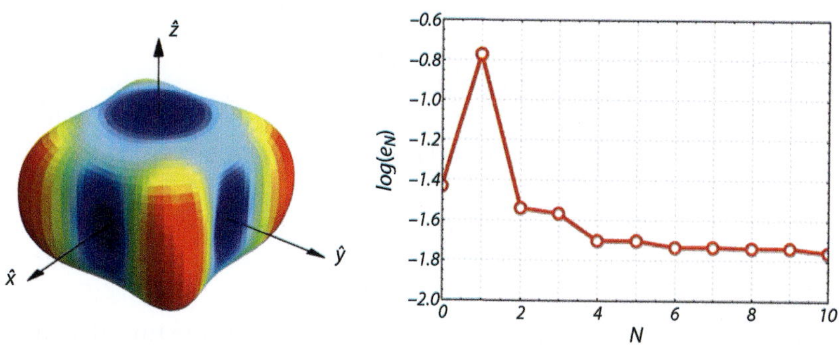

Figure 12.3 The solution of the Robin problem for the Laplace equation in a shell S with fourfold symmetry in XY direction. Left: Boundary behaviour along the inner shell surface of the partial sum U_N of order $N = 11$. Right: relative boundary error e_N as function of the order N of the truncated spherical harmonic expansion for the super-shaped shell on the left. (Reproduced with permission from De Tommasi *et al.* [12.27].)

12.4 Going Through the Wall: Simulating Light Propagation in the Frustule

Describing the propagation of light through the intricate but still regular diatom frustules, ultimately means to solve Maxwell equations in the presence of a periodic dielectric medium. In their macroscopic formulation, Maxwell equations can be expressed as [12.64]:

$$\begin{aligned} \nabla \cdot \mathbf{B} &= 0 \\ \nabla \times \mathbf{E} + \frac{\partial \mathbf{B}}{\partial t} &= 0 \\ \nabla \cdot \mathbf{D} &= \rho \\ \nabla \times \mathbf{H} - \frac{\partial \mathbf{D}}{\partial t} &= \mathbf{J} \end{aligned} \tag{12.15}$$

where **E** and **H** are the macroscopic electric and magnetic fields, **D** and **B** are the displacement and magnetic induction fields and finally ρ and **J** stand for free charge and current densities. In absence of sources of light both ρ and **J** can be set equal to zero. In linear regime, for isotropic media, and neglecting material dispersion, we have:

$$\begin{aligned} \mathbf{D}(\mathbf{r}) &= \epsilon_0\, \epsilon(\mathbf{r})\mathbf{E}(\mathbf{r}) \\ \mathbf{B}(\mathbf{r}) &= \mu_0\, \mu(\mathbf{r})\mathbf{H}(\mathbf{r}) \simeq \mu_0\, \mathbf{H}(\mathbf{r}) \end{aligned} \tag{12.16}$$

where ϵ_0 stands for vacuum permittivity, ϵ is the dielectric function of the material (purely real and positive for transparent media, like diatom biosilica in visible spectral range), μ_0 is the vacuum permeability and the magnetic permeability $\mu(\mathbf{r})$ is very close to unity for most dielectric materials (thus the refractive index n equals $\sqrt{\epsilon\mu} \simeq \sqrt{\epsilon}$). With all these assumptions, Maxwell equations can be rewritten as follows:

$$\begin{aligned} \nabla \cdot \mathbf{H}(\mathbf{r},t) &= 0 \\ \nabla \times \mathbf{E}(\mathbf{r},t) + \mu_0 \frac{\partial \mathbf{H}(\mathbf{r},t)}{\partial t} &= 0 \\ \nabla \cdot [\epsilon(\mathbf{r})\mathbf{E}(\mathbf{r},t)] &= 0 \\ \nabla \times \mathbf{H}(\mathbf{r},t) - \epsilon_0 \epsilon(\mathbf{r}) \frac{\partial \mathbf{E}(\mathbf{r},t)}{\partial t} &= 0 \end{aligned} \tag{12.17}$$

Exploiting the linearity of Maxwell equations, we can separate the time dependence from the spatial dependence by expanding the fields into a set of harmonic modes:

$$\begin{aligned} \mathbf{H}(\mathbf{r}, t) &= \mathbf{H}(\mathbf{r})\, e^{-i\omega t} \\ \mathbf{E}(\mathbf{r}, t) &= \mathbf{E}(\mathbf{r})\, e^{-i\omega t} \end{aligned} \tag{12.18}$$

We are thus considering field patterns that vary sinusoidally with time at a frequency ω, which is useful since any solution of the Maxwell equations can be expressed as a combination of harmonic modes. Inserting the equations 12.18 in 12.17 we obtain:

$$\begin{gathered}\nabla \cdot \mathbf{H}(\mathbf{r}) = 0 \\ \nabla \cdot [\epsilon(\mathbf{r})\,\mathbf{E}(\mathbf{r})] = 0 \\ \nabla \times \mathbf{E}(\mathbf{r}) - i\omega\mu_0\mathbf{H}(\mathbf{r}) = 0 \\ \nabla \times \mathbf{H}(\mathbf{r}) + i\omega\epsilon_0\,\epsilon(\mathbf{r})\mathbf{E}(\mathbf{r}) = 0\end{gathered} \tag{12.19}$$

The first two equations express the transversality of the electromagnetic waves, while a rearrangement of the last two equations leads to:

$$\nabla \times \left(\frac{1}{\epsilon(\mathbf{r})} \nabla \times \mathbf{H}(\mathbf{r}) \right) = \left(\frac{\omega}{c} \right)^2 \mathbf{H}(\mathbf{r}) \tag{12.20}$$

which is known as the *master equation* ($c = 1/\sqrt{\epsilon_0\mu_0}$ stands for the vaccum speed of light). Thus, for a given structure characterized by the dielectric function $\epsilon(\mathbf{r})$ (in our case the three-dimensional structure of the diatom frustule), describing how the optical field interacts with it means to solve the master equation to find the modes $\mathbf{H}(\mathbf{r})$ and the corresponding frequencies. The electric fields $\mathbf{E}(\mathbf{r})$ can finally be retrieved by the last equation in 12.19. Identifying the left side of Eq. 12.20 with a differential operator $\hat{\Theta}$ allows expressing the master equation as an eigenvalue problem:

$$\hat{\Theta}\mathbf{H}(\mathbf{r}) = \left(\frac{\omega}{c} \right)^2 \mathbf{H}(\mathbf{r}) \tag{12.21}$$

Let's consider a dielectric structure which is periodic in three-dimensions (similar observations can be conducted for one- and two-dimensional systems), i.e. which is invariant under translations through a *lattice vector* $\mathbf{R} = l\mathbf{a}_1 + m\mathbf{a}_2 + n\mathbf{a}_3$, where $(\mathbf{a}_1, \mathbf{a}_2, \mathbf{a}_3)$ are the *primitive lattice vectors* and l, m, and n are integers (in other words we can say that $\mathbf{a}_1$, $\mathbf{a}_2$, and $\mathbf{a}_3$ *span* the space of lattice vectors). The *primitive reciprocal lattice vectors* $(\mathbf{b}_1, \mathbf{b}_2, \mathbf{b}_3)$, defined as

$$\mathbf{a}_i \cdot \mathbf{b}_j = 2\pi\delta_{ij} \tag{12.22}$$

span the so-called *reciprocal lattice*. In this scenario the eigenvectors of $\hat{\Theta}$ corresponding to eigenfrequencies $\omega(\mathbf{k})$ are the *Bloch states*:

$$\mathbf{H}_\mathbf{k}(\mathbf{r}) = e^{i\mathbf{k}\cdot\mathbf{r}}\,\mathbf{u}_\mathbf{k}(\mathbf{r}) \tag{12.23}$$

where $\mathbf{k} = k_1\mathbf{b}_1 + k_2\mathbf{b}_2 + k_3\mathbf{b}_3$ is the *Bloch wave vector* (with k_1, k_2, and k_3 integer numbers) and $\mathbf{u}_\mathbf{k}(\mathbf{r}) = \mathbf{u}_\mathbf{k}(\mathbf{r} + \mathbf{R})$ is a periodic function on the lattice. In other words, the eigensolutions of Eq. 12.21 can be expressed as a product of plane waves and lattice periodic functions (*Bloch-Floquet theorem*). The region of non-redundant values of **k** (corresponding to identical Bloch states) in the reciprocal lattice is known as the (first) *Brillouin zone.*[1] It can be easily shown that inserting the Bloch state described by 12.23 in the master equation 12.20 leads to the following eigenvalue problem:

$$\hat{\Theta}_{\mathbf{k}}\mathbf{u}_{\mathbf{k}}(\mathbf{r}) = \left(\frac{\omega(\mathbf{k})}{c}\right)^2 \mathbf{u}_{\mathbf{k}}(\mathbf{r}) \tag{12.24}$$

where the differential operator $\hat{\Theta}_{\mathbf{k}}$ is defined as:

$$\hat{\Theta}_{\mathbf{k}} = (i\mathbf{k} + \nabla)\times \frac{1}{\epsilon(\mathbf{r})}(i\mathbf{k} + \nabla)\times \tag{12.25}$$

The mode profiles satisfying the eigenvalue equation 12.24 are subject to the transversality and periodicity conditions defined by:

$$\begin{gathered}(i\mathbf{k} + \nabla)\cdot \mathbf{u}_{\mathbf{k}} = 0 \\ \mathbf{u}_{\mathbf{k}}(\mathbf{r}) = \mathbf{u}_{\mathbf{k}}(\mathbf{r} + \mathbf{R})\end{gathered} \tag{12.26}$$

The periodic boundary condition allows us to restrict the eigenvalue problem to a single cell of the lattice. It can be demonstrated [12.66] that, for each value of **k**, this leads to a discrete spectrum of eigenvalues that we can label by a band index n. Analogous with the energy band structure of a lattice of atoms subjected to a periodic electric potential $V(\mathbf{r})$, the functions $\omega_n(\mathbf{k})$ describe the frequency of the modes belonging to the nth band of a photonic structure characterized by a periodic refractive index. Photonic crystals (PhCs), in particular, are characterized by a geometry and a refractive index contrast such that there exist frequency intervals for which no modes can propagate in the structure. In analogy with the energy band gaps present in semiconductors, these bands are known as *photonic band gaps* (PBGs).

Several species of birds, insects, flora, and protists are provided, as constituent parts of their organisms, with periodic structures at the sub-micron scale acting as photonic crystals [12.119] [12.94] [12.67]. This is at the basis of the so-called *structural color* [12.114] [12.26] which is not due to the presence of pigments absorbing visible radiation but is rather based on spectrally selective reflectance induced by interference processes, which in turn are caused by the periodic spatial arrangement of nanostructured biomaterials. As an example we can mention the flashing, intense blue wings of several species of *Morpho* butterflies [12.118] [12.69] [12.55]. Their bright coloration, which is mainly used by males to mark their territory and scare off potential rivals in mating, is mostly due to multilayer interference originating from the *lamellae* of the tree-like ridges observed in the single scales of the

[1]For more details on the properties of the reciprocal lattice and the Brillouin zone it can be useful to consult a solid-state physics handbook such as Kittel *et al.* [12.71], since the mathematical formalism of lattices is the same.

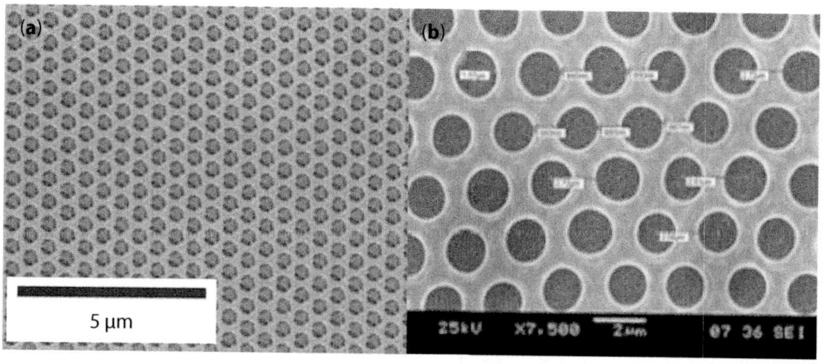

Figure 12.4 Artificial PhC (a) and a detail of a *Coscinodiscus wailesii* valve (b). (Reproduced with permission from De Tommasi *et al.* [12.27].)

wings. In other words, we can claim that the periodically nanostructured wing scales are characterized by a PBG corresponding to the blue spectral range of the visible spectrum.

Nanostructured photonic structures present in nature are not only exploited by animals for intraspecies and interspecies signaling or for camouflage, but also by flowers for attracting specific pollinators [12.123] [12.91] and by plants to enhance photosynthetic efficiency [12.65]. Nevertheless the most impressive resemblance with artificial PhCs is observed in diatom frustules (see Figure 12.4), which allowed Fuhrmann and coworkers to define diatoms as "living photonic crystals" in their pioneering paper in 2004 [12.45].

In the following sections we will briefly review the main numerical methods aimed at the simulation of the electromagnetic field propagation in a periodic photonic structure, both in frequency and time domain and including techniques for PBGs retrieval in a PhC. For every described algorithm, several examples of applications to diatom frustules will be reported.

12.4.1 Plane Wave Expansion (PWE) Method

In order to introduce the plane wave expansion (PWE) method, it can be useful to start from Eq. (12.24) in one dimension.[2] Taking into account the periodicity condition $u_k(x) = u_k(x + a)$, with a period of the lattice, we can express $u_k(x)$ as an infinite sum of sines and cosines, i.e., as a *Fourier series*:

$$u_k(x) = \sum_{n=-\infty}^{\infty} c_n(k) e^{i\frac{2\pi n}{a}x} \tag{12.27}$$

with

$$c_n(k) = \frac{1}{a}\int_0^a u_k(x) e^{-i\frac{2\pi n}{a}x} dx \tag{12.28}$$

[2]For a more detailed discussion on PWE method, refer to Busch and John [12.18] and Johnson and Joannopoulos [12.68].

The coefficients c_n decay with $|n|$, thus the sum in Eq. (12.27) can be truncated and reduced to a finite number of N terms. The problem is thus transformed into a set of linear equations for N unknown c_n coefficients. It is worth noticing that the quantity $\frac{2\pi n}{a}$ in the exponential is a reciprocal lattice vector for a one-dimensional lattice. For a three-dimensional lattice we have, analogously:

$$\mathbf{u_k}(\mathbf{r}) = \sum_{\mathbf{G}} \mathbf{c_G}(\mathbf{k}) e^{i\mathbf{G}\cdot\mathbf{r}} \qquad (12.29)$$

$$\mathbf{c_G} = \frac{1}{V}\int \mathbf{u_k}(\mathbf{r}) e^{-i\mathbf{G}\cdot\mathbf{r}} d^3 r$$

where $\mathbf{G} = l'\mathbf{b}_1 + m'\mathbf{b}_2 + n'\mathbf{b}_3$ is a reciprocal lattice vector and V is the volume of the unit cell. Applying the transversality constraint $(i\mathbf{k} + \nabla) \cdot \mathbf{u_k} = 0$ to the first equation in (12.29), we obtain:

$$(\mathbf{k} + \mathbf{G}) \cdot \mathbf{c_G} = 0 \qquad (12.30)$$

For each $\mathbf{G}$, we can choose two perpendicular unit vectors $\hat{\mathbf{e}}_{\mathbf{G}}^{(1)}$ and $\hat{\mathbf{e}}_{\mathbf{G}}^{(2)}$ orthogonal to $\mathbf{k} + \mathbf{G}$ such that:

$$\mathbf{c_G} = \mathbf{c_G}^{(1)}\hat{\mathbf{e}}_{\mathbf{G}}^{(1)} + \mathbf{c_G}^{(2)}\hat{\mathbf{e}}_{\mathbf{G}}^{(2)} \qquad (12.31)$$

thus reducing the problem to two unknown coefficients $\mathbf{c_G}^{(1)}$ and $\mathbf{c_G}^{(2)}$ for every $\mathbf{G}$.

We can now substitute $\mathbf{u_k}(\mathbf{r})$ in Eq. (12.24) with the sum reported in Eq. (12.29) and then take the Fourier transform of both sides, obtaining the following equation:

$$\sum_{\mathbf{G}} \left[-\epsilon_{\mathbf{G'}-\mathbf{G}}^{-1} \cdot (\mathbf{k}+\mathbf{G'}) \times (\mathbf{k}+\mathbf{G}) \times \right] \mathbf{c_G} = \left(\frac{\omega}{c} \right)^2 \mathbf{c_{G'}} \qquad (12.32)$$

$\epsilon_{\mathbf{G}}^{-1}$ standing for the Fourier transform of $\epsilon^{-1}(\mathbf{r})$. We can approximate $\epsilon_{\mathbf{G}}^{-1}$ using the *discrete Fourier transform* (DFT), i.e., replacing it with a finite sum of N terms. After the truncation to a finite set of $\mathbf{G}$ values, the calculation is reduced to a finite matrix eigenequation of the form $Ax = \omega^2 x$. In terms of *time complexity*, the iterative process used to solve the eigenequation generally takes $O(N^2)$ time (i.e., the number of operations of the algorithm scales with N^2). A possible way to reduce time complexity is to make use of the *fast Fourier transform* (FFT) algorithm, which allows computing the multidimensional DFT over N points in $O(N \log N)$ time. We can thus proceed as follows: *i)* we compute the product $(\mathbf{k} + \mathbf{G}) \times \mathbf{c_{G}}$, which takes $O(N)$ time; *ii)* we calculate the inverse FFT to pass to $\mathbf{r}$ space and multiply by $\epsilon^{-1}(\mathbf{r})$ in $O(N)$ time; finally, *iii)* we pass back to $\mathbf{G}$ space by FFT and perform the cross-product $(\mathbf{k} + \mathbf{G'}) \times \mathbf{c_{G'}}$. The overall computation time is $O(N \log N)$ plus $O(N)$ for storage, which noticeably improves the speed of the iteration. Each step of the iteration can be further accelerated by using so-called *preconditioners*, i.e., approximate solutions of the eigenproblem. For PWE method an efficient preconditioner is obtained by considering only the diagonal entries of A, which dominate the problem for large values of $|\mathbf{G}|$ [12.66].

The first application of the PWE method to diatom frustules seen as photonic periodic structures was introduced in the aforementioned work by Fuhrmann *et al.* [12.45]. The authors looked at *Coscinodiscus granii* valves and girdles as *photonic crystal slabs*, i.e., photonic structures characterized by a two-dimensional periodicity but a finite thickness. In this case, the properties of a two-dimensional photonic band structure combine with that of a planar waveguide. For the considered species, valves and girdles present an hexagonal and a square pattern of holes, respectively (see Figure 12.5). In particular, the valves are characterized by holes with 450 nm radius, a lattice constant of 900-950 nm and a thickness of about 700 nm while the girdles present holes with 90 nm radius, a lattice constant of 250 nm and a thickness which varies in the range of 200-600 nm. PWE was exploited to reconstruct the band structure of the girdle in air (see Figure 12.5d) while a two-dimensional calculation based on the effective-index method [12.99] was used to investigate the properties of a frustule immersed in water (low refractive index contrast, see Figure 12.5b). Let us consider in detail the band diagram reported in Figure 12.5d, referring to a girdle with a thickness of 200 nm, a hole radius r/lattice constant a ratio of 0.33, and a refractive index of 1.43. The labels Γ, X, and M indicate the vertices of the irreducible part of the Brillouin zone of a square lattice, while the normalized frequencies of the modes are expressed in units of speed of light divided by the lattice constant. Circles label TE modes (i.e., modes with the electric field parallel to the slab plane) while squares mark TM modes (i.e., modes with the electric field parallel to the holes' axis). The shaded area is relative to non-guided modes radiating into air. Let us now focus on the spectral range of interest for photosynthesis ($\lambda = 400 - 700$ nm). The frequencies at which the resonances occur and the number of guided modes depend on the shell thickness (in this case 200 nm) and on the refractive index of the surrounding medium. For a frustule immersed in air, the X-resonances are found in green-yellow spectral range while the M-resonances lie in the blue spectral range. Passing to water as the surrounding environment, the resonating wavelengths are found at 665 nm and 470 nm at the X- and M-points, respectively. In general, in correspondence of these points the group velocity $\frac{\partial\omega}{\partial k}$ is low, implying a long interaction between light and matter. In the case of the valve (Figure 12.5b), the lattice constant is higher than the wavelength of visible light, thus the guided modes lie in infrared region. For visible light, higher order resonances have to be considered which stand above the light line (defined as $\omega(k) = \frac{ck}{\sqrt{\epsilon}}$), thus being partially coupled to radiative modes. This means that the valve can be regarded as a two-dimensional grating coupler of higher order. When immersed in water, as it is the case reported in figure, several resonances are distributed in the visible range. The authors made also calculations on the *penetration depth* associated to the modes supported by the silica walls, starting from the evaluation of the evanescent wave:

$$E(z) = E_0 e^{-\frac{2\pi}{\lambda_0}\sqrt{n_{eff}^2 - n_m^2}} \tag{12.33}$$

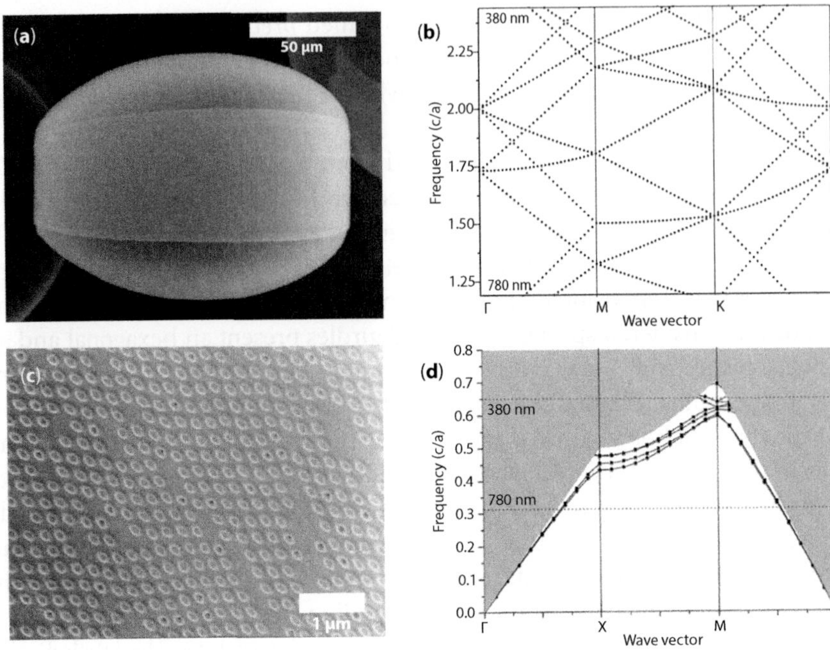

Figure 12.5 SEM image of a *C. granii* whole frustule (a) and two-dimensional band structure of a valve immersed in water as retrieved by effective-index method calculations (b). Detail of the girdle (c) and relative, numerically reconstructed three-dimensional band structure obtained by PWE method, considering air as surrounding medium (d). Reproduced with permission from Fuhrmann *et al.* [12.45].)

with n_{eff} effective refractive index of the frustule, n_m refractive index of the surrounding medium, λ_0 vacuum wavelength, and z direction perpendicular to the plane of the slab. It is worth noticing how, for small refractive index contrasts (like the one between porous silica and water and/or cytoplasm), the evanescent wave can reach far into the living cell, allowing an efficient optical coupling with the plastids, which have been observed to migrate close to the frustule walls under dim lighting conditions [12.46].

PWE was also applied by Yamanaka *et al.* [12.124] in the calculation of the band structure of the inner shell of *Melosira variance* diatom, viewed as a PhC slab characterized by an hexagonal lattice of pores. Simulations have been carried out considering the frustule immersed in air and in water. In both cases, looking at the spectral range between 400 and 500 nm and considering the M-K direction (Γ, M and K being the vertices of the irreducible part of the Brillouin zone of an hexagonal lattice), the dispersion curves $\omega = \omega(k)$ are parabolic, meaning an enhanced interaction between light and the silica shell. One of the possible advantages of this behavior lies in the ability of the frustule to weaken an excess supply of blue light, which would be harmful for the cell because of the inducted production of active oxygen species such as superoxide, hydrogen peroxide, and hydroxyl radical [12.8].

12.4.2 Finite Difference Time Domain (FDTD) Method

Finite difference time domain (FDTD) method represents the most common computational technique used to numerically solve Maxwell equations in time domain [12.66], simulating the propagation in time of both the electric and magnetic fields. A detailed description of the algorithm can be found, for example, in ref. Taflove and Hagness [12.116]. Basically, space and time are divided into a (usually) uniform grid of discrete points and the derivatives present in the Maxwell equations are approximated by finite differences, as the name of the method suggests. The electric fields **E** at time t are computed starting from the stored values of the **E** fields at time $t - \Delta t$ and from the **H** fields at time $t - \Delta t/2$, and vice versa for **H** at $t + \Delta t/2$. Like all the time-domain methods, FDTD allows estimating the transmission or reflection of a structure at many frequencies with a single computation, starting from the Fourier transform of its response to a short pulse (broad bandwidth excitation). Frequency eigenvalues of the system, corresponding to resonant or leaky modes, can be retrieved as peaks in the spectrum of the response, while the imposition of Bloch-periodic boundary conditions allows for the computation of band structures. Loss rates can be calculated as easily as resonant frequencies. A separate simulation performed with a narrow-bandwidth source can be used to reconstruct the field pattern corresponding to a specific eigenfrequency. For high spatial resolution simulations, a high temporal resolution is required as well, thus the computational time of a 3D FDTD simulation scales to the fourth power of the resolution.

A precise morphological characterization based on scanning electron microscopy (SEM) allowed Goessling and co-workers [12.56] to perform FDTD calculations on accurate 3D CAD reconstructions of *C. granii* girdle, which, as we saw in the previous subsection, can be described as a square lattice of cylindrical holes perforating a silica slab (see Figures 12.6a,b). These simulations showed, for a girdle immersed in water and considering in-plane propagation, the presence of a so-called *pseudogap* (i.e., a photonic band gap existing only for particular directions of propagation) spectrally located in near infrared (with a central wavelength $\lambda_c \simeq 780$ nm). The pseudogap has also been observed experimentally and undergoes a blue-shift as the incidence angle θ_i increases (see Figure 12.6c). A secondary reflectance pattern, due to in-plane diffraction of the modes guided by the girdle, appears in green spectral region for $\theta_i \simeq 50°$. Since both pseudogap and guided modes resonate far from the absorption maxima of chlorophylls, the authors hypothesize that the photonic properties of the girdle evolved in such a way to reduce the interference of photosynthetically unproductive wavelengths with pigment light absorption.

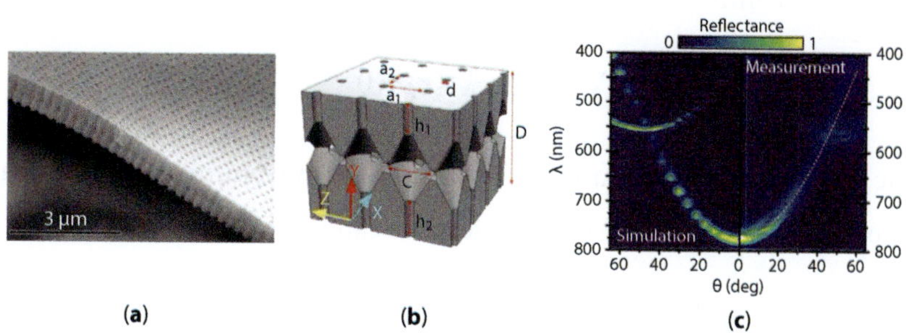

Figure 12.6 (a) Cross-sectional view of a *C. granii* girdle (SEM image). (b) 3D CAD reconstruction over four unit cells of the girdle. Numerical values derived by SEM characterization: $a_1 = 285 \pm 5$ nm; $a_2 = 279 \pm 11$ nm; $d = 124 \pm 21$ nm; $h_1 = 225 \pm 21$ nm; $h_2 = 250 \pm 32$ nm; $C = 290 \pm 18$ nm; $D = 745 \pm 43$ nm. (c) Reflectance of the girdle immersed in water as a function of incidence angle (for in-plane propagation) and for different wavelengths as retrieved by FDTD simulations (left) and measured by micro-scatterometry (right). (Reproduced with permission from Goessling *et al.* [12.56].)

In ref. D'Mello *et al.* [12.36], the photonic band structure associated with the triangular and square lattice configurations of *Nitzschia filiformis* frustules has been calculated by 3D FDTD, revealing the presence, in both cases, of resonances at symmetry points around 430 and 660 nm respectively, coinciding with the main peaks of absorption of chlorophyll a (Chl-a), one of the chromophores typically found in diatoms. In resonant condition the group velocity lowers, the propagation time within the frustule becomes longer and the probability of light-matter interaction in the proximity of the silica walls, where the chloroplasts lie, increases as well. The resonant photonic behavior of *N. filiformis* shells has also been demonstrated experimentally by means of near-field optical microscopy (SNOM) and correlated atomic force microscopy (AFM) [12.35] [12.36].

12.4.3 Wide-Angle Beam Propagation Method (WA-BPM)

Besides the evaluation of the band structure of a diatom frustule seen as a photonic crystal, it can be useful to study how the light normally impinging on a valve interacts with its porous matrix and how the transmitted radiation is spatially distributed inside the cell. The typical dimensions of the valve pores are indeed of the same order of magnitude of the visible wavelength, thus giving rise to diffraction. Diffracted light contributions coming from the pores interact with each other by means of interference along the direction of propagation of the field, giving rise to intensity maxima which look like intense hot-spots. This phenomenon has been observed in several centric species [12.23] [12.24] [12.32] [12.113] [12.40] [12.82] [12.103] and has been exploited in combination with far-field sub-diffraction techniques in order to obtain a sort of bio-derived superlens [12.25]. In order to simulate light propagation through a nano-patterned diatom valve, a properly adjusted version of beam propagation method (BPM) can be used. Indeed, BPM is usually employed in the study of light propagation in waveguides and optical fibers, i.e. in conditions of paraxiality (small angles in respect to the optical axis) and of uniformity of refractive index along the direction of propagation of the field. This is not the case of a diatom frustule, where the pores induce diffraction of light (leading to non-paraxial conditions) and where an abrupt variation in the refractive index between the relatively thin valve and the surrounding environment (water or cytoplasm) takes place.

Starting from the Helmholtz equation for a scalar field:

$$\nabla^2 E(\mathbf{r}) + k^2(\mathbf{r}) = 0 \tag{12.34}$$

with E electric field, $k = nk_0$ wavenumber (with $k_0 = \frac{2\pi}{\lambda}$ wavenumber in free space), and $n = n(x, y, z)$ refractive index spatial distribution, we can write the solution as:

$$E(\mathbf{r}) = E(x, y, z) = U(x, y, z)e^{-ik_r z} \tag{12.35}$$

The electric field can thus be expressed as the product of a slowly varying envelope factor $U(x, y, z)$ and a rapid varying phase factor $e^{-ik_r z}$, with $k_r = n_r k_0$ reference wavenumber (expressed in terms of the reference refractive index n_r), which takes into account the average phase variation of the field. We are assuming that the considered wave propagates primarily along z (i.e., we are considering, at first, paraxial conditions). We will also suppose, for now, that the amplitude varies slowly along z axis too. Inserting $U(x,y,z)e^{-ik_r z}$ into Eq. (12.34) we obtain:

$$\frac{\partial^2 U}{\partial z^2} + 2ik_r \frac{\partial U}{\partial z} + \frac{\partial^2 U}{\partial x^2} + \frac{\partial^2 U}{\partial y^2} + \left(k^2 - k_r^2\right)U = 0 \tag{12.36}$$

Making use of the *slowly varying envelope approximation*:

$$\left|\frac{\partial^2 U}{\partial z^2}\right| \ll \left|2k_r \frac{\partial U}{\partial z}\right| \tag{12.37}$$

we obtain the basic BPM equation:

$$\frac{\partial U}{\partial z} = \frac{i}{2k_r}\left[\frac{\partial^2 U}{\partial x^2} + \frac{\partial^2 U}{\partial y^2} + \left(k^2 - k_r^2\right)U\right] \tag{12.38}$$

Specifying $U(x, y, z)$ at a plane $z = z_0$, we can iterate U along the z-axis using finite differences for the x and y derivatives.

The most popular BPM variant which can take into account non-paraxial conditions (wide-angle beam propagation method, WA-BPM), is known as the *multistep Padé-based technique* [12.58] [12.59]. We can denote $\frac{\partial}{\partial z}$ with D, and, consequently, $\frac{\partial^2}{\partial z^2}$ with D^2. Eq. (12.36) can now be viewed as a quadratic equation to be solved for the differential operator D. This yields the following solution for a first order equation in z:

$$\frac{\partial U}{\partial z} = ik_r\left(\sqrt{1+P} - 1\right)U \tag{12.39}$$

with

$$P \equiv \frac{1}{k_r^2}\left(\frac{\partial^2}{\partial x^2} + \frac{\partial^2}{\partial y^2} + \left(k^2 - k_r^2\right)\right) \tag{12.40}$$

Even though it is restricted to forward propagation of the field ($z > 0$), the above equation is exact in that no paraxiality approximation has been introduced. The radical in Eq. (12.39) can be evaluated by using a Taylor expansion. The first order of the expansion leads to the standard, paraxial BPM, while higher orders lead to more accurate representations of the propagating field. However, expansion via Padé approximants [12.59] is more accurate than Taylor expansion for the same order of terms. This approach leads to the following equation:

Table 12.2 Low-order Padé approximants expressed in terms of the operator P defined in Eq. (12.40).

Padé order (m,n)	$\mathbf{N_m}$	$\mathbf{D_n}$
(1,0)	$\frac{P}{2}$	1
(1,1)	$\frac{P}{2}$	$1+\frac{P}{4}$
(2,2)	$\frac{P}{2}+\frac{P^2}{4}$	$1+\frac{3P}{16}+\frac{P}{2}$
(3,3)	$\frac{P}{2}+\frac{P^2}{2}+\frac{3P^3}{32}$	$1+\frac{5P}{4}+\frac{3P^2}{8}+\frac{P^3}{64}$

$$\frac{\partial U}{\partial z} = ik\frac{N_m(P)}{D_n(P)}U \tag{12.41}$$

where N_m and D_n are polynomials in the operator P, and (m, n) is the order of approximation. Some of their low-order values are reported in Table 12.2.

WA-BPM has been applied with success to *Coscinodiscus wailesii* [12.24] [12.28], *Arachnoidiscus* sp. [12.40], and *Ctenophora pulchella* [12.29] (in this last case the algorithm has also been tested on simplified models of whole live cells immersed in water). In Figure 12.7 both experimental and numerical results are reported for a valve of *Arachnoidiscus* sp. immersed in air and illuminated by red radiation ($\lambda = 640$ nm). A CAD model of the valve has been obtained by a SEM morphological characterization of the frustule. In particular, the gray-scale SEM image of the valve has been converted into a binary image, so to obtain a refractive index map which has then been properly extruded (so as to obtain a three-dimensional model) and used in simulations. Diffraction caused by the ultrastructure of the valve induces coherent superposition of radiation along the optical axis, giving rise to a train of hot-spots. Even though the refractive index map is retrieved starting from the image of a real valve (thus giving rise to a more accurate representation in respect to simplified valve CAD models used in other numerical approaches), the position of the hot-spots as obtained by WA-BPM does not coincide exactly with that observed in experiments, mainly because the CAD model used for simulations is basically a perforated disk and does not take into account the actual curvature of the valve. Still, the efficient light focusing exerted by the valve is reproduced with a reasonable accuracy. The train of hot-spots extends in a region of hundreds of micrometers. Since, for the considered species, mature frustules are characterized by multiple girdles [12.17], i.e., the height of the frustule is of the same order of magnitude of the diameter of the valve ($\simeq 200\ \mu$m), it is plausible to hypothesize that part of the confined light lies inside the living cell and that valves contribute to an effective collection of PAR even in environmental conditions where sunlight is not so easily accessible.

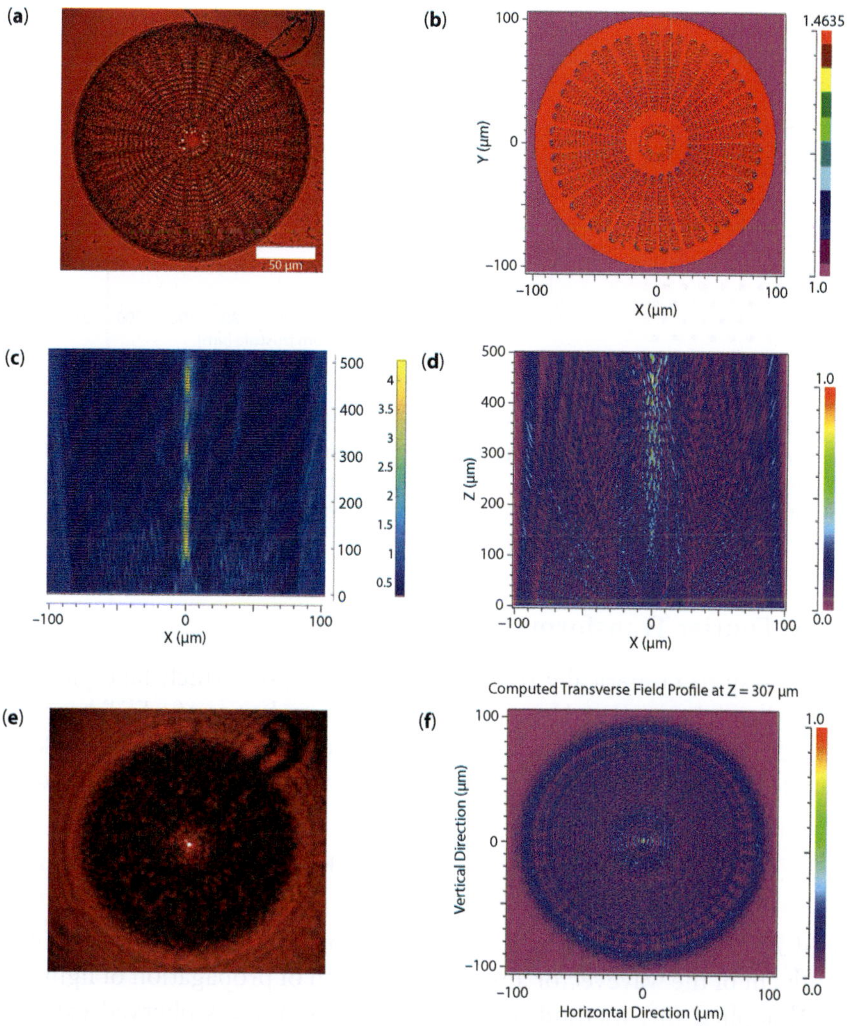

Figure 12.7 (a) Single valve of an *Arachnoidiscus* sp. diatom illuminated by radiation at $\lambda = 640$ nm; (b) refractive index map (XY plane) of the valve immersed in air retrieved from a SEM characterization of the frustule; (c) experimental transmitted intensity in XZ plane; (d) transmitted intensity in XZ plane as evaluated by WA-BPM; (e) hot-spot acquired at $z = 320$ μm; (f) transmitted intensity evaluated by WA-BPM in XY plane at $z = 307$ μm.

Furthermore, it has been observed both experimentally and numerically that, in the case of UVB radiation (detrimental to DNA mainly through formation of dimeric photoproducts between adjacent pyrimidines), the angle of divergence of diffracted light is such that the train of hot-spots takes place very far from the valve (i.e., outside the living cell) or, for sufficiently low wavelengths, it does not take place at all [12.40]. Finally, WA-BPM allows easily evaluating light propagation in whatever environment (e.g., water [12.24] or cytoplasm [12.28]), simply by assigning the right value of refractive index in the algorithm.

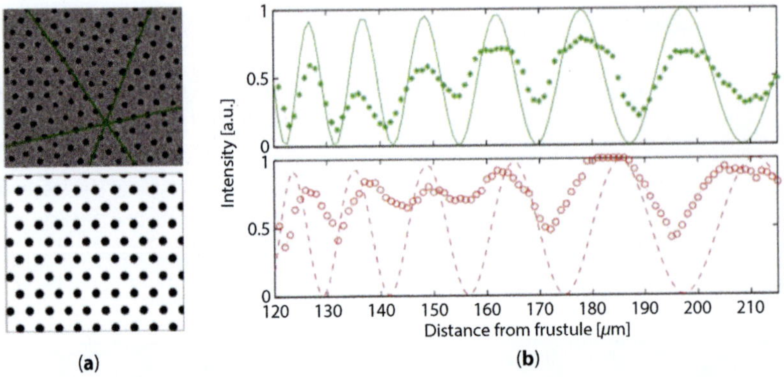

Figure 12.8 (a) SEM image of a detail of a *C. granii* valve with indication of the hexagonal symmetry of the lattice (top) and corresponding mask used in numerical simulations (bottom). (b) transmitted intensity as a function of the distance from the frustule for λ = 532 nm (top) and λ = 633 nm (bottom). Dots represent experimental data, while continuous (top) and dashed (bottom) lines refer to numerical simulations. (Reproduced with permission from Maibohm *et al.* [12.82].)

12.4.4 Fast Fourier Transform (FFT) Approach

An alternative way to numerically reconstruct the hot-spots which take place beyond a diatom valve when it is irradiated by a plane wave, is to make use of a FFT-based approach within a full diffraction algorithm like the one developed by Delen and Hooker [12.30]. Field amplitude $U(x, y, z)$ at a distance $z = L$ from the valve can indeed be expressed as follows:

$$U(x,y,L) = \mathcal{F}^{-1}\{\mathcal{F}\{U(x,y,0)\}e^{ik_zL}\} \tag{12.42}$$

where $\mathcal{F}$ and $\mathcal{F}^{-1}$ stand for the Fourier and inverse Fourier transform, respectively, and k_z is the component of the wavevector along the direction of propagation of light. Maibohm *et al.* [12.82] applied this method to reproduce the hot-spots oberved experimentally when a valve of *C. granii* diatom is exposed to coherent (laser) radiation (see Figure 12.8). Starting from the quasi-hexagonal symmetry of the valves, a hexagonal pattern of holes has been used as model for the algorithm, assuming them to be completely transparent to radiation. An absorbing boundary has been implemented in order to avoid undesired reflections.

12.5 Fractional Calculus for Diatoms

The interaction of electromagnetic fields and biological tissues and cells has become a topic of increasing interest for new research activities combining knowledge of electromagnetic theory, modeling, and simulations, physics, material science, cell biology, and medicine [12.14]. The cytoplasm of diatoms contains heterogeneous mixtures of various materials and quantities such as water, ions, membranes, and macromolecules, with a broad variety of profiles. Therefore, their interaction with electromagnetic radiation takes place in different relaxation processes, including (*i*) reorientation of water and protein-bound water molecules, (*ii*) interfacial polarization arising by the presence of two or more regions with different electrical properties, (*iii*) ionic diffusion, (*iv*) tumbling motion of the protein molecules, and (*v*) relaxations due to the non-spherical shape. These phenomena can cause complex frequency dispersion patterns of permittivity and conductivity [12.14]. Conventional FDTD methods cannot handle such materials, and there is a need for dedicated customized procedures.

Taking into consideration that the study of fractional calculus opens the mind to entirely new branches of applications, we propose to apply such concept to the modeling of diatoms, i.e., as a novel scheme for simulating electromagnetic pulse propagation in this class of biological structures. Just as fractional exponents may find their way into innumerable equations and applications, the use of fractional order can find practical use in many modern problems. The concept of fractional exponents is an outgrowth of exponents with integer value. In the same way, non-integral order of integration is a generalization of the mathematical operations of differentiation and integration to arbitrary, general, non-integer order. Although it better describes the higher complexity of nature, it is still easy to physically represent its meaning. The method is based on the fractional calculus theory and a general series expansion of the permittivity function, dealing with spatial dispersion effects as well. The resulting formulation is explicit, has a second-order accuracy, and the need for additional storage variables is minimal.

The macroscopic dielectric properties of diatoms are defined by the interaction of the light/electromagnetic energy with the material constituents at microscopic and mesoscopic scale. As a result, the permittivity and electrical conductivity depend on the working frequency. In particular, the characteristics of a general dispersive medium exhibiting multi–relaxation processes can be modeled by using the following relationship [12.44]:

$$\epsilon_r(\omega) = \epsilon_{r_\infty} + \sum_{l=1}^{N} \frac{\Delta\epsilon_{rl}}{\Gamma_l(i\omega\tau_l)} + \frac{\sigma}{i\omega\epsilon_0}, \tag{12.43}$$

where $\omega = 2\pi f$ is the angular frequency, ϵ_{r_∞} denotes the asymptotic relative permittivity at high frequency (for $\omega \to +\infty$), $\Delta\epsilon_{r_l}$ is a dimensional constant (also called the dielectric increment in some cases), and $\Gamma_l(i\omega\tau_l)$ is a heuristically derived function depending on the characteristic time τ_l for $l = 1, 2, \ldots, N$, with N being the number of relaxation processes occurring in the considered dielectric material. In (12.43), σ is the static ionic conductivity, and ϵ_0 denotes the permittivity of free space. Furthermore, we assume that the following condition holds true:

$$\lim_{\omega \to +\infty} |\Gamma_l(i\omega\tau_l)| = +\infty, \tag{12.44}$$

so to preserve the consistency of the representation.

12.5.1 Fractional Calculus Based Dielectric Dispersion Model

The following approximated fractional power expansion is adopted here:

$$\Gamma_l(i\omega\tau_l) \simeq \sum_{n=0}^{K_l} \chi_{n,l}(i\omega\tau_l)^{\zeta_{n,l}} = \Gamma_l^{(a)}(i\omega\tau_l, \zeta_{n,l}, \chi_{n,l}), \tag{12.45}$$

where $\zeta_{n,l}$ and $\chi_{n,l}$ denote suitable real-valued parameters. In particular, in order to avoid model singularities, the exponents $\zeta_{n,l}$ are required to be non-negative.

The parameters K_l, $\zeta_{n,l}$, and $\chi_{n,l}$ can be evaluated, for $l = 1, 2, \ldots, N$, by minimizing the following relative error function over the frequency band of operation:

$$e_\eta(\tau_l, \zeta_{n,l}, \chi_{n,l}) = \sqrt{\frac{\int_{\omega_{\min}}^{\omega_{\max}} \left|\Gamma_l(i\omega\tau_l) - \Gamma_l^{(a)}(i\omega\tau_l, \zeta_{n,l}, \chi_{n,l})\right|^2 d\omega}{\int_{\omega_{\min}}^{\omega_{\max}} |\Gamma_l(i\omega\tau_l)|^2 \, d\omega}}. \tag{12.46}$$

To this end, the numerical procedure based on the enhanced weighted quantum particle swarm optimization (*EWQPSO*) in [12.85] [12.13] can be conveniently used. The *EWQPSO* technique does not rely on complex operators or require gradient information and, thanks to that, is characterized by reduced computational time, and low memory usage [12.50].

It is worth stressing the fact that the non-linear fitting of the dispersive terms $\Gamma_l(i\omega\tau_l)$ ($l = 1, 2, \ldots, N$) is carried out in such a way as to prevent outbreaks of the model parameters $\chi_{n,l}$ and $\zeta_{n,l}$ ($n = 0, 1, \ldots, K_l$) that could lead to instabilities of the time-marching scheme adopted for the numerical solution of Maxwell's equations within the considered dielectric medium. To this end, the search domain is properly constrained, so that the inequality:

$$\operatorname{Im}\left\{\Gamma_l^{(a)}(i\omega\tau_l, \zeta_{n,l}, \chi_{n,l})\right\} \geqslant 0, \tag{12.47}$$

is satisfied for any angular frequency $\omega \in [\omega_{\min}, \omega_{\max}]$ and index $l = 1, 2, \ldots, N$. As known from theory [12.63], the enforcement of the inequality (12.47) is essential to ensure passivity and, in this way, preserve the physical consistency of the approximate model (12.45).

12.5.2 Basic Time-Marching Scheme

The evaluation of the electromagnetic field distribution excited by light impinging on a given diatom is here performed by using an extended formulation of the aforementioned finite-difference time-domain (*FDTD*) scheme described in Taflove *et al.* [12.117] and Caratelli *et al.* [12.20], which is useful to model the ohmic losses as well as the multi-relaxation response of media with fractional-power-law frequency dispersion as per (12.43).

Let us consider a non-magnetic dispersive medium with complex relative permittivity described by (12.43). Under such assumption, the differential version of the Ampere's law in time domain, within said material, can be written as:

$$\nabla \times \mathbf{H} = \epsilon_0 \epsilon_{r\infty} \frac{\partial \mathbf{E}}{\partial t} + \sigma \mathbf{E} + \sum_{l=1}^{N} \mathbf{J}_l, \tag{12.48}$$

where the auxiliary displacement current density terms $\mathbf{J}_l$ ($l = 1, 2, \ldots, N$) have been introduced. It is straightforward to find out that the k–th term ($1 \leq k \leq N$) is such to satisfy the equation:

$$\mathcal{D}_t^{(k)} \mathbf{J}_k = \epsilon_0 \Delta \epsilon_{rk} \frac{\partial \mathbf{E}}{\partial t}, \tag{12.49}$$

involving the generalized fractional derivative operator:

$$\begin{aligned} \mathcal{D}_t^{(k)} &= \mathcal{F}^{-1}\{\Gamma_k(i\omega\tau_k)\} \simeq \mathcal{F}^{-1}\left\{\Gamma_k^{(a)}(i\omega\tau_k, \zeta_{n,k}, \chi_{n,k})\right\} \\ &= \sum_{n=0}^{K_k} \chi_{n,k} \tau_k^{\zeta_{n,k}} \mathrm{D}_t^{\zeta_{n,k}}. \end{aligned} \tag{12.50}$$

Upon substituting (12.49) in (12.48), and applying a second-order accurate finite-difference scheme, one readily obtains, at the time instant $t = m\Delta t$:

$$(\nabla \times \mathbf{H})\big|^m - \frac{\epsilon_{r\infty}}{\Delta \epsilon_{r_k}} \left(\mathcal{D}_t^{(k)} \mathbf{J}_k\right)\Big|^m = \sum_{l=1}^{N} \mathbf{J}_l\big|^m + \sigma\, \mathbf{E}\big|^m \tag{12.51}$$

where the vector terms appearing on the right-hand side of the equation are evaluated by means of the semi-implicit approximation:

$$\left\{\begin{matrix} \mathbf{J}_l \\ \mathbf{E} \end{matrix}\right\}\Bigg|^m = \frac{1}{2}\left(\left\{\begin{matrix} \mathbf{J}_l \\ \mathbf{E} \end{matrix}\right\}\Bigg|^{m-\frac{1}{2}} + \left\{\begin{matrix} \mathbf{J}_l \\ \mathbf{E} \end{matrix}\right\}\Bigg|^{m+\frac{1}{2}}\right) \tag{12.52}$$

In a similar way, from Eq. (12.49) it follows that:

$$\mathbf{E}\big|^{m+\frac{1}{2}} = \mathbf{E}\big|^{m-\frac{1}{2}} + \frac{\Delta t}{\epsilon_0 \Delta \epsilon_{r_k}} \left(\mathcal{D}_t^{(k)} \mathbf{J}_k\right)\Big|^m \tag{12.53}$$

Let $v_{n,k}$ be the integer number such that $v_{n,k} - 1 \leq \zeta_{n,k} \leq v_{n,k}$. So, applying the Riemann–Liouville theory-based procedure detailed in Mescia *et al.* [12.85] for the finite-difference approximation of the operator $\mathcal{D}_t^{(k)}$ yields, after some mathematical manipulations which are omitted here:

$$\left(\mathcal{D}_t^{(k)}\mathbf{J}_k\right)\Big|^m \simeq \sum_{n=0}^{K_k} \frac{\chi_{n,k}}{(\nu_{n,k}-\zeta_{n,k})!}\left(\frac{\tau_k}{\Delta t}\right)^{\zeta_{n,k}} \sum_{p=0}^{\nu_{n,k}} \left[A_{n,k,p}\, \mathbf{J}_k\Big|^{m-p+\frac{1}{2}} + \sum_{q=1}^{Q_{n,k}} B_{n,k,p,q}\, \mathbf{\Psi}_{n,k,q}\Big|^{m-p} \right] \tag{12.54}$$

with:

$$\left\{ \begin{array}{c} A_{n,k,p} \\ B_{n,k,p,q} \end{array} \right\} = (-1)^p \left(\begin{array}{c} \nu_{n,k} \\ p \end{array} \right) \left\{ \begin{array}{c} \sum_{q=1}^{Q_{n,k}} a_{n,k,q} \\ e^{-b_{n,k,q}} \end{array} \right\}, \tag{12.55}$$

the positive real-valued coefficients $a_{n,k,q}$ and $b_{n,k,q}$ for $q = 1, 2, \ldots, Q_{n,k}$ being derived through the application of the *EWQPSO* method to the non linear minimization of the function:

$$\Delta_{n,k}(p) = \left| (p+1)^{\nu_{n,k}-\zeta_{n,k}} - p^{\nu_{n,k}-\zeta_{n,k}} - \sum_{q=1}^{Q_{n,k}} a_{n,k,q} e^{-b_{n,k,q} p} \right|, \tag{12.56}$$

depending on the integer index p. In (12.54), $\mathbf{\Psi}_{n,k,q}$ denotes the auxiliary current density vector defined by the following recursion formula:

$$\mathbf{\Psi}_{n,k,q}\Big|^m = \begin{cases} a_{n,k,q}\, \mathbf{J}_k\Big|^{m-\frac{1}{2}} + e^{-b_{n,k,q}}\, \mathbf{\Psi}_{n,k,q}\Big|^{m-1}, & m \geqslant 1, \\ 0, & m \leqslant 0. \end{cases} \tag{12.57}$$

Finally, by combining (12.51) with (12.52), (12.53), (12.54), one can readily obtain:

$$\left[\frac{1}{2} + \left(\epsilon_{r_\infty} + \frac{\sigma \Delta t}{2\epsilon_0} \right) \frac{C_k}{\Delta \epsilon_{r_k}} \right] \mathbf{J}_k\Big|^{m+\frac{1}{2}} + \frac{1}{2} \sum_{l=1, l\neq k}^{N} \mathbf{J}_l\Big|^{m+\frac{1}{2}} = \eta_k\Big|^m, \tag{12.58}$$

where:

$$\eta_k|^m = (\nabla \times \mathbf{H})|^m - \sigma \mathbf{E})|^{m-\frac{1}{2}} - \frac{1}{2}\sum_{l=1}^{N} \mathbf{J}_l|^{m-\frac{1}{2}} - \left(\epsilon_{r_\infty} + \frac{\sigma\,\Delta t}{2\epsilon_0}\right)$$

$$\cdot \frac{1}{\Delta\epsilon_{r_k}} \sum_{n=0}^{K_k} \frac{\chi_{n,k}}{(\nu_{n,k} - \zeta_{n,k})!} \left(\frac{\tau_k}{\Delta t}\right)^{\zeta_{n,k}} \left[\sum_{p=1}^{\nu_{n,k}} A_{n,k,p}\, \mathbf{J}_k|^{m-p+\frac{1}{2}}\right.$$

$$\left. + \sum_{p=0}^{\nu_{n,k}} \sum_{q=1}^{Q_{n,k}} B_{n,k,p,q}\, \Psi_{n,k,q}|^{m-p} \right] \tag{12.59}$$

and:

$$C_k = \sum_{n=0}^{K_k} A_{n,k,0} \frac{\chi_{n,k}}{\nu_{n,k} - \zeta_{n,k}} \left(\frac{\tau_k}{\Delta t}\right)^{\zeta_{n,k}}, \tag{12.60}$$

for $k = 1, 2, \ldots, N$. It is apparent from (12.58) that the evaluation of the displacement current density entails solving a symmetric system of N linear equations, this reflecting the multi–relaxation characteristics of the dielectric material under analysis. As a matter of fact, Eq. (12.58) can be recast in the more compact matrix form:

$$\underbrace{\left[\frac{1}{2}\underline{\mathbf{U}} + \left(\epsilon_{r_\infty} + \frac{\sigma\Delta t}{2\epsilon_0}\right)\underline{\mathbf{D}}\right]}_{\underline{\mathbf{T}}} \cdot \mathbf{J}|^{m+\frac{1}{2}} = \eta|^m \tag{12.61}$$

with $\underline{\mathbf{U}}$ being the unit matrix of order N, and $\underline{\mathbf{D}} = \text{diag}\,\{D_1, D_2, \ldots, D_N\}$ the diagonal matrix with nonzero entries $D_k = C_k/\Delta\epsilon_{r_k}$. In (12.61), $\mathbf{J}|^{m+\frac{1}{2}}$ denotes the vector of the unknown current densities at the time instant $t = \left(m + \frac{1}{2}\right)\Delta t$, namely:

$$\mathbf{J}|^{m+\frac{1}{2}} = \begin{bmatrix} \mathbf{J}_1|^{m+\frac{1}{2}} \\ \mathbf{J}_2|^{m+\frac{1}{2}} \\ \vdots \\ \mathbf{J}_N|^{m+\frac{1}{2}} \end{bmatrix} \tag{12.62}$$

Similarly, the column vector $\boldsymbol{\eta}|^m$ is built up by arraying the auxiliary electromagnetic field quantities $\boldsymbol{\eta}_k|^m$ ($k = 1, 2, \ldots, N$) defined in (12.59). It is worth noting that the inverse of the coefficient matrix $\underline{\mathbf{T}}$ of the linear system (12.61) can be conveniently computed only one time before the time–marching scheme is initiated. In this way, the algorithmic implementation of the technique proposed in this research study actually results in a reduced additional computational cost of $O(N^2)$ floating–point operations useful to determine the solution of (12.58) as:

$$\mathbf{J}|^{m+\frac{1}{2}} = \underline{\mathbf{T}}^{-1} \cdot \boldsymbol{\eta}|^m. \tag{12.63}$$

Once the current density terms $\mathbf{J}_l|^{m+\frac{1}{2}}$ ($l = 1, 2, \ldots, N$) are evaluated, the electric field distribution within the considered dielectric medium can be derived from (12.48) as:

$$\mathbf{E}|^{m+\frac{1}{2}} = \frac{2\epsilon_0\epsilon_{r_\infty} - \sigma\Delta t}{2\epsilon_0\epsilon_{r_\infty} + \sigma\Delta t}\mathbf{E}|^{m-\frac{1}{2}} + \frac{2\Delta t}{2\epsilon_0\epsilon_{r_\infty} + \sigma\Delta t}\left[(\nabla \times \mathbf{H})|^m - \frac{1}{2}\sum_{l=1}^{N}\left(\mathbf{J}_l|^{m-\frac{1}{2}} + \mathbf{J}_l|^{m+\frac{1}{2}}\right)\right] \tag{12.64}$$

where judicious use of (12.52) has been made. Finally, by carrying out a second–order accurate finite–difference approximation of Faraday's law in the time domain, the following update equation for the magnetic field is readily obtained:

$$\mathbf{H}|^{m+1} = \mathbf{H}|^m - \frac{\Delta t}{\mu_0}(\nabla \times \mathbf{E})|^{m+\frac{1}{2}} \tag{12.65}$$

with μ_0 denoting the magnetic permeability of free space.

12.5.3 Uniaxial Perfectly Matched Layer Boundary Conditions

In order to truncate the *FDTD* computational volume and solve electromagnetic problems involving diatoms in domains with open boundaries, dedicated uniaxial perfectly matched layer (*UPML*) conditions [12.48] have to be derived and implemented numerically, accounting for the electrical conductivity and the multi–relaxation characteristics of the dielectric material under analysis. To this end, let us first introduce the auxiliary electric field vector $\mathbf{e}$ as:

$$\mathbf{e} = \left(\kappa_x + \frac{\sigma_x}{i\omega\epsilon_0}\right)\mathbf{E}, \tag{12.66}$$

with κ_x, σ_x denoting the *UPML* material parameters in accordance with the complex coordinate stretching approach Taflove *et al.* [12.117]. Multiplying both sides of (12.66) by $i\omega$ and transforming into the time domain immediately yields:

$$\frac{\partial \mathbf{E}}{\partial t} = \kappa_x \frac{\partial \mathbf{E}}{\partial t} + \frac{\sigma_x}{\epsilon_0} \mathbf{E}. \tag{12.67}$$

In this way, it is not difficult to find out that Ampere's law can be written within the *UPML* region, as:

$$\nabla \times \mathbf{H} = \epsilon_0 \epsilon_{r_\infty} \frac{\partial \mathbf{E}}{\partial t} + \sigma \mathbf{e} + \sum_{l=1}^{N} \mathbf{j}_l, \tag{12.68}$$

where the l–th displacement current density term ($1 \leq l \leq N$) satisfies the fractional derivative equation:

$$\mathcal{D}_t^{(l)} \mathbf{j}_l = \epsilon_0 \Delta \epsilon_{r_l} \frac{\partial \mathbf{E}}{\partial t}. \tag{12.69}$$

The discretization of the considered equations on the Yee lattice can be conveniently carried out by adopting the usual leapfrog scheme in time, wherein the loss terms are averaged according to the semi–implicit approximation [see (12.52)]. Overall, as it can be easily figured out by comparison with (12.48) and (12.49), this leads to time–stepping expressions which are formally equivalent to (12.58) and (12.64) with the stretched vectors **e** and $\mathbf{j}_l$ replacing the quantities **E** and $\mathbf{J}_p$, respectively, for $l = 1, 2, \ldots, N$. On the other hand, the electric field update equation directly follows from (12.67) as:

$$\mathbf{E}\big|^{m+\frac{1}{2}} = \frac{2\epsilon_0 \kappa_x - \sigma_x \Delta t}{2\epsilon_0 \kappa_x + \sigma_x \Delta t} \mathbf{E}\big|^{m-\frac{1}{2}} + \frac{2\epsilon_0}{2\epsilon_0 \kappa_x + \sigma_x \Delta t} \cdot \left(\mathbf{e}\big|^{m+\frac{1}{2}} - \mathbf{e}\big|^{m+\frac{1}{2}} \right) \tag{12.70}$$

By Fourier–transforming the Faraday's law $\nabla \times \mathbf{E} = -\mu_0 (i\omega\kappa_x + \sigma_x/\epsilon_0)\,\mathbf{H}$ and discretizing at the time instant $t = (m + \frac{1}{2})\,\Delta t$, one can readily derive the time–stepping expression for the magnetic field within the *UPML* termination:

$$\mathbf{H}\big|^{m+1} = \frac{2\epsilon_0 \kappa_x - \sigma_x \Delta t}{2\epsilon_0 \kappa_x + \sigma_x \Delta t} \mathbf{H}\big|^{m} - \frac{2Y_0^2 \Delta t}{2\epsilon_0 \kappa_x + \sigma_x \Delta t} (\nabla \times \mathbf{E})\big|^{m+\frac{1}{2}} \tag{12.71}$$

with $Y_0 = \sqrt{\epsilon_0 / \mu_0}$ being the wave admittance in free space.

The developed numerical procedure can be used to characterize the electromagnetic behavior of diatoms, modeled as multi-layered dielectric structure, under plane–wave excitation.

12.6 Beyond the Glass Cage: The Fate of Light Inside the Cell

The last leg of the light journey is the cell. The absorbed photons represent sources of information about the environment and environmental signals controlling different physiological, adaptive, and biochemical processes [12.31]. Pigments are responsible for capturing solar energy. The absorbed light can have three possible fates: generate photochemical reactions leading to production of organic matter; when in excess, can be dissipated as heat [12.7] or be emitted back to the environment as fluorescence [12.81]. Diatoms contain two types of pigments involved in light harvesting and photoprotection: chlorophylls and carotenoids (fucoxanthin, β-carotene and the xanthophylls, diatoxanthin, diadinoxanthin, violaxanthin, antheraxanthin, and zeaxanthin). Two forms of chlorophylls are found in diatoms: Chl-*a* and Chl-*c*, identified in various algae. Chlorophyll absorbs light in the red (long wavelength) and the blue (short wavelength) portions of the electromagnetic spectrum, which are used in photosynthesis while carotenoids are engaged mainly in photoprotection, with the exception of fucoxanthin that transfers excitation energy very efficiently to Chl-*a* and plays a major role in the so-called LHC (light-harvesting complex) photosystems [12.75]. The LHC system contains Chl-*a* and *c*, and the fucoxanthin (*Fx*) and is thus referred to as FCP, for fucoxanthin-chlorophyll protein. The pigment composition in FCP reflects the adaptation of diatoms to the blue-green light of the marine environment [12.49]. Moreover, the high amount of the fucoxanthin that masks the other carotenoids, is responsible for the characteristic brown color of diatoms [12.12]. However, due to the mixtures of pigments, the absorption spectra of phytoplankton are complex, and it is difficult to quantify the contribution of each pigment. In the last few years, different studies have been reported in order to find a method that can be successfully used to identify the individual absorption spectra of component pigments such as different derivative analysis and the Gaussian-Lorentzian technique. However, to date, it is still difficult to match the information from the pigments with the absorption spectrum [12.3].

Like for all living organisms, variation in light/dark cycles due to Earth rotation influences the circadian rhythms of diatoms as well [12.4]. In addition, the Moon gives off some light of its own and a few papers reported its influence on photosynthesis [12.100]. Very recently, Breitler and co-workers [12.16] reported for the first time a transcriptomic analysis of the plant *Coffea arabica* under full moonlight conditions. This additional source of photons plays a key role in the light reactions, however, at the moment we can only speculate a possible influence of moonlight on rates of photosynthesis also in diatoms.

Thus, having discussed how the light gets through the water and the cell wall, we now consider what happens inside the cell and which components are involved.

12.6.1 The Diatom Chloroplast and its Evolution

Unlike plants, green and red algae, the diatom plastid is an evolutionary mosaic, as the one of other photosynthetic members of the stramenopile algae group that are believed to have arisen by secondary endosymbiosis, whereby a non-photosynthetic eukaryote acquired a chloroplast by engulfing a photosynthetic eukaryote, probably a red algal endosymbiont [12.39]. More recent studies suggest a tertiary endosymbiotic origin, in which a red alga was acquired through secondary endosymbiosis through another algal group which was in turn engulfed by the photosynthetic stramenopile [12.33] [12.34]. Each endosymbiotic event led to new combinations of genes from the hosts and the endosymbionts. The thylakoid membrane inside chloroplasts, capable of forming elaborate networks, harbors the protein complexes that are necessary for the photosynthesis. Plastid genomes have retained

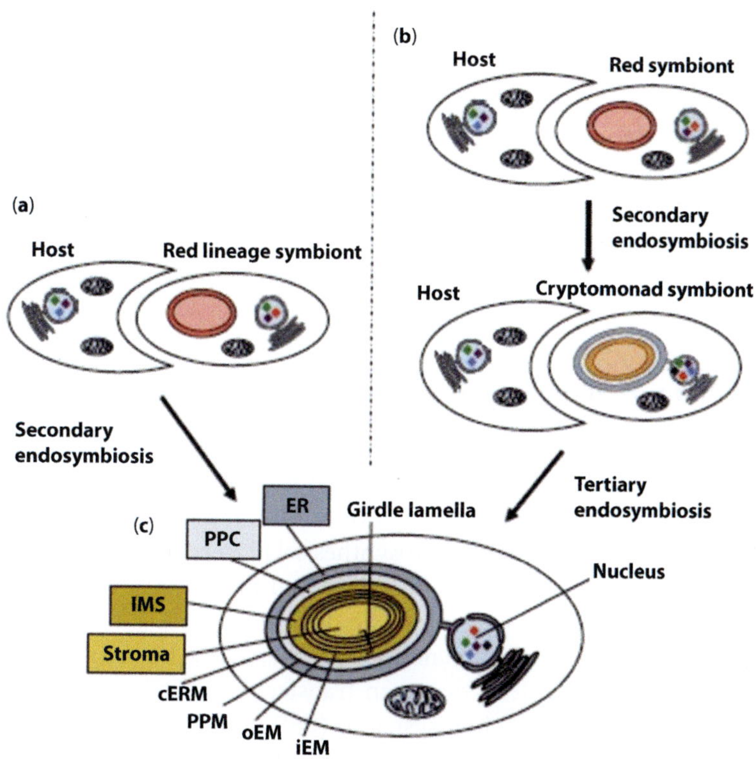

Figure 12.9 Origin and structure of the diatom chloroplast. The upper panel shows the schematic representation of the secondary (a) and tertiary (b) endosymbiont hypothesis of diatom evolution. The lower panel (c) shows a schematic diagram of the four membranes surrounding the diatom chloroplast. Abbreviations are as follows: cERM, chloroplast endoplasmic reticular membrane; ER, endoplasmic reticulum; iEM, inner envelope membrane; IMS, intermembrane space; oEM, outer envelope membrane; PPC, periplastid compartment; PPM, periplastid membrane. (Reproduced with permission from [12.92].)

a limited number of genes related to the chloroplast function, and other genes have either been lost or transferred to the nuclear genome [12.92]. For this reason, the thylakoids of diatoms are different from the chlorophyte one, with an envelope surrounded by four annular membranes instead of two, and arranged in stacks of three, that are disposed in parallel in a highly ordered manner, where the grana stacking is absent [12.79]. The four membranes are, from inside out, the inner envelope membrane (iEM), the outer envelope membrane (oEM), the periplasmic membrane (PPM), and the chloroplast endoplasmic reticular membrane (cERM) [12.92] (Figure 12.9).

Depending on the species, diatom plastids can be highly diverse in number and size. Centric diatoms possess plastids ranging from 2 to 3 per cell (as in *Thalassiosira pseudonana*) up to several hundreds (as in *Odontella sinensis*) [12.107]. Interestingly, diatom plastids are characterized by a pronounced motility which strongly depends on both illumination conditions (i.e. photons availability) and spectral content of the impinging radiation. In particular, as already mentioned in section 12.4.1, in several centric species and for weak illumination conditions, it has been observed a migration of plastids towards frustule walls, well within the region of influence of the evanescent field associated with the modes coupled to valves and girdles (see again Eq. 12.33). On the other hand, an excess of light intensity induces plastid relocation towards the center of the cell [12.46], suggesting the presence of a mechanism useful to nucleus protection from photodamage and/or for an efficient coupling with the hot-spots described in section 12.4.3. Migration of plastids towards the nucleus is not only triggered by exposure to high light intensity, but it results prominent also for irradiation in the blue spectral range [12.110], where one of the maxima of the absorption spectrum of chlorophylls is located.

12.6.2 The Photosynthetic and Electron Transport Chain

Photosynthesis is a unique process which, in the case of diatoms, takes place with an extremely high efficiency, explaining the ecological success of these microalgae and how they are able to produce approximately 20% of all the oxygen on Earth, being directly responsible for 40% of total marine photosynthesis [12.41]. Despite distinctions in the architecture of thylakoid membranes, the fundamental machinery responsible for photosynthetic electron transfer is highly conserved in diatoms [12.79].

Like all oxygenic organisms, light energy is captured by pigments bound to light-harvesting antenna proteins (LHCs) that then transfer the energy to the two multi-subunit membrane protein complexes photosystem II (PSII) and PSI, to initiate photochemical reactions (Figure 12.10). Recently, thanks to a sophisticated analysis, Flori and colleagues reconstructed the 3D structure of the diatom chloroplast using focused ion beam scanning electron microscopy (FIB-SEM). They demonstrated that, unlike previously thought and as in the case of plants, also in diatoms PSI and PSII are separated and that this separation is necessary to avoid any "short circuits" that would make photosynthesis less effective [12.42]. Even though photosynthetic complexes share fundamental structures, they differ in composition, ratio, and dynamic control of the photosynthetic reactions due to the substantial differences in the light field of marine environments. To cope with the various light environments which different photosynthetic organisms inhabit, different types and numbers of light-harvesting antenna proteins are attached to the PSII core to enhance the

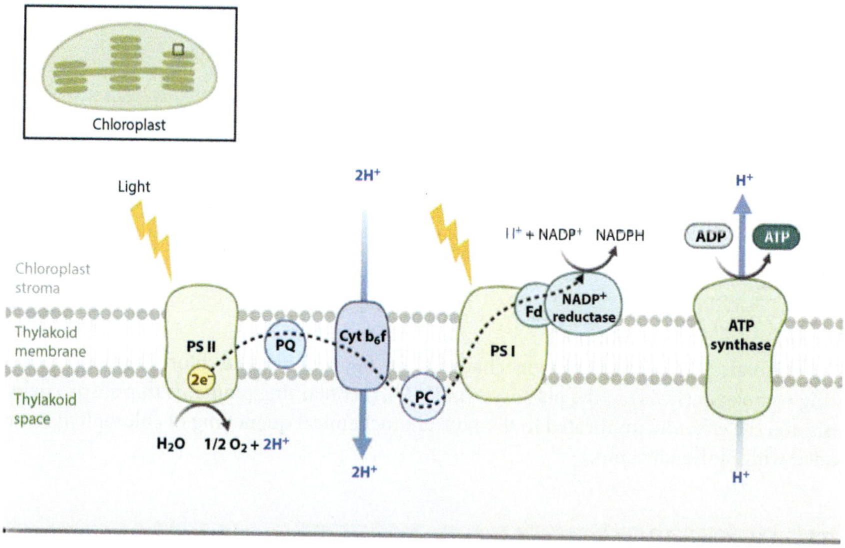

Figure 12.10 Schematic diagrams of the photosynthetic electron transport chain. (Created with BioRender.com.)

light-harvesting and energy-quenching capacities [12.38] [12.121]. Very recently it also has been demonstrated that different diatom classes have various architectures of PSII as an adaptation strategy, whilst a convergent evolution occurred concerning PSI and the overall plastid structure [12.5].

12.6.3 The Photoprotection Mechanism

An excess of light energy may induce damage to the photosynthetic apparatus. A photoprotective response consists in the dissipation of this excess as heat, which occurs in PSII whenever the light absorption exceeds the maximum CO_2 assimilation rate. These processes are at the basis of the so-called non-photochemical quenching (NPQ) and are typical of almost all the photosynthetic eukaryotes, regulating and protecting photosynthesis. The ecological dominance of diatoms also relies on their very efficient photoprotective mechanisms. Diatoms, compared to other photosynthetic organisms, possess a high capacity to dissipate excess light energy as heat through high energy quenching (qE) that, together with the photoinhibitory quenching (qI), can be visualized via the NPQ of Chl-*a* fluorescence [12.78] [12.57]. NPQ takes place in the light-harvesting complex antennae (LHC) of PSII. These include the xanthophyll cycle, which allows diatoms to activate NPQ through the conversion of the accessory pigment diadinoxanthin into diatoxanthin. Along with pigments of the xanthophyll cycle, FCPs and in particular an expanded set of LHCXs family, light-harvesting proteins also participate in the formation of NPQ to avoid photo-oxidation [12.115] (Figure 12.11). Moreover, it has been recently demonstrated that the excess in chloroplast reducing potential is translocated into the mitochondria and dissipated through respiratory electron transport as an additional photoprotective strategy [12.34].

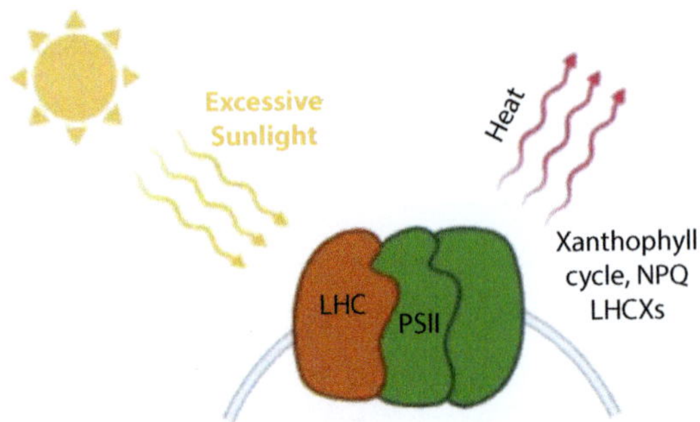

Figure 12.11 Schematic of photoprotection mechanism. The figure shows the chloroplastic membranes, the light-harvesting complexes (LCH) of the photosystem II (PSII) containing pigments that absorb light for photosynthesis and the effectors implicated in the non-photochemical quenching of chlorophyll fluorescence (NPQ). (Created with BioRender.com.)

12.6.4 The Diatom Photoreceptors

Light consists of a spectrum of colors and, as mentioned in section 12.2, the spectral properties of the underwater light field are very distinct from the terrestrial ones. Light undergoes absorptive and scattering processes throughout the water column, and is also affected by the presence of different photosynthetic organisms [12.31]. The blue-green (400–500 nm) spectral components dominate at high depths, while the red and infrared wavebands are absorbed by the water, and are present at low intensities, mostly derived from the chlorophyll *a* fluorescence due to the presence of photosynthetic organisms.

Photoreceptors are light-sensing molecules able to perceive the light at specific wavelengths and propagate downstream signal pathways in response, converting the physical light signal into a biochemical signal. They are photon detectors that use visible light to learn about the world around them. These proteins are usually modular in their architecture. They are provided with photoreceptive domains binding chromophores to absorb light signals at specific wavelengths. Chromophores undergo physicochemical and structural changes upon light absorption, which regulate the signal propagation [12.31] (Figure 12.12). In diatoms genome many types of photoreceptors have been identified and partially characterized, e.g., the red light receptors, phytochrome (DPH) [12.43], different cryptochromes [12.21] [12.73] and aureochromes [12.84] for the blue and also heliorhodopsins, a new family of rhodopsins recently discovered [12.97].

12.6.5 Chlorophyll Optical Signals for Satellite Population Monitoring

After being absorbed, light can be re-emitted at higher wavelengths through a process called fluorescence. Approximately 2-6% of the solar energy absorbed by chlorophyll is re-emitted as fluorescence [12.98]. Chlorophyll fluorescence is a detectable signal that provides a way to directly estimate actual photosynthesis from space, reflecting phytoplankton presence and primary productivity (see Figure 12.13). In 2016, Li and colleagues calculated the quantum yield of chlorophyll fluorescence from space by using the images coming from the moderate resolution imaging spectroradiometer (MODIS) and medium resolution

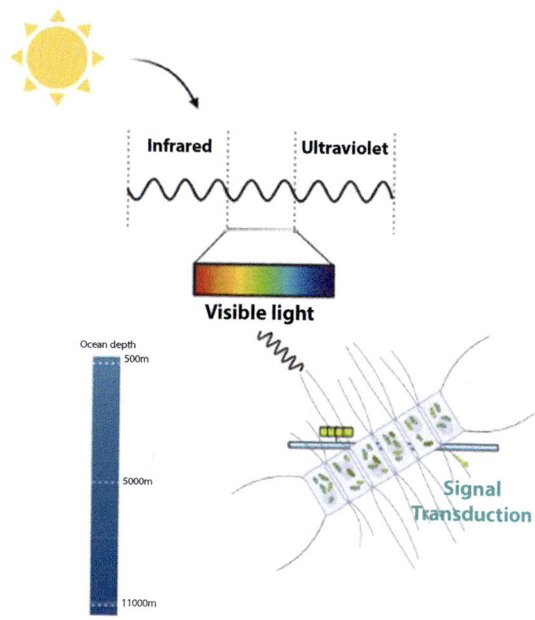

Figure 12.12 Schematic representation of a simplified view of the phototransduction cascades mediated by photoreceptors. Light in the visible spectrum colors is absorbed by specific photoreceptors which trigger signal transduction cascades that generate physiological responses. (Created with BioRender.com.)

imaging spectrometer (MERIS) satellites, which possess the capability of remotely detecting solar induced chlorophyll fluorescence signals from the global ocean [12.80]. Chl-*a* is routinely derived from satellite-based reflectance using regionally optimized algorithms. The OLCI Neural Network Swarm (ONNS) is an example of a bio-geo-optical algorithm

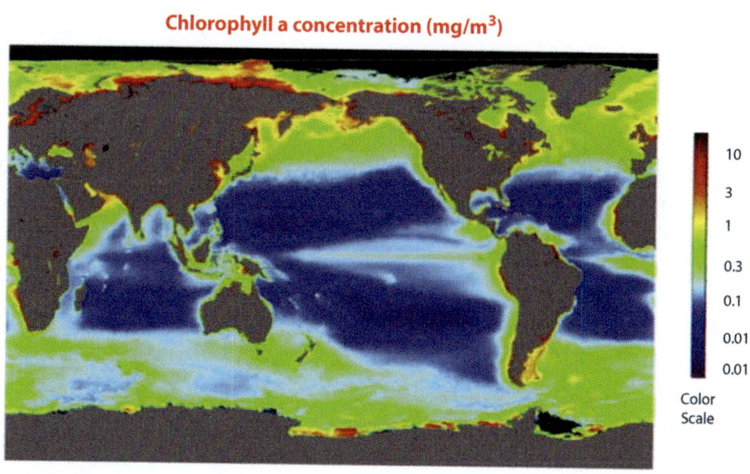

Figure 12.13 This map shows the average spatial concentration of chlorophyll *a* in the oceans. (Adapted from Rogato *et al.*, [12.102].)

for the retrieval of water quality parameters from atmospherically corrected satellite imagery or *in situ* radiometric measurements. The aim of the development is to provide a single algorithm that is suitable to all natural waters, from oligotrophic ocean waters to very turbid coastal or highly absorbing inland waters. However, total phytoplankton biomass, as indexed by Chl-*a* concentration, tells only part of the story and differences in optical properties of diatoms can be exploited to develop computational instruments able to distinguish them from other types of phytoplankton. Starting from the observation that diatom populations exhibit much lower specific absorption coefficients than other phytoplankton populations for a given region [12.112], in 2004 Sathyendranath and colleagues introduced a bio-optical algorithm able to distinguish their contribution and applied it to SeaWiFS data on ocean color [12.106]. The SeaWiFS instrument was launched from NASA by Orbital Sciences Corporation on the OrbView-2 (a.k.a. SeaStar) satellite in August 1997, and collected data from September 1997 until December 2010. SeaWiFS had 8 spectral bands from $\lambda = 412$ nm to $\lambda = 865$ nm. Reflectance at a specific wavelength λ and for a given depth z can be defined as the ratio of upwelling irradiance (E_u) to the downwelling one (E_d):

$$R(\lambda, z) = \frac{E_u(\lambda, z)}{E_d(\lambda, z)} \tag{12.73}$$

The contribution of elastic scattering to reflectance at sea surface, $R^E(\lambda, 0)$, is related to total absorption and backscattering coefficients $a(\lambda)$ and $b_b(\lambda)$ by:

$$R^E(\lambda, 0) = r\frac{b_b(\lambda)}{a(\lambda) + b_b(\lambda)} \tag{12.74}$$

with r proportionality factor. Total absorption coefficient $a(\lambda)$ is given by the sum of the absorption coefficients of pure water (a_w), dissolved organic matter (a_{dom}) and phytoplankton (a_p):

$$a(\lambda) = a_w(\lambda) + a_{dom}(\lambda) + a_p(\lambda) \tag{12.75}$$

The coefficient a_p can be parametrized as a non-linear function of Chl-*a* concentration C [12.105] derived from *in-situ* high performance liquid chromatography (HPLC) data:

$$a_p(\lambda) = U(\lambda)\{1 - \exp[-S(\lambda)C)]\} + Ca_2^*(\lambda) \tag{12.76}$$

with U [m^{-1}], S [m^3 (mg Chl-a)$^{-1}$] and a_2^* [m^2 (mg Chl-a)$^{-1}$] fitting parameters. HPLC data can indeed be used to separate diatom-dominated samples from mixed populations starting from the concentration of Chl-c_3 and fucoxanthin relative to Chl-*a* (values of Chl-c_3 to Chl-*a* ratio less than 0.02 and fucoxanthin to Chl-*a* ratio greater than 0.4 correspond to diatom-dominated regions). In the algorithm, the ratio of reflectances at two wavelengths is always used in order to reduce the effect of any potential errors in the factor r of Eq. (12.74), since the variations in r are independent by the wavelengths used in the ratio. For each pixel of a satellite image we can define the *normalized remote-sensing reflectance* R_{RS}:

$$R_{RS}(\lambda) = \frac{L(0,\lambda)}{E_d(0,\lambda)} \tag{12.77}$$

where $L(0, \lambda)$ is the upwelling radiance from the sea surface in the zenith direction at wavelength λ and $E_d(0, \lambda)$ is the downwelling irradiance at the same wavelength. The chlorophyll concentration is first estimated using both the R_{RS} (510)/R_{RS} (555) and R_{RS} (490)/R_{RS} (670) ratios, assuming a diatom population. The difference between the two computed chlorophyll values is then estimated and normalized to the mean of the two values. The operation is repeated assuming a mixed phytoplankton population. The correct model yields smaller differences in the concentrations retrieved using two different waveband pairs if compared to the wrong model. Therefore, each pixel can be assigned to diatom or mixed population on the basis of the computed normalized differences. In 2018, Kramer and co-workers [12.74] refined and improved the algorithm through successive re-parameterizations and reformulations of the absorption and backscattering coefficients, with data collected in different sites.

12.7 Conclusions

The main analytical models describing the underwater optical field, the essential numerical algorithms for the study of the photonic properties of the diatom seen as a natural metamaterial and methods for complex EM propagation problems and wave propagation in dispersive materials with multiple relaxation times are described. In the near future, the use of fractional calculus-based methods will be extended to study diatoms and their substructures, in those cases where conventional FDTD techniques fail. In Bia *et al.* [12.14], the plane-wave propagation in a variety of dispersive media has been examined, whereby the comparison between simulated results and those evaluated by using an analytical method based on Fourier transformation demonstrated the accuracy and effectiveness of the developed FDTD model. The dispersive models might be related further to the absorption spectra of the chromophores described in section 12.6 and to biosilica dispersion.

Gielis transformations are presented as a unifying geometrical model framework. They allow for great flexibility in modeling and for computational efficiency to be achieved, and have been used for automated detection and determination of diatoms (see Figure 12.14) [12.109].

In presenting a new viewpoint, beyond the classical Euclidean and Riemannian, one enters the realm of Minkowski-Finsler geometry, basically the geometry where the fundamental metric goes beyond the quadratic one. Today, Finsler geometry is the full-sized research theme in mathematics, but it turns out that the simplest of metrics, actually the one proposed by Lamé and by Riemann for the infinitesimals, is extremely useful to study natural shapes from a systems biological and evolutionary perspective.

Finally, the presented discussion ends up inside the cell, describing the pigments and the processes involved in light harvesting, photoprotection, and, ultimately, in photosynthesis. Examples of numerical analysis of satellite images for remote sensing of diatom populations

Figure 12.14 Jewels of the sea. https://maxseeger.de/jewels-of-the-sea. (Reproduced with permission of the author.)

are also described, confirming how mathematical modeling not only allows the understanding of the interaction of light with frustules, but also the retrieval of fundamental information about diatom-related primary production.

References

[12.1] Yogesh C Agrawal. The optical volume scattering function: Temporal and vertical variability in the water column off the new jersey coast. *Limnology and oceanography*, 50(6):1787–1794, 2005.

[12.2] Luis Ever Aguirre, Liangqi Ouyang, Anders Elfwing, Mikael Hedblom, Angela Wulff, and Olle Inganäs. Diatom frustules protect DNA from ultraviolet light. *Scientific reports*, 8(1):1–6, 2018.

[12.3] R Aguirre-Gomez, AR Weeks, and SR Boxall. The identification of phytoplankton pigments from absorption spectra. *International Journal of Remote Sensing*, 22(2-3):315–338, 2001.

[12.4] Rossella Annunziata, Andrés Ritter, Antonio Emidio Fortunato, Alessandro Manzotti, Soizic Cheminant-Navarro, Nicolas Agier, Marie JJ Huysman, Per Winge, Atle M Bones, Francois-Yves Bouget, *et al.* bHLH-PAS protein RITMO1 regulates diel biological rhythms in the marine diatom *Phaeodactylum tricornutum*. *Proceedings of the National Academy of Sciences*, 116(26):13137–13142, 2019.

[12.5] Rameez Arshad, Claudio Calvaruso, Egbert J Boekema, Claudia Büchel, and Roman Kouřil. Revealing the architecture of the photosynthetic apparatus in the diatom *Thalassiosira pseudonana*. *Plant Physiology*, 186(4):2124–2136, 2021.

[12.6] RW Austin. The remote sensing of spectral radiance from below the ocean surface. In NG Jerlov and ES Nielsen, editors, *Optical Aspects of Oceanography*. Academic Press, 1974.

[12.7] B. Bailleul, A. Rogato, A. de Martino, S. Coesel, P. Cardol, C. Bowler, A. Falciatore, and G. Finazzi. LHCX1 an atypical member of the LHCSR protein family, modulates diatoms response to light. *Proceedings of the National Academy of Sciences*, 107: 18214–9, 2010.

[12.8] C Jacyn Baker and Elizabeth W Orlandi. Active oxygen in plant pathogenesis. *Annual review of phytopathology*, 33(1):299–321, 1995.

[12.9] Karen S Baker and Robert Frouin. Relation between photosynthetically available radiation and total insolation at the ocean surface under clear skies 1. *Limnology and Oceanography*, 32(6):1370–1377, 1987.

[12.10] TMWJ Bandara, M Furlani, I Albinsson, Angela Wulff, and B-E Mellander. Diatom frustules enhancing the efficiency of gel polymer electrolyte based dyesensitized solar cells with multilayer photoelectrodes. *Nanoscale Advances*, 2 (1):199–209, 2020.

[12.11] HS Barlow. The scientific basis of illuminating engineering. *Physics Bulletin*, 14(1):16, 1963.

[12.12] Martine Bertrand. Carotenoid biosynthesis in diatoms. *Photosynthesis research*, 106(1):89–102, 2010.

[12.13] Pietro Bia, Diego Caratelli, Luciano Mescia, R Cicchetti, Guido Maione, and F Prudenzano. A novel FDTD formulation based on fractional derivatives for dispersive Havriliak-Negami media. *Signal Processing*, 107:312–318, 2015.

[12.14] Pietro Bia, Luciano Mescia, and Diego Caratelli. Fractional calculus-based modeling of electromagnetic field propagation in arbitrary biological tissue. *Mathematical Problems in Engineering*, 2016, 2016.

[12.15] Max Born and Emil Wolf. *Principles of optics: electromagnetic theory of propagation, interference and diffraction of light*. Elsevier, 2013.

[12.16] Jean-Christophe Breitler, Doaa Djerrab, Sophie Leran, Lucile Toniutti, Charlie Guittin, Dany Severac, M Pratlong, Alexis Dereeper, Hervé Etienne, and Benoît Bertrand. Full moonlight-induced circadian clock entrainment in *Coffea arabica*. *BMC Plant Biology*, 20(1):1–11, 2020.

[12.17] Nicholas Edward Brown. *Arachnoidiscus: An Account of the Genus, Comprising its History, Distribution, Development and Growth of the Frustule, Structure and its Examination and Purpose in Life, and a Key to and Descriptions of All Known Species, Illustrated*. W. Watson, 1933.

[12.18] Kurt Busch and Sajeev John. Photonic band gap formation in certain self-organizing systems. *Physical Review E*, 58(3):3896, 1998.

[12.19] Diego Caratelli, Bruna Germano, Johan Gielis, Matthew He, Pierpaolo Natalini, and Paolo Emilio Ricci. Fourier solution for the Dirichlet problem for the Laplace and Helmholtz equations in starlike domains. In *Lecture Notes of Tbilisi International Centre of Mathematics and Informatics*, volume 10. Tbilisi University Press, 2009.

[12.20] Diego Caratelli, Alessandro Massaro, Roberto Cingolani, and Alexander G. Yarovoy. Accurate time-domain modeling of reconfigurable antenna sensors for non-invasive melanoma skin cancer detection. *IEEE Sensors Journal*, 12 (3):635–643, mar 2012.

[12.21] Sacha Coesel, Manuela Mangogna, Tomoko Ishikawa, Marc Heijde, Alessandra Rogato, Giovanni Finazzi, Takeshi Todo, Chris Bowler, and Angela Falciatore. Diatom PtCPF1 is a new cryptochrome/photolyase family member with DNA repair and transcription regulation activity. *EMBO Reports*, 10(6):655–661, 2009.

[12.22] Charles Cox and Walter Munk. Measurement of the roughness of the sea surface from photographs of the sun's glitter. *Journal of the Optical Society of America*, 44(11):838–850, 1954.

[12.23] Luca De Stefano, Ilaria Rea, Ivo Rendina, Mario De Stefano, and Luigi Moretti. Lensless light focusing with the centric marine diatom *Coscinodiscus walesii*. *Optics Express*, 15(26):18082–18088, 2007.

[12.24] E De Tommasi, I Rea, V Mocella, Luigi Moretti, Mario De Stefano, I Rendina, and L De Stefano. Multi-wavelength study of light transmitted through a single marine centric diatom. *Optics Express*, 18(12):12203–12212, 2010.

[12.25] E De Tommasi, AC De Luca, L Lavanga, P Dardano, M De Stefano, Langella De Stefano, C Langella, I Rendina, K Dholakia, and M Mazilu. Biologically enabled sub-diffractive focusing. *Optics Express*, 22(22):27214–27227, 2014.

[12.26] E De Tommasi, E Esposito, S Romano, A Crescitelli, V Di Meo, V Mocella, G Zito, and I Rendina. Frontiers of light manipulation in natural, metallic, and dielectric nanostructures. *La Rivista del Nuovo Cimento*, pages 1–68, 2021a.

[12.27] Edoardo De Tommasi, Johan Gielis, and Alessandra Rogato. Diatom frustule morphogenesis and function: a multidisciplinary survey. *Marine genomics*, 35:1–18, 2017.

[12.28] Edoardo De Tommasi, Roberta Congestri, Principia Dardano, Anna Chiara De Luca, Stefano Managò, Ilaria Rea, and Mario De Stefano. UV-shielding and wavelength conversion by centric diatom nanopatterned frustules. *Scientific reports*, 8(1):1–14, 2018.

[12.29] Edoardo De Tommasi, Ilaria Rea, Maria Antonietta Ferrara, Luca De Stefano, Mario De Stefano, Adil Y Al-Handal, Marija Stamenković, and Angela Wulff. Underwater Light Manipulation by the Benthic Diatom *Ctenophora pulchella*: From PAR Efficient Collection to UVR Screening. *Nanomaterials*, 11(11):2855, 2021b.

[12.30] Nuri Delen and Brian Hooker. Free-space beam propagation between arbitrarily oriented planes based on full diffraction theory: a fast Fourier transform approach. *JOSA A*, 15(4):857–867, 1998.

[12.31] Frauke Angelique Depauw, Alessandra Rogato, Maurizio Ribera d'Alcalá, and Angela Falciatore. Exploring the molecular basis of responses to light in marine diatoms. *Journal of Experimental Botany*, 63(4):1575–1591, 2012.

[12.32] Giuseppe Di Caprio, Giuseppe Coppola, Luca De Stefano, Mario De Stefano, Alessandra Antonucci, Roberta Congestri, and Edoardo De Tommasi. Shedding light on diatom photonics by means of digital holography. *Journal of Biophotonics*, 7(5):341–350, 2014.

[12.33] Richard G Dorrell and Chris Bowler. Secondary plastids of stramenopiles. *Advances in Botanical Research*, 84:57–103, 2017.

[12.34] Richard G Dorrell, Gillian Gile, Giselle Mccallum, Raphaël Méheust, Eric P Bapteste, Christen M Klinger, Loraine Brillet-Guéguen, Katalina D Freeman, Daniel J Richter, and Chris Bowler. Chimeric origins of ochrophytes and haptophytes revealed through an ancient plastid proteome. *eLife*, 6:e23717, 2017.

[12.35] Yannick D'Mello, Dan Petrescu, James Skoric, Melissa Campbell, Mark Andrews, and David Plant. Characterization of the photonic response in *Nitzschia filiformis* phytoplankton. In *CLEO: QELS_Fundamental Science*, pages JTh2A–96. Optical Society of America, 2018.

[12.36] Yannick D'Mello, Santiago Bernal, James Skoric, Dan Petrescu, Mark Andrews, and David V Plant. Photonic crystal behavior of *Nitzschia filiformis* phytoplankton for chlorophyll A photosynthesis. In *CLEO: QELS_Fundamental Science*, pages JW2A–121. Optical Society of America, 2019.

[12.37] Steven Emerson and John Hedges. *Chemical oceanography and the marine carbon cycle*. Cambridge University Press, 2008.

[12.38] Paul G Falkowski and Yi-Bu Chen. Photoacclimation of light harvesting systems in eukaryotic algae. In BR Green and WW Parson, editors, *Light-harvesting antennas in photosynthesis*, pages 423–447. Springer, Dordrecht, 2003.

[12.39] Paul G Falkowski, Miriam E Katz, Andrew H Knoll, Antonietta Quigg, John A Raven, Oscar Schofield, and FJR Taylor. The evolution of modern eukaryotic phytoplankton. *Science*, 305(5682):354–360, 2004.

[12.40] Maria Antonietta Ferrara, Principia Dardano, Luca De Stefano, Ilaria Rea, Giuseppe Coppola, Ivo Rendina, Roberta Congestri, Alessandra Antonucci, Mario De Stefano, and Edoardo De Tommasi. Optical properties of diatom nanostructured biosilica in *Arachnoidiscus* sp: micro-optics from Mother Nature. *PLoS One*, 9(7):e103750, 2014.

[12.41] Christopher B Field, Michael J Behrenfeld, James T Randerson, and Paul Falkowski. Primary production of the biosphere: integrating terrestrial and oceanic components. *Science*, 281(5374):237–240, 1998.

[12.42] S Flori, PH Jouneau, B Bailleul, *et al.* Plastid thylakoid architecture optimizes photosynthesis in diatoms. *Nat. Commun.*, 8:15885, 2017.

[12.43] Antonio Emidio Fortunato, Marianne Jaubert, Gen Enomoto, Jean-Pierre Bouly, Raffaella Raniello, Michael Thaler, Shruti Malviya, Juliana Silva Bernardes, Fabrice Rappaport, Bernard Gentili, *et al.* Diatom phytochromes reveal the existence of far-red-light-based sensing in the ocean. *Plant Cell*, 28 (3):616–628, 2016.

[12.44] Kenneth R Foster, Herman P Schwan, *et al.* Dielectric properties of tissues. In *CRC Handbook of Biological Effects of Electromagnetic Fields*, pages 27–96. CRC Press, Boca Raton, FL, 1986.

[12.45] T Fuhrmann, S Landwehr, M El Rharbi-Kucki, and M Sumper. Diatoms as living photonic crystals. *Applied Physics B*, 78(3):257–260, 2004.

[12.46] T Furukawa, M Watanabe, and I Shihira-Ishikawa. Green-and blue-light-mediated chloroplast migration in the centric Diatom *Pleurosira laevis*. *Protoplasma*, 203(3-4):214–220, 1998.

[12.47] David M Gates. Energy exchange in the biosphere. *Soil Science*, 96(1):76, 1963.

[12.48] S.D. Gedney. An anisotropic perfectly matched layer-absorbing medium for the truncation of FDTD lattices. *IEEE transactions on Antennas and Propagation*, 44(12):1630–1639, 1996.

[12.49] Andrius Gelzinis, Ramūnas Augulis, Claudia Büchel, Bruno Robert, and Leonas Valkunas. Confronting FCP structure with ultrafast spectroscopy data: evidence for structural variations. *Physical Chemistry Chemical Physics*, 23(2): 806–821, 2021.

[12.50] A Giaquinto, L Mescia, G Fornarelli, and F Prudenzano. Particle swarm optimization-based approach for accurate evaluation of upconversion parameters in Er3+ -doped fibers. *Optics Letters*, 36(2):142–144, 2011.

[12.51] Johan Gielis. *Inventing the circle: The geometry of nature*. Geniaal, 2003a.

[12.52] Johan Gielis. A generic geometric transformation that unifies a wide range of natural and abstract shapes. *American journal of botany*, 90(3):333–338, 2003b.

[12.53] Johan Gielis. *The geometrical beauty of plants*. Atlantis Press, Paris, 2017.

[12.54] Johan Gielis, Stefan Haesen, and Leopold Verstraelen. Universal natural shapes: From the super eggs of Piet Hein to the cosmic egg of Georges Lemaître. *Kragujevac Journal of Mathematics*, 28:57–68, 2005.

[12.55] MA Giraldo, S Yoshioka, C Liu, and DG Stavenga. Coloration mechanisms and phylogeny of morpho butterflies. *Journal of Experimental Biology*, 219(24): 3936–3944, 2016.

[12.56] Johannes W Goessling, William P Wardley, and Martin Lopez-Garcia. Highly reproducible, bio-based slab photonic crystals grown by diatoms. *Advanced Science*, 7(10):1903726, 2020.

[12.57] Reimund Goss and Bernard Lepetit. Biodiversity of NPQ. *Journal of Plant Physiology*, 172:13–32, 2015.

[12.58] G Ronald Hadley. Multistep method for wide-angle beam propagation. *Optics Letters*, 17(24):1743–1745, 1992a.

[12.59] G Ronald Hadley. Wide-angle beam propagation using Padé approximant operators. *Optics Letters*, 17(20):1426–1428, 1992b.

[12.60] Michelle S Hale and James G Mitchell. Functional morphology of diatom frustule microstructures: hydrodynamic control of brownian particle diffusion and advection. *Aquatic Microbial Ecology*, 24(3):287–295, 2001.

[12.61] Christian E Hamm, Rudolf Merkel, Olaf Springer, Piotr Jurkojc, Christian Maier, Kathrin Prechtel, and Victor Smetacek. Architecture and material properties of diatom shells provide effective mechanical protection. *Nature*, 421(6925):841–843, 2003.

[12.62] Christian Hill, Iouli E Gordon, Roman V Kochanov, Lorenzo Barrett, Jonas S Wilzewski, and Laurence S Rothman. Hitranonline: An online interface and the flexible representation of spectroscopic data in the hitran database. *Journal of quantitative spectroscopy and radiative transfer*, 177:4–14, 2016.

[12.63] AR Von Hippel. *Dielectrics and Waves*. Artech House, Norwood, MA, 1994.

[12.64] John David Jackson. *Classical electrodynamics*. Wiley, 3rd edition, 1999.

[12.65] Matthew Jacobs, Martin Lopez-Garcia, O-Phart Phrathep, Tracy Lawson, Ruth Oulton, and Heather M Whitney. Photonic multilayer structure of begonia chloroplasts enhances photosynthetic efficiency. *Nature Plants*, 2(11):1–6, 2016.

[12.66] John D Joannopoulos, Steven G Johnson, Joshua N Winn, and Robert D Meade. Molding the flow of light. *MIT Physics Annual*, pages 32–43, 2001.

[12.67] Villads Egede Johansen, Olimpia Domitilla Onelli, Lisa Maria Steiner, and Silvia Vignolini. Photonics in nature: from order to disorder. *Functional surfaces in biology III*, pages 53–89, 2017.

[12.68] Steven G Johnson and John D Joannopoulos. Block-iterative frequency-domain methods for Maxwell's equations in a planewave basis. *Optics Express*, 8(3): 173–190, 2001.

[12.69] Shuichi Kinoshita, Shinya Yoshioka, Yasuhiro Fujii, and Naoko Okamoto. Photophysics of structural color in the morpho butterflies. *Forma*, 17(2):103–121, 2002.

[12.70] John TO Kirk. *Light and photosynthesis in aquatic ecosystems*. Cambridge University Press, 1994.

[12.71] Charles Kittel, Paul McEuen, and Paul McEuen. *Introduction to solid state physics*, volume 8. Wiley New York, 1996.

[12.72] Miyuki Koiso and Bennett Palmer. Equilibria for anisotropic surface energies and the Gielis formula. *Forma*, 23(1):1–8, 2008.

[12.73] Sarah König, Matthias Juhas, Stefanie Jäger, Tilman Kottke, and Claudia Büchel. The cryptochrome—photolyase protein family in diatoms. *Journal of Plant Physiology*, 217:15–19, 2017.

[12.74] Sasha J Kramer, Collin S Roesler, and Heidi M Sosik. Bio-optical discrimination of diatoms from other phytoplankton in the surface ocean: Evaluation and refinement of a model for the northwest atlantic. *Remote Sensing of Environment*, 217:126–143, 2018.

[12.75] Paulina Kuczynska, Malgorzata Jemiola-Rzeminska, and Kazimierz Strzalka. Photosynthetic pigments in diatoms. *Marine drugs*, 13(9):5847–5881, 2015.

[12.76] Francesca Romana Lamastra, Roberta De Angelis, Alessandra Antonucci, Damiano Salvatori, Paolo Prosposito, Mauro Casalboni, Roberta Congestri, Sonia Melino, and Francesca Nanni. Polymer composite random lasers based on diatom frustules as scatterers. *RSC Advances*, 4(106):61809–61816, 2014.

[12.77] Karine Leblanc, Bernard Queguiner, Frederic Diaz, Veronique Cornet, Mónica Michel-Rodriguez, Xavier Durrieu de Madron, Chris Bowler, Shruti Malviya, Melilotus Thyssen, Gérald Grégori, *et al.* Nanoplanktonic diatoms are globally overlooked but play a role in spring blooms and carbon export. *Nature Communications*, 9(1):1–12, 2018.

[12.78] Bernard Lepetit, Sabine Sturm, Alessandra Rogato, Ansgar Gruber, Matthias Sachse, Angela Falciatore, Peter G Kroth, and Johann Lavaud. High light acclimation in the secondary plastids containing diatom *Phaeodactylum tricornutum* is triggered by the redox state of the plastoquinone pool. *Plant Physiology*, 161(2):853–865, 2013.

[12.79] Orly Levitan, Muyuan Chen, Xuyuan Kuang, Kuan Yu Cheong, Jennifer Jiang, Melissa Banal, Nikhita Nambiar, Maxim Y Gorbunov, Steven J Ludtke, Paul G Falkowski, *et al.* Structural and functional analyses of photosystem II in the marine diatom *Phaeodactylum tricornutum*. *Proceedings of the National Academy of Sciences*, 116(35):17316–17322, 2019.

[12.80] Wenkai Li, Weidong Guo, Yongkang Xue, Congbin Fu, and Bo Qiu. Sensitivity of a regional climate model to land surface parameterization schemes for east asian summer monsoon simulation. *Climate Dynamics*, 47(7):2293–2308, 2016.

[12.81] Hanzhi Lin, Fedor I Kuzminov, Jisoo Park, SangHoon Lee, Paul G Falkowski, and Maxim Y Gorbunov. The fate of photons absorbed by phytoplankton in the global ocean. *Science*, 351(6270):264–267, 2016.

[12.82] Christian Maibohm, Søren Michael Mørk Friis, Marianne Ellegaard, and Karsten Rottwitt. Interference patterns and extinction ratio of the diatom *Coscinodiscus granii*. *Optics Express*, 23(7):9543–9548, 2015.

[12.83] Stefano Managò, Gianluigi Zito, Alessandra Rogato, Maurizio Casalino, Emanuela Esposito, Anna Chiara De Luca, and Edoardo De Tommasi. Bioderived three-dimensional hierarchical nanostructures as efficient surface enhanced raman scattering substrates for cell membrane probing. *ACS applied materials & interfaces*, 10(15):12406–12416, 2018.

[12.84] Marcus Mann, Manuel Serif, Thomas Wrobel, Marion Eisenhut, Shvaita Madhuri, Samantha Flachbart, Andreas PM Weber, Bernard Lepetit, Christian Wilhelm, and Peter G Kroth. The aureochrome photoreceptor PtAUREO1a is a highly effective blue light switch in diatoms. *iScience*, 23(11):101730, 2020.

[12.85] Luciano Mescia, Pietro Bia, and Diego Caratelli. Fractional derivative based FDTD modeling of transient wave propagation in Havriliak–Negami media. *IEEE Transactions on Microwave Theory and Techniques*, 62(9):1920–1929, 2014.

[12.86] Kathleen J Meyers-Schulte and John I Hedges. Molecular evidence for a terrestrial component of organic matter dissolved in ocean water. *Nature*, 321 (6065):61–63, 1986.

[12.87] Curtis D Mobley. A numerical model for the computation of radiance distributions in natural waters with wind-roughened surfaces, part 2: User's guide and code listing. Technical report, Washington Univ Seattle Joint Inst for Study of Atmosphere and Ocean, 1988.

[12.88] Curtis D Mobley. A numerical model for the computation of radiance distributions in natural waters with wind-roughened surfaces. *Limnology and oceanography*, 34(8):1473–1483, 1989.

[12.89] Curtis D Mobley, Bernard Gentili, Howard R Gordon, Zhonghai Jin, George W Kattawar, Andre Morel, Phillip Reinersman, Knut Stamnes, and Robert H Stavn. Comparison of numerical models for computing underwater light fields. *Applied Optics*, 32(36):7484–7504, 1993.

[12.90] Curtis D Mobley, Lydia K Sundman, and Emmanuel Boss. Phase function effects on oceanic light fields. *Applied optics*, 41(6):1035–1050, 2002.

[12.91] Edwige Moyroud, Tobias Wenzel, Rox Middleton, Paula J Rudall, Hannah Banks, Alison Reed, Greg Mellers, Patrick Killoran, M Murphy Westwood, Ullrich Steiner, *et al.* Disorder in convergent floral nanostructures enhances signalling to bees. *Nature*, 550(7677):469–474, 2017.

[12.92] Tomomi Nonoyama, Elena Kazamia, Hermanus Nawaly, Xia Gao, Yoshinori Tsuji, Yusuke Matsuda, Chris Bowler, Tsuyoshi Tanaka, and Richard G Dorrell. Metabolic innovations underpinning the origin and diversification of the diatom chloroplast. *Biomolecules*, 9(8):322, 2019.

[12.93] N Ogura and T Hanya. Nature of ultra-violet absorption of sea water. *Nature*, 212(5063):758–758, 1966.

[12.94] AR Parker. Natural photonic structures: an overview. *Optical Biomimetics*, pages 1–20e, 2012.

[12.95] Rudolph W Preisendorfer. Eigenmatrix representations of radiance distributions in natural waters with wind-roughened surfaces. Technical report, Washing- ton Univ Seattle Joint Inst for Study of Atmosphere and Ocean, 1988.

[12.96] Rudolph W Preisendorfer and Curtis D Mobley. Albedos and glitter patterns of a wind-roughened sea surface. *Journal of Physical Oceanography*, 16(7): 1293–1316, 1986.

[12.97] Alina Pushkarev, Keiichi Inoue, Shirley Larom, José Flores-Uribe, Manish Singh, Masae Konno, Sahoko Tomida, Shota Ito, Ryoko Nakamura, Satoshi P Tsunoda, *et al.* A distinct abundant group of microbial rhodopsins discovered using functional metagenomics. *Nature*, 558(7711):595–599, 2018.

[12.98] Bo Qiu, Yongkang Xue, Joshua B Fisher, Weidong Guo, Joseph A Berry, and Yongguang Zhang. Satellite chlorophyll fluorescence and soil moisture observations lead to advances in the predictive understanding of global terrestrial coupled carbon-water cycles. *Global Biogeochemical Cycles*, 32(3):360–375, 2018.

[12.99] Min Qiu. Effective index method for heterostructure-slab-waveguide-based two-dimensional photonic crystals. *Applied physics letters*, 81(7):1163–1165, 2002.

[12.100] JA Raven and CS Cockell. Influence on photosynthesis of starlight, moonlight, planetlight, and light pollution (reflections on photosynthetically active radiation in the universe). *Astrobiology*, 6(4):668–675, 2006.

[12.101] Alessandra Rogato and Edoardo De Tommasi. Physical, chemical, and genetic techniques for diatom frustule modification: Applications in nanotechnology. *Applied Sciences*, 10(23):8738, 2020.

[12.102] Alessandra Rogato, Alberto Amato, Daniele Iudicone, Maurizio Chiurazzi, Maria Immacolata Ferrante, and Maurizio Ribera d'Alcalà. The diatom molecular toolkit to handle nitrogen uptake. *Marine genomics*, 24:95–108, 2015.

[12.103] J Romann, J-C Valmalette, A Røyset, and M-A Einarsrud. Optical properties of single diatom frustules revealed by confocal microspectroscopy. *Optics letters*, 40(5):740–743, 2015.

[12.104] Frank Eric Round, Richard M Crawford, and David G Mann. *Diatoms: biology and morphology of the genera*. Cambridge university press, 1990.

[12.105] S Sathyendranath, G Cota, V Stuart, H Maass, and T Platt. Remote sensing of phytoplankton pigments: a comparison of empirical and theoretical approaches. *International Journal of Remote Sensing*, 22(2-3):249–273, 2001.

[12.106] Shubha Sathyendranath, Louisa Watts, Emmanuel Devred, Trevor Platt, Carla Caverhill, and Heidi Maass. Discrimination of diatoms from other phytoplankton using ocean-colour data. *Marine ecology progress series*, 272:59–68, 2004.

[12.107] Alexander F Schober, Carolina Río Bártulos, Annsophie Bischoff, Bernard Lepetit, Ansgar Gruber, and Peter G Kroth. Organelle studies and proteome analyses of mitochondria and plastids fractions from the diatom *Thalassiosira pseudonana*. *Plant and Cell Physiology*, 60(8):1811–1828, 2019.

[12.108] Joseph Seckbach and Patrick Kociolek. *The diatom world*, volume 19. Springer Science & Business Media, 2011.

[12.109] Maximilian Seeger. Digitale narrative in mikrokosmos - jewels of the sea. Hochschule für angewandte Wissenschaften W Schweinfurt, Germany, 7 2021. BSc. Thesis.

[12.110] Ikuko Shihira-Ishikawa, Takanori Nakamura, Sho-ichi Higashi, and Masakatsu Watanabe. Distinct responses of chloroplasts to blue and green laser microbeam irradiations in the centric diatom *Pleurosira laevis*. *Photochemistry and photobiology*, 83(5):1101–1109, 2007.

[12.111] John P Smol and Eugene F Stoermer. *The diatoms: applications for the environmental and earth sciences*. Cambridge University Press, 2010.

[12.112] V Stuart, S Sathyendranath, EJH Head, T Platt, B Irwin, and H Maass. Bio-optical characteristics of diatom and prymnesiophyte populations in the labrador sea. *Marine Ecology Progress Series*, 201:91–106, 2000.

[12.113] Yanyan Su, Nina Lundholm, Søren MM Friis, and Marianne Ellegaard. Implications for photonic applications of diatom growth and frustule nanostructure changes in response to different light wavelengths. *Nano Research*, 8(7):2363–2372, 2015.

[12.114] Jiyu Sun, Bharat Bhushan, and Jin Tong. Structural coloration in nature. *RSC Advances*, 3(35):14862–14889, 2013.

[12.115] Lucilla Taddei, Giulio Rocco Stella, Alessandra Rogato, Benjamin Bailleul, Antonio Emidio Fortunato, Rossella Annunziata, Remo Sanges, Michael Thaler, Bernard Lepetit, Johann Lavaud, *et al.* Multisignal control of expression of the LHCX protein family in the marine diatom *Phaeodactylum tricornutum*. *Journal of Experimental Botany*, 67(13):3939–3951, 2016.

[12.116] Allen Taflove and Susan C Hagness. Computational electromagnetics: the finite-difference time-domain method. *Artech House*, 3, 2000.

[12.117] Allen Taflove, Susan C. Hagness, and Melinda Piket-May. Computational electromagnetics: The finite-difference time-domain method. In Wai-Kai Chen, editor, *The Electrical Engineering Handbook*, pages 629–670. Elsevier, 2005.

[12.118] P Vukusic, JR Sambles, CR Lawrence, and RJ Wootton. Quantified interference and diffraction in single morpho butterfly scales. *Proceedings of the Royal Society of London. Series B: Biological Sciences*, 266(1427):1403–1411, 1999.

[12.119] Pete Vukusic and J Roy Sambles. Photonic structures in biology. *Nature*, 424 (6950):852–855, 2003.

[12.120] Anya Waite, Anne Fisher, Peter A Thompson, and Paul J Harrison. Sinking rate versus cell volume relationships illuminate sinking rate control mechanisms in marine diatoms. *Marine Ecology Progress Series*, 157:97–108, 1997.

[12.121] Wenda Wang, Songhao Zhao, Xiong Pi, Tingyun Kuang, Sen-Fang Sui, and Jian-Ren Shen. Structural features of the diatom Photosystem II–light-harvesting antenna complex. *FEBS journal*, 287(11):2191–2200, 2020.

[12.122] Charles H Whitlock, David S Bartlett, and Ernest A Gurganus. Sea foam reflectance and influence on optimum wavelength for remote sensing of ocean aerosols. *Geophysical Research Letters*, 9(6):719–722, 1982.

[12.123] Heather M Whitney, Mathias Kolle, Piers Andrew, Lars Chittka, Ullrich Steiner, and Beverley J Glover. Floral iridescence, produced by diffractive optics, acts as a cue for animal pollinators. *Science*, 323(5910):130–133, 2009.

[12.124] Shigeru Yamanaka, Rei Yano, Hisanao Usami, Nobuaki Hayashida, Masakatsu Ohguchi, Hiroyuki Takeda, and Katsumi Yoshino. Optical properties of diatom silica frustule with special reference to blue light. *Journal of Applied Physics*, 103(7):074701, 2008.

13

A Generalized Model for the Light Response of the Nonphotochemical Quenching of Chlorophyll Fluorescence of Diatoms

João Serôdio[1]* and Johann Lavaud[2]

[1]*Department of Biology, Centre for Environmental and Marine Studies (CESAM), University of Aveiro, Aveiro, Portugal*
[2]*Laboratoire des Sciences de l'Environnement Marin – LEMAR (UMR 6539), CNRS/Univ Brest/ Ifremer/IRD, Institut Européen de la Mer-IUEM, Plouzané, France*

Abstract
As photoautotrophs, diatoms have developed photoprotective mechanisms to deal with excessive ambient light and to minimize photodamage to the photosynthetic apparatus. Most of these mechanisms rely on energy-dissipation processes, collectively named non-photochemical quenching of fluorescence (NPQ). The functioning of these photoprotective processes is commonly characterized by generating light-response curves of NPQ ("NPQ light curves"), which provide a detailed and ecologically relevant description of how the energy dissipating pathways are activated or relaxed as ambient light intensity varies. The light curves (LCs) are constructed by evaluating NPQ at different light intensities, by measuring the emission of chlorophyll *a* fluorescence *in vivo*, typically using a noninvasive technique called pulse amplitude modulation (PAM) fluorometry. NPQ LCs have been quantitatively described by the three-parameter mathematical model of Serôdio and Lavaud (2011), which adequately describes the monotonic rise of NPQ with light intensity, observed in many conditions and for a wide range of taxa. However, in some diatoms (also in other groups of algae, although less frequently), the NPQ LC is not always monotonic but shows an initial decrease, reaching a minimum under low-intermediate light intensities, before increasing under high light, a pattern not describable by the existing model. This work presents a new model, which generalizes the previous one of Serôdio and Lavaud and permits the description of the type of low light-response pattern observed in some diatom species and growth conditions. The study further illustrates the general applicability of the model by fitting it to a wide range of data sets measured on diatoms, both grown in unialgal cultures or as part of natural communities.

***Keywords*:** Chlorophyll fluorescence, light curves, photoprotection, photoinhibition, xanthophyll cycle

**Corresponding author*: jserodio@ua.pt
João Serôdio: www.cesam.ua.pt/joaoserodio, https://orcid.org/0000-0002-9047-1740, https://www.researchgate.net/profile/Joao-Serodio
Johann Lavaud: https://www.researchgate.net/profile/Johann-Lavaud, https://scholar.google.fr/citations?user=VJojlMoAAAAJ&hl=fr, https://www-iuem.univ-brest.fr/lemar/equipe/lavaud_johann/

Janice L. Pappas (ed.) The Mathematical Biology of Diatoms, (393–412) © 2023 Scrivener Publishing LLC

13.1 Introduction

One of the main reasons for the interest in diatoms is their enormous importance as primary producers. Through their photosynthesis, diatoms contribute up to 40% of marine primary productivity and to a massive 20-25% of the Earth's global photosynthetic carbon fixation and oxygen released into the atmosphere [13.1]–[13.3]. Being the dominant group of microalgae in many oceanic, coastal, estuarine and freshwater ecosystems, the photosynthetic activity of diatoms represents a key source of carbon for trophic webs that support a wide range of animal populations, from small crustacean to fishes, seabirds and marine mammals [13.4].

The high productivity attained by diatoms has been attributed, at least partially, to their capacity to effectively cope with the systematic exposure to high and fluctuating levels of solar radiation, a potential source of photoinhibitory damage to the photosynthetic apparatus, and of limitation of carbon fixation and growth [13.5] [13.6]. The reduction of the risk of photoinhibition is achieved by the operation of photoprotective energy-dissipating pathways, known to be particularly efficient in diatoms [13.7]–[13.9]. The various existing energy-dissipation processes are collectively grouped and named as non-photochemical quenching of fluorescence (NPQ). The main photoprotective processes contributing to NPQ rely on the regulated thermal dissipation of absorbed light energy (q_E, "energy-dependent quenching") [13.10] [13.11], that operates in the light-harvesting antenna of the photosystem II (PSII) and involves specific proteins (Lhcx in the case of diatoms [13.12]–[13.14]) and, under high light, the formation of a transthylakoid proton gradient (ΔpH), and the subsequent triggering of the xanthophyll cycle (XC) [13.15]. In diatoms, the XC consists in the enzyme-mediated, reversible conversion between the pigments diadinoxanthin and diatoxanthin, promoting the dissipation of absorbed light energy as heat and the reduction of excitation pressure on PSII [13.9] [13.16] [13.17]. Under certain conditions, there is a close relationship between NPQ and the XC, demonstrated by the observation of a linear relationship between NPQ and diatoxanthin concentration [13.17] [13.18].

Due to its importance as a key set of mechanisms used by diatoms to cope with environmental stress, the study of NPQ has attracted considerable research interest for a long time [13.9] [13.10] [13.19]–[13.23]. Typically, NPQ is studied quantitatively using variable *in-vivo* chlorophyll fluorescence, often through pulse amplitude modulation (PAM) fluorometry [13.9] [13.10] [13.24]. One of the main forms of studying the operation of NPQ, and of related XC, is through the generation of light-response curves (LCs), by measuring chlorophyll fluorescence indices under various (often sequentially increasing) actinic light intensities [13.25]–[13.28]. These NPQ LCs are a valuable and often used tool employed to characterize how the energy dissipating pathways are activated or relaxed as a response to changes in ambient light [13.29]. NPQ LCs can be quantitatively described by a three-parameter function, shown to adequately trace the monotonic rise of NPQ with light intensity, under a wide range of conditions and for a large range of taxa [13.30].

However, in some diatom species and diatom-dominated communities, as well as in some other algal groups (i.e., for instance brown algae [13.31]), it has been observed that NPQ is not only formed under high light, during the generation of the LC, but may be present at the start of a LC, formed during prolonged darkness [13.9] [13.26] [13.28] [13.32]–[13.34]. When NPQ has been formed in the dark in a sustained manner, the resulting NPQ

LCs are no longer monotonic but rather show a distinctive bi-phasic pattern, with an initial phase of NPQ decrease from initial high values measured in the dark, a minimum reached under low-intermediate light levels, and the generally observed increase under high light (e.g., [13.18] [13.35] [13.36]).

This biphasic pattern is not describable by Serôdio and Lavaud's 2011 model and, even though it has been observed in numerous studies, no mathematical description has yet been proposed. The present study proposes a generalized model to describe NPQ LCs that includes dark-induced NPQ. It further aims at extending the use of the proposed model to an alternative NPQ index increasingly used in the literature (Y(NPQ) (see below), and for which LC models have not been explicitly tested. The applicability of the proposed model is tested by fitting it to a wide range of NPQ LC data sets measured on diatoms, published or original, either grown in unialgal cultures or as part of natural communities.

13.2 Model Formulation

13.2.1 Nonphotochemical Quenching Indices NPQ and Y(NPQ)

The nonphotochemical quenching of chlorophyll fluorescence is most commonly quantified by the Stern-Volmer NPQ index, based on the relative difference between fluorescence parameters F_m and F'_m, the maximum fluorescence measured on dark-adapted and on illuminated samples, respectively (see Table 13.1 for notation) [13.20] [13.37]:

$$\mathrm{NPQ} = \frac{F_m - F'_m}{F'_m} \tag{13.1}$$

This index is frequently used as a measure of the operation of photoprotective energy-dissipation processes which, in the case of diatoms, are mainly based on the diadinoxanthin-diatoxanthin XC [13.9] [13.17], although it also englobes the effects of nonreversible photoinhibitory damage to the photosynthetic apparatus [13.9] [13.10]. The NPQ index was originally formulated in the context of the study of plants, for which always $F_m > F'_m$, ensuring NPQ > 0 in all cases. However, it was later observed in diatoms and in other algal groups that, when exposing dark-adapted samples to low light, the values of F'_m may increase above the "maximum" F_m value measured in dark conditions. In these cases, to avoid the obtention of negative values due to this "dark NPQ," the following formulation of the NPQ index was proposed [13.38]:

$$\mathrm{NPQ} = \frac{F'_{m,m} - F'_m}{F'_m} \tag{13.2}$$

where F'_{mm} represents the maximum F'_m value, higher than F_m, that is reached under low irradiance.

More recently, an alternative formulation for the reversible component of NPQ was proposed [13.39]:

Table 13.1 Notations.

E = PAR irradiance (μmol m^{-2} s^{-1})
ETR = Electron transport rate in PSII
E_{50} = Irradiance level corresponding to 50% of NPQ$_m$ in a NPQ vs. E curve
E_{NPQmin} = Irradiance level at which NPQ reaches a minimum in a NPQ vs E curve
F_o, F_m = Minimum and maximum fluorescence of a dark-adapted sample
F_s, F'_m = Steady state and maximum fluorescence of a light-adapted sample
F'_{mm} = Maximum F'_m value, higher than F_m, reached under low irradiance in a NPQ vs E curve
k_D = Coefficient of NPQ$_{Dark}$ dissipation under low irradiance in a NPQ vs E curve
NPQ = Non-photochemical quenching; Stern-Volmer coefficient
NPQ$_0$ = Maximum NPQ value reached in a NPQ vs E curve
NPQ$_{Dark}$, NPQ$_{HL}$ = Dark- and high light-induced components of NPQ
NPQ$_m$ = Maximum NPQ value reached in a NPQ vs E curve
n = Sigmoidicity coefficient of the NPQ vs E curve
PAR = Photosynthetically Active Radiation
PSII = Photosystem II
q_E = Energy-dependent nonphotochemical quenching
XC = Xanthophyll cycle
Y(NPQ) = Regulated thermal energy dissipation related to NPQ
Zx = Zeaxanthin

$$Y(NPQ) = \frac{F}{F'_m} - \frac{F}{F_m} \tag{13.3}$$

where F is the steady-state fluorescence, measured under ambient light. This formulation has been increasingly used in recent years [13.13] [13.40]–[13.44], as it refers more exclusively to reversible thermal dissipatory mechanisms, having the additional operational advantage of yielding results bounded between 0 and 1, while NPQ (Eq. 13.1) has no upper limit. Following the rationale behind Eq. (13.2), when $F_m < F'_m$ this equation may be adapted by replacing F_m by F'_{mm}.

13.2.2 Standard Model for NPQ LCs

The induction of NPQ under high light (hereafter called NPQ_{HL}) in the course of the generation of a light-response curve, can be modeled by a simple function, adapted from the Hill equation, originally proposed to describe the cooperativity of enzymatic reactions [13.30]:

$$NPQ_{HL}(E) = \mathrm{NPQ}_m \frac{E^n}{E_{50}^n + E^n} \tag{13.4}$$

where NPQ_m is the maximum NPQ value, E_{50} is the irradiance level corresponding to 50% of NPQ_m, and n characterizes the sigmoidicity of the curve. This model is valid under conditions when NPQ is dominated by the fluorescence quenching associated with the build-up of the transthylakoid proton gradient (ΔpH), or energy-dependent quenching (q_E), and to the operation of the XC. The proposal of this model was supported by the excellent fit to experimental data observed for a large range of taxa and physiological conditions, and since its introduction it has been applied in numerous studies (e.g., [13.18] [13.41] [13.45]–[13.49]), becoming the standard model for describing light-response curves of NPQ.

13.2.3 Generalized Model for NPQ LCs

The model here proposed is based on the idea that the NPQ measured under a certain light level during the generation of a LC is caused by the simultaneous and addictive influence of two types of processes: i) the well-known high light induction of NPQ (NPQ_{HL}), based on the build-up of a ΔpH caused by the accumulation of protons in the lumen of the thylakoid due to photosynthetic activity, inducing a monotonic rise in NPQ as irradiance increases, and well described by Eq. (13.4); ii) the less known dark-induced NPQ (NPQ_{Dark}), formed during prolonged darkness, and dissipated under low light, resulting in a decrease of NPQ values during the first steps of the curve. We thus propose a simple formulation consisting of the sum of the two components NPQ_{Dark} and NPQ_{HL}, and describing the decrease of NPQ_{Dark} with irradiance E as a first order negative exponential function:

$$\mathrm{NPQ}(E) = NPQ_{Dark} + NPQ_{HL} = \mathrm{NPQ}_0 e^{-k_D E} + \mathrm{NPQ}_m \frac{E^n}{E_{50}^n + E^n} - C \tag{13.5}$$

where NPQ_0 represents the NPQ level measured in the dark ($E = 0$), k_D is the coefficient of NPQ dissipation upon exposure to increasing light levels during initial stages of the light curve, and C is a factor to allow the light curve to reach the minimum value of NPQ = 0. As discussed in detail below, in most cases, $C \sim 0$, but in the case of $C > 0$, then NPQ_0 and NPQ_m must be corrected by subtracting the value of C (Figure 13.1). This model is a generalization of the standard model (Eq. 13.4) since it can also be used to describe LCs without the presence of NPQ_{Dark}, because in that case $NPQ_0 = 0$ and Eq. (13.5) readily collapses into Eq. (13.4).

As also depicted in Figure 13.1, fitting Eq. (13.5) to experimental data further allows the estimation of an emerging parameter, E_{NPQmin}, that describes the light intensity for which NPQ is minimum. Although not supported by a mathematical model, this parameter was

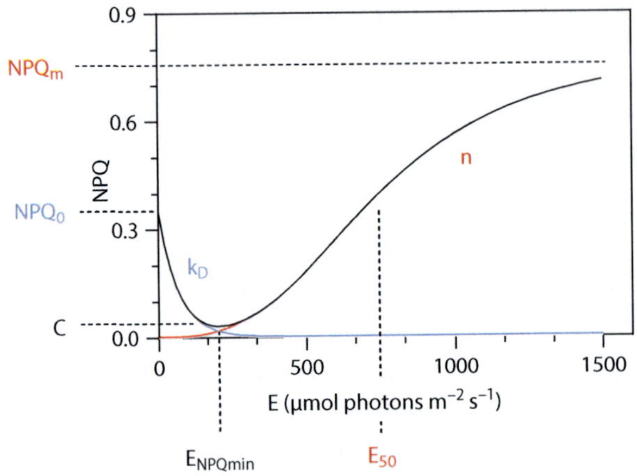

Figure 13.1 Schematic description of the model proposed to describe light curves of chlorophyll fluorescence NPQ index (Eq. 13.6) and graphical depiction of its parameters. Lines and letters in blue and red refer to the functions describing relaxation of dark-induced NPQ (NPQ_{Dark}; Eq. 13.5) and build-up of high light-induced NPQ (NPQ_{HL}, Eq. 13.4), respectively. Black line represents the sum of the two components NPQ_{Dark} and NPQ_{HL} following Eq. 13.6. For clarity, the value of C was exaggerated relative to the values typically obtained from fitting the model to experimental data. The figure and the parameters refer to NPQ but are readily applicable to light curves of Y(NPQ).

introduced before [13.36]. We did not search for a mathematical expression for the determination of this parameter, and it was estimated numerically (see below).

Eq. (13.5) expresses the model in terms of the index NPQ. However, the model was also tested as a representation of light-response curves of index Y(NPQ). For simplicity, we did not define a duplicated set of parameters for Y(NPQ) vs. E curves but relied on the assumption that the discussion of the results can be understood under the context of the type of LC being analyzed.

13.2.4 Model Fitting and Parameter Estimation

The model of Eq. (13.5) was fitted to NPQ vs. E curves measured on diatoms, both grown in unialgal cultures or as part of natural communities. Published as well as original, yet unpublished, data sets were used, as summarized in Table 13.2. When not available, LC data points were estimated from the published figures using WebPlotDigitizer software (http://automeris.io/WebPlotDigitizer). The light-response curve models were fitted using a procedure written in Microsoft Visual Basic and based on Microsoft Excel Solver as described in [13.30]. Parameters of the model were estimated iteratively by minimizing a least-squares function, forward differencing, and the default quasi-Newton search method. E_{NPQmin} was also estimated using MS Excel Solver by searching the value of E that minimizes NPQ. Because of the emphasis of this study on the NPQ_{Dark} component of NPQ LCs, only data sets presenting a minimum of 3 data points measured for $E < E_{NPQmin}$ were considered. Punctually, LCs not showing (or not substantial) NPQ_{Dark} were used when useful for establishing comparisons with light curves of the same dataset showing important NPQ_{Dark}.

Table 13.2 Results of fitting of Eq. (13.5) to NPQ vs. *E* and Y(NPQ) vs. *E* curves measured in diatoms or diatom-dominated communities. Growth conditions/Treatment: sampling dates, conditions applied during growth or before the measurement of the light curves, when more than one LC available for the same taxon (HL: high growth light; LL: low growth light). LC: light curve protocol (LCx: steady state LC; RLCx: rapid light curve, x seconds of light step; UP, Down: sequential increase or decrease light levels, respectively).

			Model parameters NPQ vs *E* curve Model parameters Y(NPQ) vs *E* curve								
Taxon/sample	**Growth conditions/ treatment**	**LC**	**NPQ_m**	**E_{50}**	**n**	**NPQ_0**	**k_D**	**C**	**E_{NPQmin}**	**r^2**	**Reference**
Phaeodactylum tricornutum		LC300-600	9.58 0.86	3524 1538	2.01 2.14	0.29 0.07	0.028 0.024	0.03 0.01	131 134	0.997 0.996	[13.35]
MPB	July	LC90	2.96 0.55	1494 917	2.33 3.00	0.36 0.10	0.014 0.015	0.03 0.01	200 209	0.998 0.994	[13.38]
	November		5.78 0.68	1295 792	2.52 2.89	2.67 0.32	0.024 0.016	0.08 0.01	194 135	0.989 0.989	
MPB	July	LC90	3.28 0.63	2285 1548	2.80 3.02	0.20 0.06	0.018 0.015	0.01 0.00	227 236	0.999 0.999	[13.36]
			2.25 0.58	1705 1287	2.40 2.50	0.12 0.04	0.014 0.013	0.02 0.01	181 168	0.996 0.997	
			2.93 0.64	1167 827	2.34 2.46	0.24 0.07	0.030 0.028	0.02 0.01	111 108	0.999 0.998	
	November		1.99 0.59	1802 1262	2.71 3.03	1.46 0.24	0.033 0.015	0.00 0.01	200 268	0.992 0.990	
			3.10 0.59	2497 1423	2.78 3.38	1.60 0.27	0.026 0.010	0.00 0.01	253 378	0.973 0.982	

(Continued)

Table 13.2 Results of fitting of Eq. (13.5) to NPQ vs. *E* and Y(NPQ) vs. *E* curves measured in diatoms or diatom-dominated communities. Growth conditions/Treatment: sampling dates, conditions applied during growth or before the measurement of the light curves, when more than one LC available for the same taxon (HL: high growth light; LL: low growth light). LC: light curve protocol (SSx: steady state LL; RLCx: rapid light curve, x seconds of light step; UP, Down: sequential increase or decrease light levels, respectively). (*Continued*)

			Model parameters NPQ vs *E* curve Model parameters Y(NPQ) vs *E* curve								
Taxon/sample	**Growth conditions/ treatment**	**LC**	NPQ_m	E_{50}	**n**	NPQ_0	k_D	**C**	E_{NPQmin}	r^2	**Reference**
			1.61 0.55	1087 777	2.23 2.69	0.70 0.19	0.026 0.018	0.02 0.02	160 175	0.993 0.990	
MPB	Control	LC90	6.13 0.70	2483 1052	1.93 2.50	0.27 0.06	0.014 0.018	0.06 0.01	166 142	1.000 0.999	[13.60]
	LatA		2.48 0.65	861 698	2.49 2.56	0.04 0.01	0.090 0.085	0.00 0.00	43 43	0.999 0.999	
MPB	15 July	RLC10	0.69 -	810 -	5.21 -	1.71 -	0.007 -	0.07 -	615 -	0.933 -	[13.28]
		RLC30	0.53 -	872 -	5.44 -	3.70 -	0.020 -	0.01 -	357 -	1.000 -	
	16 July	RLC10	1.28 -	942 -	3.16 -	2.29 -	0.002 -	0.98 -	696 -	0.999 -	
		RLC30	0.32 -	742 -	20.81 -	3.35 -	0.018 -	0.00 -	515 -	0.999 -	

(*Continued*)

Table 13.2 Results of fitting of Eq. (13.5) to NPQ vs. *E* and Y(NPQ) vs. *E* curves measured in diatoms or diatom-dominated communities. Growth conditions/Treatment: sampling dates, conditions applied during growth or before the measurement of the light curves, when more than one LC available for the same taxon (HL: high growth light; LL: low growth light). LC: light curve protocol (SSx: steady state LL; RLCx: rapid light curve, x seconds of light step; UP, Down: sequential increase or decrease light levels, respectively). (*Continued*)

			Model parameters NPQ vs *E* curve **Model parameters Y(NPQ) vs *E* curve**								
Taxon/sample	**Growth conditions/ treatment**	**LC**	**NPQ_m**	**E_{50}**	**n**	**NPQ_0**	**k_D**	**C**	**E_{NPQmin}**	**r^2**	**Reference**
	17 July	RLC10	1.14 -	1100 -	25.93 -	1.50 -	0.003 -	0.19 -	927 -	0.985 -	
		RLC30	14.47 -	7247 -	1.99 -	2.68 -	0.015 -	0.05 -	362 -	0.999 -	
	18 July	RLC10	4.54 -	1250 -	18.38 -	1.30 -	0.005 -	0.00 -	857 -	0.993 -	
		RLC30	0.47 -	720 -	5.83 -	2.56 -	0.012 -	0.00 -	404 -	0.988 -	
Navicula sp.	HL	LC120	2.76 0.70	1061 628	2.31 2.66	0.31 0.13	0.087 0.078	0.00 0.00	60 64	0.998 0.998	J. Serôdio, unpublished
			2.11 0.63	1057 0.71	2.86 3.34	0.54 0.19	0.096 0.076	0.00 0.00	75 86	0.998 0.998	
			1.62 0.57	1177 760	2.31 2.68	0.52 0.21	0.079 0.056	0.00 0.00	78 94	0.997 0.997	
	LL		0.96 0.49	748 615	2.64 2.64	0.05 0.02	0.785 0.788	0.00 0.00	0 0	0.997 0.997	

(*Continued*)

Table 13.2 Results of fitting of Eq. (13.5) to NPQ vs. *E* and Y(NPQ) vs. *E* curves measured in diatoms or diatom-dominated communities. Growth conditions/Treatment: sampling dates, conditions applied during growth or before the measurement of the light curves, when more than one LC available for the same taxon (HL: high growth light; LL: low growth light). LC: light curve protocol (SSx: steady state LL; RLCx: rapid light curve, x seconds of light step; UP, Down: sequential increase or decrease light levels, respectively). (*Continued*)

			Model parameters NPQ vs *E* curve Model parameters Y(NPQ) vs *E* curve								
Taxon/sample	**Growth conditions/ treatment**	**LC**	**NPQ_m**	**E_{50}**	**n**	**NPQ_0**	**k_D**	**C**	**E_{NPQmin}**	**r^2**	**Reference**
			0.83 0.43	884 647	2.05 2.35	0.09 0.04	0.147 0.134	0.00 0.00	50 77	0.996 0.997	
			0.78 0.42	718 0.55	2.18 2.49	0.05 0.02	0.271 0.253	0.00 0.00	19 41	0.998 0.998	
Phaeodactylum tricornutum		LC120	3.54 0.68	1037 667	2.41 2.45	0.18 0.06	0.062 0.054	0.01 0.00	67 68	1.000 1.000	J. Serôdio, unpublished
			3.13 0.66	1090 741	2.72 2.80	0.33 0.11	0.074 0.059	0.00 0.00	77 83	0.999 0.999	
Thalassiosira weissflogii		LC120, Up	0.85 0.50	1375 1173	3.30 3.15	0.66 0.39	0.009 0.006	0.04 0.05	434 443	0.992 0.988	J. Serôdio, unpublished
		LC120, Down	1.89 0.66	1153 778	2.14 2.15	0.06 0.04	0.039 0.036	0.00 0.00	72 72	1.000 1.000	
Fragilariopsis cylindrus		RLCs 30s	0.96 0.47	161 126	1.95 2.08	0.13 0.08	0.153 0.145	0.02 0.01	18 18	0.997 0.995	J. Lavaud, unpublished
			0.51 0.35	107 97	2.59 2.42	0.05 0.04	0.081 0.122	0.02 0.01	20 17	0.963 0.954	
			0.43 0.30	171 154	2.02 2.06	0.04 0.03	0.189 0.182	0.01 0.00	15 15	0.997 0.997	

13.3 Results

The fitting of the model to experimental data revealed in all cases an excellent capacity to trace and describe the main features of the LCs. Considering all tested LCs, the fitting of the model explained on average 99.3% of the data variability, the minimum value being 95.4% (Table 13.2). Although the adequacy of Eq. (13.4) to describe the high light-induced NPQ (NPQ_{HL}) was to be expected, the results show that this model can be efficiently combined with a negative exponential function and that the generalized model of Eq. (13.5) can be successfully fit to a large range of NPQ LCs.

Figure 13.2 illustrates the goodness of fit of the model when describing LCs of both NPQ and Y(NPQ), measured in cultures of a benthic diatom acclimated to contrasting light regimes. Very good fits ($r^2 \geq 0.954$) were obtained for the two indices, with the model being capable of adequately tracing both the low light and the high light part of the response curves. These results further illustrate how the model can cope with various relative magnitudes of low- and high-light NPQ (Figure 13.2a vs. Figure 13.2b), with different degrees of sigmoidicity under high light (Figure 13.2a vs. Figure 13.2c), and with uncorrelated variation in the NPQ_{Dark} and NPQ_{HL} parts of the curve.

The large flexibility of the model can be further illustrated by the good fitting obtained for Y(NPQ) light-response curves measured on diatom-dominated biofilms, grown under high light and cultures of a polar diatom (*Fragilariopsis cylindrus*), adapted to very low light regimes, and exhibiting very different curve characteristics (Figure 13.3). While showing relatively similar high light-induction features (NPQ_m varying between 0.47 and 0.59),

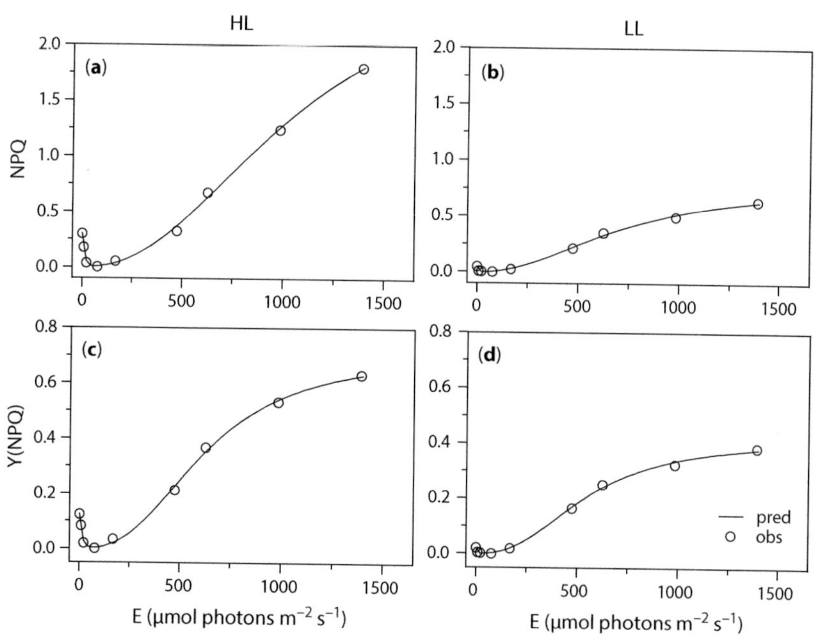

Figure 13.2 Fit of the model (Eq. 13.6) to light curves of NPQ (a, b) and Y(NPQ) (c, d) of contrasting magnitude of NPQ_{Dark}, as induced by different photoacclimation states (light step: 120 s). Data are unpublished results from cultures of the benthic diatom *Navicula* sp. (see Table 13.2 for details), grown in f/2 medium photoacclimated to 15 μmol m^{-2} s^{-1} (LL) and 160 μmol m^{-2} s^{-1} (HL).

the contribution of dark-induced NPQ varied substantially, especially between the data of Figure 13.3a and of Figure 13.3b and Figure 13.3c, with NPQ_0 varying between 0.24 and 0.10 and 0.08, respectively. For the biofilms represented in Figure 13.3a, a rather high dark-induced NPQ_{Dark} was observed, the dissipation of which was very closely followed by the model. A particularly striking difference between these LCs was the substantially lower

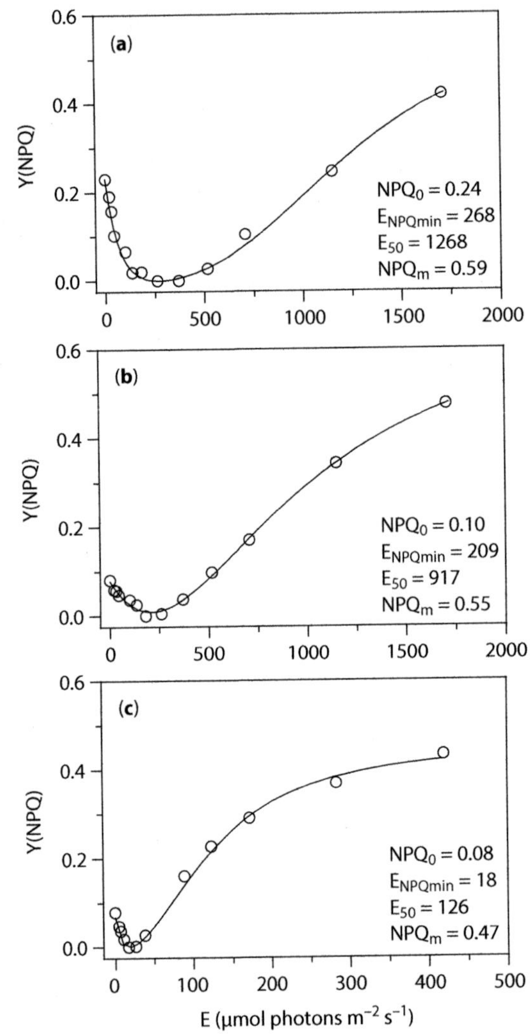

Figure 13.3 Further examples illustrating the fit of the model (Eq. 13.6) to experimental data. Three Y(NPQ) light-response curves show contrasting relationships between NPQ_{Dark} and NPQ_{HL} components. While having similar magnitude of NPQ_{HL} (as quantified by NPQ_m estimates), the three light curves show very different contributions of NPQ_{Dark}. (a) High light-acclimated natural diatom-dominated biofilm [13.35] showing a high dark-induced NPQ_{Dark} (NPQ_0), dissipating gradually with light increase, as well as high E_{50}, resulting in very high E_{NPQmin} (light step: 90 s). (b) High light-acclimated natural diatom-dominated biofilm [13.37] showing a lower NPQ_0 level, while maintaining light saturation at high light levels (high E_{50}) and resulting high E_{NPQmin} (light step: 90 s). (c) Culture of the low light-adapted artic diatom *Fragilaria* sp. (J. Lavaud, unpublished data) showing NPQ_0 and NPQ_m similar to panel b, but dramatically lower values of E_{50} and E_{NPQmin} (light step: 30 s; note x-axis scale).

light levels for saturation of NPQ_{HL}, registered for the polar diatom, quantified by E_{50}, and also resulting in much lower E_{NPQmin}. Regardless of these large differences in the relative contribution of NPQ_{Dark} and NPQ_{HL}, the model was able to trace with great detail the LCs in all cases.

To test the fit of the model to LCs with and without the influence of NPQ_{Dark}, NPQ LCs were generated starting with dark-acclimated samples, by applying increasing light levels and, immediately after, by applying the same levels in reverse, decreasing order, to the same samples (now light acclimated). This caused the dissipation of the NPQ formed in the dark while not affecting the high light part of the NPQ light-response curve. Figure 13.4 shows the response of F_m', the parameter on which NPQ is based (Eq. 13.1), and NPQ to the sequential increase and subsequent decrease in light intensity, revealing marked hysteresis patterns resulting from the dissipation of NPQ_{Dark} by light exposure. Following the increase in light intensity, F_m' increased from the level measured in the dark (F_m) until $E = 500$ μmol photons m^{-2} s^{-1}, after which it began to decrease as expected due to high light exposure (Figure 13.4a). This resulted in the inverted variation of NPQ, first increasing under low light, then gradually decreasing under high light intensities. During the exposure to decreasing light intensities, the high light part of the curve matched almost perfectly the previous measurements, but when irradiance reached below $E < 750$ μmol photons m^{-2} s^{-1}, the curve trajectories started to depart: F_m' continued to increase, and NPQ to decrease, as NPQ formed in the dark was fully dissipated and no longer present. Despite the large

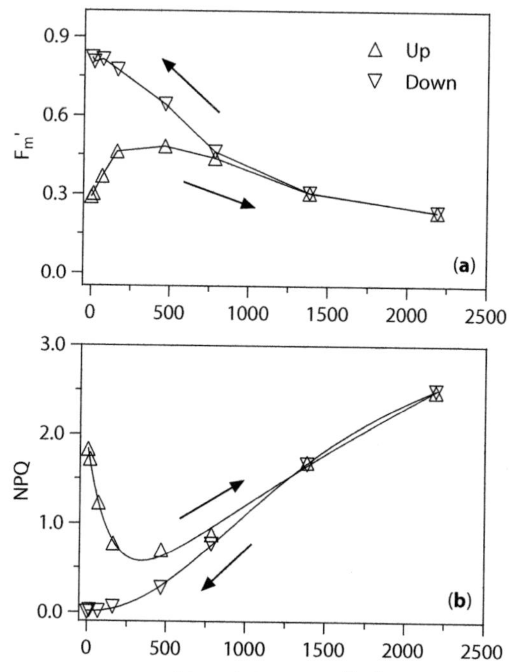

Figure 13.4 Dependence of NPQ_{Dark} on light history, as measured on the diatom *Thalassiosira weissflogii*. Variation of parameter F_m' (a) and NPQ (b) during sequential increase (up) and decrease (down) of light intensity (light step: 120 s). Lines in panel b represent the fit of the proposed model (Eq. 13.6). Arrows represent the direction of increasing light intensity.

differences in the contribution of NPQ_{Dark} and overall shape of the two LCs, the model was able to fit very well to the experimental data (Figure 13.4b), explaining more than 99.2% of the variability.

13.4 Discussion

13.4.1 Model Assumptions

The main original part of this work regards the description of the variation (decrease) of sustained, dark-induced NPQ_{Dark} under low light, during the initial steps of a light-response curve. The resulting general model was based on two main assumptions.

First, the modeling of the dissipation of NPQ_{Dark} was assumed to follow a simple first-order kinetics, characterized by only two parameters, the initial value NPQ_0, quantifying the NPQ level formed prior to the start of the LC, and k_D, a coefficient defined to quantify the rate of NPQ dissipation under exposure to low light levels. This assumption is supported by the experimental observation of exponential decrease in NPQ formed in the dark in diatoms [13.18]. However, the situation under the low light part of the LC can be expected to be inherently more complex than that predicted by a simple exponential decay. This is because first order processes are often observed for constant conditions, while during the construction of the LC dissipation occurs under multiple light intensities, usually increasing over time. In fact, especially when generated sequentially, light-response curves implicitly represent a variation of the photosynthetic activity over time. The dissipation of NPQ_{Dark} under low light, or the induction of NPQ_{HL} under high light, depends not only on the light intensity applied at each light step, but also on the duration of each light step, and on the order of their application [13.25] [13.28] [13.36] [13.50]. The dependence of time and light history is well illustrated by the hysteresis patterns shown in Figure 13.4 that highlight how the NPQ response varies markedly with the sequence of light exposure, especially for low light levels: the NPQ_{Dark} present when the LC is started is later completely dissipated as the light intensity returns to initial low levels. The model here proposed, as may others that have been proposed and extensively used to describe light-response curves (for instance [13.28] [13.51]), lacks the explicit parameterization of the influence of time. One possible consequence of this simplification is that the apparent rate of NPQ dissipation, as characterize by parameter k_D, may not depend on the light levels applied but also on the duration of the light step and their sequence.

Another important assumption of the model is that the processes contributing to NPQ_{Dark} and NPQ_{HL} act independently from each other. That is, that their contribution to overall NPQ under each light level can be considered as addictive, without interaction, thus represented as a sum of two independent equations (Eq. 13.5). The independent operation of the two NPQ components is supported by the observation of uncorrelated variation of the dark- and high light-induced NPQ in a wide range of LCs, as illustrated in Figure 13.3, and in the absence of significant correlations between the parameters NPQ_0 and NPQ_m: from the data compiled in Table 13.2, $r^2 = 0.015$ ($N = 33$, $P = 0.495$) for index NPQ, and $r^2 = 0.022$ ($N = 25$, $P = 0.474$) for index Y(NPQ). In practice, there is some degree of overlap between the dissipation of NPQ_{Dark} and the light-induction of NPQ_{HL}, justifying the use of the parameter C in the model, although in most cases the estimated value for C was very low (Table 13.2).

13.4.2 Fitting to Experimental Data

The simple formulation of Eq. (13.5) reflects a parsimonious approach, aiming for the use of the minimum possible number of model parameters. This seems particularly advantageous considering the fact that NPQ LCs are often based on a small number of data points, and that the fitting of the model becomes problematic when the number of data points is relatively small compared to the number of model parameters. Even keeping the number of model parameters to a minimum, as in Eq. (13.5), the proposed model has six parameters (considering C), while published LCs are often based on eight or fewer data points, twelve at maximum. Although possibly oversimplifying the description of the LC, the use of a reduced set of parameters may help in avoiding fitting issues and allowing for more robust fitting and estimation of model parameters.

The fitting of Eq. (13.5) to experimental data may also encounter difficulties arising from the fact that NPQ LCs may not include enough data points in the low light range. Light-response curves of NPQ are commonly employed to study the response to high light, sacrificing the allocation of light steps to the low light region. This issue may be worsened by the fact that LCs of NPQ are often a "secondary product" of LCs designed to cover the response of ETR, the electron transport rate of PSII, which tends to saturate at lower intensities than NPQ [13.30]. The lack of detailed measurements in the low light region, and along the whole curve, makes the fitting of the model much more dependent on the iterative procedure to use and, in particular, on the choice of adequate start values.

13.4.3 Application

Despite its simplicity, the proposed model was found to be able to closely describe light-response curves of NPQ and of Y(NPQ), displaying a wide range of combinations of relative magnitudes of NPQ_{Dark} and NPQ_{HL} components, light saturation levels, sigmoidicity, and with as few as eight light levels (LCs with fewer light levels were not tested). The model preserves the ability to describe the high light part of the LC, through the estimation of parameters NPQ_m, E_{50} and n. The first two parameters convey quantitative information to characterize the photoprotective capacity through heat dissipation of excess absorbed light energy (NPQ_m) and to locate the light range under which q_E and XC are activated, through the conversion of pigment diadinoxanthin into diatoxanthin (E_{50}). The sigmoidicity parameter, n, consistently attaining values above 1, confirms previous results indicating the allosteric regulation of NPQ by light exposure [13.30]. The same fitting procedure can also be used to parametrize LCs of diatoxanthin synthesis, allowing cross analysis between ETR, NPQ and diatoxanthin-related parameters [13.41] [13.48].

The main novelty of the proposed model is the possibility of quantifying the relative magnitude of dark-induced NPQ_{Dark}, through parameter NPQ_0, and the rate at which it is dissipated under low light, through parameter k_D. The build-up and maintenance of sustained NPQ during prolonged darkness, and its subsequent dissipation under exposure to low light is well-known in diatoms (e.g., [13.9] [13.33] [13.36] [13.52] [13.53]). This is due to the formation of a transthylakoidal ΔpH in the dark, involving the conversion of energy-dissipating pigments, as usually triggered during the operation of the XC under high light [13.9] [13.33]. The causes for this process have been debated thoroughly and attributed to chlororespiratory [13.32] [13.33] [13.52] or, more recently, to the reversed operation of chloroplastic ATP synthase, hydrolyzing ATP produced in the mitochondria, resulting in the transport of stroma protons into the lumen of the thylakoid [13.53]. Parameter NPQ_0 can thus serve as an indicator of the magnitude of ΔpH formed in the dark, allowing the direct comparison of LCs measured under different conditions or for different taxa or growth regimes. NPQ_{Dark} is thought to support the fast resumption of photochemistry when diatoms are exposed to light after a long period of darkness, as is often the case, for instance, for intertidal benthic species, which can spend several days buried in the sediments before being exposed again to light during low tide, or for arctic diatoms to respond to the return of light in the spring after the winter polar night [13.54] [13.55].

The relaxation of NPQ_{Dark} under low light is thought to be due to the dissipation of the transthylakoidal ΔpH due to the gradual activation of ATP synthase and carbon metabolism, promoting the removal of protons from the lumen of the thylakoid. Under the proposed model, the pace of the relaxation of NPQ_{Dark}, as followed by the decrease of NPQ from the initial NPQ_0 level is quantified by a single parameter, k_D. However, because NPQ indices are largely determined by the relative variation of fluorescence parameter $F_m{}'$ during the generation of the LCs, and because its variation (increase under low light levels and subsequent decrease under high light levels) is simultaneously affected by the processes underlying NPQ_{Dark} and NPQ_{HL}, both parameters NPQ_0 and k_D can be limited by the range of variation of $F_m{}'$ in relation to the maximum value registered during the generation of the LC. Thus, as the increase of $F_m{}'$ under low light may be, after some intermediate light levels, masked by the decrease induced by high light, the values reached by NPQ_0 and by NPQ_m are only interpretable in relative terms. For the same reason, the observable dissipation of NPQ_{Dark}, measured by k_D, may not be complete, which may limit the accuracy of its description by this parameter.

Another situation where NPQ_{Dark} is present is when it develops at light but does not relax during subsequent prolonged darkness exposure. This was especially observed in species that thrive in the polar cold waters ([13.34]) and was compared with the sustained ("locked") NPQ of overwintering evergreen trees [13.56]. It is based on high concentrations of de-epoxidized xanthophylls (diatoxanthin in diatoms) that are not converted back to their initial form (diadinoxanthin in diatoms) [13.57], and consequently can maintain NPQ in darkness, even in the absence of a transthylakoidal ΔpH [13.9]. Consequently, this type of NPQ_{Dark}, and in contrast to the one that is induced in the dark, cannot relax during the course of a LC [13.34].

Although the occurrence of dark-induced NPQ is best known for diatoms, it has also been observed for other photoautotrophs, such as cyanobacteria [13.58], brown algae [13.31], and other stramenopile [13.59]. The proposed model can be readily applied to these other groups, contributing to a better characterization of the somewhat elusive and ill-described NPQ_{Dark} component, which may include establishing under which conditions it is developed, describing its transient nature under variable light, and untangling the dependence on light intensity and of duration of exposure.

Acknowledgments

We thank A. Bastos, S. Frankenbach, J. Frommlet, D. Croteau and S. Guérin for allowing the use of their unpublished data. This work was supported by project SearchLight (PTDC/BIA-BMA/1058/2021) and CESAM (UIDP/50017/2020+UIDB/50017/2020+ LA/P/0094/2020), funded by national funds (OE) through FCT/MCTES, and by the CNRS. JL also thanks support funding from the NSERC (Discovery Grant), the Canada First ULaval-Sentinelle Nord Research Program, and the IUEM (Polar Research Program).

References

[13.1] Field, C.B., Behrenfeld, M.J., Randerson, J.T., and Falkowski, P.G. (1998) Primary production of the biosphere: Integrating terrestrial and oceanic components. *Science (80-.).*, 281 (5374), 237–240.
[13.2] Sarthou, G., Timmermans, K.R., Blain, S., and Tréguer, P. (2005) Growth physiology and fate of diatoms in the ocean: A review. *J. Sea Res.*, 53 (1–2), 25–42.
[13.3] Armbrust, E.V. (2009) The life of diatoms in the world's oceans. *Nature*, 459 (7244), 185–192.
[13.4] Benoiston, A., Ibarbalz, F.M., Bittner, L., Guidi, L., Jahn, O., Dutkiewicz, S., Bowler, C., and Bowler, C. (2017) The evolution of diatoms and their biogeochemical functions. *Philos. Trans. R. Soc. B Biol. Sci.*, 372 (1728), 20160397.
[13.5] Wagner, H., Jakob, T., Fanesi, A., and Wilhelm, C. (2017) Towards an understanding of the molecular regulation of carbon allocation in diatoms: the interaction of energy and carbon allocation. *Philos. Trans. R. Soc. Lond. B. Biol. Sci.*, 372 (1728), 20160410.
[13.6] Raven, J.A. (2011) The cost of photoinhibition. *Physiol. Plant.*, 142 (1), 87–104.
[13.7] Wilhelm, C., Jungandreas, A., Jakob, T., and Goss, R. (2014) Light acclimation in diatoms: From phenomenology to mechanisms. *Mar. Genomics*, 16 (1), 5–15.
[13.8] Ruban, A., Lavaud, J., Rousseau, B., Guglielmi, G., Horton, P., and Etienne, A.L. (2004) The super-excess energy dissipation in diatom algae: Comparative analysis with higher plants. *Photosynth. Res.*, 82 (2), 165–175.
[13.9] Lavaud, J., and Goss, R. (2014) *Non-Photochemical Quenching and Energy Dissipation in Plants, Algae and Cyanobacteria*, Springer Netherlands, Dordrecht.
[13.10] Müller, P., Li, X.P., and Niyogi, K.K. (2001) Non-Photochemical Quenching. A Response to Excess Light Energy. *Plant Physiol.*, 125 (4), 1558–1566.
[13.11] Demmig-Adams, B., and Adams, W.W. (2006) Photoprotection in an ecological context: the remarkable complexity of thermal energy dissipation. *New Phytol.*, 172 (1), 11–21.
[13.12] Blommaert, L., Vancaester, E., Huysman, M.J.J., Osuna-Cruz, C.M., D'hondt, S., Lavaud, J., Lepetit, B., Winge, P., Bones, A.M., Vandepoele, K., Vyverman, W., and Sabbe, K. (2020) Light Regulation of LHCX Genes in the Benthic Diatom Seminavis robusta. *Front. Mar. Sci.*, 7 (April), 1–10.
[13.13] Buck, J.M., Sherman, J., Bártulos, C.R., Serif, M., Halder, M., Henkel, J., Falciatore, A., Lavaud, J., Gorbunov, M.Y., Kroth, P.G., Falkowski, P.G., and Lepetit, B. (2019) Lhcx proteins provide photoprotection via thermal dissipation of absorbed light in the diatom Phaeodactylum tricornutum. *Nat. Commun.*, 10 (1), 4167.
[13.14] Taddei, L., Stella, G.R., Rogato, A., Bailleul, B., Fortunato, A.E., Annunziata, R., Sanges, R., Thaler, M., Lepetit, B., Lavaud, J., Jaubert, M., Finazzi, G., Bouly, J.-P.P., and Falciatore, A. (2016) Multisignal control of expression of the LHCX protein family in the marine diatom Phaeodactylum tricornutum. *J. Exp. Bot.*, 67 (13), 3939–3951.

[13.15] Goss, R., and Jakob, T. (2010) Regulation and function of xanthophyll cycle-dependent photoprotection in algae. *Photosynth. Res.*, 106 (1–2), 103–122.

[13.16] Olaizola, M., and Yamamoto, H.Y. (1994) Short-term response of the diadinoxanthin cycle and fluorescence yield to high irradiance in Chaetoceros muelleri (Bacillariophyceae). *J. Phycol.*, 30, 606–612.

[13.17] Lacour, T., Babin, M., and Lavaud, J. (2020) Diversity in Xanthophyll Cycle Pigments Content and Related Nonphotochemical Quenching (NPQ) Among Microalgae: Implications for Growth Strategy and Ecology. *J. Phycol.*, 56 (2), 245–263.

[13.18] Blommaert, L., Chafai, L., and Bailleul, B. (2021) The fine - tuning of NPQ in diatoms relies on the regulation of both xanthophyll cycle enzymes. *Sci. Rep.*, 11, 12750.

[13.19] Bassi, R., and Dall'Osto, L. (2021) Dissipation of Light Energy Absorbed in Excess: The Molecular Mechanisms. *Annu. Rev. Plant Biol.*, 72 (1), 47–76.

[13.20] Bilger, W., and Bjorkman, O. (1990) Role of the xanthophyll cycle in photoprotection elucidated by measurements of light-induced absorbance changes, fluorescence and photosynthesis in leaves of Hedera canariensis. *Photosynth. Res.*, 25 (3), 173–185.

[13.21] Demmig-Adams, B., Adams III, W., and Adams, W.W. (1996) The role of xanthophyll cycle carotenoids in the protection of photosynthesis. *Trends Plant Sci.*, 1 (1), 21–26.

[13.22] Ruban, A. V (2016) Nonphotochemical Chlorophyll Fluorescence Quenching: Mechanism and Effectiveness in Protecting Plants from Photodamage. *Plant Physiol.*, 170 (4), 1903–16.

[13.23] Goss, R., and Lepetit, B. (2015) Biodiversity of NPQ. *J. Plant Physiol.*, 172, 13–32.

[13.24] Schreiber, U., Schliwa, U., and Bilger, W. (1986) Continuous recording of photochemical and nonphotochemical chlorophyll fluorescence quenching with a new type of modulation fluorometer. *Photosynth. Res.*, 10, 51–62.

[13.25] Perkins, R.G., Mouget, J.L., Lefebvre, S., and Lavaud, J. (2006) Light response curve methodology and possible implications in the application of chlorophyll fluorescence to benthic diatoms. *Mar. Biol.*, 149 (4), 703–712.

[13.26] Perkins, R.G., Kromkamp, J.C., Serôdio, J., Lavaud, J., Jesus, B., Mouget, J.L., Lefebvre, S., and Forster, R.M. (2010) Chlorophyll a Fluorescence in Aquatic Sciences: Methods and Applications, in *Chlorophyll a Fluorescence in Aquatic Sciences: Methods and Applications* (eds.Suggett, D.J., Prášil, O., and Borowitzka, M.A.), Springer Netherlands, Dordrecht, pp. 237–275.

[13.27] Frankenbach, S., and Serôdio, J. (2017) One pulse, one light curve: Fast characterization of the light response of microphytobenthos biofilms using chlorophyll fluorescence. *Limnol. Oceanogr. Methods*, 15 (6), 554–566.

[13.28] Lefebvre, S., Mouget, J.-L., and Lavaud, J. (2011) Duration of rapid light curves for determining the photosynthetic activity of microphytobenthos biofilm in situ. *Aquat. Bot.*, 95 (1), 1–8.

[13.29] Rohacek, K., Bertrand, M., Moreau, B., Jacquette, B., Caplat, C., Morant-Manceau, A., Schoefs, B., Roháček, K., Bertrand, M., Moreau, B., Jacquette, B., Caplat, C., Morant-Manceau, A., & Schoefs, B. (2014). Relaxation of the non-photochemical chlorophyll fluorescence quenching in diatoms: kinetics, components and mechanisms. *Phil. Trans. R. Soc.* B, 369 (1640), 20130241–20130241.

[13.30] Serôdio, J., and Lavaud, J. (2011) A model for describing the light response of the nonphotochemical quenching of chlorophyll fluorescence. *Photosynth. Res.*, 108 (1), 61–76.

[13.31] Mouget, J., and Tremblin, G. (2002) Suitability of the Fluorescence Monitoring System (FMS, Hansatech) for measurement of photosynthetic characteristics in algae. *Aquat. Bot.*, 74 (3), 219–231.

[13.32] Cruz, S., Goss, R., Wilhelm, C., Leegood, R., Horton, P., and Jakob, T. (2011) Impact of chlororespiration on non-photochemical quenching of chlorophyll fluorescence and on the regulation of the diadinoxanthin cycle in the diatom *Thalassiosira pseudonana*. *J. Exp. Bot.*, 62 (2), 509–519.

[13.33] Jakob, T., Goss, R., and Wilhelm, C. (1999) Activation of diadinoxanthin de-epoxidase due to a chlororespiration proton gradient in the dark in the diatom *Phaeodactylum tricornutum*. *Plant Biol.*, 1, 76–82.

[13.34] Lacour, T., Larivière, J., Ferland, J., Bruyant, F., Lavaud, J., and Babin, M. (2018) The role of sustained photoprotective non-photochemical quenching in low temperature and high light acclimation in the bloom-forming arctic aiatom *Thalassiosira gravida*. *Front. Mar. Sci.*, 5 (October), 1–16.

[13.35] Geel, C., Verluis, W., and Snel, J.F.H. (1997) Estimation of oxygen evolution by marine phytoplankton from measurement of the efficiency of photosystem II electron flow. *Photosynth. Res.*, 51, 61–70.

[13.36] Serôdio, J., Vieira, S., Cruz, S., and Coelho, H. (2006) Rapid light-response curves of chlorophyll fluorescence in microalgae: relationship to steady-state light curves and non-photochemical quenching in benthic diatom-dominated assemblages. *Photosynth. Res.*, 90 (1), 29–43.

[13.37] Schreiber, U., Bilger, W., Neubauer, C., Shulze, E.D., and Caldwell, M.M. (1994) Chlorophyll fluorescence as a nonintrusive indicator for rapid assessment of *in vivo* photosynthesis, Springer-Verlag, Berlin, pp. 49–70.

[13.38] Serôdio, J., Cruz, S., Vieira, S., and Brotas, V. (2005) Non-photochemical quenching of chlorophyll fluorescence and operation of the xanthophyll cycle in estuarine microphytobenthos. *J. Exp. Mar. Bio. Ecol.*, 326 (2), 157–169.

[13.39] Klughammer, C., and Schreiber, U. (2008) Complementary PS II quantum yields calculated from simple fluorescence parameters measured by PAM fluorometry and the Saturation Pulse method. *PAM Appl. Notes*, **1** (1), 27–35.

[13.40] Xu, K., Lavaud, J., Perkins, R., Austen, E., Bonnanfant, M., and Campbell, D.A. (2018) Phytoplankton σPSII and Excitation Dissipation; Implications for estimates of primary productivity. *Front. Mar. Sci.*, 5 (August), 281.

[13.41] Croteau, D., Guérin, S., Bruyant, F., Ferland, J., Campbell, D.A., Babin, M., and Lavaud, J. (2021) Contrasting nonphotochemical quenching patterns under high light and darkness aligns with light niche occupancy in Arctic diatoms. *Limnol. Oceanogr.*, 66 (S1), S231–S245.

[13.42] Perkins, R., Williamson, C., Lavaud, J., Mouget, J.-L., and Campbell, D.A. (2018) Time-dependent upregulation of electron transport with concomitant induction of regulated excitation dissipation in Haslea diatoms. *Photosynth. Res.*, 0 (0), 1–12.

[13.43] Miyata, K., Ikeda, H., Nakaji, M., Kanel, D.R., and Terashima, I. (2015) Rate constants of PSII photoinhibition and its repair, and PSII fluorescence parameters in field plants in relation to their growth light environments. *Plant Cell Physiol.*, 56 (9), 1841–1854.

[13.44] Schrameyer, V., Krämer, W., Hill, R., Jeans, J., Larkum, A.W.D., Bischof, K., Campbell, D.A., and Ralph, P.J. (2016) Under high light stress two Indo-Pacific coral species display differential photodamage and photorepair dynamics. *Mar. Biol.*, 163 (8), 1]–[13.

[13.45] Nitschke, U., Connan, S., and Stengel, D.B. (2012) Chlorophyll a fluorescence responses of temperate Phaeophyceae under submersion and emersion regimes: a comparison of rapid and steady-state light curves. *Photosynth. Res.*, 114 (1), 29–42.

[13.46] Bozzato, D., Jakob, T., and Wilhelm, C. (2019) Effects of temperature and salinity on respiratory losses and the ratio of photosynthesis to respiration in representative Antarctic phytoplankton species. *PLoS One*, 14 (10).

[13.47] Trsková, E.K., Belgio, E., Yeates, A.M., Sobotka, R., Ruban, A. V., and Kaňa, R. (2018) Antenna proton sensitivity determines photosynthetic light harvesting strategy. *J. Exp. Bot.*, 69 (18), 4483–4493.

[13.48] Barnett, A., Méléder, V., Blommaert, L., Lepetit, B., Gaudin, P., Vyverman, W., Sabbe, K., Dupuy, C., and Lavaud, J. (2015) Growth form defines physiological photoprotective capacity in intertidal benthic diatoms. *ISME J.*, 9 (1), 32–45.
[13.49] Lepetit, B., Gélin, G., Lepetit, M., Sturm, S., Vugrinec, S., Rogato, A., Kroth, P.G., Falciatore, A., and Lavaud, J. (2017) The diatom *Phaeodactylum tricornutum* adjusts nonphotochemical fluorescence quenching capacity in response to dynamic light via fine-tuned Lhcx and xanthophyll cycle pigment synthesis. *New Phytol.*, 214 (1), 205–218.
[13.50] Herlory, O., Richard, P., and Blanchard, G.F. (2007) Methodology of light response curves: application of chlorophyll fluorescence to microphytobenthic biofilms. *Mar. Biol.*, 153 (1), 91–101.
[13.51] Platt, T., Gallegos, C.L., and Harrison, W.G. (1980) Photoinhibition of photosynthesis in natural assemblages of marine phytoplankton. *J. Mar. Res.*, 38, 678–701.
[13.52] Ting, C.S., and Owens, T.G.C. (1993) Photochemical and nonphotochemical fluorescence quenching processes in the diatom *Phaeodactylum tricornutum. Plant Physiol*, 101 (185405), 1323–1330.
[13.53] Bailleul, B., Berne, N., Murik, O., Petroutsos, D., Prihoda, J., Tanaka, A., Villanova, V., Bligny, R., Flori, S., Falconet, D., Krieger-Liszkay, A., Santabarbara, S., Rappaport, F., Joliot, P., Tirichine, L., Falkowski, P.G., Cardol, P., Bowler, C., and Finazzi, G. (2015) Energetic coupling between plastids and mitochondria drives CO_2 assimilation in diatoms. *Nature*, 524 (7565), 366–369.
[13.54] Morin, P.I., Lacour, T., Grondin, P.L., Bruyant, F., Ferland, J., Forget, M.H., Massicotte, P., Donaher, N., Campbell, D.A., Lavaud, J., and Babin, M. (2019) Response of the sea-ice diatom *Fragilariopsis cylindrus* to simulated polar night darkness and return to light. *Limnol. Oceanogr.*, lno.11368.
[13.55] Lacour, T., Morin, P.I., Sciandra, T., Donaher, N., Campbell, D.A., Ferland, J., and Babin, M. (2019) Decoupling light harvesting, electron transport and carbon fixation during prolonged darkness supports rapid recovery upon re-illumination in the Arctic diatom Chaetoceros neogracilis. *Polar Biol.*, 42 (10), 1787–1799.
[13.56] Verhoeven, A. (2014) Sustained energy dissipation in winter evergreens. *New Phytol.*, 201 (1), 57–65.
[13.57] Fernández-Marín, B., Roach, T., Verhoeven, A., and García-Plazaola, J.I. (2021) Shedding light on the dark side of xanthophyll cycles. *New Phytol.*, 230 (4), 1336–1344.
[13.58] Schreiber, U., Endo, T., Mi, H., and Asada, K. (1995) Quenching analysis of chlorophyll fluorescence by the saturation mulse Method: particular aspects relating to the study of eukaryotic algae and Cyanobacteria. *Plant Cell Physiol.*, 36 (5), 873–882.
[13.59] Berne, N., Fabryova, T., Bistaz, F., Cardol, P., and Bailleul, B. (2018) The peculiar NPQ regulation in the stramenopile *Phaeomonas sp.* challenges the xanthophyll cycle dogma. *Biochim. Biophys. Acta - Bioenerg.*, 1859 (7), 491–500.
[13.60] Cartaxana, P., & Serôdio, J. (2008). Inhibiting diatom motility: a new tool for the study of the photophysiology of intertidal microphytobenthic biofilms. *Limnol. Oceanogr. Meth.* 6 (9), 466–476.

14

Coscinodiscus wailesii as Biogenic Charge-Based Sensors for Heavy Metal Ion Contamination Detection

Rajeshwari Taruvai Kalyana Kumar, Diem-Thuy Le, Antra Ganguly and Shalini Prasad*

Department of Bioengineering, University of Texas at Dallas, Richardson, Texas, USA

Abstract

The potential of marine diatoms as a naturally abundant ion absorbent present in the environment as a water quality indicator was investigated in this study. The effect of ion-based surface charge characterization using heavy metals on diatoms has been described in detail in this work. Cationic effect (lead) and anionic effect (arsenic) heavy metals were studied on *Coscinodiscus wailesii* diatoms to understand the effect of sorption using infrared characterization and shift in zeta potential. The role of water pH was investigated and transient measurements of sorption behavior over 24 hours and 1 week are presented. Additional data based on morphological changes of diatoms is presented to provide a unique understanding of heavy metal effects on marine diatoms and their use as water quality indicators. Data reveal information on intricate changes in the surface charge accumulation and potential shift at the solid-liquid interface. Zeta potential on diatom samples can provide useful information on the environmental factors affecting the growth and biological mechanism which can provide basic knowledge on sample contamination. This chapter provides a unique perspective through a surface charged-based approach to using *C. wailesii* diatoms as an early indicator of heavy metal ion contamination in water samples.

Keywords: Diatoms, water quality detection, surface charge, charge-based sensor, zeta potential, Fourier transform infrared spectroscopy

14.1 Introduction

Heavy metals are highly toxic, high-density chemical metals with atomic weights between 63.5 and 200.6 g mol^{-1} and a specific gravity greater than 5 g cm^{-3} [14.1]. Heavy metal ions in water sources have a sustained long-term adversarial impact on human health [14.2].

**Corresponding author*: Shalini.prasad@utdallas.edu

Rajeshwari Taruvai Kalyana Kumar: https://www.researchgate.net/profile/Rajeshwari-Taruvai-Kalyana-Kumar-2

Antra Ganguly: https://www.researchgate.net/profile/Antra-Ganguly, https://orcid.org/0000-0002-0776-0008, https://scholar.google.com/citations?user=99cjXbQAAAAJ&hl=en&oi=ao

Shalini Prasad: https://www.researchgate.net/profile/Shalini-Prasad-3, https://orcid.org/0000-0002-2404-3801, https://scholar.google.com/citations?user=cFphA2cAAAAJ&hl=en

Janice L. Pappas (ed.) The Mathematical Biology of Diatoms, (413–426) © 2023 Scrivener Publishing LLC

They accumulate in the biosphere and are absorbed by plants, thereby becoming a part of the food chain [14.3]. As a result, water contamination, and aquatic life endangerment are some of the key issues that are constantly being tackled by the United States Environmental Protection Agency (US EPA) [14.4]. In addition, the leaching of inorganic compounds into water sources through man-made chemical wastes is one of the most prevalent risks associated with water contamination. Chemicals, such as heavy metals like arsenic (As) [14.5], lead (Pb) [14.6], copper (Cu) [14.7], zinc (Zn) [14.8], and cadmium (Cd) [14.9], threaten aquatic life and the human population.

The two metals most commonly detected in water samples are arsenic and lead. These heavy metals are considered environmental health hazards, as they top the list from the Agency for Toxic Substances and Disease Registry's Priority List of Hazardous Substances, based on toxicity when exposed to water, air, and soil [14.10]. A key factor in arsenic and lead toxicity is its chemical speciation, distribution, and cycling, along with biological activity as in biogeochemistry. According to the US EPA, the maximum contamination level of arsenic is 10 μg/L and for lead is 15 ppb [14.11]. In most cases, trace heavy metals are important for biological lifeform cell signaling/functioning. However, if these metal ions exceed optimum concentration and/or move out of the mechanistic pathway, they become toxic and produce adverse effects on biological cells. Therefore, with the constant increase in the industrial usage of these metals leading to high levels of toxic waste disposal, it is important to detect the levels of heavy metals in water to prevent further environmental damage [14.12].

Various techniques such as water sampling through colorimetric testing to electrochemical biosensors provide detailed insights into the detection and presence of heavy metals [14.1]. Because, though metals are important for carrying out functions, their concentration range has a huge impact on human health. However, a deeper understanding of heavy metal adsorption and adverse effects on biological cells is still an emerging area of study that needs novel approaches to characterize water samples. Table 14.1 lists various sources, toxicity levels (according to the World Health Organization – WHO), and limit of detection offered by some of the novel sensing technologies for As and Pb metal ions (according to data referenced in [14.1]).

In addition to the detection of heavy metals, understanding their adsorption behavior and fluctuations caused at the ionic level will provide greater insights into the food chain and material behavior of these toxins. Hence, to study the adsorption and interfacial effects of heavy metals, marine biological cells are the perfect entities for material characterization. Biological freshwater algae, otherwise known as marine diatoms, are important contributors to the primary production in aquatic ecosystems. The common unicellular algae exist as filamentous colonies and are good indicators of environmental integrity. The diatoms are made up of inorganic frustules encasing the organic lifeform that provide a large functional surface area for interfacial charge characterization. Upon increasing exposure to such chemical contaminants, depending on the degree of exposure, the frustule either absorbs these chemicals or significant changes can be observed in the cell's morphology and metabolism [14.13]. Hence, diatoms can be used as biogenic sensors to monitor environmental effects. In addition, diatoms have also been used in several applications, including optical biosensors, filtration, and biofilm applications [14.14-14.16].

Fluctuations in salinity, temperature, as well as the chemical composition of the water in which the diatoms are suspended, can cause tremendous changes in them. Chemicals

Table 14.1 Toxicity limits, common sources, sensing modality, and limit of detection.

Metal ion	WHO/ US EPA limits	Common sources and adverse effects	Sensing modality	Limit of detection and sensitivity
Arsenic (As)	0.05 mg/L 10 μg/L	*Source of contamination:* Fertilizers, chemical wastes, untreated effluents, and arsenic-based salts – result in water and soil-based contamination *Health implications:* Nervous system damage, cardiovascular disease, gastrointestinal disease, acute heavy metal toxicity	Electrochemical interface studies (Fe3O4microspheres–RTIL composite)	LoD: $8X10^{-4}$ ppb Sensitivity: 4.91 μA ppb^{-1}
			Nanotechnology-based techniques (Au-decorated tripod-shaped Te hybrids)	LoD: 0.0096 ppb Sensitivity: 6.35 $\mu Appb^{-1}$
Lead (Pb)	0.05 mg/L 15 ppb	*Source of contamination:* Paints, jewelry, lead batteries, pipelines, plastics, and chemical wastes – result in water, air, and soil contamination *Health implications:* Blood-brain barrier damage, Dementia, Alzheimer's disease, kidney damage, bone damage	Co-polyaniline nanoparticles	LoD: 2.2E-11 M Sensitivity: 29.3 mV $decade^{-1}$
			Calixarenes/Multiwalled Carbon nanotubes	LoD: 4E-11 $molL^{-1}$ Sensitivity: $10^{-6}AV^{-1}$

are made up of either charged or neutral ions. Continuous exposure to these chemicals can cause charge-based fluctuations on the solid surfaces of the diatoms [14.17]. The phenomenon of charges at the interface of the solid and liquid material is known as the zeta potential (ζ). The zeta potential indicates the amount of charges absorbed at the interface. It can reveal unique charge-based signatures about an analyte that are either physically or chemically absorbed onto a surface [14.18]. In this study, we used two different types of heavy metals to study the absorption behavior of diatoms that can be used to characterize the presence of heavy metals in water samples: (1) anion–arsenic-based salt and (2) cation–lead-based salt.

14.2 Materials and Methods

14.2.1 Chemicals and Reagents

All chemicals and reagents were of analytical grade and used as received without further purification. Confluent stock cultures of *C. wailesii* and natural seawater were purchased from Provasoli-Guillard National Center for Marine Algae (MA, USA). The F/2 enrichment solution and sodium metasilicate nonahydrate ($Na_2SiO_3.9H_2O$), hydrochloric acid (HCl), arsenic trioxide (As_2O_3), and lead nitrate ($Pb(NO_3)_2$) were purchased from Sigma-Aldrich (MO, USA). Deionized water and 0.15 M Phosphate buffered saline (PBS, pH 7.0) was purchased from Corning (NY, USA). Ethyl alcohol was purchased from Fisher Scientific (MA, USA).

14.2.2 Cell Culture

The daughter culture of *C. wailesii* was grown in Guillard's medium (f/2 medium). The recipe for f/2 medium consisted of natural seawater, F/2 enrichment solution, and sodium metasilicate nonahydrate salt solution. The salinity of natural seawater purchased was maintained between 32 to 33 ppt. 20 mL of f/2 enrichment solution was used for 1 L volume of seawater followed by the addition of 1 mL of $Na_2SiO_3.9H_2O$ to the stock solution. F/2 enrichment solution consisted of salt solutions, trace metal solution, and f/2 vitamin solution. These solutions are critical for the formation of the silica exoskeleton of the diatoms. 10 mL of confluent stock culture was cultivated in 20 mL of prepared f/2 medium in a 500 mL Erlenmeyer flask with a large air column for aerating the culture. A 1:2 ratio was chosen to prevent the diatoms from undergoing shocks during cultivation. The daughter culture was generated from the cultivation solution after an 8-week transfer cycle. During the culture period, the diatoms were exposed to optimal conditions of light intensity, duration of light and dark cycle, air supply, temperature, and culture medium. It is essential to optimize growth conditions as sub-optimal growth can lead to changes in pore dimensions and exoskeletal alterations. The optimal culture conditions include 13 hours of light and 11 hours of dark cycle with a light intensity of 700 lumens maintained at a temperature ~19 to 20 °C with sufficient air column in the flask.

14.2.3 Heavy Metal Doping and Characterization

Known concentrations of heavy metal salts, arsenic trioxide (As_2O_3) and lead nitrate ($Pb(NO_3)_2$), were added to 2 mL of confluent diatom cultures each. The final concentrations of the sample were adjusted to 1, 5, 10, 50, 100, and 150 ppm. The diatom samples were characterized 1 day and 1 week post heavy metal doping. To determine the effect of heavy metals on the growth of *C. wailesii*, 500 μL samples were taken from flasks for each heavy metal treatment (24 hours and 1-week post-treatment of Arsenic and Lead) under sterile conditions. From this sample, 100 μL was used and counted using a hemocytometer under a phase-contrast microscope. Similarly, each sample was loaded onto a sterile microscopic glass slide to measure cell size. Visual morphological changes were observed and analyzed through an inverted optical light microscope. The electrical conductivity of the samples was measured using an Accumet conductivity meter (IL, USA). Fourier transform infrared measurements were performed on liquid diatom samples doped with arsenic (As) and lead (Pb) to characterize the silica diatomic frustule. FTIR characterization was used to analyze the material present on the diatom frustule through sorption (Nicolet iS50 FTIR with Harrick VariGATR sampling stage, Thermo Scientific (MA, USA)). The operational conditions in FTIR analysis were scanned 256 times from wavenumber 600-4000 cm^{-1}.

14.2.4 Electrophoretic Measurements

Zeta potential (ZP) measurements were obtained using a Malvern Zetasizer (Malvern, UK), which measures the electrophoretic mobility and converts the value to the respective charge potential, ZP, based on the Smoluchowski approximation, which is valid when $\kappa a > 1$. Seawater used for diatom culture was used as a background electrolyte to maintain ionic strength between measurements. After adsorption equilibrium, the diatom solutions with varying heavy metal doping concentrations were measured to obtain the mean range of the surface charge potential on the diatom frustules. As control measurements, boiled dead diatoms were used across the same doping conditions to establish a baseline. The effect of pH on zeta potential was monitored by measuring the potential of doped diatoms in varying pH solutions ranging from 2 to 12. Time points of 24 hours and 1 week were chosen to observe charge-based changes at the diatom interface in the water samples for an early indication of contamination. All measurements were obtained in triplicate.

14.3 Results and Discussion

14.3.1 Effect of Heavy Metal Doping on Cell Culture

The effect of the different concentrations of arsenic and lead (1, 5, 10, 50, 100, 150 ppm) on the growth and biochemical composition of *C. wailesii* varied significantly. Cell growth per given sample volume, also known as cell density, was measured for varying heavy metal concentration diatom samples, as shown in Figure 14.1. The number of cells per volume is significantly reduced with an increase in the concentration of arsenic and lead. As a control, stock diatom samples suspended in seawater in the absence of heavy metal supplements were measured. These measurements were taken after the diatoms were incubated in heavy metal salts for 24 hours and one week. Cell density for the control population was observed to be constant at around 200 and 220 diatomic cells per 100 μL of sample volume respectively. Unlike control samples, cell density of diatoms exposed to arsenic and lead for the first 24 hours of treatment, decreased by 70% and 88% of cells per 100 μL of sample volume respectively.

In addition, a consistent monotonic decrease in the trend was observed in diatom growth for one-week post heavy metal treatment. Cell density significantly decreased to 110 and 130 cells from 220 cells per 100 μL of sample volume for arsenic and lead respectively. An increase in dose concentration resulted in reduced cell density to less than 40 cells per 100 μL of sample volume for both arsenic and lead salts. Replication was such that coefficients of variation within treatments were generally less than 10%. This behavior indicates rapid uptake of heavy metal salts by the cells during their growth process and metabolic pathway implementation. Generally, the toxicity of heavy metals increases with an increase

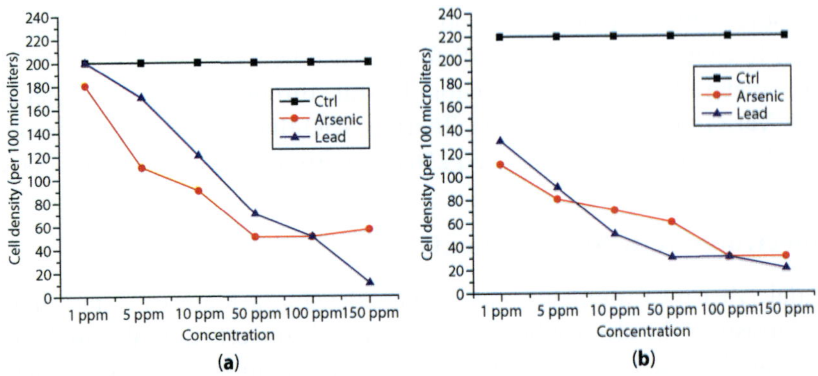

Figure 14.1 Results of diatoms doped with arsenic and lead heavy metal ions: (a) cell density after 24-hour incubation, (b) cell density after 1-week incubation.

in dosage. The supporting hypothesis is that these anionic and cationic heavy metal salts tend to be physisorbed onto the diatom silica frustule. Further surface characterization analyses are shown in subsequent sections.

Along with cell density, cell size was also observed for all samples that were incubated with heavy metals for a week. For each diatom sample, 10 diatomic cells were considered for counting to obtain the average cell size of the batch per sample (corresponding to $p < 0.05$ with no significant differences in the rest of the sample). Although no trend was observed with the varying cell sizes with increasing concentration of heavy metals, it was observed that the range of cell size was well-defined for all the samples tested. Cell size ranged between 74 and 79 μm for samples treated for 24 hours and one week. No consistency was observed for cell sizes from samples treated with either of the heavy metals, indicating that the absorption of heavy metals did not have a significant impact on the external size of the diatom. These results may suggest that the toxicity due to heavy metals could potentially affect the cell metabolism but have minimal to negligible effect on silica frustule.

14.3.2 Effect of Heavy Metal Doping on Zeta Potential

To understand surface charge changes due to the addition of arsenic and lead, zeta potential studies were carried out on four different types of sample solutions. Diatoms were treated with arsenic- and lead-based salt along with pristine live and dead diatoms as control samples. Results shown in Figure 14.2 indicate the overall zeta potential observed for varying concentrations of heavy metal doping on diatoms.

The results indicate a monotonic increase in zeta potential with the increase in heavy metal concentration over 24 hours of incubation. Zeta potential values increased approximately from -8 mV to -3.5 mV for As doped diatoms and -9 mV to -4 mV for Pb doped diatoms. The dynamic range of the surface potential does not significantly vary for an anion- or cation-based dopant within the post 24 hours of treatment. Control live diatoms cells' zeta potential remained averagely constant without any addition of dopant, while for dead cells the surface charge potential reflects the physisorbed ions onto the silica frustule. The values for the dead control sample increased to approximately -5.5 mV throughout the time study and showed a significant difference from that of live control samples.

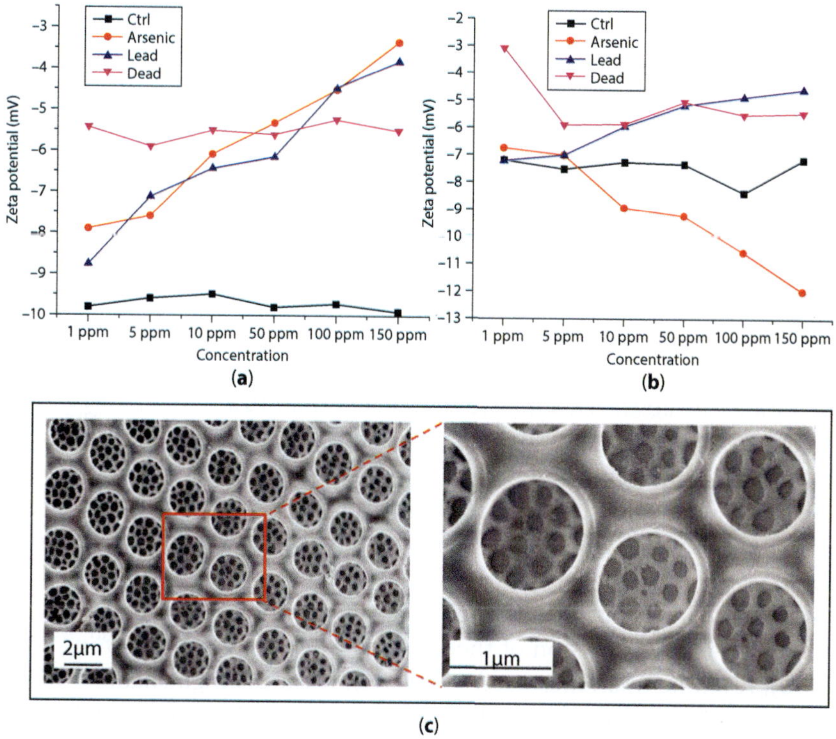

Figure 14.2 Zeta potential of diatoms at varying concentrations of heavy metal salts after (a) 24-hour incubation and (b) 1-week incubation with control sample as diatoms without any dopant along with dead diatoms (silica frustule only). (c) SEM micrographs of diatom frustules.

However, as indicated in Figure 14.2, the surface potential of diatoms after 1-week incubation with the dopants shows significant differences in ζ values. Zeta potential for As doped diatoms decreased from approximately -7 mV to -12 mV and increased for lead doped diatoms from -7 mV to -5 mV. This behavior could very well reflect on the surface absorption phenomenon of ions by silica frustules. In an aqueous solution, the predominant arsenic species is the neutral H_3AsO_3, however, its anionic form becomes more evident upon prolonged exposure. Accumulation of arsenite (AsO_2^-) anions resulted in increased negative zeta potential, while the addition of cationic lead (Pb^{2+}) compound resulted in decreased potential values. The shift in the zeta potential at higher concentrations indicates charge-based influence at the diatom-dopant interfacial double layer, as shown in Figure 14.3. The surface morphological structures were captured by SEM, as shown in Figure 14.2c. Larger segments and smoother surfaces were observed along with visible concurrent porous structures of the diatoms. No significant morphological changes were observed in heavy metal-treated diatom samples for the time period of 24 hours and one-week of treatment.

14.3.3 Dependency of pH on Surface Charge Potential

For the diatom frustule to function as a sensory element it is essential to gain insight into the mechanism of inorganic heavy metal intake; therefore, the effect of solution physical-chemical conditions such as salinity-based conductivity and pH on the adsorption capacity were studied. The zeta potentials of the diatoms loaded with arsenic and lead salts at varying pH solutions were observed and indicated in Figure 14.3. As pH drastically affects the adsorption characteristics of these materials, the inorganic heavy metal uptake capacity of the diatoms was measured across different pH solutions. The surface charge behavior inherently depends on the impregnation behavior of the substrate and its charge density at varying pH. For silica, this is illustrated in Figure 14.3f. At lower pH, silica exhibits positive charges on surfaces with the presence of surface OH and H^+ ions due to protonation. At pH values optimal for charge balance, the so-called point of zero charge (PZC) is experienced, leading to electrically neutral surfaces. However, at higher pH, silica H^+ ions are removed and, as a result, the surface is negatively charged. Therefore, for anions to be adsorbed onto the carrier surface, pH values below the PZC offer suitable condition for deposition, whereas the cations are adsorbed at pH values greater than PZC.

The data represented in Figure 14.3g indicates the charge-based behavior of silica surfaces when treated with arsenic anions and lead cations. Under acidic conditions (pH 3), the zeta potential values of As doped diatoms were observed to be between -4 mV and -6 mV over a period of 1 week. However, under weakly acidic or weakly basic conditions (pH 5–9), the presence of anions at the neutrally charged silica surfaces increased as a result of the slight increase in negative zeta potential observed from -5 mV to -8 mV. A further significant increase in the repulsion of these ions was observed under strong basic conditions (pH >10), resulting in an increased stern layer. Clear distinguishability was observed at higher pH values for diatom at 1-week post treatment, leading to zeta potential value of approximately -20 mV.

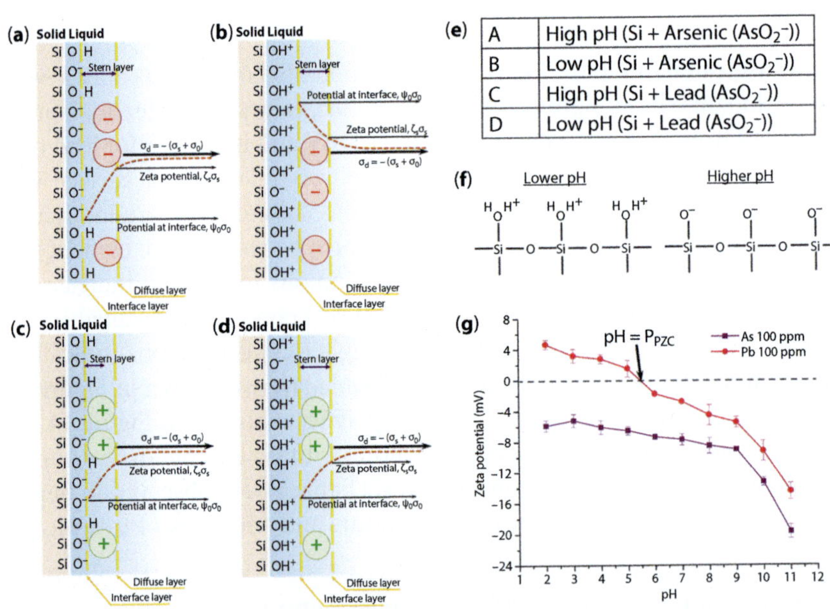

Figure 14.3 Schematic representation of Stern layer model of electric double layer on (a-b) silica and anionic charges and (c-d) silica and cationic charges. (e-f) Protonated and deprotonated silanol groups are located at the interface layer at higher and lower pH values respectively. (g) Zeta potential of diatoms treated with arsenic- and lead-treated diatoms at 100 ppm salt concentration with varying solution pH post treatment.

It is well known that inorganic arsenic and lead exist in different ionic forms in solution. When changing the solution from acidic to basic, the dominant form of arsenic trioxide, i.e., arsenite (As(III)) exists in ionic AsO_2^- at greater pH levels, thereby indicating improved physisorption of ions or even electrostatic interactions at the solid diatom surface. With a greater possibility of oxidation of these ions, arsenic could also be present in solution as arsenate ions (AsO_4^-), leading to an increase in negative zeta potential value. After being treated with 100 ppm of As salt, no zero-point charge pH (pH_{pzc}) was observed, indicating a strong negative impact of electrostatic interactions on the As adsorption process. Meanwhile, at any given pH, the zeta potentials of 1 week incubated As-loaded sample was lower than that of lead-treated samples. Longer exposure to arsenic ions results in the formation of negatively charged surface complexes, as shown in Figure 14.3. These negatively charged complexes strongly electrostatically repulse the anionic As(III) ions and result in increased zeta potential with increase in the Stern layer. Although this could account for weak to negligible adsorption of As ions, a plausible explanation is that some As(III) is quickly oxidized to As(V) and enhanced deprotonation of surface As(III) can occur. This has further been explored and explained using FTIR characterization in the following section.

A similar monotonic decrease trend in zeta potential can be observed when treated with lead salt. Clearly, with the addition of positive cations to the solution, the zeta potential at lower pH was positive. The protonated silica substrate further enhances the positive zeta potential values. With the increase in pH, the zeta potential became more negative. However, at pH 5.5, the sample exhibited neutral charge behavior with no protonation and deprotonation of silica substrate. At weakly acidic and basic pH values between 6 and 9, a linear trend in the zeta potential values was observed changing from -2 mV to -6mV. At higher pH values, the silica substrate becomes silanol, leading to negatively charged SiO^- terminations, allowing for electrostatic and chemical reaction-based bonding between the cations and surface charges. At pH > p_{PZC}, zeta potential values follow a similar trend to that of arsenic-treated diatoms, reaching up to -14.8 mV at 100 ppm for 1 week of incubation. The error bar indicates covariance from three replicates.

14.3.4 FTIR Characterization

Fourier transform infrared spectroscopy can be employed to identify the functional groups present on the cell surface of diatoms because each group has a unique energy adsorption band. The FTIR spectrum of pristine diatoms exhibits distinctive peaks, suggesting the presence of amide, carboxyl, nitrile, hydroxyl, and halide groups from the sample mixture. The broad mixed stretching vibrations frequency of N-H and O-H are observed in the region of 3500-3300 cm^{-1} and those for strong silane (Si-H) groups can be observed at 2134 cm^{-1}. The formation of silane groups due to biogenic silica suspended in water provides a strong surface area for charge-based interaction with heavy metal ions. Similarly strong and broad Si-OR stretching band frequency can be observed at 1100 cm^{-1} for all samples. The presence of amides and nitriles can be attributed to the strong peaks at 862 and 773 cm^{-1}. A small presence of halides in the seawater can be observed at lower wavelengths between 600 and 700 cm^{-1}. The shift of around 2 cm^{-1} in the FTIR analysis may be due to the resolution of the device. Figure 14.4 shows the FTIR spectra in which (a) arsenic-doped diatoms and (b) lead-doped diatoms are compared with control live diatom samples. The concentration

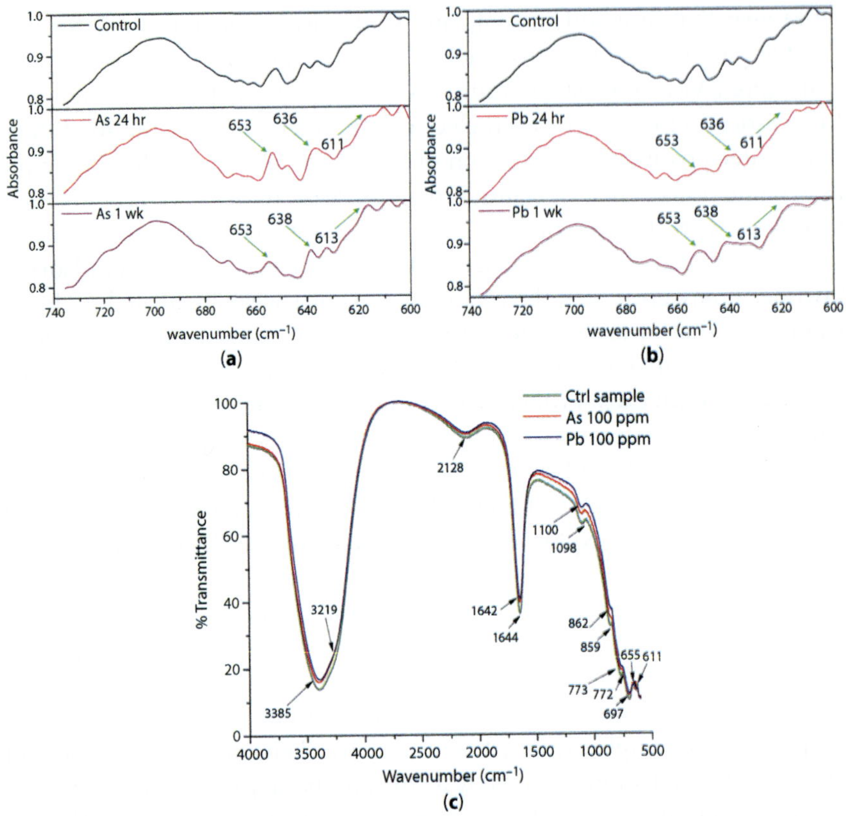

Figure 14.4 FTIR spectra of (a) arsenic- and (b) lead-treated diatoms with 100 ppm salt concentration at 24 hours and 1 week exposure time. (c) Full range FTIR spectra of heavy metal-doped diatom samples.

of heavy metal salts used for this study is 100 ppm and samples were suspended in seawater with constant ionic content.

As discussed above, the main adsorption bands observed at 3400, 2127, 1640, 1100, 860 to 611 cm^{-1} are in close agreement with the reported spectrum of silica in water solution as shown in Table 14.2. The presence of strong peaks corresponding to silanes and Si-OR stretching can also be observed for samples treated with heavy metals. The shift in the silane and silica stretching bonds may indicate the formation of a counter ion layer around the diatom silica. The mechanism of heavy metal ions adsorbed onto solid oxide surfaces is essential for a full understanding of their presence and removal from water samples. The vibration of the Si-OH bond results in a band at around 3400 cm^{-1}, together with a doublet at approximately 860 cm^{-1}, which can be attributed to the adsorption of ions from the surrounding medium. However, upon treatment with As the bands at 860 and 772 cm^{-1} in the spectrum are characteristic of As-O bonds of the adsorbed oxidized arsenate species at the silica surface. The bands at 653 cm^{-1} and 636 cm^{-1} also matched well with the stretching frequencies of As-O bonds in AsO_2^- groups and are closely consistent with the υ_3 values reported by [14.19].

A shift in the band and intensity of the absorbance indicates negative electrostatic interaction with the silica frustule of the diatoms. Decoupled spectra bands at 638 cm^{-1} indicate

Table 14.2 FTIR spectra and band assignment for their corresponding functional groups.

Wavenumber (cm^{-1})	Peak assignment	Sample type	Bond details
3391	N-H, O-H	All sample diatoms	Strong
2134	Si-H silane		Strong and broad
1643	C=C (in ring), C=O		Strong
1100	Si-OR (silica)		Strong
862	ρ(C-N) (amine)	Control sample	Medium
773	CNC, NCN		
860	As-O (arsenate)	Arsenic treated	Medium to weak
772			
697	NH_2	All sample diatoms	Medium to weak
655	CNC		
638	Si-H/Pb (accumulation and stretching)	Lead treated	Weak
611–613			
653	As-O (AsO_2^- stretching)	Arsenic treated	Medium to weak
636			

the uptake of lead metal ions over a longer exposure time. The adsorption band at 613 cm^{-1}, which is not originally observed for the control sample but is shifted at 1-week post treatment, further corroborates the hypothesis. A small shift in the wavenumber between 640 and 610 cm^{-1} is noted after the treatment with lead salt, which may be due to electrostatic adsorption of lead cation onto the silica surface. The resultant shift observed at the 860 cm^{-1} wavenumber may be due to oxygen atom in the hydroxyl groups present, which could be involved in lead adsorption. Adsorption of metal ions mainly involves the functional groups present on the cell surface. The functional groups such as carboxyl, amides, hydroxyl, and silanes were chemically treated with heavy metals to understand their role in physical or chemical adsorption that could potentially be used as an early indicator for detection of contamination in water samples.

Similar spectral behavior was observed for both heavy metals, indicating physisorption and electrostatic interactions between ions and silica frustule over 24 hours and 1 week incubation time. The surface area analysis based on the visual interpretation of the diatom samples indicates that heavy metal ion uptake could vary depending on the available area for adsorption. According to the literature, pore volume and pore size play a vital role in ion adsorption rather than available enhanced surface area. The experimental results for the As and Pb interaction with *C. wailesii* diatoms provide an understanding of anion- and cation-based doping of biogenic silica. Continuous changes in zeta potential and solution pH shows the presence of a slow step in the mechanism, and possible diffusion of reactant species interacting with solid surfaces through the double layer. Such behavior can provide valuable information regarding the ionic content of a solution, in this case, freshwater quality.

As seen in the results presented, doping of diatoms with small concentrations of heavy metals can alter the adsorption capacity, leading to varied surface charge potential. Similar behavior has been observed by research groups where they have shown adsorption of heavy metal ions onto inorganic and organic adsorbents for the removal and filtration processes [14.20-14.22]. Applications that require early detection of heavy metal presence are critical and need to be addressed. This study has shown the potential of using surface charge potential as an early indicator of evaluating sample content and quality control. The continuous zeta potential change observed with an increase in concentrations of anions (As) and cations (Pb) over 24 hours and 1 week indicate the sensitivity of these green freshwater algae to their surrounding environment. Increasing specific ion adsorption by diatomaceous earth due to chemical waste or other forms of pollution can be studied and characterized based on surface charge potential as early as 24 hours of exposure. This work can potentially be translated to develop diatom-based biosensors for early detection of metals in water samples by transducing the surface charge effects readable outputs as a function of the levels of the target contaminant in the water sample. This can pave the way for the development of novel, non-faradaic impedance-based electrochemical biosensors which can detect and quantify subtle charge changes on the diatom surface in a sensitive and label-free manner. Further detailed study of the time effects and electrokinetic behavior changes along with biochemical characterization to explore metal ion uptake and metabolism changes would be necessary to ascertain the role and fate of heavy metal adsorption in diatoms.

14.4 Conclusion

We demonstrated a viable option for early indication of heavy metal contamination using surface charge-based analysis. Arsenic and lead being the most commonly found heavy metals in water sources, we have presented their ionic behavior in aqueous solution and their responses to biogenic silica based on physisorption and electrostatic interaction at the double layer interface. We have observed a 70 to 88% change in cell density of diatom culture upon 24-hour and 1-week treatment with the heavy metal salts respectively. Although diatom frustule size remained within 74 to 79 μm, significant changes in cell density were observed. Zeta potential values varied for varying concentrations of As- and Pb-treated diatoms. The data also suggests that there is non-uniform adsorption of ions, which results in multilayer charge accumulation rather than uniform monolayer adsorption. Further characterization with FTIR indicates clear distinguishability in the presence of anions and cations, however, a minimal shift was observed due to weak physisorption or electrostatic interaction of inorganic heavy metal ions. The FTIR data also suggest that the interaction of heavy metal ions could be due to the presence of functional groups present on the silica surface. In addition, we have also presented pH dependency of ionic charge-based behavior as water samples to mimic real-life contaminated samples. The possibility of identifying the presence of heavy metal contamination in water at the earliest could potentially help prevent adverse health complications. Further studies using electrokinetic techniques are being investigated concerning the development of point-of-care technology for ease of use in the field.

Acknowledgments

The authors would like to thank Dr. Kai-Chun Lin (Biomedical Microdevices and Nanotechnology Lab, BMNL) for diatom cultures and David Kinnamon from BMNL for the support in FTIR experiments. The authors acknowledge the funding resources from BMNL UT Dallas for this work.

References

[14.1] Gumpu, M.B., *et al.*, A review on detection of heavy metal ions in water–An electrochemical approach. *Sensors and Actuators B: Chemical*, 2015. 213: p. 515-533.

[14.2] Khraisheh, M.A., Y.S. Al-degs, and W.A. Mcminn, Remediation of wastewater containing heavy metals using raw and modified diatomite. *Chemical Engineering Journal*, 2004. 99(2): p. 177-184.

[14.3] Mehta, S. and J. Gaur, Use of algae for removing heavy metal ions from wastewater: progress and prospects. *Critical reviews in biotechnology*, 2005. 25(3): p. 113-152.

[14.4] Jaafarzadeh, N., *et al.*, Biosorption of heavy metals from aqueous solutions onto chitin. *International Journal of Environmental Health Engineering*, 2015. 4(1): p. 7.

[14.5] Foster, S., D. Thomson, and W. Maher, Uptake and metabolism of arsenate by anexic cultures of the microalgae Dunaliella tertiolecta and Phaeodactylum tricornutum. *Marine Chemistry*, 2008. 108(3): p. 172-183.

[14.6] Bairagi, H., *et al.*, Adsorption profile of lead on Aspergillus versicolor: a mechanistic probing. *Journal of hazardous materials*, 2011. 186(1): p. 756-764.

[14.7] Yan, H. and G. Pan, Toxicity and bioaccumulation of copper in three green microalgal species. *Chemosphere*, 2002. 49(5): p. 471-476.

[14.8] Morin, S., *et al.*, Long-term survey of heavy-metal pollution, biofilm contamination and diatom community structure in the Riou Mort watershed, South-West France. *Environmental Pollution*, 2008. 151(3): p. 532-542.

[14.9] Arkas, M. and D. Tsiourvas, *Organic/inorganic hybrid nanospheres based on hyperbranched poly (ethylene imine) encapsulated into silica for the sorption of toxic metal ions and polycyclic aromatic hydrocarbons from water.* Journal of hazardous materials, 2009. 170(1): p. 35-42.

[14.10] Zhang, J., *et al.*, Enhanced adsorption of trivalent arsenic from water by functionalized diatom silica shells. *PloS one*, 2015. 10(4): p. e0123395.

[14.11] Anantharaj, K., *et al.*, Effect of heavy metals on marine diatom Amphora coffeaeformis (Agardh. Kutz). *Global Journal of Environmental Research*, 2011. 5(3): p. 112-117.

[14.12] Fisher, N. and D. Frood, Heavy metals and marine diatoms: influence of dissolved organic compounds on toxicity and selection for metal tolerance among four species. *Marine Biology*, 1980. 59(2): p. 85-93.

[14.13] Hernández-Ávila, J., *et al.*, Diatoms and Their Capability for Heavy Metal Removal by Cationic Exchange. *Metals*, 2017. 7(5): p. 169.

[14.14] Viji, S., *et al.*, Diatom-based label-free optical biosensor for biomolecules. *Applied biochemistry and biotechnology*, 2014. 174(3): p. 1166-1173.

[14.15] Henderson, R.K., S.A. Parsons, and B. Jefferson, Surfactants as bubble surface modifiers in the flotation of algae: dissolved air flotation that utilizes a chemically modified bubble surface. *Environmental science & technology,* 2008. 42(13): p. 4883-4888.
[14.16] Yu, Y., J. Addai-Mensah, and D. Losic, Functionalized diatom silica microparticles for removal of mercury ions. *Science and Technology of Advanced Materials,* 2012. 13(1): p. 015008.
[14.17] Ontiveros-Ortega, A., *et al.*, Effect of heavy metals on the surface free energy and zeta potential of volcanic glass: implications on the adhesion and growth of microorganisms. *Journal of Materials Science*, 2014. 49(9): p. 3550-3559.
[14.18] Zhao, C., *et al.*, Extracting local surface charges and charge regulation behavior from atomic force microscopy measurements at heterogeneous solid-electrolyte interfaces. *Nanoscale,* 2015. 7(39): p. 16298-16311.
[14.19] Carabante, I., *et al.*, *In situ* ATR–FTIR studies on the competitive adsorption of arsenate and phosphate on ferrihydrite. *Journal of colloid and interface science,* 2010. 351(2): p. 523-531.
[14.20] Erdemoğlu, M. and M. Sarıkaya, Effects of heavy metals and oxalate on the zeta potential of magnetite. *Journal of colloid and interface science,* 2006. 300(2): p. 795-804.
[14.21] Zhang, J., R. Stanforth, and S. Pehkonen, Proton–arsenic adsorption ratios and zeta potential measurements: Implications for protonation of hydroxyls on the goethite surface. *Journal of colloid and interface science,* 2007. 315(1): p. 13-20.
[14.22] Somasundaran, P., Zeta potential of apatite in aqueous solutions and its change during equilibration. *Journal of Colloid and Interface Science*, 1968. 27(4): p. 659-666.

Index

Absorption coefficient, 349, 382
Adhesion complex, 292
Adsorption, 414, 417, 419–424
Agency for toxic substances and disease registry's priority list of hazardous substances, 414
Age-structured models, 29
Aging, 24
Algae, 121–122
Algorithmic information theory (AIT), 230, 240, 245–247, 249–252
Allometric, 130
Amphitetras genus, 355
Anabaena catenula, 281
Aneural cognition, 230–240, 244, 247–252
Annulus, 105–107, 115, 116
Apical pore field, 107
Aquaculture, 64, 91, 96
Aquaphy model, 134
Arachnoidiscus sp., 354, 366–367
Araphid pennates, 106, 107
Areola, 106, 110, 112, 114
Areolae, 74, 80, 92
Arrhenius relation, 129
Arsenic, 414–424
Asymmetric division, 28
Asymptote, 148, 152–153
Atomic force microscopy (AFM), 75, 77, 79–81, 301
Aulacoseira, 22, 24–25
Auxospore, 23
Axial area, 108, 114, 115

Bacillaria, 230–233, 235–240, 242, 246–252, 316
Bacillaria paradoxa, 195
Bacillaria paxillifer, 195–198
Baranyi-Roberts, 150–152, 154–157, 160
Basal araphid, 106, 116
Basic time-marching scheme, 370–374
Bayesian information criterion, 41
Bellman-Harris process, 40, 48
Benthic, 67, 72, 73, 90
Biofuel, 122, 136
Biogenic sensors, 414
Biogenic silica, 421, 424
Biomass, 122, 134
Bio-optic model, 130
Bipolar centric, 116
Biraphid, 106, 116
Bloch states, 357
Bloom phase, 52
Branching process, 39–41, 48–49
Brillouin zone, 358

C. wailesii, 416–417, 423
C:N ratio, 123
Canal raphe, 293
Carbohydrate, 122–123, 134
Cardinal points, 21
  initial size, 21
  minimal size, 21
  sexual size threshold, 21
Cell, 132, 135
  cycle, 133, 135
  division, 134
  mass, 128
  membrane, 123, 127, 132, 135–136
  volume, 125, 128–130, 135–136
  wall, 121
Cell colony, 230–233, 236–242, 244, 246–249, 251, 252
Cell cycle checkpoint, 22
Cell density, 417–418, 424
Cell division, 108, 109, 206, 265–270
Cellulose synthase, 292
Central pattern generator (CPG), 241–242, 249
Centric, 73, 85, 86

Centric diatoms, 105, 106
Chain,
colony, 108, 109
of annulus-like rings, 107, 116
of small rings, 110, 111, 115
Characterization of diatoms genomes,
allocation of gene functions, 169
annotation methods, 168
available genomics data, 167
subcellular localization, 169
Chlorophyll fluorescence, 393–398, 408
Chlorophylls, 86, 91
Chl-*a*, 376, 381–382
Chl-*c*, 376, 382
fluorescence, 376, 379–381
Chloroplast, 364, 377–380
Clonal chain, 266, 274, 281
Clonal colony, 266, 274, 281
Collective pattern generator (CoPG), 230, 232, 241–242
Colonies, 64–70, 72–74, 90
Connectionist, connective components, 232, 236, 239–241, 243, 247–250
connectionist learning, 237
connectionist model, 238, 242, 244, 248, 251–252
Continuous dynamical systems, 29
Core araphid, 105, 106, 116
Coscinodiscus, 71, 74–76, 78–81, 85, 92, 93
granii, 361–363, 368
wailesii, 359, 366
Coupled oscillators, 197–198, 203–208
Cribellum, 74, 79, 80
Cribrum, 74, 79
Critical coupling, 212–218
Ctenophora pulchella, 366
Cycle length, 21
Cytoplasm, 70, 74
Cytoplasmic streaming, 292

D'Alelio *et al.* model, 25–27
Delay differential equations, 37
Developmental Braitenberg vehicle (dBV), 230, 251
Braitenberg vehicles, 252
Diadinoxanthin, 394, 395, 407, 408
DiaLCM model, 46
Diatom colony, 266–268
Diatom life cycle, 20–21, 25
Diatom trail, 293, 306, 309
Diatoma tenuis, 286
Diatoma vulgaris, 266
Diatomite, 82–84
Diatoms, 230–232, 234–236, 238–243, 246–248, 251, 252, 413–424
benthic, 4–5, 15
frustules, 4–16
marine, 4–5, 14–15
Diatotepum, 300
Diatoxanthin, 394, 305, 407, 408
Difference equations, system of, 29–33
Differential equations, system of, 35–37
Diffusion constant, 302
Diploneis,
central nodule, 5, 8–16
Diploneis aff. *chersonensis*, 4–16
Diploneis sp., 4–6, 8, 10
Doppel-T-Träger, 14, 16
girdle bands, 13
longitudinal canals, 5, 8–14, 16
raphe sternum, 4–8, 12–14, 16
Director direction, 311
Discrete dynamical systems, 29–33
Discrete Fourier transform, 219–221
Ditylum brightwellii, 28
Division delay, 26
Division rate, 22
Domino effect, 304
diatom speed, 304, 305
Dopant, 418–419
Double layer, 419–420, 424
Down feather-like structure, 112, 113, 115
Dynamical system,
continuous, 35
discrete, 29

Eigenfrequency, 85, 86
Elastic modulus, 76, 77, 79–81, 85
Electric current, 298
Electrochemical, 414–415, 424
Electrokinetic, 424–425
Electrolyte, 417
Electron transport chain, 378–379
Electrophoretic, 417
Electrostatic, 421–424
Ellerbeckia arenaria, 26
Empirical data, 130
Enzyme, 132, 135

Epitheca, 74
Eppley curve, 129
Equilibrium,
continuous system, 35
discrete system, 30
Exoskeleton, 63, 64, 77, 83
Explosive hydration, 302
External coupling, 53

Fast Fourier transform approach, 368
FIB,
milling, 4, 6, 8–12
sequential milling, 4–6, 8
x axis, 6, 8, 10, 16
y axis, 4, 12–13
z axis, 4, 6, 8–9, 13
Fibonacci growth, 26
Fibulae, 293
Filter feeders,
Manila clam, 14–15
Finite difference time domain method, 362–364
Firefly, 198, 223–225
Fisher information, 43
Fluid mechanics, 298
Fluorescence, 76, 90, 91
Fold-change, 230, 239
Foramen, 74
Fourier series, 54
Fourier transform infrared spectroscopy (FTIR), 421–424
Fractional calculus, 368–370
Fragilaria mesolepta, 105, 107, 108, 110–112
Free energy, 234, 236–237, 249
Fresnel equation for unpolarized fields, 347
Frustule, 63–66, 68–70, 72–77, 79–83, 85–87, 89, 92–96, 414, 417–419, 422–424
Fuhrmann-Lieker *et al.* model, 49–52

Gaussian mixture model, 41
Generative density (G-density), 230, 240
Gielis transformations,
three-dimensional, 354
two-dimensional, 353
Girdle band, 74, 76, 79–81
Gompertz, 126–127, 130–131, 148, 150, 152–159, 161
Gradient, 292
Growth curve, 142–144, 146–150, 157–158, 160
Growth function, 142, 149, 151, 160
Growth kinetics, 63, 64, 90, 91, 96
Growth models,
nonparametric, 147–149, 152, 157–158, 160–161
parametric, 147–149, 152–155, 157-161
Growth parameter, 143, 147
Growth rate, 22, 143, 147–152, 156, 158–161

Habituation, 233, 235
Half saturation constant, 127, 132–133
Hardness, 75–80
Haslea ostrearia, 23
Heavy metal, 413–424
Hebbian learning, 232, 235, 237, 248, 251
Helictoglossa, 293
Helmholtz equation, 364
Hense-Beckmann model, 45–48
Heuristic (approach), 149, 152
Hierarchical, 63–65, 74–77, 80, 83, 84, 92, 93
Hoff-Arrhenius relation, 129
Homology, 105, 115
Hydrated silicon dioxide, 121
Hydration, 298
Hylalin area, 293
Hypotheca, 74

Incremental size diminution, 22
Individual-based models, 39, 46, 48
Initial cell, 20
Initial cell formation,
automixis, 23
biparental auxosporulation, 24
vegetative enlargement, 23
Initial cell size range, 22
Initial value problem,
continuous system, 35
discrete system, 30

Jacobian, 33
Janus-faced causation, 303
Jury test, 32

Kuramoto model, 207–208

Lag phase, 142–143, 146–147
Lambert-Beer law, 348
Laney model, 28
Lattice model, 295
Lead, 414–424

Left-censoring, 147
Life cycle of diatoms, 20–21, 25
Light,
  confinement, 87
  focusing, 74, 87
  harvesting, 74, 75, 87
  microscopy, 67, 70, 71
Light-harvesting complex (LHC), 376, 378–380
Limit of detection (LoD), 414–415
Lindenmayer system (L-system), 281–284
Linking spine, 107–110, 114
Lipid, 122–123
Local polynomial regression (loess), 148–149, 152–157, 160
Logistic, 148, 150–153, 161
Loop, 105, 112–116

MacDonald-Pfitzer rule, 20
Macronutrient, 121
Mantle, 107–110
Markov blankets, 247
Markov process, 39
Maximum contamination level, 414
Maxwell equations, 356
Mean squared error (MSE), 151–157
Mechanical,
  behavior, 77
  performance, 63, 73
  properties, 65, 75–77, 79–82, 84, 87
  response, 64
  stability, 75
  strength, 81, 82
  stresses, 81
  test, 75
Mechanistic approach, 123
Mechanistic model, 123, 134
Membrane speed, 297
MEMS, 63, 64, 85, 87, 96
Metabolic modeling,
  genomes, 173
  horizontal gene transfer, 180
  light acclimation, 178
  model constraints, 173, 176–177
  production of valuable metabolites, 181
  proteome allocation, 183
Metabolic rate, 123, 128–129, 134
Michaelis-Menten function, 132
Michaelis-Menten model, 132–133, 135–136
Microfilament bundles, 292
Minimal cognition, 230, 235
Modeling,
  displacement, 4, 12–13, 16
  elasticity, 5, 14
  finite element method, 4, 7
  fracture, 13, 15
  geometric model, 4, 7, 9–10
  normalized displacement, 13
  numerical model, 4, 7
  reference models, 7
  stiffness, 12–16
Monod expression, 131–132
Monoraphid, 106
Monte Carlo methods, 150
Monte Carlo simulation, 215–218, 281–282
Morphogenesis, 105, 106, 109, 114, 116
Morphology, 77, 82, 83, 86, 87
Müller model, 26
Multiple input model, 123, 136
Multiple variables, 130

Nanoindentation, 75, 76, 78, 80
Nanostructured, 64, 85, 96
Nematic, 293
NEMS, 63, 64, 85, 96
Non-photochemical quenching (NPQ),
  dark NPQ, 397–398, 404–408
  energy-dependent quenching (q_E), 394, 396–397, 407
  regulated thermal energy dissipation index (Y(NPQ)), 395–396, 398, 403–404, 406–407
  Stern-Volmer index (NPQ), 395–396

Ontogenetic, 105, 107, 110, 115, 116
Oscillation, oscillatory units, 230, 233, 237, 240–248, 251
Oscillations, 31–32, 36, 52–55

Partial differential equation, 38
Pascal's triangle, 20
Pattern center, 105, 106, 108, 115, 116
Pattern formation, 267–288
PDMPO, 109
Peleg, 152–156, 160
Pennate, 73, 74, 85, 86
Pennate diatoms, 105, 114

Periodic light source, 198, 221–223
Periodic solution,
  continuous system, 35
  discrete system, 30
Periphyton, 66–68, 71, 72
Perron-Frobenius theorem, 33–34
pH, 416–417, 419–421, 424
Phaeodactylum tricornutum, 20, 23
Phase transition, 247–248
Phenotypic, 143
Photoacoustic, 86, 90
Photoinhibition, 394–395
Photosynthesis, 122–123, 129–130, 134, 361, 376–380, 394, 397, 406
Photosystem II (PSII), 394, 396, 407
Physisorption, 421, 423–424
Phytoplankton, 121, 130
Picard-Lindelöf theorem, 35
Plane wave expansion method, 359–362
Planktonic, 67, 68, 71, 73
Plasmolysis, 300
Plasticity, 105, 116
Polar centrics, 106, 115, 116
Population dynamics, 28–33
Power law, 128
PPND model, 45
Predictive processing, 232, 235–239, 247, 251, 252
Probability generating function, 40, 48
Production rules, 281–283
Propagator matrix, 30
Proton gradient (ΔpH), 394, 397, 407, 408
Pseudo-annulus, 106
Pseudo-nitzschia delicatissima, 41–43
Pseudo-nitzschia multistriata, 43–45, 52–53
Psychophysics, 231, 232, 235–236, 238, 243–244, 247, 251

Q10, 129
Quota model, 134

Radial centrics, 106
Radiance, 350
Radiative transfer equation for the underwater optical field, 349
Random walk, biased, 298
Raphan, 292
Raphan swelling, 306
Raphan synthase, 292
Raphan, concatenated, 295
Raphe, curved, 293
Raphid pennates, 106, 107
Rayleigh instability, 299
Reaction-diffusion equation, 38
Receiver operator characteristic (ROC), 230, 251
Reciprocal lattice, 357
Recognition density (R-density), 230, 240
Reductive cell cycle, 147
Replacement rules, 281–283
Reptation, 303
Retardation, 207–217
Richards, 148, 150, 153
Ricker function, 50
Right-censoring, 146, 151
Rimoportula, 107, 108, 110, 112, 287
Rotational Brownian motion, 292
  raphe cross section, 293
Routh-Hurwitz criterion, 37

Scattering,
  coefficient, 349–350, 382
  function, 350
  phase function, 350
Schwarz *et al.* model, 23–25
Seawater, 416–417, 421–422
Seminavis robusta, 23–24
Senescence, 144
Sensation, 234, 239
Sensitivity, 415, 424
Sensomotoric system, 204–205
Sex clock, 19
Sexual period, 45
Sexual reproduction, 20
Shear, 303
Sigmoid function, 127
Sigmoidal, 143, 146, 148, 150–151, 160
Signal detection theory (SDT), 230, 239, 250–251
Silica, 65, 74, 75, 77, 79–85, 88, 89, 93, 96
Silica deposition vesicle (SDV), 65, 89, 96, 105, 115, 116
Silica spherules, 110, 111, 114
Silicate, 121
Siliceous,
  exoskeleton, 4–5, 14
  mechanical properties, 4–6, 8, 14–15
Siliceous threads, 112

Silicic acid, 133
Silicon, 65, 81–83, 85, 86
Simulation of the synchronous movement, 211–212, 214–221, 223
Size class, 20
Size distribution, 20
Size reduction-restitution cycle, 20
Skeletonema subsalsum, 144–146, 158–159
Slime mold (*Physarum*), 231, 233, 250
Smectic, 293
Smoluchowski, 417
Snell's law, 347
Splines, 148, 152, 159–160
State transitions, 146
State variable, 134–135
Staurosira binodis, 105, 107–109, 112–114
Stephanodiscus neoastreaea, 23–24
Sternum, 105–110, 112, 114–116
Stimulus intensity, 230, 234–235, 238–241, 248
  light intensity, 234, 235, 240, 241
Stochastic model, 281–282
Surface charge, 417–421, 424
Surface to volume ration, 127, 129
Swarm, 225
Symmetry, 67, 68, 70, 72
Synchronization of the movement, 197–198, 202–225
Synchronous cell division, 267–268, 274–275, 281
Synchrony, synchronizing, 230, 237, 246–249
Synedra tabulata, 22

Taxis, 232, 250
  chemotaxis, 231, 235, 236
  phototaxis, 231
  thermotaxis, 231
Terzieva-Terziev model, 48–49
Thalassiosira pseudonana, 20
Threads,
  with longer virgae (TLV), 112, 113
  with shorter virgae (TSV), 112, 113
Tools to predict cellular localization, 170
Toxicity, 414–415, 417–418
Tracking, 198–199, 201

Ulnaria acus, 75, 85
Uniaxial perfectly matched layer boundary conditions, 374–375
United States Environmental Protection Agency (US EPA), 414–415
Unorganized silica, 112, 113, 115
Unstable equilibrium, 303
Uptake rate, 130, 132–133

Valve,
  face, 107–109
  morphogenesis, 105, 106, 109, 114, 116
Vegetative division, 20
Vimens, 110, 112, 114
Virgae, 105, 107, 110–116
Voltage controlled oscillator (VCO), 205

Water column model, 45–47
Weber-Fechner law, 236, 238–239, 241, 250–251
Wide-angle beam propagation method, 364–367
Wolfram pattern, 309
Wulff shape, 354

Xanthophyll cycle (XC), 394–397, 407–408

Young's modulus, 75–78, 86

Zeitgeber rhythm, 54
Zeta potential, 415, 417–421, 424
Zooplankton,
  copepods, 14
  grazing, 4–5, 14–15

Also of Interest

Check out these published and forthcoming titles in "Diatoms: Biology and Applications" series from Scrivener Publishing

Diatom Ecology
Edited by Eduardo Morales and Nora I. Maidana
Forthcoming 2024

Diatom Cultivation for Biofuel, Food and High Value Products
Edited by Vandana Vinayak and Richard Gordon
Forthcoming 2023

Mathematical Macroevolution in Diatom Research
By Janice L. Pappas
Forthcoming 2023

Diatom Photosynthesis: From High Value Molecules to Primary Production
Edited by Johannes Wilhelm Goessling, João Serôdio and Johann Lavaud
Forthcoming 2023

The Mathematical Biology of Diatoms
Edited by Janice L. Pappas
Published 2023. ISBN 978-1-119-74985-1

Diatom Microscopy
Edited by Nirmal Mazumder and Richard Gordon
Published 2022. ISBN 9781119711537

Diatoms Morphogenesis
Edited by Vadim V. Annenkov, Joseph Seckbach and Richard Gordon
Published 2021. ISBN 978-1-119-48795-1

Diatoms Motility
Edited by Stanley A. Cohn, Kalina M. Manoylov and Richard Gordon
Published 2021. ISBN 978-1-119-52635-3

Diatoms
Fundamentals and Applications
Edited by Joseph Seckbach and Richard Gordon
Published 2019. ISBN 978-1-119-37021-5

www.scrivenerpublishing.com

Printed and bound by CPI Group (UK) Ltd, Croydon, CR0 4YY

17/09/2026

14964249-0002